Sustainable Agriculture Systems and Technologies

Sustainable Agriculture Systems and Technologies

Edited by

Pavan Kumar

College of Horticulture and Forestry
Rani Lakshmi Bai Central Agricultural University
Jhansi, UP, India

A.K. Pandey

College of Horticulture and Forestry
Rani Lakshmi Bai Central Agricultural University
Jhansi, UP, India

Susheel Kumar Singh

College of Agriculture
Rani Lakshmi Bai Central Agricultural University
Jhansi, UP, India

S.S. Singh

Extension Education
Rani Lakshmi Bai Central Agricultural University
Jhansi, UP, India

V.K. Singh

ICAR-Central Research Institute for Dryland Agriculture
Hyderabad, Telangana, India

WILEY

This edition first published 2022

Registered Offices
John Wiley & Sons, Inc., 111 River Street, Hoboken, NJ 07030, USA
John Wiley & Sons Ltd, The Atrium, Southern Gate, Chichester, West Sussex, PO19 8SQ, UK

Editorial Office
The Atrium, Southern Gate, Chichester, West Sussex, PO19 8SQ, UK

For details of our global editorial offices, customer services, and more information about Wiley products visit us at www.wiley.com.

Wiley also publishes its books in a variety of electronic formats and by print-on-demand. Some content that appears in standard print versions of this book may not be available in other formats.

Library of Congress Cataloging-in-Publication Data

Names: Kumar, Pavan (Professor of forestry), editor. | Pandey, A. K. (Scientist), editor. | Singh, S. K. (Professor of Soil science), editor. | Singh, S. S. (Sati Shankar), editor. | Singh, V. K. (Vinod Kumar) (Director of ICAR-CRIDA), editor.
Title: Sustainable agriculture systems and technologies / edited by Pavan Kumar, A. K. Pandey, Susheel Kumar Singh, S. S. Singh, V. K. Singh.
Description: Hoboken, NJ : Wiley, 2022. | Includes bibliographical references and index.
Identifiers: LCCN 2021970023 (print) | LCCN 2021970024 (ebook) | ISBN 9781119808534 (cloth) | ISBN 9781119808541 (adobe pdf) | ISBN 9781119808558 (epub)
Subjects: LCSH: Sustainable agriculture. | Agricultural innovations.
Classification: LCC S494.5.S86 S837 2022 (print) | LCC S494.5.S86 (ebook) | DDC 338.1–dc23/eng/20220118
LC record available at https://lccn.loc.gov/2021970023
LC ebook record available at https://lccn.loc.gov/2021970024

Cover Design: Wiley
Cover Image: © MONOPOLY919/Shutterstock

Set in 9.5/12.5pt STIXTwoText by Straive, Pondicherry, India
Printed and bound by CPI Group (UK) Ltd, Croydon, CR0 4YY

C9781119808534_210222

Contents

List of Contributors

Aishwarya
College of Agriculture
Rani Lakshmi Bai Central Agricultural University
Jhansi, Uttar Pradesh, India

Safik Ahamad
GSM Division
ICAR-IGFRI
Jhansi, Uttar Pradesh, India

S. Babu
Division of Agronomy
ICAR-IARI
New Delhi, India

Boubacar Siddighi Balde
Integrated Research System for Sustainability Sciences (IR3S), The University of Tokyo
Kisohigashi, Tokyo, Japan

B.P. Bhatt
NRM Division
Indian Council of Agricultural Research
New Delhi, India

Ramesh Chand
Department of Plant Pathology and Mycology
Institute of Agricultural Sciences, Banaras Hindu University
Varanasi, Uttar Pradesh, India

Bhagwan Singh Chaudhary
Department of Geophysics
Kurukshetra University
Kurukshetra, Haryana, India

Shib Kinkar Das
Department of Aquaculture
Faculty of Fishery Sciences
West Bengal University of Animal and Fishery Sciences (WBUAFS)
Kolkata, India

A. Dey
ICAR Research Complex for Eastern Region
Patna, Bihar, India

Anwesha Dey
Department of Agricultural Economics
Institute of Agricultural Sciences
Banaras Hindu University
Varanasi, Uttar Pradesh, India

Pavel Dmitriev
The Botanical Garden
Southern Federal University
Rostov-on-Don, Russia

Mohamed Esham
Department of Agribusiness Management
Faculty of Agricultural Sciences
Sabaragamuwa University of Sri Lanka
Belihuloya, Sri Lanka

Shilan Felegari
Department of Soil Science
Faculty of Agriculture
University of Zanjan
Zanjan, Iran

Avijit Ghosh
GSM Division
ICAR-IGFRI
Jhansi, Uttar Pradesh, India

Ahmad Golchin
Department of Soil Science
Faculty of Agriculture
University of Zanjan
Zanjan, Iran

J.J. Gupta
ICAR Research Complex for Eastern Region
Patna, Bihar, India

Avijit Haldar
Department of Animal Reproduction
ICAR-Agricultural Technology Application Research Institute (ATARI)
Kolkata, West Bengal, India

H.M. Halli
ICAR-Indian Grassland and Fodder Research Institute
Jhansi, Uttar Pradesh, India

Harshi Jain
Forest Research Institute Deemed to be University (FRIDU)
Dehradun, Uttarakhand, India

Pawan Jeet
Division of Land & Water Management
ICAR-Research Complex for Eastern Region
Patna, Bihar, India

Sachin Onkar Khairnar
Department of Aquaculture
College of Fisheries, Guru Angad Dev Veterinary and Animal Sciences University (GADVASU)
Ludhiana, Punjab, India

Bal Krishna
Department of Plant Breeding & Genetics
Bihar Agricultural University
Sabour, Bihar, India

Manoj Kumar
GIS Centre
Forest Research Institute (FRI)
Dehradun, Uttarakhand, India

Parveen Kumar
Natural Resource Management Division
ICAR-Central Coastal Agricultural Research Institute
Old Goa, Goa, India

Pavan Kumar
Department of Forest Biology and Tree Improvement
College of Horticulture and Forestry
Rani Lakshmi Bai Central Agricultural University
Jhansi, Uttar Pradesh, India

Pradeep Kumar
Department of Forestry
North Eastern Regional Institute of Science Technology
Nirjuli, Arunachal Pradesh, India

Rakesh Kumar
Division of Crop Research
ICAR-Research Complex for
Eastern Region
Patna, Bihar, India

R.V. Kumar
GSM Division
ICAR-IGFRI
Jhansi, Uttar Pradesh, India

Sunil Kumar
GSM Division
ICAR-IGFRI
Jhansi, Uttar Pradesh, India

Narendra Kumawat
AICRP for Dry land Agriculture
College of Agriculture
Indore, Madhya Pradesh, India

Bharat Lal
College of Agriculture
Rani Lakshmi Bai Central Agricultural
University
Jhansi, Uttar Pradesh, India

Amit Mandal
Department of Aquaculture
College of Fisheries
Guru Angad Dev Veterinary and Animal
Sciences University (GADVASU)
Ludhiana, Punjab, India

Manjanagouda S. Sannagoudar
GSM Division
ICAR-IGFRI
Jhansi, Uttar Pradesh, India

Riya Mehrotra
Central Institute of Medicinal and
Aromatic Crops
Lucknow, Uttar Pradesh, India

Tatiana Minkina
Academy of Biology and Biotechnology
Southern Federal University
Rostov-on-Don, Russia

Kamran Moravej
Department of Soil Science
Faculty of Agriculture
University of Zanjan
Zanjan, Iran

I. Wayan Nampa
Department of Agribusiness
Faculty of Agriculture
Nusa Cendana University
Kupang, Indonesia

Rajiv Nandan
College of Agriculture
Rani Lakshmi Bai Central Agricultural
University
Jhansi, Uttar Pradesh, India

V. Paramesha
Natural Resource Management Division
ICAR-Central Coastal Agricultural
Research Institute
Old Goa, Goa, India

Shubhi Patel
Department of Agricultural Economics
Institute of Agricultural Sciences
Banaras Hindu University
Varanasi, Uttar Pradesh, India

Amlan Kumar Patra
Department of Animal Nutrition
West Bengal University of Animal and
Fishery Sciences
Kolkata, West Bengal, India

G.A. Rajanna
ICAR-Directorate of Groundnut Research
Regional Station
Anantapur, Andhra Pradesh, India

Vishnu D. Rajput
Academy of Biology and Biotechnology
Southern Federal University
Rostov-on-Don, Russia

Meenu Rani
Department of Geography
Kumaun University
Nainital, Uttarakhand, India

Sanjay S. Rathore
Division of Agronomy
ICAR-IARI
New Delhi, India

Sapna Rawat
Department of Botany
University of Delhi
New Delhi, Delhi, India

Indranil Samanta
Department of Veterinary Microbiology
West Bengal University of Animal and Fishery Sciences
Kolkata, West Bengal, India

M.S. Sannagoudar
ICAR-Indian Grassland and Fodder Research Institute
Jhansi, Uttar Pradesh, India

Martiwi Diah Setiawati
Research Center for Oceanography
National Research and Innovation Agency (BRIN)
Jakarta, Indonesia

Alireza Sharifi
Department of Surveying Engineering
Faculty of Civil Engineering
Shahid Rajaee Teacher Training University
Tehran, Iran

Ragini Sharma
Department of Zoology
Punjab Agricultural University
Ludhiana, Punjab, India

Kapila Shekhawat
Division of Agronomy
ICAR-IARI
New Delhi, India

Abhishek Kumar Shukla
College of Agriculture
Rani Lakshmi Bai Central Agricultural University
Jhansi, Uttar Pradesh, India

Abhishek Singh
Department of Agricultural Biotechnology
Sardar Vallabhbhai Patel University of Agriculture and Technology
Meerut, Uttar Pradesh, India

Amit K. Singh
GSM Division
ICAR-IGFRI
Jhansi, Uttar Pradesh, India

Anil Kumar Singh
Division of Land & Water Management
ICAR-Research Complex for Eastern Region
Patna, Bihar, India

Anil Kumar Singh
University of Allahabad Senate House Campus
Allahabad, Uttar Pradesh, India

Awani Kumar Singh
Centre for Protected Cultivation and Technology
Indian Agricultural Research Institute
New Delhi, Delhi, India

H.P. Singh
Department of Agricultural Economics
Institute of Agricultural Sciences
Banaras Hindu University
Varanasi, Uttar Pradesh, India

Omkar Singh
Department of Soil Science & Agril. Chemistry
College of Agriculture, Sardar Vallabhbhai Patel University of Agriculture and Technology
Meerut, Uttar Pradesh, India

Prashant Deo Singh
GSM Division
ICAR-IGFRI
Jhansi, Uttar Pradesh, India

Rakesh Singh
Department of Agricultural Economics
Institute of Agricultural Sciences
Banaras Hindu University
Varanasi, Uttar Pradesh, India

R.K. Singh
Division of Agronomy
ICAR-IARI
New Delhi, India

Ram Kumar Singh
Department of Natural Resources
TERI School of Advanced Studies
New Delhi, India

Rudra Pratap Singh
College of Agriculture
Acharya Narendra Dev University of Agriculture and Technology
Ayodhya, Uttar Pradesh, India

Shashank Singh
Department of Agricultural Biotechnology
Sardar Vallabhbhai Patel University of Agriculture and Technology
Meerut, Uttar Pradesh, India

Susheel Kumar Singh
Department of Soil Science and Agricultural Chemistry
College of Agriculture
Rani Lakshmi Bai Central Agricultural University
Jhansi, Uttar Pradesh, India

V.K. Singh
Division of Agronomy
ICAR- Central Research Institute for Dryland Agriculture
Hyderabad, Telangana, India

Prem K. Sundaram
Division of Land & Water Management
ICAR-Research Complex for Eastern Region
Patna, Bihar, India

Aqil Tariq
State key Laboratory of Information Engineering in Surveying Mapping and Remote Sensing (LIESMARS)
Wuhan University
Wuhan, China

Ram Sewak Tomar
College of Horticulture and Forestry
Rani Lakshmi Bai Central Agricultural University
Jhansi, Uttar Pradesh, India

Keshav Tyagi
Forest Research Institute Deemed to be University (FRIDU)
Dehradun, India

Pravin Kumar Upadhyay
Division of Agronomy
ICAR- Indian Agricultural Research Institute
New Delhi, Delhi, India

Preface

Technological change has been the major driving force for increasing agricultural productivity and promoting agriculture development across the globe. In the past, the choice of technologies and their adoption was to increase production, productivity, and farm incomes. However, with changing agrarian economy, food habits, and climate scenario, demand for nutritious food and ecofriendly cultivation practices are becoming a major concern. Over many decades, policies for agriculture, trade, research and development, education, and training have been strong influences on technology adoption, agricultural production, and farm management. Agriculture is one of the most important sources of food nutrition, income, and employment in most of the developing world, including India. With passes of time, predominant rice, wheat, and other grains producing tracts have started showing symptoms of fatigue due to several eco-physical and socioeconomic constraints. Effects are witnessed as frequent drought occurrence, soil carbon depletion and degradation, and reduced farm income. Under these circumstances, cultivators, advisors, and policy makers are facing technological complexities, which are either available or under development, causing pressure on agricultural research and advisory services. Although, few attempts have been made in establishing the role of climate on crop productivity in current and future scenarios. But it does not consider non-climatic factors such as land use, technological advancement, change in irrigation pattern, soil fertility, and crop destruction due to insects, pests and diseases. Integrating all these may become robust tools for decision- and policy-making to prioritize the vulnerable zones that need immediate attention.

This book covers significant and updated contribution in the field of sustainable agriculture systems and technologies linked to climate change. The updated knowledge from countries like India, Indonesia, Japan, Sri Lanka, Iran, and China, is presented in this book through selected case studies for major thematic areas that have basic preliminary concepts and elaborates the scientific understanding of the relationship between sustainable agriculture systems and climatic drivers. The book has been separated into four major themes, each having subject-specific chapters to develop the concept and to present the findings in a lucid way that is useful for a wide range of readers. While the range of applications and innovative techniques is constantly increasing, this book provides a summary of key case studies to provide the most updated information. Chapters incorporate multisource data and information that offer critical understanding to explain the causes and effects of environmental changes linked to sustainable agriculture systems. This book will be of interest to researchers and practitioners in the field of agriculture, remote sensing, geographical information, and policy studies, etc., related to agricultural systems. Also, researchers, graduate, and postgraduate students of various disciplines, planners, and policy makers will find valuable information in this book. We believe that the book will be read by people with a common interest in sustainable development and other diverse backgrounds within earth observation.

The scientific quality of the book was ensured by a rigorous review process where leading researchers from India, Indonesia, Japan, Sri Lanka, Iran, and China, participated to provide constructive comments to improve the chapters. Due to the confidentiality of the review process, we are unable to provide their name; however, we are deeply indebted and thankful for their voluntary support. On behalf of the team of authors, we express our gratitude to the entire crew of Wiley for all kind of assistance to make this successful endeavor.

Pavan Kumar
A. K. Pandey
Susheel Kumar Singh
S. S. Singh
V. K. Singh

About the Editors

Dr. Pavan Kumar is an Assistant Professor at the College of Horticulture and Forestry, Rani Lakshmi Bai Central Agricultural University, Jhansi, Uttar Pradesh, India. He obtained his PhD degree from Faculty of Natural Sciences, Jamia Millia Islamia, New Delhi, India. He did a BSc (Botany) and MSc (Environmental Science) from Banaras Hindu University, Varanasi, India, and subsequently obtained his master's degree in Remote Sensing (MTech) from Birla Institute of Technology, Mesra Ranchi, India. His current research interests include resilient agriculture and climate change studies. He is recipient of Innovation China National academy award for Remote Sensing. Dr. Kumar has published 50 research papers in international journals and authored several books. He has visited countries including USA, France, the Netherlands, Italy, China, Indonesia, Brazil, and Malaysia for various academic/scientific assignments, workshops, and conferences. Dr. Kumar is a member of the International Associations for Vegetation Science, USA, and Institution of Geospatial and Remote Sensing, Malaysia.

Dr. A.K. Pandey is currently Dean at the College of Horticulture and Forestry, Rani Lakshmi Bai Central Agricultural University, Jhansi, Uttar Pradesh, India. Prior to joining RLBCAU, Dr. Pandey served for almost six years as Dean, College of Horticulture and Forestry, Pasighat, Arunachal Pradesh, under Central Agricultural University, Imphal, Manipur, India. He is an ARS Scientist of 1985 batch. Dr. Pandey obtained his MSc (Ag.) and PhD. degrees in Horticulture from C.S. Azad University of Agriculture and Technology, Kanpur. Dr. Pandey participated in the 1st International Post Graduate Course on Protected Agriculture in Arid and Semi-arid Region at Hebrew University, Jerusalem, Israel. He has published more than 80 research papers, 167 popular articles and review articles in journals of national and international repute. Dr. Pandey has authored 14 books and has participated in more than 70 national conferences/seminars and symposia and in the position of Organizing Secretary, organized several national seminars/symposia and one International Symposium on Minor Fruits, Medicinal & Aromatic Plants (ISMF, M&AP) at Pasighat, Arunachal Pradesh. Dr. A.K. Pandey has been conferred several awards and honors for his distinguished academic contributions. Hon'ble President of India has conferred on him the prestigious Rajiv Gandhi Gyan-Vigyan Purskar for his book Dalhani Sabjiya during 2012. Defense Research and Development Organization (DRDO) honored him four times for his significant contribution. Dr. Pandey is a prolific writer and for his significant contribution, Scientific and Applied Research Centre, Meerut, India has conferred on him the Outstanding Authorship in Science and

Technology Award (2010). He was also awarded the Life Time Achievement Award (2016) by BSRD, Allahabad. Dr. Pandey is a board member of a number of Scientific Societies and Academic panels. He is a Fellow of Indian Society of Vegetable Science (ISVS), Varanasi and Society of Biological Sciences and Rural Development, Allahabad, India.

Dr. Susheel Kumar Singh is currently Assistant Professor at the College of Agriculture, Department of Soil Science and Agricultural Chemistry, Rani Lakshmi Bai Central Agricultural University, Jhansi, Uttar Pradesh, India. He obtained his PhD degrees from Faculty of Soil Sciences, Sardar Vallabhbhai Patel University of Agriculture and Technology, Meerut (UP), India. He did a Bachelor of Science in Agriculture and Master of Science in Soil Science and Agricultural Chemistry from Tilak Dhari Post Graduate College Jaunpur affiliated to Veer Bahadur Singh Purvanchal University, Jaunpur, UP, India. He was also awarded as Research Fellow through IRRI-CSISA project during his PhD research program. Dr. Singh primarily works in the field of climate change, and related interdisciplinary fields with wider applications of Information Technology, Remote Sensing, and GIS tools with working experience of more than eight years. His current research interests include conservation agriculture and precision nutrient management studies. Dr. Singh has published several research papers including book chapters, as well as popular articles also.

Dr. S.S. Singh is the Director, Extension Education, Rani Laxmi Bai Central Agricultural University, Jhansi, Uttar Pradesh, India. He has served as Director, ICAR – Agricultural Technology Application Research Institute, Kolkata, WB (2017–2020). He was Head of Division, Crop Production at Indian Institute of Pulses Research, Kanpur, UP (2014–2017) and Head, Crop Research Division at ICAR Research Complex for Eastern Region, Patna, Bihar (2009–2014). Dr. Singh has also worked in RWC- CIMMYT, India, from 2004 to 2006. Dr. Singh has served RAU Pusa, Bihar, from 1986 to 1998 as Junior Scientist cum Assistant Professor. He is BSc (Ag) from CSAUAT Kanpur and MSc (Ag.) and PhD (Agronomy) from NDUAT, Faizabad, UP, India. Dr. Singh has handled 16 foreign/external funded projects on Natural Resource Management, Crop Management, Livelihood Development and Crop Improvement funded by DFID, IFAD, USAID, BMGF, IRRI, CIMMYT, Ford Foundation, and European Union. As an agronomist, he has contributed to the development of five rice varieties, which have been released by CVRC and Bihar SVRC and are suitable for aerobic drought prone, late direct seeding, contingency cropping, and rainfed lowland conditions. He has also guided and monitored ICAR/DAC extension projects like ARYA, Farmers FIRST, Skill Development, MGMG, TSP, SCSP, CFLD Pulses & Oilseeds, NICRA TDC, DAMU, and Seed Hub program from 2017 to 2020. He has published 115 research papers, 6 books, 20 book chapters, 15 technical bulletins, 135 papers in proceedings/symposium/seminar, 50 popular articles, and 40 extension folders. He has visited USA, UK, Australia, Mexico, Thailand, Philippines, Bangladesh, and Nepal. He is recipient of Rajeev Gandhi Gyan Vigyan Award from Ministry of Home Affairs, FAI award, Senior Research Fellowship of ICAR, Excellent Team Research Award of ICAR in Social Science.

Dr. V.K. Singh is currently, Director, Central Research Institute for Dryland Agriculture, Hyderabad, India. Dr. Singh has made valuable contributions in the area of soil fertility appraisal and soil health restoration through site-specific nutrient management (SSNM) and inclusion of legumes in intensive cropping systems. His effort on soil fertility appraisal using geo-statistical tools in different agro-ecologies revealed widespread multinutrient deficiencies. The extensive studies by him at cultivators' fields underlined the significance of SSNM for addressal of multinutrient deficiencies, improving yields, nutrient use efficiency, and profits under different cropping systems. He also explored different options for inclusion of legumes in rice–wheat system (RWS) to reduce subsoil compaction, enhanced organic matter accumulation, and minimized NO_3-N leaching. Dr. Singh has standardized conservation agriculture practices for improving soil health, nutrient and water use efficiency, and net returns. The Integrated Farming Systems models developed by him have great potential to raise the income and employment to small holders. Besides publishing research recognized peer-reviewed journals, he also published his work in popular language for the advantage of farmers and extension personnel. He possesses an illustrious academic record, with several awards and distinctions, viz. Fellow of NAAS, ISNS, ISA, SEE; NAAS Young Scientist Award, NAAS Associate, PS Deshmukh Young Agronomist Award, UPCAR Young Agricultural Scientist Award, IPNI-FAI award, FAI Golden Jubilee Award, PPIC-FAI award, Dr. J.S.P. Yadav Memorial Award of ISSS, Sriram Award, and Dhiru Morarji Memorial Award for his outstanding contributions in the area of efficient agronomic input management research.

Foreword 1

डॉ अरविन्द कुमार
कुलपति

Rani Lakshmi Bai
Central Agricultural University
Jhansi, UP, India

भाकृअनुप
ICAR

Dr. Arvind Kumar
Vice-Chancellor
Source: RLB Central Agricultural University

Modern agriculture depends heavily on engineering and technology and on the biological and physical sciences. Agriculture not only contributes to overall growth of the economy but also reduces poverty by providing employment and food security to the majority of the population in the continent, and thus it is the most inclusive growth sectors of the economy. In addition, growth in agriculture significantly influences the growth of nonagriculture sectors, also. Within the agricultural sector, smallholder farmers remain central to agricultural development and continue to play important roles promoting an ecologically rational and socially available food system. The ultimate goal or the ends of sustainable agriculture is to develop farming systems that are productive and profitable, conserve the natural resource base, protect the environment, and enhance health and safety, and to do so over the long-term. In recent past, satellite technologies available for agricultural applications promise to offer multiple benefits to the growers like estimating the timing of harvest, predicting in-season yields, understanding water and nutrient status, planning crop nutrition programmes and informing in-season irrigation, forecasting in diseases and pests, etc. Advances in satellite constellations, payloads, and launch are enabling increased connectivity and observational capability. Coupling these developments with "smarter" computing, data infrastructures, and analytics is increasing the possibilities for the use of satellite technologies for multiple uses in agriculture. While this creates new possibilities for products, services, and decision support, it also requires proper planning to ensure that the latest technology is linked appropriately with production challenges and, therefore, can be used to deliver the gains required to meet the societal, economic, political, and environmental needs.

The compiled text encircles updated information on sustainable agriculture systems and technologies addressing a variety of areas related to food security within context of sustainable practices, crop modeling, irrigation practices, micro-irrigation, agricultural statistics, agricultural economics, climate change scenario, flood routing, spatial modeling, farmers income, and agricultural policy in the twenty-first century. This book would serve as a hand book encompassing several scopes of interests on sustainable technologies toward reliable practices and income generation in areas agriculture, livestock, and fishery resources for sustainable agriculture as a whole.

This book would be beneficial for academics, scientists, environmentalists, meteorologists, environmental consultants, computing experts working in the areas of agricultural sciences.

Arvind Kumar

Foreword 2

Dr. Suresh Kumar Chaudhari
Indian Council of Agricultural Research
New Delhi, India
Deputy Director General (Natural Resource Management)
Source: Indian Council of Agricultural Research

08.06.2021

Agriculture is not only a key driver for inclusive growth of the economy in Asian countries but also has become means to elevate poverty, employment generation, and food security to the millions of growers, consumers, and other stakeholders in the continent. In addition, the forward and backward linkage effects of agriculture growth have also scaled up the incomes in the nonagriculture enterprises. Within the agricultural sector, smallholder farmers remain central to agricultural development and continue to play important roles for promoting an ecologically rational and socially available food system.

The primary aim of this book on "Sustainable Agriculture Systems and Technologies" is to advance the scientific understanding and application of technologies addressing a variety of areas related to food security within the context of sustainable productivity, system diversification, irrigation practices, precision agriculture, climate change, crop modeling, big data analytics, farmer's livelihood, and agricultural policy framework in the twenty-first century. This book will serve as a hand book encompassing several scopes of interests on sustainable agricultural technologies toward improved livelihoods, income generation and addressing sustainable development goals through agriculture, livestock, and fishery resources. A variety of scholars will also be benefited to explore risks and potential solutions in different agricultural systems under changing climate scenario, which can further be transmitted to the producers, policy makers, and other stakeholders. Further, this book will extend knowledge support for academics, researchers, and environmental consultant working within the framework of agricultural sciences.

I compliment the contributors and the editors for this worthy publication.

S.K. Chaudhari

Section 1

Food Security and Agrarian Livelihood

1

Agriculture and Nutritional Security in India

Shubhi Patel[1], Anwesha Dey[1], Rakesh Singh[1], and Ramesh Chand[2]

[1]*Department of Agricultural Economics, Institute of Agricultural Sciences, Banaras Hindu University, Varanasi, Uttar Pradesh, India*
[2]*Department of Plant Pathology and Mycology, Institute of Agricultural Sciences, Banaras Hindu University, Varanasi, Uttar Pradesh, India*

CONTENTS

1.1 Introduction

India is a young nation with the advantage of demographic dividend. Demographic dividend is accompanied with economic growth, provided, good health, education, and employment opportunities exist (Bloom and Williamson 1998; Ross 2004). At present conditions, India will sustain demographic dividend for 37 more years and thus, the health of children is an important aspect as it tells about the past and also the future of the workforce (UNICEF 1998). In 2020, India ranked 94 out of 107 countries in Global Hunger Index, third economy out of 181 in GDP (PPP) and 131 out of 189 countries in Human Development Index. The level of hunger in India is categorized as serious (Global Hunger Index 2020). The child wasting rate is 17.3%, child stunting rate is 34.7%, and under-five mortality rate is 3.7%; there is a 14% prevalence of undernourished children and a decreasing trend in the global hunger index score since 2000. Although, India ranked 71 out of 113 countries in food security index, and it is noteworthy that it was ranked 37 in food availability, 76 in affordability, 74 in quality and safety, and 98 in natural resources and

Sustainable Agriculture Systems and Technologies, First Edition. Edited by Pavan Kumar, A.K. Pandey, Susheel Kumar Singh, S.S. Singh, and V.K. Singh.

resilience (Global Food Security Index 2020). Being a leading producer of food grains make it obvious to score well in availability of food but this merely is not enough as quality and safety are also important. Food security is composed of four dimensions, i.e. availability, access, utilization, and stability of the three dimensions over time (FAO 2008). Equitable distribution of food ensures that the good quality, available food reaches the table of every household in the nation. Failing to do so leads to hunger which means distress related to lack of sufficient calories (Global Hunger Index 2019). This leads up to another complex dimension that is malnutrition. Malnutrition refers to deficiencies, excesses, or imbalances in a person's intake of energy and/or nutrients (WHO 2021). The term malnutrition covers two broad groups of conditions one is "undernutrition" and other is overweight, obesity, and diet-related noncommunicable diseases (WHO 2021). Undernutrition is defined as the outcome of insufficient food intake and repeated infectious diseases, it includes being underweight for one's age, too short for one's age (stunted), dangerously thin for one's height (wasted), and deficient in vitamins and minerals (micronutrient malnutrition) (UNICEF). Stunting is an indicator of chronic undernutrition, the result of prolonged food deprivation and/or disease or illness; wasting is an indicator of acute undernutrition, the result of more recent food deprivation or illness; underweight is used as a composite indicator to reflect both acute and chronic under nutrition, although it cannot distinguish between them (WHO 1995). Undernutrition has been a cause of health issues in developing countries (Rice et al. 2000; Schofield and Ashworth 1996; WHO 2002). This is a matter of concern because; stunting affects the mental development of the child and reduces their productive efficiency (Mendez and Adair 1999). Thus, posing a risk on the future of nations and emerging as a public health challenge. Not only this, the economic losses caused by malnutrition are 11% of GDP every year in Africa and Asia, whereas the rate of return on investment in prevention is $16 for each dollar invested (Global Nutrition Report 2016). Thus, preventing malnutrition is an economically viable option.

India is opting for a multipronged approach to tackle the burden of malnutrition through nutritional programmes like Mid-day meals, Integrated Child Development Programme, Public Distribution System (PDS), National Nutrition Mission, Bio-fortification, etc. Biofortification is the process of using conventional plant breeding techniques to enrich staple food crops with higher level of vitamin A, zinc, and iron. PDS in India covers a large amount of beneficiaries who are unable to afford a minimum dietary requirement. Distribution of bio-fortified wheat though PDS can help uplift the nutritional status of the majority. Sustainable development goals setup in 2012 have given a blueprint of actions needed to be taken for ensuring a safe and sound future for the upcoming generation. Zero hunger, as the second goal says, it targets to end hunger, ensure food security, and achieve sustainable agriculture development. And our performance in the SDG index is important on global scale as one-sixth of world population resides in India. In the SDG India Index 2020, nutrition and gender equality have a long road to take to perform well in the score. Occurrence of pandemic like COVID-19, where complete closure of economic activities has been witnessed, also threatens the nutrition security of the population (specially the unorganized sector). The slow economic growth rate due to COVID-19 will lead to reduced aggregate demand, consumption expenditure, and hamper income as well as food security (IFPRI 2015). Thus, it becomes important to point out where we are at the nutritional level, how much have we have improved and what are the factors responsible and associated with the prevalence of under-nutrition.

1.2 Growth of Agriculture in India

India is a leading producer of food grains, milk, wool, and other agricultural products. But what is the situation at present and how has India become what it is today, the journey to achieve this position was not a child's play. It began in the 1960s with the advent of the green revolution. During the independence period, India was left with own national governance and food insecurity. We were importing rice and wheat to feed the starving population. In 1966, India was the leading importer of rice and wheat. This was the time of introduction of green revolution. Introduction of dwarf wheat variety, subsidies on fertilizers, irrigation expansion, procurement at minimum support price, and chemicals for crop protection were done to increase the production (Mandal et al. 2020; Pandey et al. 2015). And the improved package of practice actually increased the production manifold. In 11 years, India's wheat production increased 205% and rice production increased 72% (Table 1.1) bringing India among the leading exporters at the global level in 1977–1978. The productivity (kg/hectare) increased 79% in wheat between 1966–1967 and 1977–1978, which was a record increment. The increase in rice productivity (kg/hectare) was 51% between 1966 and 1977. Since then, the production of wheat has but rice has not seen a setback. Area under wheat grew at an annual rate of 1.68 while production speeded up with Compound Annual Growth Rate (CAGR) of 4.3. Wheat has been a winner in green revolution with record increase in production and ensuring that the country is food secure. Area under rice grew at an annual rate of 0.55 while production speeded up with CAGR of 2.45.

Wheat and rice are among the staple food of the country. Apart from this the government of India procures wheat and rice at Minimum support price and distributes it to the below poverty line population at subsidized price under National food security programme (Bisht et al. 2014; Singh et al. 2013; Tomar et al. 2014; Yadav et al. 2014). It is known that wheat and rice have a lion's share in procurement under Minimum Support Price. This has led to

Table 1.1 Percentage change in area production and productivity of wheat and rice in India.

	Area		Production		Productivity	
Year	Wheat	Rice	Wheat	Rice	Wheat	Rice
1965–1966	17.7	13.4	29.6	8.4	10.3	−4.4
1977–1978	70.7	13.6	205.5	72.2	79.0	51.7
1989–1990	9.5	4.7	57.0	39.7	43.3	33.4
2001–2002	12.1	6.5	46.0	26.9	30.2	19.1
2013–2014	15.7	−1.7	31.7	14.3	13.9	16.2

Source: Indiastat.com,www.fao.org/faostat.

cereal dominance in food grain production. Per capita availability of food grains has shown increase in rice, wheat, and cereals, while it has declined in the case of gram, pulses, and other cereals (Figure 1.1). This reveals that the advent of green revolution has shifted India from food insecure to food secure country. This raises up a question that whether being food secure ensures zero hunger?

1.2.1 Role of Agriculture in Nutrition

Agriculture plays a pivotal role in nutrition (Kadiyala et al. 2011). Agriculture provides the major food items required for the balanced dietary intake. As already discussed in the above section, India has made significant growth in agricultural production indicating that the availability of food is not a constraint in ensuring nutrition. But is this the only way through which agriculture ensures nutrition security? Nutrition security has three dimensions – access to adequate food, care and feeding practices, and sanitation and health (UNSCN). Here, the emphasis is on access to adequate food. The access to adequate food comes from the ability to pay for the demand of food. Source of livelihood here plays an important role in this ability to purchase apart from the subsidized food grains obtained from the NFSA. In 2020, agriculture employed around 58% of the Indian population and was the primary occupation of the country (IBEF 2021). In monetary terms, the Rs.19.48 lakh crores of Gross Value Added (GVA) was for agriculture, forestry, and fishing in 2020. Thus, a major source of income for a major section of Indian population. The year 2014 was recognized as the year of family farming emphasizing the role of agriculture in food and nutrition security. Moving to the care and feeding determinant of nutrition security it can be covered under agriculture by ensuring crop diversity and development of bio-fortified food grains which have higher nutritional content, diversification of cropping pattern in order to ease the availability of a variety of food items at local level. And for the

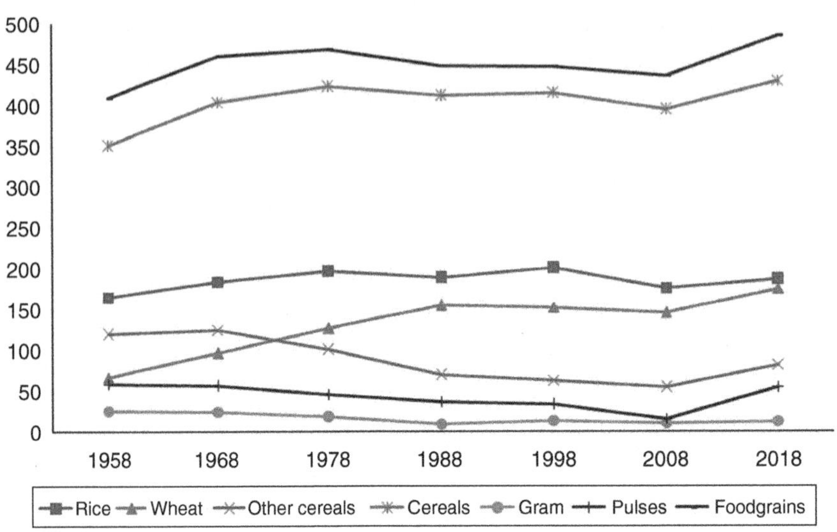

Figure 1.1 Per capita per day availability of food grains 1958–2018 in gm/day.

third determinant a model named "Farming System for Nutrition" model was developed (Das et al. 2014). This model suggested the location-specific nutrition focused agriculture encompassing the diet need and surrounding environment development. Thus, it can be seen that the upliftment of nutritional security revolves around agriculture.

1.3 Dynamics of Under Nutrition in India

Hunger is measured considering children under five years who are under nourished, stunted, wasted, underweight, and under-five mortality rate. Stunting means the percentage of children under five years who are stunted (height to age) is presented in Figure 1.2a from data obtained from National Family Health Survey (NFHS) 4. It shows that the highest prevalence of stunting is seen in Bihar, Uttar Pradesh, Jharkhand, Meghalaya, and Madhya Pradesh with more than 40% of children stunted, while Kerala, Goa, and Tripura have less than 25% of children stunted. Lessons can be taken from these states to improve the status in other states. Also, regional disparities were observed with higher percentage of stunting in rural area as compared to urban area. This is subject to the reasons that sanitation, antenatal care, and literacy are higher in the urban side. This means, emphasis on improvement of these parameters can help uplift the children from malnutrition.

Child wasting means low weight for height and indicates the acute food shortage or prevalence of disease in children (UNICEF). As per NFHS 4 Jharkhand, Gujarat, Karnataka, Madhya Pradesh, and Maharashtra, more than 23% of children are wasted (Figure 1.2b), while Mizoram and Manipur have less than 10% wasted children. The reasons can be similar to the reasons of stunting. There is 50% chance that a stunted child will also be wasted. The urban rural divide is prevalent in this case too. Hunger also leads to underweight children and if prolonged it leads to mortality of children. As seen in Figure 1.2c, in Madhya Pradesh, Jharkhand, and Bihar more than 40% of children are underweight, while Manipur, Mizoram, and Sikkim have less than 15% underweight population of children under five years. This shows strong association between stunting, wasting, and underweight of children as Madhya Pradesh, Jharkhand, and Bihar have the highest prevalence of all these parameters. Under five infant mortality is highest in Uttar Pradesh, Madhya Pradesh, and Chhattisgarh, i.e. more than 60 deaths per 1000 lives (Figure 1.2d). This reveals that Madhya Pradesh, Bihar, Uttar Pradesh, and Jharkhand have the highest level of malnourishment in India. Combining all the four parameters, we see that Madhya Pradesh, Bihar, Uttar Pradesh, and Jharkhand have the highest level of undernourishment in India. An insight into the economy of these states shows that Madhya Pradesh has 62% of labor force participation in agriculture and agriculture contributes 45% to the Gross State Value Added (GSVA) (Madhya Pradesh Budget Analysis 2019–20). Uttar Pradesh, the most populous state of India (17.11%) is the leading producer of food grains in India (Agricultural Statistics at a Glance 2018). Jharkhand is a rural state with 76% of its population living in rural area. And almost 49% of its population lives below poverty line. Bihar has 23% of its GSVA to the state's economy by agriculture (Bihar Budget Analysis 2019-20). Apart from this out of total labor force the casual labor force of Bihar is 32.2%, Jharkhand 23.6%, Madhya Pradesh 28.2%, and Uttar Pradesh 21.3% and they are not able to find job after COVID-19 outbreak. Percentage of population below poverty line of Bihar is 33.7%, Jharkhand 36.7%, Madhya Pradesh 31.7%, and Uttar Pradesh

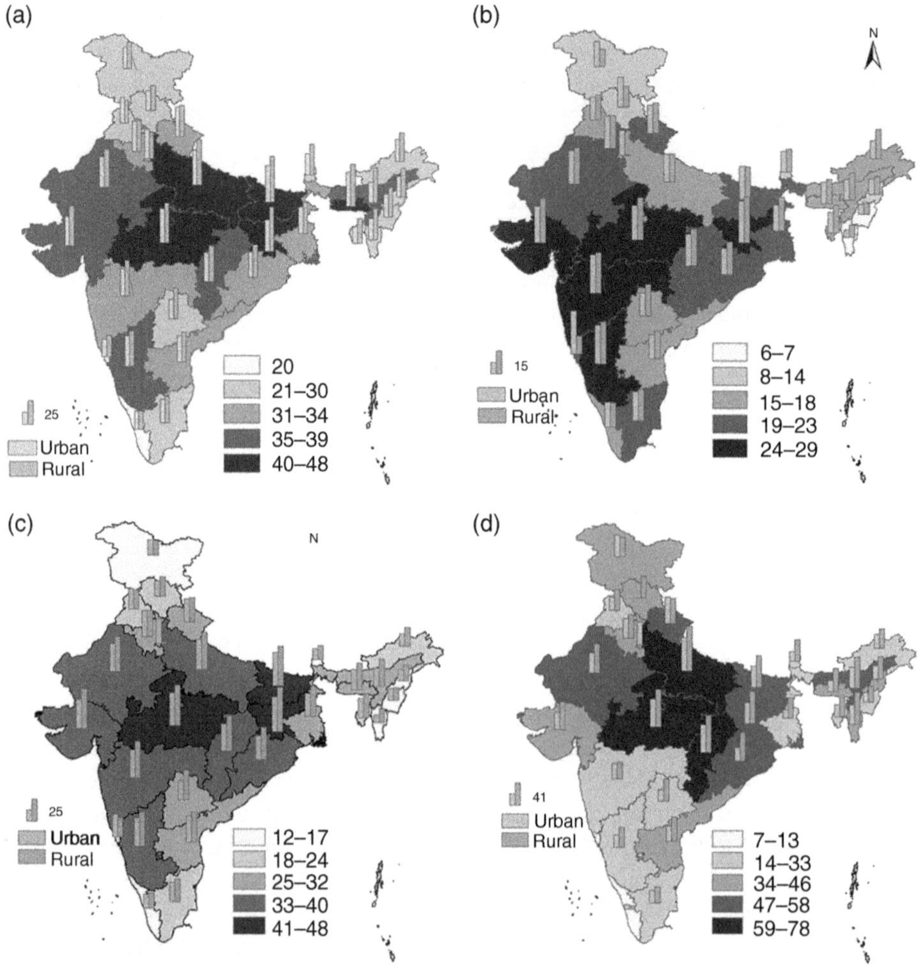

Figure 1.2 Status of undernutrition in India. (a) Prevalence of stunted children. (b) Prevalence of child wasting. (c) Prevalence of underweight children. (d) Under-five mortality rate. *Source:* Based on NFHS 2015–16. (*See insert for color representation of the figure.*)

29.4% and Uttar Pradesh has 17.12% of population, while Bihar, Madhya Pradesh, and Jharkhand comprise of 9, 6, and 3% of population, i.e., 35% of the population of India. These states are mainly agriculture dependent making the population more vulnerable. Madhya Pradesh and Uttar Pradesh being highly populous states in spite of having high number of beneficiaries under Integrated Child Development Scheme are still among the poor performers. This implies that unequal distribution of income might be the reason that despite economic growth and food production, prevalence of hunger is there.

1.3.1 Trends Over Time

The performance of the states over two National Family Health Survey (NFHS) reveals that the situation has improved over the years in most of the cases (Figure 1.3). NFHS 3 conducted in 2005–2006 and NFHS 4 conducted in 2015–2016 although not strictly comparable, have substantial results to show. The reduction in percentage of children who are stunted was highest in Arunachal Pradesh (32%), Tripura was followed by Himachal Pradesh and Punjab. It is noteworthy that Tripura, Himachal Pradesh, and Punjab are among the top five states with the least percentage of stunted children. While Uttar Pradesh,

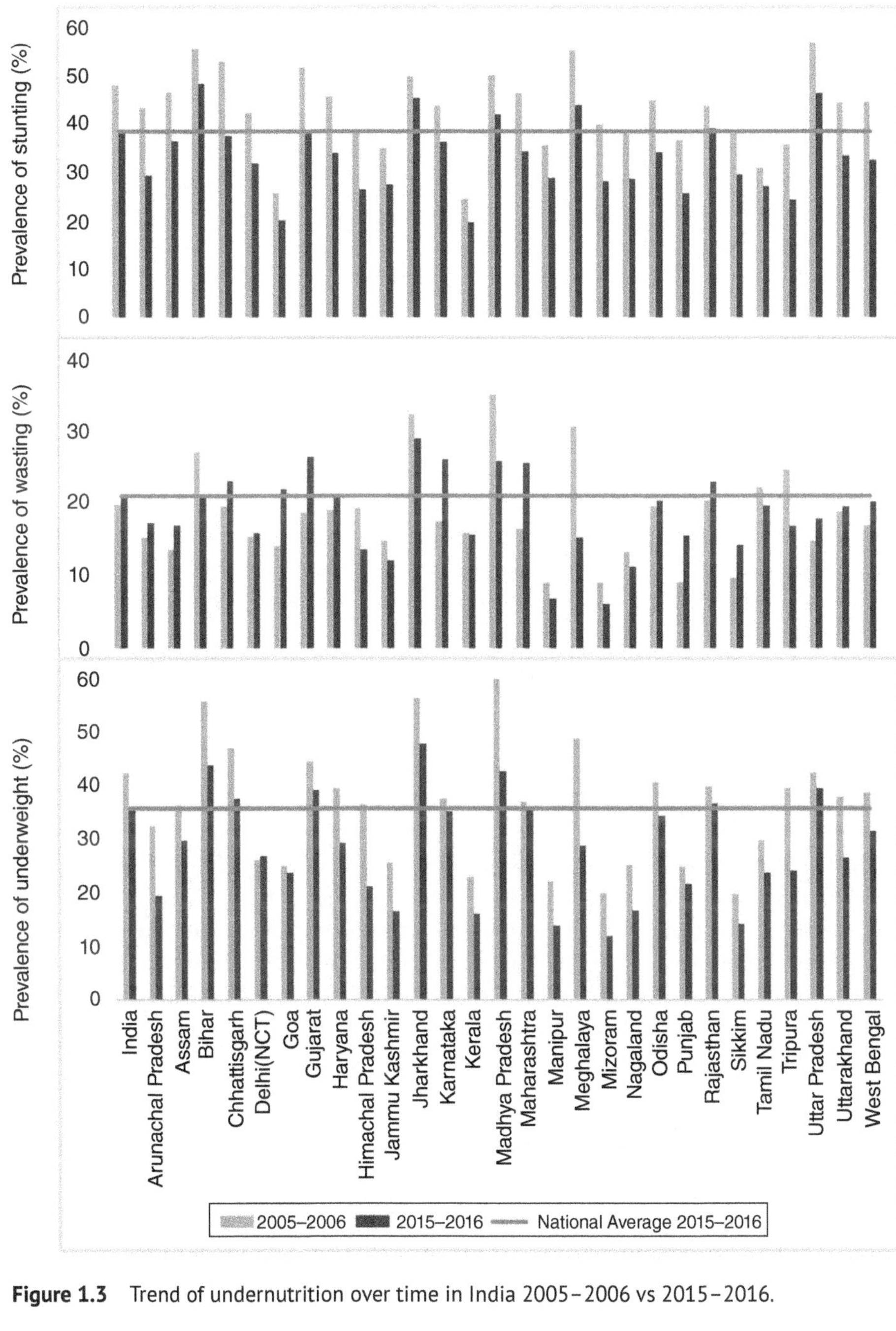

Figure 1.3 Trend of undernutrition over time in India 2005–2006 vs 2015–2016.

Madhya Pradesh, and Jharkhand could not make substantial efforts to reduce the stunted children percentage and remain at the top five states with the highest percentage of stunted children. In case of child wasting, the results have been striking. Meghalaya, Mizoram, and Tripura have shown the highest level of reduction in percentage of population wasted, while increase in child wasting has been observed in Orissa, Haryana, Andhra Pradesh, Chhattisgarh, West Bengal, Uttar Pradesh, Assam, Gujarat Sikkim, Karnataka, Maharashtra, Goa, and Punjab with highest increase in the percentage. Madhya Pradesh has reduced by 26% indicating efforts done by the government. Despite this fact Madhya Pradesh, Jharkhand, Gujarat, Karnataka, and Maharashtra have higher percentage than the national average. Child underweight situation means weight-to-age situation Mizoram has reduced by 39% and is the state with the least percentage of underweight children. Under-five infant mortality again is highest in Madhya Pradesh, Jharkhand although Madhya Pradesh has reduced it by 30% but the highest decrease of 62% was witnessed in Arunachal Pradesh.

Average Annual Rate of Reduction – It is the average relative percentage decrease per year in prevalence or rate (WHO). A positive sign indicated reduction in the prevalence and negative sign indicates increase in the prevalence. It is calculated by the formula:

$$Y_{t+n} = Y_t\left(1-b\%\right)^n$$

where Y_{t+n} is the prevalence of the next year, Y_t is the prevalence of the given year, and $b\%$ is the annual rate of reduction (UNICEF 2015).

The calculation of Average Annual Rate of Return revealed that the AARR of stunting in India is 2.2% which means till 2030 the prevalence will be reduced up to 33% if the situation remains same. This is a positive pace but slower in speed because the target of SDG is to reduce the stunting prevalence by 50% by 2030 from the level of 2012. The AARR of wasting (Table 1.2) shows that there are states like Arunachal Pradesh, Tripura, Himachal Pradesh, Punjab, Mizoram, Chhattisgarh, West Bengal, and Nagaland which have an AARR of more than 3%. These states will reach around 50% reduction by 2030. But at the same time, states like, Bihar, Tamil Nadu, Madhya Pradesh, Rajasthan, and Jharkhand have AARR lower than 1.5% which brings down the national average and makes reaching the SDG target tough.

In conclusion, we can say that the most vulnerable states in terms of hunger security are Madhya Pradesh, Uttar Pradesh, Jharkhand, and Bihar. The states which have performed well in reducing the prevalence are Tripura and Himachal Pradesh. Both the states have least prevalence of hunger. Giving us a lesson, although centrally sponsored schemes are there, still regional disparity makes the difference. States with higher dependence on agriculture and tribal population have higher level of hunger, indicating unequal income distribution and lack of improvement in influencing factors.

1.3.2 Association with SocioEconomic Indicators

Poverty affects nutrition (Nelson 2000) leading to undernutrition in the poor population. Stunting was the most dominant phenomenon among children under five years. So the percentage of stunting was associated with socioeconomic factors like percentage of

Table 1.2 Average Annual Rate of Return of stunting and wasting from 2005–2006 to 2015–2016.

State	AARR in stunting	AARR in wasting
India	2.21	−0.59
Arunachal Pradesh	3.80	−1.24
Assam	2.42	−2.18
Bihar	1.40	2.61
Chhattisgarh	3.36	−1.71
Delhi(NCT)	2.76	−0.32
Goa	2.39	−4.50
Gujarat	2.90	−3.51
Haryana	2.91	−1.05
Himachal Pradesh	3.76	3.37
Jammu Kashmir	2.42	1.99
Jharkhand	0.94	1.07
Karnataka	1.87	−4.02
Kerala	2.16	0.13
Madhya Pradesh	1.73	3.00
Maharashtra	2.93	−4.49
Manipur	2.06	2.76
Meghalaya	2.27	6.73
Mizoram	3.42	3.81
Nagaland	3.00	1.62
Odisha	2.74	−0.40
Punjab	3.50	−5.42
Rajasthan	1.11	−1.21
Sikkim	2.54	−3.88
Tamil Nadu	1.30	1.19
Tripura	3.77	3.74
Uttar Pradesh	2.02	−1.92
Uttarakhand	2.78	−0.37
West Bengal	3.12	−1.85

Source: Calculated by author from data obtained from NFHS 4.

population below poverty line, per capita GDP, and growth rate of the state. Undernutrition is a result of poor dietary intake, poor maternal health combined with lack of safe water and sanitation along with poor health services (UNICEF 1998). Undernutrition is responsible for poor mental health (Martins et al. 2011), higher vulnerability to ill health, and a reduced physical work capacity (Non et al. 2016), which is making the workforce inefficient and thus posing a problem in the economic development of nations. Further, under nutrition which is caused by poverty also leads back to undernutrition which again causes poverty and the cycle continues. The marginal propensity to consume is higher for poor people because they spend higher proportion of their income in consumption. Thus, decline in income will hamper the consumption expenditure. Bivariate analysis was done to work out relationship between prevalence of stunting and each economic factor under consideration. The association between the percentage of stunted children and the percentage of the population below poverty line is strong and positive (Figure 1.4a). This is anticipated because poverty leads to insufficient food intake, less prenatal care, child malnutrition, and unhealthy diet. A few states however deviate from the predicted line. Meghalaya, Rajasthan, and Gujarat are clear negative outliers with a much higher percentage of stunted children as compared to their poverty level. Goa, Kerala, Manipur, and Arunachal Pradesh, on the other hand, are positive deviants, i.e. they have lower percentage of stunting in comparison to the level of poverty. Improved drinking water sources, improved sanitation facility, use of iodized salt, literacy among women, antenatal care, and anemia among children and women also impact the level of undernutrition in India (Ghosh 2020).

A scatterplot of the percentage of stunting and net state domestic product (NSDP) per capita, with the latter serving as a proxy for each state's per capita income in Figure 1.4b. In this case, the two variables show a negative association, with poorer states having a significantly higher percentage of stunting as compared with more prosperous states. The association though has a number of outliers. For instance, Cluster 1- Bihar, Uttar Pradesh, Jharkhand, Meghalaya, Uttar Pradesh, and Madhya Pradesh have much higher level of stunting as expected from states of their income level. While states like Manipur, Mizoram, Tripura, and Jammu Kashmir are positive deviants with much lower percentage of stunting as compared with states with similar income level. These are states with better sanitation, literacy, and care for pregnant women. This indicates that not only income but also other socioeconomic factors might also be major contributors. The association between the percentage of stunting and the rate of economic growth for each state is shown in Figure 1.4c, which shows not much strong relationship between the two variables. Meghalaya with negative growth rate in 2014–2015 has a high level of stunting 42% but so did states like Jharkhand, Bihar, Uttar Pradesh, and Madhya Pradesh with much higher growth rate as compared to Meghalaya. Goa and Mizoram have a much higher growth rate but the percentage of stunting is not low in these states. This gives a conclusion that growth rate of a state is weakly associated with states prevalence of malnutrition. Thus, it can be concluded that despite economic progress, India has to struggle to combat malnutrition that adversely affects the country's socioeconomic progress. This is not due to food scarcity but due to the lack of equitable distribution of the available food. Food security depends not only on crop production, but also on the policies that affect food systems, from farm to table (World Food Conference 1974). To improve the diet consumption of whole population, when there is an increasing growth rate and unbalanced growth of working sectors, injecting money is not the solution. There are other challenges like inequitable distribution of income, lack of awareness and access to healthy food, climate change, slowdown of economy, and lower income levels that threaten the hunger of the countrymen.

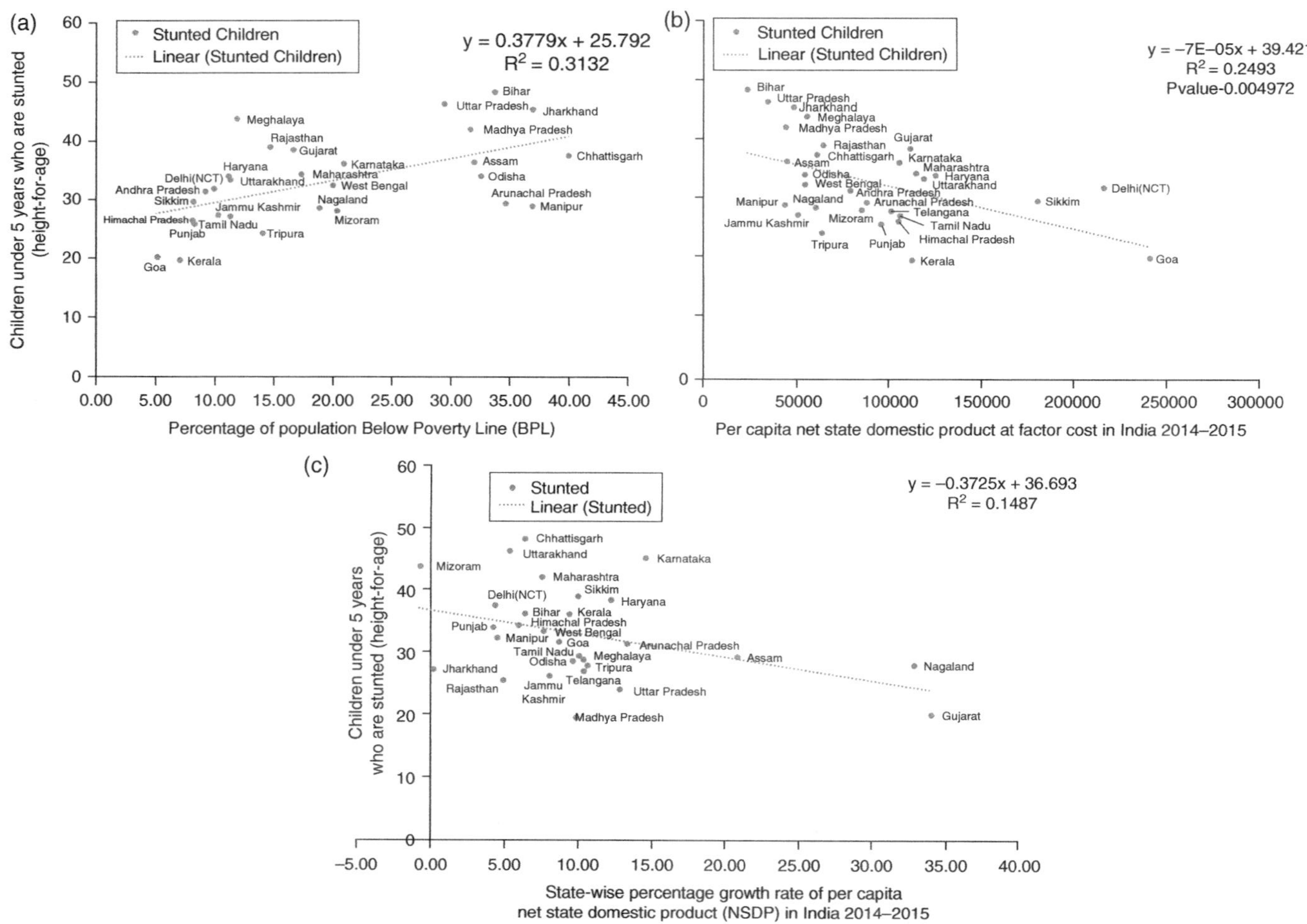

Figure 1.4 Association of stunting prevalence with socio-economic indicators. Note- (a) State-wise percentage of BPL (based on MRP-consumption) in rural and urban areas of India, 2011–2012 (b) Measurement at current prices based on 2011–2012 series, (c) State-wise, (at constant prices based on 2011–2012 series).

Comparison of the states with lower stunting percentage and those with higher stunting percentage, based on seven indicators, was done viz. improved drinking water sources, improved sanitation facility, use of iodized salt, literacy among women, antenatal care, and anemia among children and women (Figure 1.5). These indicators also impact level of undernutrition in India (Ghosh 2020; Martorell and Young 2012). The interstate comparison showed that states who have higher prevalence of stunting, underweight and wastage have lower percentage of improved sanitation facility, literacy among women, antenatal care, and higher percentage of anemia among children and women as compared to states having lower prevalence of the same factors. These are the factors which influence absorption of nutrients, access to nutritious food, and care of feeding right diet to the child at the right time.

1.4 Institutional Interventions to Cope Up with Malnutrition

Nutrition is accredited as one of the important characters of human development, dealing with poverty reduction and scaling up the economy. As discussed above, a nutrition policy is needed because the attainment of food self-sufficiency does not ensure nutrition security. India, being the leader in production of a number of food products, lags behind when it comes to the nutrition status of children. This gap between production and consumption urged a need to devise a policy that could ensure the upliftment of the nutrition status. In such situations, it is the institutional intervention that emerges as a problem solving resort to cope up with the issue. Since independence, the Government of India has taken various steps to deal with the burden of malnutrition which can also be called as the double burden

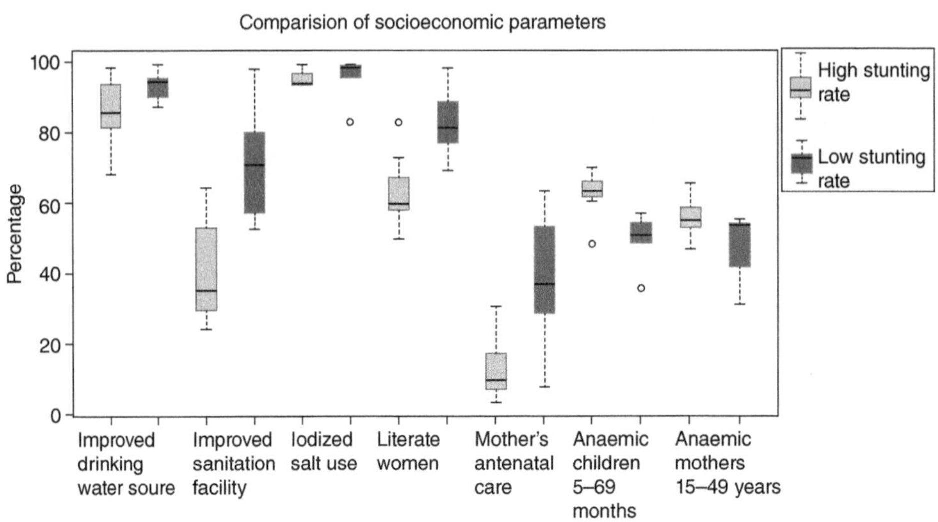

Figure 1.5 Comparison of socioeconomic indicators of states with higher rate of stunting with those having lower prevalence.

of malnutrition. But, before moving to the policies adopted and executed, it is imperative to understand the mechanism of the causes of malnutrition. Malnutrition is not something that is caused by a single causal factor or habit or reason. Occurrence of malnutrition is a complex activity and is caused by a set of factors which can be majorly divided into (i) personal, (ii) socioeconomic, and (iii) administrative. Among the personal factors, the problem associated with the individual are insufficient dietary intake, inadequate absorption of nutrients, inadequate care of child and mother, and mismanagement of disease prevention and care. Factors relating to the habit of the child and mother can be rectified through women empowerment, awareness, proper diet, and medical care. The socioeconomic factors are poverty, unorganized livelihood, lack of resources, lack of access to health care, unavailability of safe drinking water, unhygienic surroundings, etc. These call for development efforts at societal level where the area administration needs to ensure the following: availability of safe drinking water, cleaning of the area, generation of employment options. The last and the foremost factor is the administration where despite the existence of suitable schemes and programmes to eradicate the existing causal effects, the implementation lags behind. Reasons like poor scheme implementation, low quality of food distributed, lack of awareness, and inadequate evaluation prevail. This implementation lacunae need to be addressed and taken care of. Policies play an important role in curbing the issue of nutrition (Hawkes 2005; Mozaffarian et al. 2018). Integrated approaches are required for solving the issues of hunger and malnutrition (Burchi et al. 2011). While formulating the policy there are three things that should be kept in mind as the policy directly increases the coverage of nutritious food, incorporates behavioral changes among mothers, targeting adolescent girls, prenatal and postnatal care and nutrition, and provision of nutrient rich food.

The indirect effect of a policy includes change in consumption pattern, food security, income security, food safety, and health care. And in order to make it effective it is necessary to do surveillance of implementation, research and development, extension of knowledge, and awareness programmes. The history of policy shows that there was Special nutrition programme which started in 1970–1971 to provide Supplementary feeding to children, pregnant, and lactating mothers. Also Balwadi Nutrition Programme started in the same time period. In 1975, came Integrated Child Development Services, that focused upon Supplementary nutrition, immunization, health check-up. Day Meal Programme 1962–1963, Integrated Child Development Services 1975, Tamil Nadu Integrated Nutrition Programme 1980, Wheat-based supplementary nutrition programme 1986, Mid Nutrition Anemia Prophylaxix Programme 1970, National Diarrhea Disease Control Programme 1978, National Goiter Control Programme 1992, National Nutrition Policy 1993, Targeted Public Distribution System 1997, National Health Policy 2002, Mahatma Gandhi National Rural Employment Guarantee Scheme 2005, National Food Security Mission 2007, National Rural Drinking Water Programme 2009, Matritva Sahyog Yojana 2010, SABLA for adolescent girls 2011, National Rural Livelihood Mission 2011, National Policy for Children 2013, National Food Security Act 2013, and National Health Mission 2013, including Janani Suraksha Yojana, Swachh Bharat Abhiyan 2014. The national food security act (NFSA) shifted the approach from welfare to right based. The act legally entitles up to 75% of the rural population and 50% of the urban population to receive subsidized food grains under Targeted Public Distribution System (nfs.gov). And under the act, the eldest woman of the house of age 18 or above is entitled as the head of the house and the ration card is issued in her name, thus empowering women too. At present, around 80 crore persons have benefitted from the act. The NFSA has a life-cycle approach in which the nutrition is provided as per the life stage of the beneficiary. It provides prenatal care and postnatal care by entitling the pregnant women and lactating mothers to receive nutritious meal free of cost through Integrated Child Development Services (ICDS) centers, called Anganwadi Centers and to children through mid-day meal scheme. Under NFSA, the foodgrains are made available at a subsidized price of rice at Rs 3/kg, wheat at Rs 2/kg, and coarse grains at Rs 1/kg. Mid-day meal scheme was launched in 1962–1963 in which meal of one time (i.e. lunch) is given to children of class 1 to eighth in government primary schools. This helps to improve the school attendance of children by avoid dropping out of school and ensuring child nutrition (Laxmaiah et al. 1999; Si and Sharma 2008; Singh and Gupta 2015). Especially in the case of situations like drought, this scheme acted as a safety net for child nutrition (Singh et al. 2014). And at present 11.59 crore children have been enrolled under mid-day meal programme and it has provided employment to 25.95 lakh cook-cum-helpers (mdm.nic). The gloomy picture of prevalence of undernutrition shows that despite such multipronged schemes, lags existed. Studies have shown that distribution of subsidized food does not ensure nutrition (Desai and Vanneman 2015). In future of 2050s, the population of India will increase and there will be need to develop transportation and storage infrastructure as, at present, the buffer stocks maintained are not evenly distributed across the country (Chakraborty and Sarmah 2019). Similar drawbacks in not able to provide proper nutrition were observed in mid-day meal scheme (Chhabra and Rao 2014; Prakasam 2019).

1.5 Policy Implication

The situation of having sufficient amount of calories for the Indian population and yet evidences of significant lack in children and women shows the gap between demand and supply. The scattered distribution of prevalence of malnutrition across the states indicates that there is a need to classify the regions as per the gravity of the situation. There was more percentage of affected children in rural area as compared to urban area, suggesting that the focus needs to be more toward the rural beneficiaries. The surrounding environmental conditions in rural area are less clean and hygienic than those of urban area, hence the implementation of Swachh Bharat and village sanitation should be done more efficiently than that at present state. The situation of other socioeconomic indicators shows that there is need of a holistic approach toward improving the nutrition as well as education, awareness, and hygiene of the vulnerable states. The vulnerable sections need to be identified and then ways to tackle them need to be devised at localized scale rather than with a single policy at national level. There is a need to develop smooth transit of quality agricultural produce to different regions as the supply is the issue and not the availability. The focus needs to be more on raising the standard of living of vulnerable sections rather than increasing the growth rate of the state, as it does not indicate equitable distribution of income. Agriculture can help in improving the situation provided the focus shifts from production centric to profit-centric agriculture. Make sure that the implementation is done properly through better monitoring and strict evaluation.

1.6 Conclusion

Green revolution ensured the food security of the nation. But if looked into details, food security and nutrition security although sound similar but are dimensionally different aspects. Nutrition security is more concerned about the nutrient intake and then the absorption of the quality nutrients by the child and mother, while food security deals with availability, access, and affordability of quality food. India has gained paramount success in increasing the production, productivity of agricultural crops, and allied activities. But, the status of nutrition in children shows that this production has not been able to fulfill the calorie requirement of the children. This indicates a disparity between the production and consumption of the food items along with its equitable distribution. The SDG 2 – Zero hunger targets to end hunger, end malnutrition, increase agricultural productivity by 2030. When observed, undernutrition in India is a serious issue that needs immediate attention. This is because it starts from the health of the mother and then the nutrition of the child, thus, improvement in this status is a time consuming task. In states like Madhya Pradesh, Uttar Pradesh, Bihar, and Jharkhand one out of every two children is stunted and is suffering from undernutrition. These are states with dominance of agriculture and tribal population (Agricultural statistics at a glance 2018). The AARR of stunting in India is 2.2% which means till 2030, the prevalence will be reduced up to 33%, if the situation remains the same. This is a positive pace but slower in speed because the target of SDGs is to reduce the stunting prevalence by 50% by 2030 from the level of 2012. There is a dire need of research intervention on effect of socioeconomic factors, climate change, etc. on malnutrition. Investments will be needed to strengthen agriculture, improve food quality and nutrition, enhance overall food availability and access to all population segments, and to improve child nutrition and mortality outcomes.

References

Agricultural Statistics at a Glance (2018). Directorate of Economics and Statistics, Ministry of Agriculture & Farmers Welfare, Government of India 2019.

Bihar Budget Analysis (2019–20). Bihar budget analysis 2019–20. https://prsindia.org/budgets/states/bihar-budget-analysis-2019-20 (accessed 20 April 2020).

Bisht, P., Kumar, P., Yadav, M. et al. (2014). Spatial dynamics for relative contribution of cropping pattern analysis on environment by integrating remote sensing and GIS. *Int. J. Plant Prod. 8* (1): 1–17.

Bloom, D.E. and Williamson, J.G. (1998). Demographic transitions and economic miracles in emerging Asia. *World Bank Econ. Rev.* 12 (3): 419–456.

Burchi, F., Fanzo, J., and Frison, E. (2011). The role of food and nutrition system approaches in tackling hidden hunger. *Int. J. Environ. Res. Public Health* 8 (2): 358–373.

Chakraborty, S. and Sarmah, S.P. (2019). India 2025: the public distribution system and national food security act 2013. *Dev. Pract.* 29 (2): 230–249.

Chhabra, I. and Rao, K.A. (2014). Challenges and interventions in meeting delivery of nutrition in mid-day meal scheme: a pilot in district Faizabad, Uttar Pradesh. *Indian J. Community Health* 26 (Supp 2): 183–192.

Das, P.K., Bhavani, R.V., and Swaminathan, M.S. (2014). A farming system model to leverage agriculture for nutritional outcomes. *Agric. Res.* 3 (3): 193–203.

Desai, S. and Vanneman, R. (2015). Enhancing nutrition security via India's National Food Security act: Using an axe instead of a scalpel?. *In India Policy Forum:[papers]. India Policy Forum. Conference 11*, New Delhi (17 Aug 2015), 11, 67–113.

FAO (2008) An introduction to the basic concepts of food security. http://www.fao.org/3/al936e/al936e00.pdf (accessed 12 November 2019).

Ghosh, S. (2020). Factors responsible for childhood malnutrition: a review of the literature. *Curr. Res. Nutr. Food Sci. J. 8* (2): 360–370.

Global Food Security Index (2020). Country rankings weblink. https://foodsecurityindex.eiu.com/Index (accessed 12 March 2021).

Global Hunger Index (2019). The challenge of hunger and climate change. https://www.globalhungerindex.org/results.html (accessed 14 February 2021).

Global Hunger Index (2020). https://www.globalhungerindex.org/india.html (accessed 25 April 2021).

Hawkes, C. (2005). The role of foreign direct investment in the nutrition transition. *Public Health Nutr.* 8 (4): 357–365.

IBEF (2021). Indian agriculture and allied industries report. https://www.ibef.org/industry/agriculture-india.aspx#login-box (accessed 25 April 2021).

IFPRI (2015). Highlights of recent IFPRI food policy research in India weblink. http://ebrary.ifpri.org/utils/getfile/collection/p15738coll2/id/129292/filename/129503.pdf (accessed 12 March 2020).

International Food Policy Research Institute (2016). *Global Nutrition Report 2016: From Promise to Impact: Ending Malnutrition by 2030*. Washington, DC: International Food Policy Research Institute.

Kadiyala, S., Joshi, P.K., Dev, S.M. et al. (2011). *Strengthening the Role of Agriculture for a Nutrition Secure India*. New Delhi: Policy note, India International Food Policy Research Institute (IFPRI).

Laxmaiah, A., Rameshwar Sarma, K.V., Rao, D.H. et al. (1999). Impact of mid day meal program on educational and nutritional status of school children in Karnataka. *Indian Pediatr.* 36 (12): 1221–1228.

Madhya Pradesh Budget Analysis (2019–20). PRS legislative research. https://prsindia.org/budgets/states/madhya-pradesh-budget-analysis-2019-2020. (accessed 19 November 2020).

Mandal, V.P., Rehman, S., Ahmed, R. et al. (2020). Land suitability assessment for optimal cropping sequences in Katihar district of Bihar, India using GIS and AHP. *Spat. Inf. Res.* 28 (5): 589–599.

Martins, V.J., Toledo Florêncio, T.M., Grillo, L.P. et al. (2011). Long-lasting effects of undernutrition. *Int. J. Environ. Res. Public Health* 8 (6): 1817–1846.

Martorell, R. and Young, M.F. (2012). Patterns of stunting and wasting: potential explanatory factors. *Adv. Nutr.* 3 (2): 227–233.

Mendez, M.A. and Adair, L.S. (1999). Severity and timing of stunting in the first two years of life affect performance on cognitive tests in late childhood. *J. Nutr.* 129 (8): 1555–1562. https://doi.org/10.1093/jn/129.8.1555.

Mozaffarian, D., Angell, S.Y., Lang, T., and Rivera, J.A. (2018). Role of government policy in nutrition—barriers to and opportunities for healthier eating. *BMJ* 361 https://doi.org/10.1136/bmj.k2426.

Nelson, M. (2000). Childhood nutrition and poverty. *Proc. Nutr. Soc.* 59 (2): 307–315. https://doi.org/10.1017/s0029665100000343.

NFHS (2015–16). International Institute for Population Sciences (IIPS) and ICF. 2017. *National Family Health Survey (NFHS-4), 2015-16*. India. Mumbai: IIPS. http://rchiips.org/nfhs/nfhs-4Reports/India.pdf (accessed 25 December 2019).

Non, A.L., Roman, J.C., Gross, C.L. et al. (2016). Early childhood social disadvantage is associated with poor health behaviours in adulthood. *Ann. Hum. Biol.* 43 (2): 144–153.

Pandey, P.C., Mandal, V.P., Katiyar, S. et al. (2015). Geospatial approach to assess the impact of nutrients on rice equivalent yield using MODIS sensors'-based MOD13Q1-NDVI data. *IEEE Sensors J.* 15 (11): 6108–6115.

Prakasam, S. (2019). Impact of mid-day meal scheme on body mass index of school children in India. the NCERT and no matter may be reproduced in any form without the prior permission of the. *NCERT* 45 (1): 115.

Rice, A.L., Sacco, L., Hyder, A., and Black, R.E. (2000). Malnutrition as an underlying cause of childhood deaths associated with infectious diseases in developing countries. *Bull. World Health Organ.* 78: 1207–1221.

Ross, J. (2004). Understanding the demographic dividend. POLICY Project Note.

Schofield, C. and Ashworth, A. (1996). Why have mortality rates for severe malnutrition remained so high? *Bull. World Health Organ.* 74 (2): 223.

Si, A.R. and Sharma, N.K. (2008). An empirical study of the mid-day meal programme in Khurda, Orissa. *Econ. Polit. Wkly.* 43 (25): 46–55.

Singh, S. and Gupta, N. (2015). Impact of mid day meal on enrollment, attendance and retention of primary school children. *Int. J. Sci. Res.* 4: 5.

Singh, K., Kumar, P., and Singh, B.K. (2013). An associative relational impact of water quality on crop yield: a comprehensive index analysis using LISS-III sensor. *IEEE Sensors J.* 13 (12): 4912–4917.

Singh, A., Park, A., and Dercon, S. (2014). School meals as a safety net: an evaluation of the midday meal scheme in India. *Econ. Dev. Cult. Chang.* 62 (2): 275–306.

Tomar, V., Mandal, V.P., Srivastava, P. et al. (2014). Rice equivalent crop yield assessment using MODIS sensors' based MOD13A1-NDVI data. *IEEE Sensors J.* 14 (10): 3599–3605.

UNICEF (1998). *State of the World's Children*, 7–87. New York: Oxford University Press.

UNICEF (2015). How to calculate average annual rate of reduction (AARR). https://data.unicef.org/wp-content/uploads/2015/12/Technical_Note_AARR_185.pdf (accessed 25 December 2019.

WHO (2021). Fact sheets - malnutrition. https://www.who.int/news-room/fact-sheets/detail/malnutrition (accessed 12 June 2021).

World Food Conference (1974). https://www.un.org/en/development/devagenda/food.shtml (accessed 19 November 2020).

World Health Organization (1995). *Expert Committee on Nutrition and Physical Status: Uses and Interpretation of Anthropometry*. Geneva: WHO.

World Health Organization (2002). *World Health Report*. Geneva: The Organization.

Yadav, M., Sharma, M.P., Prawasi, R. et al. (2014). Estimation of wheat/rice residue burning areas in major districts of Haryana, India, using remote sensing data. *J. Indian Soc. Remote Sens.* 42 (2): 343–352.

2

Diversification for Restoration of Ecosystems and Sustainable Livelihood

Sanjay S. Rathore[1], *Kapila Shekhawat*[1], *R.K. Singh*[1], *S. Babu*[1], *and V.K. Singh*[2]

[1] *Division of Agronomy, ICAR-IARI, New Delhi, India*
[2] *Division of Agronomy, ICAR-Central Research Institute for Dryland Agriculture, Hyderabad, Telangana, India*

CONTENTS

2.1 Introduction

The agriculture is the predominant means of livelihoods of majority of the people around the world and especially in the third world countries and also land use type occupying about 40% of the terrestrial surface of the Earth (FAO 2009). It has been widely perceived that Agroecosystems are both providers and consumers of ecosystem services. Along with provisioning services (the produce, e.g. food, timber, and other raw materials), the other components of the ecosystem like plants, animals, fungi, microbials ensure indispensable regulating services such as pollination, prevention of soil erosion and water purification, and an enormous range of cultural services, as recreation and aesthetic values. Agroecosystems is the largest ecosystem, which also produces diverse ecosystem services, such as regulation of soil and water quality, carbon sequestration, support for biodiversity, and cultural services, and these are well maintained under diverse production scenarios

Sustainable Agriculture Systems and Technologies, First Edition. Edited by Pavan Kumar, A.K. Pandey, Susheel Kumar Singh, S.S. Singh, and V.K. Singh.

including well-designed crop diversification (Polasky 2008). On the planet earth, there exist half a million-plant species but we are depending only upon rice, wheat, maize, and soybean for food security, globally. Depending on fewer crops for our food and other needs may lead to serious ecological threats in terms of increasing biotic and abiotic stresses in crop production. In spite of the vast plant biodiversity across the world, unfortunately at present 150 crops species are being used widely as now commercially important crops. More worrisome is the fact that the cereals, rice, wheat, and maize account for 60% of the world's food supply. Over seven billion people depend on the productivity of these major crops and in the near future the global population will reach to 9.0 billion and meeting the food and nutritional requirement of burgeoning population is a daunting task. The mono-cropping constraints the provisioning services of the ecosystem. Therefore, the major crops alone may not be able to meet the world's food and nutritional requirements. The higher crop productivity of food grain crops can meet the energy demands of a mounting population; they may not provide adequate nutrition. The double-burden of over and under nutrition (Hidden Hunger) is a major challenge. Nutrient-poor and energy-rich diets are linked with lack of dietary diversity. In this context, it is very pertinent to mention the significance of the cropping systems diversification to promote ecosystem services, thereby reducing dependency on external inputs while sustaining the high crop productivity. Overall, diversification of cropping ensures diversity of beneficial organisms in the ecosystem, better pollination, pest management, nutrient cycling, and water regulation without compromising crop yields (Rathore et al. 2019). Hence, aboveground biodiversity reduces pest menace and helps in water regulation, while belowground biodiversity enhances nutrient cycling, soil resilience, and water regulation. It is very common that the diversification practices resulted in enhancing the support services of the ecosystem and enhances crop productivity (Tamburini et al. 2020).

2.2 The Need for Agricultural Diversification for Sustained Livelihood

The size of holdings is declining across the world, more especially in developing countries. In India, almost 86% of the farmers are small and marginal (<1.0 ha) for whom agriculture has become cost-intensive due to escalating prices of agro-inputs, labor, and energy, etc. The decreasing landholding size and mono-cropping types of crop cultivation poses grave challenges to the profitability and overall livelihood of the farmers. Achieving economic sustainability in such agricultural production system necessitates coherent linkage of climate-smart agriculture with the farmer's needs. In this context, integration of cropping with other farming enterprises under agricultural diversification as integrated farming system (IFS) can ensure enhanced system productivity, resource-use efficiency, and improved livelihood. Research evidences across the country clearly reveals that the complementary coexistence of diverse crops and other farm enterprises can provide an array of benefits in meeting out the growing demands for food and nutrition rich dietary, income stabilization, and livelihood upliftment while conserving ecological balance and environment, particularly for small holders with limited resources (Singh et al. 2020).

This is of utmost important to maintain ecosystem services adequately for ensuring sustainability in farming. The direct and indirect aids of ecosystems to plant–animal–human continuum are included mainly under the ecosystem services (TEEB 2010). Crop diversification, a type of cropping now has been widely recognized to maintain balance between human needs and ecosystem system services. It is a type of cropping with more number of the crop species. Agricultural diversification can safe guard greater food and nutritional security, minimizing environmental degradation, lessening poverty, higher use efficiency of land resources, and also preventing the impact of degradation forces and helping in contesting desertification. Diversification of agriculture opens new vistas for additional environmentally safe, sustainable systems, enhanced livelihood options for farmers. Some of the crops which are presently considered as underutilized can become a potential tool and hence will ensure more ecologically sustainable agricultural production systems and provide newer livelihood options for small holder farmers. As a matter of the fact global food supply relies on 150 crop species. And the most gruesome challenge is that out of the 150 crops, there are only 12 crops which provide around 75% of the world's food demand. This indicates toward one of the most unsustainable land use types for food production practices in the world. The simple fact that three cereal crops like rice, wheat, and maize are the principal source of more than half of the global food energy and that too from a limited number of varieties of the three "mega-crops" made the agriecosystems very unsustainable. These cereal crops are also very exhaustive in their demand for nutrients and water, which further complicates the problem of resource degradation.

2.3 Crop Diversification and Ecosystem Services

Generally, crop diversification is perceived as a change from the regional supremacy of one crop to regional production of number of the crops to meet ever increasing demand of cereals, pulses, vegetables, fruits, oilseeds, fibers, fodder, grasses, etc., of the human and livestock population. Crop diversification also guarantees higher profits for the growers. It also aims to improve soil health and to maintain dynamic equilibrium of the agroecosystem. Furthermore, in India crop diversification is generally viewed as a shift from traditionally grown less remunerative crops to more remunerative crops. In the present context, the unsustainability of the modern agricultural system, there lies high stack on crop diversification mainly due to disturbance in the ecosystem. The sustainability of agricultural production system can be achieved by maintaining ecosystem services, with the design of appropriate crop diversification plans specific to agro climatic conditions. From Figure 2.1, it can be well understood that among all the agricultural operations/practices, the crop diversification including the agricultural diversification has a greater role to play in maintaining the ecosystem services.

The agroecosystems provide ecosystem services through improved human interventions which have no or very little pressure on environmental resources, ensure recycling of waste products and support large biodiversity as maintained under natural, unmanaged ecosystems. The ecosystems are vital for survival of all living beings. The genetic diversity of crops and livestock, soil health maintenance, and the provision of water

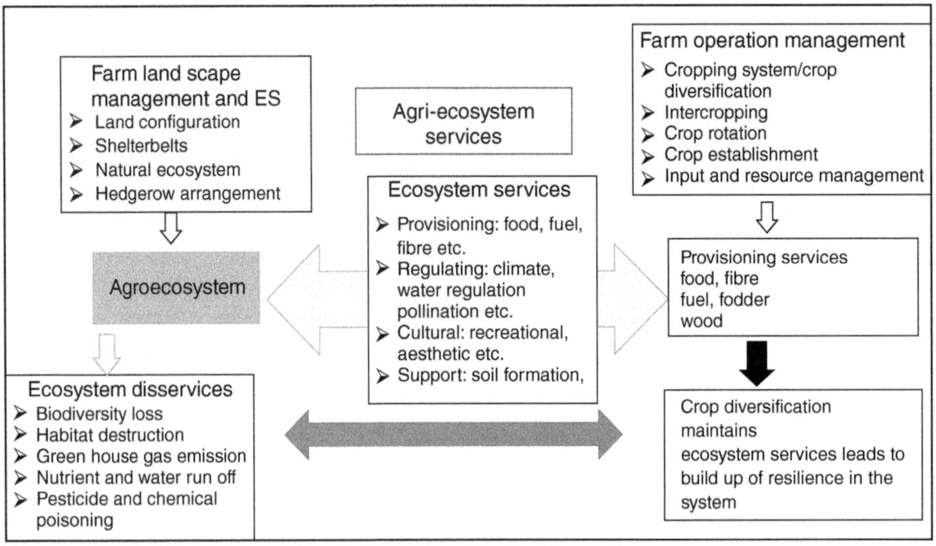

Figure 2.1 Crop diversification maintains many of the ecosystem services and hence also maintains the resilience in the production systems.

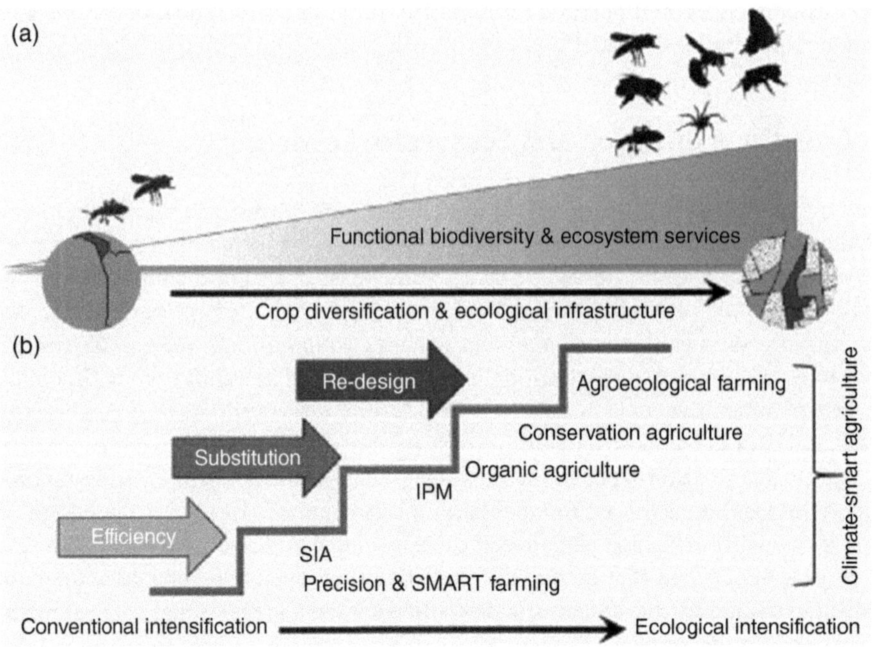

Figure 2.2 (a) Functional biodiversity delivering ecosystem services (e.g. pollination, natural biocontrol) in support of crop yields along a gradient of crop diversification and ecological infrastructure and (b) alternative options (efficiency, substitution, redesign) for farm management systems situated on a continuum from conventional agricultural intensification to ecological intensification. (*See insert for color representation of the figure.*)

sources are some of the very important supporting ecosystem services. Whereas, the regulating services of the ecosystems are in the form of pollinators, which ensures better seed setting, and the presence of the natural enemies control the crop pest within agro-ecosystems. Natural ecosystems also filter water and normalize its flow into agricultural systems. Therefore, crop diversification create favourable conditions for smooth functioning of the ecosystem services. This ultimately improves the overall health of the ecosystem which are essential for survival and well-being of mankind. It also ensures enhanced floral and faunal diversity, better pollination, natural pest management, efficient nutrient cycling, soil fertility, and water regulation without compromising on crop yields. Sustainable agronomic management targets above and below ground biodiversity, ensures better pest management, improved water regulation, enhanced nutrient cycling, and improved soil health.

Even under stressed ecologies (poor soil fertility, dry land scenario, abiotic, and biotic stresses), the crop diversification and growing of large number of crops are widely recommended to reduce the risk factor of crop failures. For the majority of farmers in different parts of the country, the gains from application of scientific tools and technologies in agriculture are yet to be realized. On realization, it will help in greater expansion of crop diversification in many areas. It is hard fact that the productivity levels of many major crops in India do not match with the yields obtained in agriculturally advanced countries even under predominant rice–wheat system. On the other hand, as a result of growing few crops in large areas, many problems in the forms of increasing biotic and abiotic stresses are emerging. Therefore, diversification of predominant cropping system in India will bring the new dawn of prosperity and resilience in agricultural production system. Therefore, efficient and effective crop diversification is a crucial for the sustainability of Indian agriculture.

2.4 Reducing Emission of Greenhouse Gases

The good agronomic practices (GAPs), such as, conservation tillage, organic nutrient sources along with inorganic nutrient sources, weeds, insect pest, and disease management along with adequate irrigation water management in diversified production system can greatly reduce greenhouse gases in the atmosphere. The selection of climate smart crops which require comparatively lesser nutrient, water from production system and have lesser impact of biotic and abiotic stresses are the essential components of the crop diversification. In this category, oilseeds, pulses are good examples; for these crops, the GHC emission is also less. Conservation tillage and deficit irrigation can also substantially minimize CO_2 and N_2O emission. Appropriate use of the crop residue and complete banning of its burning reduces the generation of CO_2, N_2O, and CH_4 to a significant extent. Crop diversification ensures greater functional biodiversity to cropping systems and it also regenerates biotic interactions underpinning yield-supporting ecosystem services (Rathore and Shekhawat 2020). It embraces a variety of practices encompassing the management of crops, noncrop habitats, soil, and landscapes (Cardinale et al. 2012; Wagg et al. 2014; Renard and Tilman 2019; Tamburini et al. 2020).

2.5 Effect of Technology-Induced Crop Diversification

The advancement in technological development especially development of high yielding and biotic, abiotic resistant varieties, of wide window of growing period, CA-based systems, land configuration, and other improved agronomic management, along with increasing value addition and better postharvest handling are the encouraging forces for promotion of crop diversification in different parts of the country. Green revolution technologies intensified cropping system in irrigated areas in favor of rice and wheat and other crops got less acreage due to low productivity and less profitability. This enhanced total output of cereals tremendously but also created second generation problems like intensive use of chemical fertilizers, pesticides, etc., led to pollution of resource base, overexploitation of groundwater resulted in depletion of ground water, salinization. This made urgent call for crop diversification to achieve sustainable high levels of productivity with conservation of natural resource base and to secure higher profitability to the farmers in the long run. Rice–wheat acreages increase substantially in Indo-Gangetic plains (IGPs) zone due to better irrigation infrastructures and higher response to the applied inputs for long time. Also, the government support system was developed over the period to safeguards the interest of the cereal growers through better pricing, input supply, and other incentives like increased availability of quality seed of improved varieties to the farmers. Later the stockholders including the farmers realized the faults in the ongoing cereal-based system and promoted crop diversification in originally green revolution areas.

The technological development in all sectors of crop production also promoted farmers awareness to explore for crop diversification. High yielding varieties, suited to different ecologies, were developed in oilseeds, pulse crops, and millets and improved agronomic practices. These also helped in harnessing higher productivity from the diversified cropping system. The better minimum support price of the millets, oilseeds, and pulses also help in increasing the farmers' awareness for bringing more and more area under crop diversification. The inclusion of leguminous crop not only helps in maintaining optimum soil fertility but also helps in minimizing the impact of biotic and abiotic stresses among the crop plants due to alteration in host–pathogen relationship. It helped in maintaining over all ecological balance in agriecosystem.

The date of sowing is considered as the simplest agronomic management to minimize the impact of climate change/variability on crop growth and productivity. It will be helpful to make contingency planning to fit proper variety to extreme weather conditions like raising and decreasing temperatures, varying rainfall and other weather components. The climate smart varieties would be used in large scale on farmers' field. The diversification of cereal-based cropping systems with the crops such as oilseed, pulse, and forage crops, is another option for minimizing biotic and abiotic stresses. The pathogenic cycles could be broken up through crop substitution (Singh et al. 2002, 2020; Rathore and Bhatt 2008). Consistent practicing of monocropping resulted in loss of soil fertility, decline of soil health, factor productivity decline and lowering crop productivity even under high productivity areas (Saha et al. 2012). Conservation tillage also promotes greater soil biodiversity thereby leading to more biotic stresse suppression. Proper pest management is a permanent challenge to the farmers, and it is a very important ecosystem service, which can be restored through crop diversification. The erosion of genetic diversity led to an increase in crop disease occurrence. Greater varietal and species richness reduce the spread of disease in agricultural systems. Habitat management in agricultural systems is an effective means to enhance natural enemies. These have been developed for effective application in crop, at farm, and few of these are reported be very economically useful for the farmers. Gurr et al. (2003). The diversity of crop species within the agro ecosystem consequently offers long-term pest suppression by building up potential natural enemies for any pest outbreaks.

2.6 Congenial Conditions for Crop Diversification

India is a highly diversified country in terms of climatic conditions, soil types, topography, and other agro climatic conditions. Therefore, the ICAR has divided the whole country into 15 agroclimatic zones. This help for wider cultivation of number of crops in the country. India is also one of the mega biodiverse countries and many of the crops originated in India. Traditionally, the farmers were practicing a diversified cropping system by inclusion of many crops during different crop season. But over the period, after green revolution and increasing government efforts and food demand, lot of emphasis has been paid only to cereal-based system. This resulted in growing of one or two crops only.

- Diverse agro-climatic conditions
- Diverse weather within a year
- Huge market
- Great potential for processing and value addition
- Large number of crops, including underutilized

Figure 2.3 indicated that diverse bioclimatic zones of India along with different agroclimate favors for promotion of crop diversification. For achieving a very useful diversification plan, there is a need to develop intensive cropping systems with diverse crops. Furthermore, these agro ecological regional classifications indicate very systematically the existence of diverse soil, climatic, and growing conditions and based on these diverse conditions, across the country the suitable crop diversification plans have been developed by various experts. There is a need to popularize these time-tested crop diversification plans vigorously in the country at farmers' fields to get all the benefits to the small and marginal farmers in particular.

2.7 Crop Diversification and Composition

It is very pertinent to suitably identify the nature of composition of crops for fool proof crop diversification plans in a specific locality. Area shifts and crop pattern changes can lead either to crop specialization or to crop diversification. Green Revolution technologies led to cereal centered specialization in large part of the country. But, later when increased productivity of food grains, especially cereals, made it possible to allocate more area to other crops such as oilseeds with a severe supply shortage, the specialization tendency witnessed earlier has given room for overall crop diversification. Crop diversification can be practiced successfully in spatial and temporal dimensions.

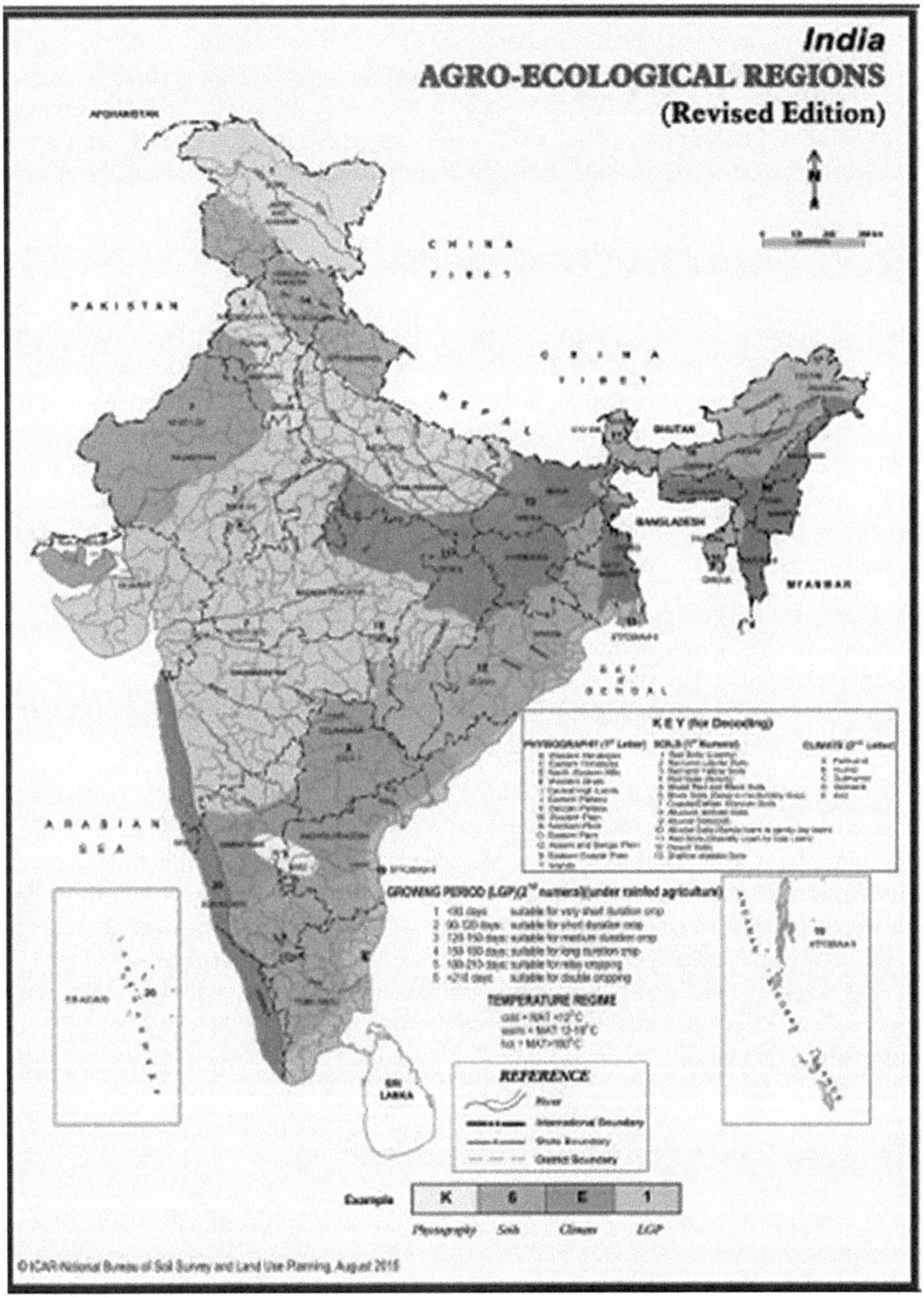

Figure 2.3 Agro-ecological regions map of the country (2015) and revised Bio-climates of India (2015) (e-Publication: ICAR-NBSS&LUP Technologies). (*See insert for color representation of the figure.*)

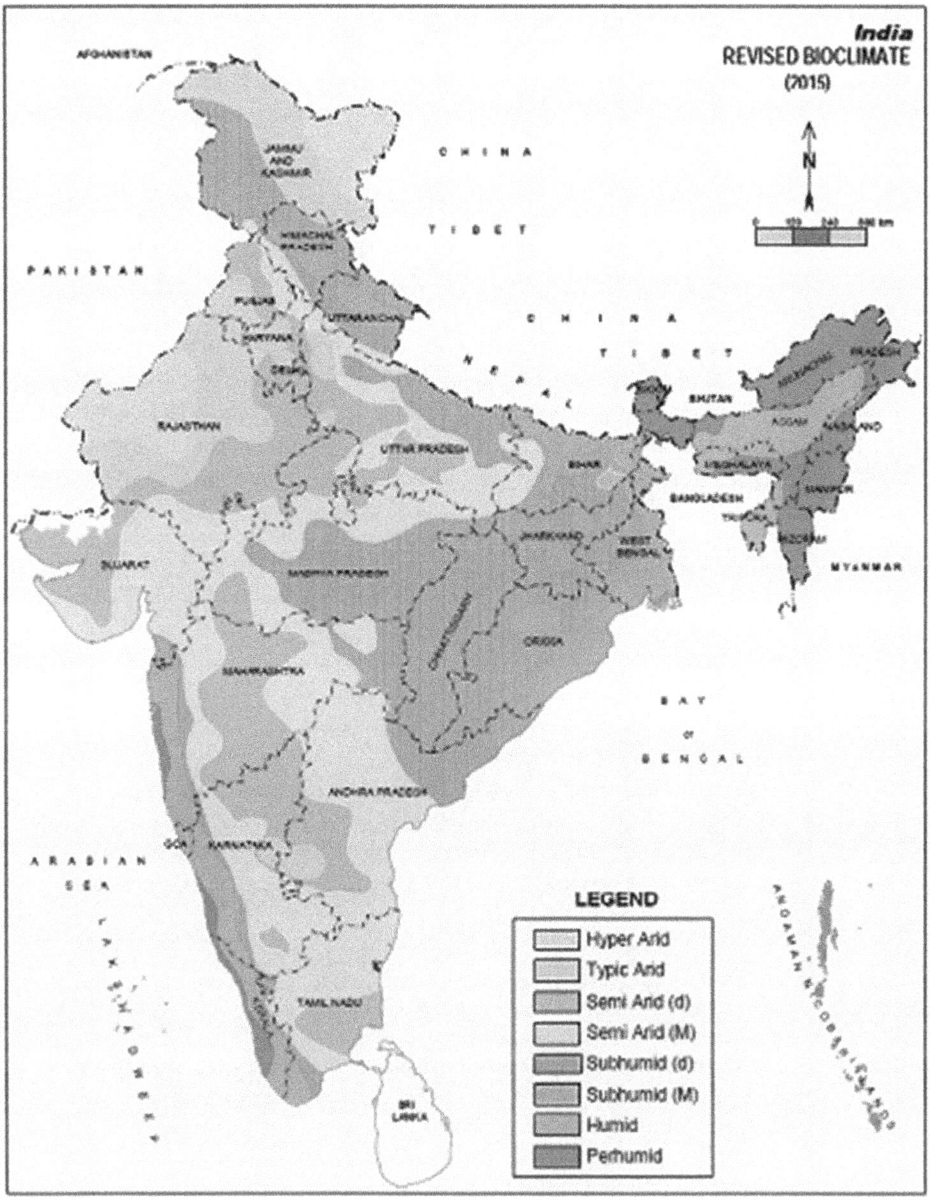

Figure 2.3 (Continued)

2.7.1 Diversification with Oilseed Crops

At present, there are around 25 Mha area under oilseed crops with 37 MT of total production, even this is not sufficient to meet domestic oil requirement, hence India is importing around 15 MT of edible oils from abroad. On the other hand, there exists enormous potential to expand oilseeds area in the country and crop diversification with oilseed crops is one of such options. And it is hugely possible in irrigated IGP regions through crop intensification, where number of computable crops are grown either in sequential or intercropping systems. There are crops which act as complementary to each other when grown together, like growing legumes with cereals, through positive allopathic effect (mustard + wheat, mustard + potato, etc.) and in tier/multistoreyed system. This type of crop combination needs to be identified and promoted for achieving desired results. The selection of crops in crop diversification depends upon many factors like available resources, prevailing weather conditions, market demand, ease of doing cultivation, and overall high productivity and profitability from the possible combinations. The established possible combination is enlisted below (Tables 2.1 and 2.2).

Inclusion of oilseed crop, Indian mustard with early maturing short duration varieties like PM 28 and PM 26 in maize (fodder)-mustard-late wheat system restyled n higher system productivity and also the profitability (Table 2.3). Maximum system productivity (49.9 t/ha) and system net return (Rs. 1.45 lakhs) were obtained from M-M(PM28)-late wheat (HD 3118) Maize-mustard (PM28)-late wheat (HD 3118) system.

Table 2.1 Crop diversification options with oilseed crops.

Crop	Area suggested for diversification
Rapeseed-mustard	The short duration, terminal heat resistant varieties like PM 25, PM 26, and PM 28
	As a substitute of meager productive rainfed wheat
	In *diara*tract in northern and eastern India
	A very good intercrop has positive allopathic effect on associated crops.
Groundnut	The short duration varieties can be introduced in many low rainfall areas during kharif season and as summer crop in the areas of adequate irrigation facilities.
	As a substitute crop for minor millets in Bihar, Jharkhand, West Bengal, and Orissa
	Substitute rice–groundnut with rice–rice system to prevent buildup of pests and diseases
Soybean	As a replacement crop for low yielding upland rice, minor millets growing areas of in Bihar, Jharkhand, and Orissa
	As a complementary rotational crop in pest endemic areas of rainfed cotton.
	In nontraditional areas of North-Eastern Hill sunder agripastoral or agri-silvicultural
	Diverting some kharif cereal area to soybean in situations of water scarcity and to restore soil health in North India.
Sunflower	As are placement crop for low yielding wheat, cotton, chickpea, sorghum crops under black cotton soils in peninsular India.
	As a spring crop in Northern India after harvest of rabi season crops.
Sesame	The short duration cv are good for raising as kharif crops in many areas. Also as a summer crop in central peninsular and Eastern India where only limited irrigation is available.
Castor	This is a nonedible but commercially very important crop and can be grown as substitute crop of cotton in Rajasthan and as bund crop in all regions.

Source: Modified from Palaniappan and Jeyabal 2002.

Table 2.2 The diversification of traditional crop base with annual oilseed crop.

Prevailing crop	Suggested crop	Region
Rice fallow	Rapeseed-mustard, soybean, and sesame	Eastern part of the country, Cauvery deltaic areas of Tamil Nadu and Coastal Andhra Pradesh
Upland rice	Groundnut (Kharif), soybean, and sunflower	Upland rice areas of Orissa, Tamil Nadu, Bihar, and A.P.
Cotton	Safflower and sunflower	Karnataka and adjoining areas of A.P.
Chickpea, dryland wheat, and coriander	Safflower (sole crop)	A.P. (rabi), Karnataka, Maharashtra, part of M.P. (Malwa region)
Linseed and barley	Safflower (sole crop)	South eastern Rajasthan (Udaipur)

Source: Reddy and Suresh 2006.

Table 2.3 System productivity and economics of maize-based diversification system.

Cropping systems	System productivity, kg/ha	WEY, kg/ha	Production efficiency, kg/ha/day		System NR, Rs./ha	System profitability, Rs./ha/day
			SP	WEY		
Maize-wheat	46556c	9179d	211.6a	42.2a	118027d	536.4a
Maize-mustard (PM26)-late wheat (HD 3118)	48901b	11834b	168.6c	50.6b	137478b	474.0c
Maize-mustard (PM28)-late wheat (HD 3118)	49922a	12252a	172.1b	51.9b	144996a	499.9b
Zn levels						
Control	44881.5c	10332d	166.5d	46.4a	117975d	435.4d
Zn 2.5	48810.8b	10828c	180.9c	46.8ab	126037c	465.9c
Zn 5.0	49570.4b	11622b	183.9b	49.7b	141019b	520.5b

Source: Data from Rathore et al. 2020.

System productivity of maize (fodder)-early mustard (PM 28)-late wheat (HD 3118) was recorded maximum with higher net return and B : C ratio. Similarly higher system productivity and profitability were recorded under 5.0 kg Zn application during kharif season crop in a calendar crop year.

2.7.2 Diversification with Pulse-Based Cropping Systems in Different Agroclimatic Zones

Legumes are the crops very vital for a successful crop diversification plan of any area. These crops through, biological N fixation, maintain optimum soil health and good soil fertility. An ideal crop diversification plan must include leguminous crops either as sequence cropping or intercropping. Recognizing the importance of pulses for meeting dietary requirements of vast vegetarian population on one hand and their role in improving soil health and conserving natural resources on the other, the efforts were made to develop high yielding, short duration and disease resistant varieties of different pulse crops which can help in diversification and intensification of popular cropping systems. The development of short duration, disease-resistant and high yielding varieties in the recent past made these crops a viable alternative to low yielding coarse cereals under rainfed conditions and also provided an opportunity for expansion in rice fallows and in double cropping systems. Due to the reasons of source of nutritious food, feed, and forage, pulses are an integral part of subsistence cropping system in large part of the country. The pulses are grown in almost all types of cropping systems as a sole crop, intercrop, catch crop, relay crop, cover crop, and green manure crop, etc., under sequential/mono-cropping in different agroecological regions. Under dryland areas pulses are predominant due to their low water requirements. The development of short duration varieties of mungbean, urdbean, and pigeonpea has paved way for crop diversification and intensification in many parts of IGP zones (Table 2.4).

2.7.3 Diversification with Horticultural Crops

The fruits, vegetables, spices, condiments, and flowering ornamental plants are very useful for developing high productive and profitable diversification plan in many areas. Depending upon the resource availability, prevailing market conditions and available technological support, different crops components may be selected. Under semiarid conditions a similar

Table 2.4 Possible new niches for pulses.

Cropping systems	Possible niches	Suitable varieties
Tur-wheat system	North-West U.P., Haryana, Punjab, and North Rajasthan	Virat, Pusa Vishal, UPAS 120, Manak, Pusa33, AL 15, AL201
Maize-*rabi* pigeonpea	Central and Eastern U.P., North Bihar, West Bengal, Assam	Pusa 9, Sharad
Maize-potato/mustard + mungbean/urdbean	Punjab, Haryana, and West U.P.	**Mungbean:** Pant Mung 2, PDM11,HUM 2, SML668, Pusa Vishal **Urdbean**: PDU 1, NarendraUrd 1, Uttara
Spring sugarcane + mungbean/urdbean	East U.P., Bihar, West Bengal	**Mungbean:** Pant Mung 2, PDM 11, Narendra Mung 1 **Urdbean:** PDU 1, Pant Urd 19
Rice-mungbean	Orissa, Part of Karnataka, Tamil Nadu, A.P.	TARM 1, Pusa 9072
Rice-urdbean	Coastal area of A.P., Karnataka, Tamil Nadu	LBG 17, LBG 402
Rice wheat mungbean	Western U.P., Haryana, Punjab	Pant Mung 2, Narendra Mung 1, PDM 139, HUM 2

Source: Based on Singh et al. 2009.

diversified fruit tree-based diversified system was developed. The IFS refers to combining one or more agriculture and allied activities with field crop cultivation. This will certainly help in making rainfed areas a stable, sustainable agroecosystem (Rathore et al. 2019). Many studies have identified the important role of perennial vegetation in supporting biodiversity in general and beneficial organisms in particular (Perfecto and Vandermeer 2008).

The agri-horti system has been developed for round the year cultivation of the crops and generation of produce for regular income and employment. During kharif season, intercropping of legume crops were taken in the rows in between the fruit crops. In fruits crops falsa (*Grewia asiatica*), karonda (*Carissa carandas*), drum stick (*Moringa oleifera*), aonla (*Phyllanthus emblica*), guava (*Psidium guajava*), and pomegranate were grown and in field crops during kharif season vegetable cowpea, mung bean was cultivated as intercrops and in this intercrop space, vegetable pea, Bengal gram, and cole crops will be taken during the rabi season. The whole system is designed to be irrigated by drip irrigated system from the pond. The well designed micro irrigation system along with fertigation device and appropriate filters ensures the risk of crop failure due to moisture shortage. The rain water is harvested and stored in the pond for life saving irrigation through micro irrigation system. Thus, every drop of water is efficiently utilized for production of different crops. The data revealed that the different agri-horti systems are economically viable, productive, and employment generator round the year (Figure 2.4). The inclusion of horticultural crops not only guarantee higher productivity and profitability but also ensure nutritional security of a family household (Table 2.5).

2.8 Constraints in Crop Diversification

The crop diversification approach has been identified as one of the potential agronomic intervention to address many of the present challenges in farming across the globe. Even out of the 17 sustainable development goals (SDGs), 6–7 SDGs have been identified as having greater involvement of CD to achieve them within set time framework.

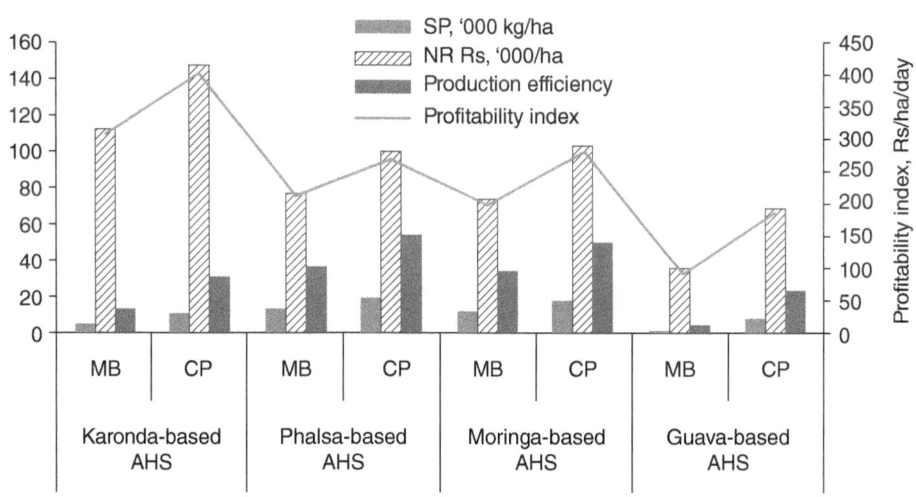

Figure 2.4 Diversified agri-horti system for higher system yield and income.

Table 2.5 Nutritional importance of vegetable crops.

S.No.	Dietary factors	Source vegetables
1	High energy	Immature seeds of broad bean and peas, lima bean, tapioca, yam, colocasiacorms, potato, brussels sprouts, onion and garlic, sweet potato
2	Proteins	Leguminous vegetable crops such as vegetable peas, beans, garlic, brussels sprouts, cowpea, lima bean seeds, amaranthus leaves, drumstick leaves, and menthe
3	Vitamin A (beta carotene)	Carrot, spinach, turnip green, palak, sarson sag, amaranth, coriander, sweet potato, pumpkin, tomato
4	Vitamin B complex	Peas, broad bean, lima bean, garlic, asparagus, colocasia and tomato
5	Vitamin C	Turnip green, green chilies, brussels sprouts, mustard green, amaranth, coriander, drumstick leaves, cauliflower, KnollKhol Spinach, cabbage, bitter gourd and reddish leaves
6	Calcium	Amaranth, parsley, palak, Chinese cabbage, kale, collard greens, broccoli, spinach
7	Iron	Amaranth, beans, peas, spinach, radish leaves, turnip greens
8	Potassium	Cowpea, peas, sword bean, colocasia, melons, potato, sweet potato, spinach, turnip green, collard greens, peas, beans

Source: Jogendra et al. 2019.

In spite of all advantages being offered due to crop diversification, its adoption is poor due to following constraints:

- Over 117.5 mha (63%) of the cropped area in the country is completely dependent on rainfall.
- Both suboptimal and over-use of resources.
- Inadequate supply of seed and planting material for crop diversification
- Fragmentation of landholdings which discourage modernization and mechanization of agriculture.
- Poor basic infrastructure like rural, roads, power, transport, communication, etc.
- Inadequate postharvest technologies and weak agro-based industries.
- Weak research-extension and farmer linkage
- Decreased investment in the agriculture sector over the years.

2.9 Conclusion and Future Perspectives

Crop diversification is of vital significance to improve livelihood, enhanced system productivity and at the same time maintain many of the ecosystem services. Diversified crop production systems are of 10 less risky, if managed proficiently, they are benefitted from synergisms among different compatible crops. The system productivity also increases by two to four times over sole cropping under diversified system of farming. The core characteristics of crop diversification is to ensure complementarily, among the selected crops in spatial and temporal dimension. Hence, the resource use efficiency is also increasing with minimal wastage of the inputs. By 2050, the food supply must double to cope for rising population pressure, and impact of climate change will further create complexity in ensuring food security. It is high time to address the constraints as explained above for wider adoption. The crop diversification has also been identified as a climate smart technology to combat the risks due to climate change in agricultural production system. Under stressed ecologies like drought and flood-prone areas, CD is an ideal approach for mitigating the negative impact of adverse weather conditions. Another vital characteristic of the agricultural diversification is efficient use of on-farm resources, better recycling of by-products which maintain ecosystem services for improving long-term system sustainability.

References

Cardinale, B.J., Duffy, J.E., Gonzalez, A. et al. (2012). Biodiversity loss and its impact on humanity. *Nature* 486: 59–67.

FAO (2009). *State of the World's Forests 2009*. Rome: FAO http://www.fao.org/docrep/011/i0350e/i0350e00.htm.

Gurr, G.M., Wratten, S.D., and Luna, J.M. (2003). Multi-function agricultural biodiversity: pest management and other benefits. *Basic and Applied Ecology* 4 (2): 107–116. https://doi.org/10.1078/1439-1791-00122.

Jogendra, S., Lyngdoh, Y.A., and Lata, S. (2019). Diversification through vegetable cultivation for doubling farmer's income. In: *Crop Diversification for Resilience in Agriculture and Doubling Farmers Income* (eds. S.S. Rathore, K. Shekhawat, G.A. Rajanna, et al.), 210. Pusa, New Delhi: ICAR –Indian Agricultural Research Institute (IARI), ISBN:978-93-83168-42-2.

Palaniappan, S.P. and Jeyabal, A. (2002). Crop diversification with oilseeds in non-traditional areas andseasons. In: *Oilseeds and Oils: Research and Development Needs* (eds. M. Rai, S. Harvir and D.M. Hegde), 94–100. Hyderabad: Indian Society of Oilseeds Research.

Perfecto, I. and Vandermeer, J. (2008). Biodiversity conservationin tropical agroecosystems: a new conservation paradigm. *The Year in Ecology and Conservation Biology* 1134: 173–200.

Polasky, S. (2008). What's nature done for you lately: measuring the value of ecosystem services. *Choices: The Magazine of Food, Farm, and Resource Issues* 23: 42–46.

Rathore, S.S. and Bhatt, B.P. (2008). Productivity improvement in jhum fields through integrated farming system. *Indian Journal of Agronomy* 53 (3): 167–171.

Rathore, S.S. and Shekhawat, K. (2020). *Climate Smart Agronomy*. Bhavya Books (BET) page 260. ISBN 978-93-83992-49-2.

Rathore, S.S., Shekhawat, K., Rajanna, G.A. et al. (2019). *Crop Diversification for Resilience in Agriculture and Doubling Farmers Income*, 210. Pusa, New Delhi: ICAR –Indian Agricultural Research Institute (IARI), ISBN:978-93-83168-42-2.

Rathore, S.S., Shekhawat, K., Singh, R.K. et al. (2020). Zinc management for higher productivity and profitability under diversified fodder maize based systems in semi-arid conditions. *Range Management and Agroforestry* 41 (2): 316–321.

Reddy, B.N. and Suresh, G. (2006). Crop diversification with oilseed crops for-maximizing productivity, profitability and resource conservation. *Indian Journal of Agronomy* 54 (2): 206–214.

Renard, D. and Tilman, D. (2019). National food production stabilized by crop diversity. *Nature* 571: 257–260.

Saha, R., Chaudhary, S., and Somasundaram, J. (2012). Soil health management under hill agro ecosystem of north east India. *Applied and Environmental Soil Science* https://doi.org/10.1155/2012/696174.

Singh, V.K., Sharma, B.B., and Dwivedi, B.S. (2002). The impact of diversification of a rice–wheat cropping system on crop productivity and soil fertility. *Journal of Agricultural Science* 139: 405–412. https://doi.org/10.1017/S0021859602002666.

Singh, K.K., Ali, M., and Venkatesh, M.S. (2009). *Pulses in Cropping Systems*. Technical Bulletin. Kanpur: IIPR.

Singh, V.K., Rathore, S.S., Singh, R.K. et al. (2020). Integrated farming system approach for enhanced farm productivity, climate resilience and doubling farmers' income. *Indian Journal of Agricultural Sciences* 90 (8): 1378–1388.

Tamburini, G., Bommarco, R., Wanger, T.C. et al. (2020). Agricultural diversification promotes multiple ecosystem services without compromising yield. *Science Advances* 6: eaba1715.

TEEB (2010). *The Economics of Ecosystems and Biodiversity. Ecological and Economic Foundations*. Abingdon: Routledge.

Vanbergena Adam, J., Aizenb, M.A., Cordeaua, S. et al. (2020). Transformation of agricultural landscapes in the Anthropocene: nature's contributions to people, agriculture and food security. *Advances in Ecological Research* https://doi.org/10.1016/bs.aecr.2020.08.002.

Wagg, C., Bender, S.F., Widmer, F., and van der Heijden, M.G. (2014). Soil biodiversity and soil community composition determine ecosystem multifunctionality. *Proceedings. National Academy of Sciences. United States of America* 111: 5266–5270.

3

Impact of Total Mixed Ration on Performance of Heifers and Homemade Concentrate Feeding on Milk Yield in Dairy Animals

A. Dey[1], *B.P. Bhatt*[2], *and J.J. Gupta*[1]

[1] *ICAR Research Complex for Eastern Region, Patna, Bihar, India*
[2] *NRM Division, Indian Council of Agricultural Research, New Delhi, India*

CONTENTS

3.1 Introduction

Small holder dairy animals thrive on crop by-products and residues with open grazing in field. Very little concentrate mixture is generally fed to the animals, which are often unbalanced based on the availability with the farmers at home. Analysis of nutritional status of animals before ration balancing revealed that 89.4 and 60.9% animals were overfed in terms of TDN and crude protein (CP), respectively and 45.8 and 76% of the animals were underfed in terms of Ca and P, respectively in Gujarat (Garg et al. 2009). In Bihar, Dey and Kaushal (2006) observed 10–27% gap in DCP intake and 5–10% gap in TDN intake in lactating animals. Traditionally in India, forage and concentrate are fed separately to the dairy animals, which sometime results in higher risk of ruminal acidosis particularly in high yielders due to improper nutrients supply and thereby low reproductive and productive performance (Maekawa et al. 2002; Gupta et al. 2016). However, in many parts of India, particularly in wheat growing areas of northern India, in contrary to this, total mixed ration (TMR) feeding is a way to improve rumen conditions by supplying balanced nutrients at a time in the rumen which are essential for better utilization of nutrients (Bargo et al. 2003).

Sustainable Agriculture Systems and Technologies, First Edition. Edited by Pavan Kumar, A.K. Pandey, Susheel Kumar Singh, S.S. Singh, and V.K. Singh.

It has been reported that supplementation of energy and protein in the diet improves average daily gain (ADG) in Holstein heifers and reduces rearing costs (Zanton and Heinrichs 2007). But at the same time, majority of the farmers in eastern India follow resource-based separate feeding of roughages and concentrates (Gupta et al. 2014a). Following scientific feeding and management practices, feed conversion efficiency and profitability could be increased and green house gases (GHGs) emission could be reduced (Makkar 2016). For production of 1 kg milk or meat, GHG emission can be reduced by 25% by 10% improvement in digestibility. However, the sustainability of precession feeding is questionable due to poor infrastructure and maintenance support in developing countries (Makkar 2016). TMR (*sani* feeding system) and preparation of homemade balanced concentrate feed are two low cost technologies that can be widely disseminated in India. Ration balancing programme has extensively been applied by NDDB in different states of India (Garg et al. 2009; Garg and Bhanderi 2011). Recently few workers have observed positive effect in *sani* feeding system. Somvanshi et al. (2017) observed improvement in milk production by 15.78% following *sani* feeding technique with supplementation of mineral mixture as compared to farmer practice i.e. roughages without minerals. However, Gupta et al. (2014b) reported that inclusion of concentrate feed to the TMR @1.5% body weight was beneficial in crossbred heifers in respect of body weight gain and nutrients digestibility as compared to concentrate mixture @1% of body weight.

Keeping in view, the positive effect on milk production and wider scope of applicability among smallholders, TMR and homemade balanced concentrate feed were evaluated through, two trials to assess their impact of feeding on milk yield, intake of nutrients, and cost of milk production in crossbred dairy animals.

3.2 Materials and Methods

3.2.1 Effect of TMR on Milk Yield and Nutrients Digestibility in Crossbred Cows

Annual cereal and legume fodders were grown during rainy and winter season under irrigated system at ICAR Research Complex for Eastern Region, Patna, under subtropical hot and humid climate. The soils were neutral in pH and alluvial-clay in nature. Multicut sorghum as cereal and rice bean as legume were grown during rainy season. Whereas, oat as cereal and berseem as legume fodder were grown during winter season on the same land on relay system. To study the effect on milk production and nutrient digestibility in first lactating crossbred cows, two feeding experiments were conducted for 60 days duration each during rainy and winter seasons on TMR. In experiment-I, six cows were distributed into two groups having three cows in each group. For preparation of TMR, wheat straw 2 kg, chapped green forages (multicut sorghum 25 kg and rice bean 2 kg) 27 kg, and concentrate feed (17% dietary CP) 5 kg were mixed together with little water to make *sani*. The above TMR was offered to each cow of T1 group daily in dividing doses i.e. morning and evening. The cows of group T2 were fed the same quantity of above feed ingredients separately. At first, wheat straw and concentrate feed was offered after mixing with water and then a gap of one to two hours un-chapped green forages were fed. A total of 10.68 kg dry matter (DM) per day per head was provided to both the groups. Similarly, experiment-II was conducted in six cows following previous feeding and management schedule. During winter season, however, wheat straw 2 kg, chapped green forages (oat 15 kg and berseem 16 kg) 31 kg and concentrate feed (16.5% dietary CP) 4.5 kg constituted the TMR. Total 11.28 kg DM per day per head was provided to the cows in both the groups. At the end of the feeding experiment, a digestion trial of five-day duration was conducted in both the experiments. Feed offered, residues and feces were collected daily for DM and CP estimation (AOAC 2005). The dry matter intake (DMI) per 100 kg body weight was calculated. The gross energy (GE) was determined in pooled dry samples. The digestibility of DM and CP, digestible energy (DE) values were calculated and analyzed statistically (Snedecor and Cochran 1994).

3.2.2 Effect of Homemade Balanced Concentrate Feed on Milk Yield and Nutrients Intake in Crossbred Cows

For the development of ration, four low-cost homemade balanced diets were prepared varying in composition of the different ingredients with locally available feed resources and crop residues available with farmers of different agro-climatic zones of Bihar. Composition of diets has been given in Table 3.1. For the evaluation of the diets, 16 crossbred cows (HF × Indigenous) of similar parity (first and second), age (26–35 months), and body weight (365.2 ± 3.66 kg) were divided into four groups and fed concentrate mixtures I–IV, respectively at rate of 1 kg concentrate per 2.5 kg of milk yield plus 2 kg for maintenance as per the schedule followed by farmers in the region. Wheat straw and Sorghum fodder were offered ad libitum as source of dry and green fodder, respectively. The study

Table 3.1 Composition of home-made balanced concentrate mixtures.

	Concentrate mixture (% fresh weight basis)			
Ingredients	**CM-I**	**CM-II**	**CM-III**	**CM-IV**
Crushed maize/wheat/broken rice	30	25	20	35
Wheat bran	15	10	10	0
Rice polish	0	0	15	20
Deoiled rice bran	15	20	10	0
Mustard cake	7	5	16	20
Linseed cake	12	15	15	15
Lentil/arhar/gram chunies	18	22	11	7
Mineral mixture	2	2	2	2
Salt	1	1	1	1
Total	100	100	100	100
CP% (DM basis)	18.04	17.99	18.60	18.63
TDN% (DM basis)	68.51	67.34	68.92	71.59

was continued for 35 days during which daily milk yield and its compositions were estimated. Before the experiment, all the animals were fed balanced concentrate feed available in the market at the same rate as used in the experiment along with similar green and dry fodder. Pre-experiment milk yield of each animal was recorded for 15 days, which was compared with the milk yield of experimental period for each group. Other feeding and management practices were the same. Daily feed offered and residues left were collected daily for DM and CP estimation (AOAC 2005). Body weight of each animal was recorded at the beginning and at the end of the experiment for two consecutive days. Based on these data, DMI per 100 kg body weight was calculated. Data were analyzed statistically (Snedecor and Cochran 1994).

3.3 Results and Discussion

3.3.1 Effect of TMR on Milk Yield and Nutrients Digestibility in Crossbred Cows

The bio-mass yield and protein content of different fodder crops used in the experiment are presented in Table 3.2. The productivity of cereal crop was observed higher than the leguminous forages in rainy season whereas opposite trend was observed in winter season. Number of cuts of fodder actually made the difference of biomass yield. The values on fodder biomass yield corroborate with the findings of Pandey and Roy (2011), Gupta and Dey (2015), and Gupta et al. (2016). The DM content of forages varied from 11.65 to 17.19% while CP content varied from 8.72 to 11.14% for cereal forages and 15.30 to 15.83% in leguminous forages. The DM and protein content of cereal (sorghum and oat) and leguminous (berseem and rice bean) fodder are in close agreement with the finding of Banerjee (2000), Gupta and Dey (2015), and Gupta et al. (2016). The minute variations in compositions may be attributed to variety of forage, soil quality, number of cuts, and management practices adopted. Feeding of TMR resulted in higher DMI (kg/100 kg body weight) by 15.92 and 2.41%, respectively in T1 group in both the experiments in comparison to group T2 where feeds were offered separately (Table 3.3). Significantly higher ($p < 0.01$) DMI was observed

Table 3.2 Season-wise production potential of different fodder.

Particulars	Total forage yield (t/ha)	Average DM (%)	Average CP (%)
Rainy season			
Multicut Sudan (threecuts at 60, 105, and 145 d)	74.78 ± 2.92	14.48	8.72
Rice bean (single cut at 90 d)	35.67 ± 1.52	17.19	15.30
Winter season			
Berseem (four cuts at 50, 85, 115, and 145 d)	67.84 ± 1.22	11.65	15.83
Oat (two cuts at 50 and 105 d)	28.23 ± 0.64	14.18	11.14

Table 3.3 Performance of crossbred cows fed on TMR.

	Experiment I		Experiment II	
Particulars	**T1**	**T2**	**T1**	**T2**
Days in lactation (d)	98 ± 28.50	84 ± 31.50	246 ± 19.50	222 ± 2.50
Av. body wt. (kg)	362 ± 20.50	385 ± 19.00	354 ± 42	360 ± 32
Av. DMI (kg/100 kg body wt.)	3.14 ± 0.06[a]	2.64 ± 0.03[a]	4.14 ± 0.28	4.04 ± 0.23
DMD (%)	68.44 ± 1.83	68.17 ± 0.62	67.54 ± 0.23[a]	60.07 ± 0.01[a]
CPD (%)	47.84 ± 1.48	43.51 ± 1.16	44.70 ± 0.53[a]	31.42 ± 1.57[a]
DE (Kcal/kg)	2754 ± 34	2687 ± 13	2542 ± 16[a]	2288 ± 08[a]
Average initial milk yield (kg/d/head)	6.13 ± 0.16	6.24 ± 0.39	5.13 ± 0.06	6.55 ± 0.44
Average final milk yield (kg/d/head)	7.17 ± 0.12	6.91 ± 0.28	6.13 ± 0.24	6.97 ± 0.09

[a] TMR, Total mixed ration; Av, Average; DMD, Dry matter digestibility; CPD, Crude protein digestibility; DE, Digestible energy.

in T1 (TMR fed) during experiment 1 (rainy season) when multicut sorghum was used as green fodder. However, total DMI did not differ significantly between groups in experiment II (winter season). This may be attributed to the succulent form of forage oat and berseem which included both in TMR and separate feeding system. Higher intake of DM, CP, and DE was also observed by Khan et al. (2010) in crossbred cows fed densified complete feed. Gupta et al. (2016) observed that feeding of concentrate feed at 1.5% of body weight in TMR increased DMI in crossbred heifers. However, Kajla et al. (2019) reported that DM intake was nonsignificant in TMR and non-TMR groups in crossbred cows. Similar results were also reported by Raja Kishore et al. (2013), who observed nonsignificant DM intake (kg/day) in TMR and non-TMR fed buffalo bulls. This may be attributed to the sorting of feed ingredients by animals fed roughage and concentrate separately.

The DM and CP digestibility and DE values of ration are presented in Table 3.2. The DM and CP digestibility and DE value of ration were nonsignificant between the groups in experiment 1, when multicut sorghum and rice bean forage were included in TMR, However, the values increased significantly ($p < 0.01$) in cows when berseem and oat were used as green fodder in experiment II. The values of DM, CP, and DE were also observed higher in crossbred cows fed densified complete feed by Khan et al. (2010). Gupta et al. (2016) observed that feeding of concentrate feed at 1.5% of body weight in TMR was beneficial and economical for higher DMI, nutrient digestibility, and growth rate in crossbred heifers. Hundal et al. (2004) reported the beneficial effect of feeding TMR than separate feeding as organic matter and neutral detergent fiber digestibility were observed significantly ($P < 0.05$) higher in TMR fed group. This may be attributed to the higher concentration of total volatile fatty acids and different nitrogen fractions in rumen of calves fed TMR as compared to the conventional feeding system. Similar trends in the rumen fermentation pattern have been reported by Reddy and Reddy (1983) in calves. Raja Kishore et al. (2013) observed that the digestibility of CP, fat, fiber, neutral detergent fiber, acid detergent fiber, hemi-cellulose, and cellulose were higher ($P < 0.01$) in buffalo bulls fed complete rations than those fed conventional ration. Kajla et al. (2019) reported significantly ($P<0.05$) higher digestibility of DM, CP, ADF, NDF, and TDN in crossbred cows fed TMR as compared to non-TMR fed cows. However, OM, EE, and CHO digestibility were not affected by mode of feeding.

In the present study, overall increase in milk production by 16.96 and 19.49% was recorded in cows fed TMR during experiment I and II, respectively in comparison to separate feeding system (Table 3.3). Similarly, an increase in milk production by 10.74 and 6.41% was observed in cows fed different feeds separately during experiment I and II, respectively. Feeding of balanced ration either in mixed form or individually had increased milk production but impact was greater when fed with TMR (Figure 3.1). This might be due to higher DMI and nutrients digestibility. The results corroborate with the observations of Khan et al. (2010) in crossbred cows fed densified complete feed. It is reported that the use of complete feeds over three lactations had no effect on herd health or milk production as compared to the conventional system in which cows were fed forage and grain separately (Larkin and Fosgate 1970). However, Bae et al. (1994) and Su and Hsieh (1999) indicated significant ($P<0.05$) impact of feeding system on milk yield and its composition. Efficient utilization of nutrients, favorable rumen environment and blood profile were responsible for higher milk production. Kajla et al. (2019) reported significantly positive ($P<0.05$) effect on milk production and milk energy yield on TMR feeding except the milk protein percent, which was observed higher in non-TMR group. Gupta et al. (2016) observed that feeding of TMR was beneficial in respect of milk production, which was attributed by proportionate intake of all feed ingredients, overall feed intake, and better digestibility of nutrients. Schraufnagel (2007) also observed positive impact on milk yield and thereby profits on TMR feeding. Similar results are reported by Bargo et al. (2002), Sarker et al. (2019), and Awlad Mohammad et al. (2017). Kerketta et al. (2020) observed that the milk yield (kg/day) was found significantly higher ($P<0.05$) in TMR (7.90 ± 0.14) than non-TMR (6.82 ± 0.11) with average increase by 1.13 l/d. TMR fed buffaloes also showed increased milk protein content (3.15% vs 3.3%). The milk fat, SNF percentage also found higher in

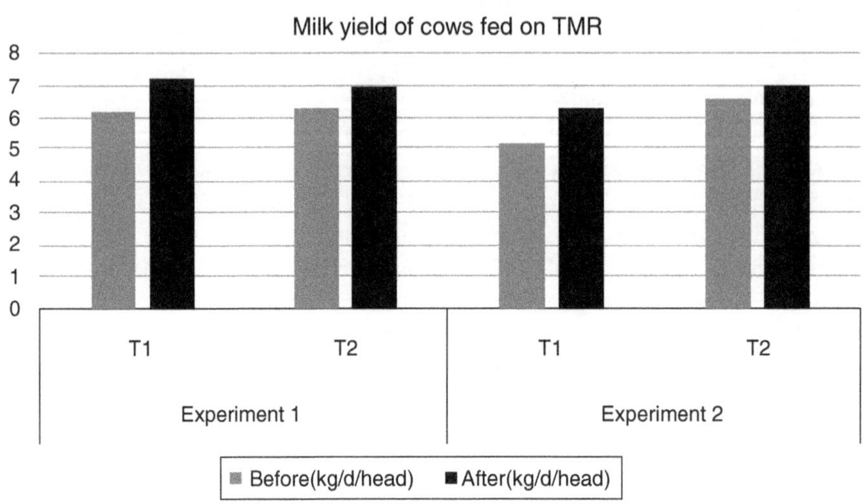

Figure 3.1 Milk yield of crossbred cows fed on TMR.

TMR with no significant difference. Percent increase in milk yield was found to be 14.3% by TMR feeding in buffalo. However, Hundal et al. (2004) reported nonsignificant impact by feeding system on milk production and its composition except lactose content, which was significantly ($P<0.05$) higher in animals fed TMR in comparison to conventional feeding system. They concluded that 15–16 kg of milk production per day could be sustained by feeding TMR based on berseem and oat hay. Several workers reported increased milk production in animals fed TMR compared to conventional method. This is because of the steady state of rumen environment conducive to the continuous rumen function and digestive flow achieved by TMR feeding. Animals fed with TMR reported a trend for more milk and higher efficiency of metabolizable energy utilization for milk production. Blending reduces the individual animal variation in feed consumption and results in better balance of nutrients than feeding the same ingredients individually (Holter et al. 1977). In on-farm trial, increased milk production was reported in lactating Murrah buffaloes fed maize stover-based TMR compared to conventional feeding system (Raja Kishore et al. 2013).

3.3.2 Effect of Homemade Balanced Concentrate Feed on Milk Yield and Nutrients Intake in Crossbred Cows

The effect of various concentrate mixtures on body weight changes and milk yield of crossbred cows has been depicted in Table 3.4. It is observed that all the groups maintained the body weight during the experimental period. The initial milk yield varied from 7.10 to

Table 3.4 Performance of crossbred lactating cows fed on homemade balanced concentrates mixtures.

	Concentrate mixtures			
Ingredients	**CM-I**	**CM-II**	**CM-III**	**CM-IV**
Initial body weight (kg)	362.1 ± 4.08	367.4 ± 4.06	364.6 ± 3.11	366.7 ± 2.68
Final body weight (kg)	363.3 ± 4.17	368.1 ± 3.96	365.7 ± 3.33	368.6 ± 3.75
Initial milk yield (kg/d/cow)	7.24 ± 0.12	7.22 ± 0.11	7.30 ± 0.15	7.10 ± 0.09
Final milk yield (kg/d/cow)	7.42 ± 0.18	7.56 ± 0.10	7.76 ± 0.18	7.94 ± 0.14
Fat content of milk (%)	3.41 ± 0.02	3.46 ± 0.02	3.21 ± 0.06	3.44 ± 0.01
Green fodder intake (kg/d/cow)	3.93 ± 0.26	4.05 ± 0.99	3.96 ± 1.01	4.14 ± 0.92
Dry fodder intake (kg/d/cow)	3.54 ± 0.88	3.53 ± 0.90	3.98 ± 0.97	3.76 ± 0.99
Concentrate intake (kg/d/cow)	5.13 ± 0.11	4.60 ± 0.21	5.62 ± 0.45	6.67 ± 0.76
Total DM intake (kg/d)	12.60 ± 0.23	12.18 ± 0.47	13.56 ± 0.65	14.57 ± 0.81
DMI (kg/100 kg body weight)	3.46 ± 0.12	3.31 ± 0.09	3.71 ± 0.16	3.95 ± 0.21
DCP intake (kg/100 kg body weight)	0.272 ± 0.06	0.250 ± 0.05	0.297 ± 0.0.9	0.341 ± 0.11
TDN intake (kg/100 kg body weight)	2.002 ± 0.18	1.882 ± 0.11	2.143 ± 0.09	2.373 ± 0.14
Cost of concentrate feed (Rs./t)	6910	6930	6940	7280
Cost of total feed (Rs./d)	62.47 ± 0.62	65.11 ± 0.84	74.52 ± 0.58	82.36 ± 0.39
Feed cost for milk production (Rs./kg of milk)	9.15 ± 0.08	8.61 ± 0.12	9.60 ± 0.18	10.37 ± 0.11

7.30 kg/day in all the animals fed concentrate feed available in the market along with ad libitum wheat straw and Sorghum fodder. After feeding of test concentrates, there was an increase of 2.5–11.85% milk yield as compared to initial milk yield of cows fed balanced concentrate feed available in the market, which indicates that the homemade balanced concentrates are comparable to the commercial concentrate available in the market. The concentration of milk constituents like fat varied from 3.21 to 3.46% and did not vary significantly among the groups. Singh et al. (2019) also evaluated cow cost concentrate feed at farmers' field in Bihar. They reported that milk yield of dairy animals increased by 14.3% per day fed on test diets and 0.71/kg of concentrate feed with increase in fat percentage by 21.3%. Farmers' income, on an average, was increased by Rs. 44 per day per animal. However, Sherasia et al. (2016) observed that balanced feeding improved daily fat corrected milk yield by 0.7 kg/cow. They also reported that balanced feeding increased the dietary protein and energy intake by 25.0 and 12.7% whereas, calcium and phosphorus intake were reduced by 30.0 and 27.0%, respectively. Atturi et al. (2018) also observed that feeding balanced ration improved body weight and milk production of dairy animals. Garg (2012) reported that balanced feeding improved milk yield and milk fat content significantly ($P < 0.05$) by 0.68 kg/day and 0.55% units, respectively, in cows and 0.19 kg/day and 0.34% units, respectively, in buffaloes in the northern region of India. In the southern and central regions of India, milk yield of cows increased by 0.42 and 0.46 kg/day, respectively.

Garg et al. (2009) reported that ration balancing improved milk production by 0.58 kg/animal/day with increment of 0.5% milk fat. After ration balancing, the improvement in milk production efficiency resulted in more milk from the same amount of feed. Milk production efficiency (kg fat corrected milk yield/kg DMI) of cows was 0.58 and 0.78 kg/kg, respectively for cows, before and after ration balancing. For buffaloes these values were 0.53 and 0.66 kg/kg. Balanced feeding can reduce the methane emissions from livestock farming, which is utilized for production purposes rather than maintenance, also known as the maintenance dilution effect (Garg et al. 2013). Vagamashi et al. (2016) observed improvement of daily milk yield by 0.95 l/day/animal and milk fat by 0.24% (3.98–4.22%) on feeding a balanced ration. This may be attributed to more efficient utilization of dietary energy and protein in lactating cow as reported by Garg and Bhanderi (2011). Findings are also similar to that of Haldar and Rai (2003), Bhanderi et al. (2016), and Mahanta (2017).

Total DMI per 100 kg body weight varied from 3.31 ± 0.09 kg to 3.95 ± 0.21 kg/d which did not differ significantly among the groups. DCP and TDN intake (kg/100 kg body weight) varied from 0.250 ± 0.05 to 0.341 ± 0.11 and 1.882 ± 0.11 to 2.373 ± 0.14 kg, respectively. The DCP and TDN intake did not differ significantly among the groups. Mahanta (2017) also reported DMI of 2.35–2.53 kg/100 kg body weight in crossbred lactating cows fed low cost balanced feed which was much lower than the present values. This may be attributed to the fibrous nature and bulkiness of the low cost feed which was responsible for low DMI. The DCP and TDN intake (g/kg $W^{0.75}$) of lactating crossbred cows were reported by Mahanta (2017) at 7.42–8.13 and 66.58–70.57, respectively. The higher DCP and TDN intake in the present study may be due to higher feed dry matter intake by crossbred cows than the values reported by Mahanta (2017). The cost of different concentrate feed varied from Rs. 6910 to 7280 per ton which was much lower than the market rate (Rs. 16 000/t) (Table 3.4). Mahanta (2017) also prepared different low cost feed costing Rs. 8.58 and 8.28 per kg for lactating cows. In the present study, the cost of total feed offered to dairy animals in a day varied from Rs. 62.47 to 82.36. The cost of milk production was also estimated to be Rs. 8.61 to 10.37 per liter (Figure 3.2). Vagamashi et al. (2016) reported an additional benefit of Rs. 38 per animal per day in cows fed on balanced ration, when milk production was increased by 0.95 l/day/animal and fat percentage by 0.24%. However, Mahanta (2017) reported the feed cost for milk production at Rs. 10.98–12.21 per kg of milk in crossbred cows fed low cost balanced feed, which are almost similar to the present study.

3.4 Conclusion and Future Prospects

TMR and homemade balanced feed are two important technologies for dairy sector in reducing the production cost as well as maintaining good health of animals. In most parts of the eastern region, undernutrition is a major constraint in animal production due to high cost of concentrate, lack of green fodder and quality dry fodder. Therefore, major share of ration is constituted by only crop residues and by-products. As a result, full potential of production and reproduction of dairy animals is not exploited. Moreover, high cost of concentrate feed results in increase in input cost and thereby, nonprofitability of dairy farming. Feeding of homemade balanced concentrate together with TMR can solve the problem, particularly in smallholder milk production system. In recent years mobile-based ration balancing programme is being implemented in many parts of India by NDDB. However, still it is not gaining up to the mark, particularly among resource poor farmers. Readymade availability of different types of homemade concentrate feed formulas may be acceptable by the smallholders. For wider dissemination, the formulas need to be handed over to the state government, KVKs and NGOs. Use of mass media and distribution of leaflets are also needed for creating general awareness among smallholders.

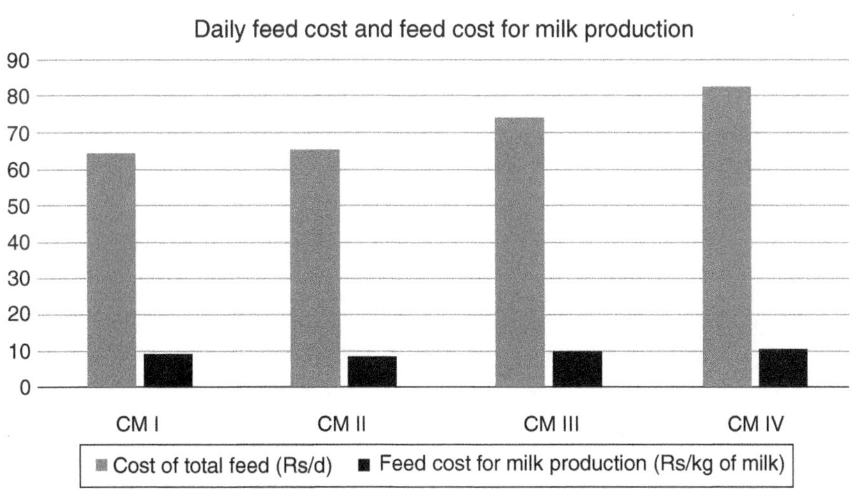

Figure 3.2 Daily feed cost and feed cost for milk production in cows fed different concentrate mixtures.

References

AOAC (2005). *Official Methods of Analysis, Association of Official Analytical Chemists*, 18e. Washington, DC: USDA.

Atturi, K.M., Singh, A., Chakravarthi Matha, K. et al. (2018). Effect of balanced ration supplementation on body weight gain and milk yield in different breeds of cattle. *International Journal of Current Microbiology and Applied Sciences* 7 (11): 2443–2446. https://doi.org/10.20546/ijcmas.2018.711.278.

Awlad Mohammad, M.E.M., Gorgulu, M., and Goncu, S. (2017). The effects of total mixed ration and separate feeding on lactational performance of dairy cows. *Asian Research Journal of Agriculture* 5 (2): 2456–561X.

Bae, O.H., Shin, C.N., and Ko, K.H. (1994). Effect of total mixed ration including apple pomace for lactating cows. *Korean Journal of Dairy Science* 16: 295–302.

Banerjee, G.C. (2000). *Feeds and Principles of Animal Nutrition*, 82–98. New Delhi: Oxford and IBH Publishing Co. Pvt. Ltd.

Bargo, F., Muller, L.D., Delahoy, J.E., and Cassidy, T.W. (2002). Performance of high producing dairy cows with three different feeding systems combining pasture and total mixed rations. *Journal of Dairy Science* 85 (11): 2948–2963.

Bargo, F., Muller, L.D., Delahoy, J.E., and Cassidy, T.W. (2003). Performance of high producing dairy cows with three different feeding systems combining pasture and total mixed rations. *Journal of Dairy Science* 86: 3237–3248.

Bhanderi, B.M., Garg, M.R., Goswami, A., and Shankhpal, S. (2016). Effect of feeding balanced ration on solid-not-fat content of milk and production performance of lactating crossbred cows. *Indian Journal of Animal Nutrition* 33: 131–137.

Dey, A. and Kaushal, D.K. (2006). Traditional Animal Husbandry and Fishery Practices and Diseases Occurrence Pattern in Some Selected Villages of Bihar. Final project (ICAR-RCER-LFIMP-22-2003) report of ICAR Research Complex for Eastern Region, Patna.

Garg, M.R. (2012). Impact of feeding balanced ration. In: *Balanced Feeding for Improving Livestock Productivity – Increase in Milk Production and Nutrientuse Efficiency and Decrease in Methane Emission* (ed. H.P.S. Makkar), 15–22. Rome, Italy: FAO Animal Production and Health Paper No. 173.

Garg, M.R. and Bhanderi B.M. (2011). Enhancing livestock productivity through balanced feeding. *Proceedings of Livestock Productivity Enhancement with Available Fedd Resources* (3–5 November 2011). Pantnagar, Uttrakhand, India: G.B. Pant University of Agriculture and Technlogy. pp. 11–21.

Garg, M.R., Biradar, S.A., and Kannan, A. (2009). Assessment of economic impact of implementing ration balancing programme in lactating cows and buffaloes under field conditions. *Indian Journal of Animal Nutrition* 26 (2): 146–150.

Garg, M., Sherasia, P., Bhanderi, B. et al. (2013). Effects of feeding nutritionally balanced rations on animal productivity, feed conversion efficiency, feed nitrogen use efficiency, rumen microbial protein supply, parasitic load, immunity and enteric methane emissions of milking animals under field conditions. *Animal Feed Science and Technology* 179: 24–35.

Gupta, J.J. and Dey, A. (2015). Nutritive value of promising cereal forages with or without combination of leguminous forages in goat. *The Indian Veterinary Journal* 92 (7): 34–37.

Gupta, J.J., Singh, K.M., Bhatt, B.P., and Dey, A. (2014a). A diagnostic study on livestock production system in eastern region of India. *The Indian Journal of Animal Sciences* 84: 198–203.

Gupta, J.J., Dey, A., Bhatt, B.P. et al. (2014b). Performance of lactating crossbred cows fed on forage based total mixed ration. *Livestock Research International* 2 (2): 30–32.

Gupta, J.J., Dey, A., Dayal, S., and Barari, S.K. (2016). Performance of heifers fed on total mixed ration. *The Indian Veterinary Journal* 93 (08): 29–31.

Haldar, S. and Rai, S.N. (2003). Effect of energy and mineral supplementation on nutrient digestibility and efficiency of milk production in lactating goats. *Indian Journal of Animal Nutrition* 20: 244–251.

Holter, J.B., Urban, W.E., Hayes, H.H. Jr., and Davis, H.A. (1977). Utilization of diet components fed blended or separately to lactating cows. *Journal of Dairy Science* 60 (8): 1288–1293.

Hundal, J.S., Gupta, R.P., Wadhwa, M., and Bakshi, M.P.S. (2004). Effect of feeding total mixed ration on the productive performance of dairy cattle. *Animal Nutrition and Feed Technology* 4: 179–186.

Kajla, S., Grewal, R.S., Kaur, J. et al. (2019). Effect of total mixed ration feeding with roughage: concentrate ratio of 60:40 on performance and residual feed intake of crossbred cows. *International Journal of Current Microbiology and Applied Sciences* 8 (04): 2866–2870. https://doi.org/10.20546/ijcmas.2019.804.334.

Kerketta, S., Sarangdevot, S.S., Naruka, P.S. et al. (2020). Production performance comparison of lactating buffaloes on feeding of total mixed ration versus conventional feeding. *Journal of Entomology and Zoology Studies* 8 (4): 1657–1660.

Khan, S.R., Singh, S.K., and Mudgal, V. (2010). Effect of feeding complete ration on the performance of lactating crossbred cows. *Indian Journal of Animal Nutrition* 27 (3): 261–264.

Larkin, J.C. and Fosgate, O.T. (1970). Comparison of two different systems of feeding dairy cows for three consecutive lactations. *Journal of Dairy Science* 53: 561.

Maekawa, M., Beauchemin, K.A., and Christensen, D.A. (2002). Effect of concentrates level and feeding management on chewing activities, saliva production and ruminal pH of lactating dairy cows. *Journal of Dairy Science* 85: 1165–1175.

Mahanta, S.K. (2017). *Economic Balanced Rations for Dairy Animals*, Research Bulletin No. 35. Indian Grassland and Fodder Research Institute, Jhansi, Uttar Pradesh, India.

Makkar, H.P.S. (2016). Smart livestock feeding strategies for harvesting triple gain – the desired outcomes in planet, people and profit dimensions: a developing country perspective. *Animal Production Science* 56: 519–534.

Pandey, K.C. and Roy, A.K. (2011). *Forage Crops Varieties*. Jhansi, India: Indian Grassland and Fodder Research Institute.

Raja Kishore, K., Srinivas Kumar, D., Ramana, J.V., and Raghava Rao, E. (2013). Field trial of maize stover based complete ration vis-à-vis conventional ration on lactation performance in graded Murrah buffaloes. *Animal Science Reporter* 7 (4): 123–127.

Reddy, D.N. and Reddy, M.R. (1983). The effect of feeding complete feeds on nutrient utilization and milk production on crossbred cows. *Indian Journal of Dairy Science* 36 (4): 421–423.

Sarker, N.R., Yeasmin, D., Tabassum, F., and Habib, M.A. (2019). An on farm study for feeding impact of total mixed ration (TMR) in milking cow. *Current Journal of Applied Science and Technology* 35 (2): 1–8.

Schraufnagel, S. (2007). *Technology Systems and Business Performance on Wisconsin Dairy Farms*. University of Wisconsin-River Falls.

Sherasia, P.L., Phondba, B.T., Hossain, S.A., and Garg, M.R. (2016). Impact of feeding balanced rations on milk production, methane emission, metabolites and feed conversion efficiency in lactating cows. *Indian Journal of Animal Research* 50: 505–511.

Singh, D.K., Sahu, S.P., and Nils, T. (2019). Increasing dairy cow productivity through new balanced concentrate feed: a study in Bihar, India. *Indian Journal of Animal Nutrition* 36 (1): 11–16.

Snedecor, G.W. and Cochran, W.G. (1994). *Statistical Methods*, 9e. New Delhi: Oxford and IBH Publishing Co. Pvt. Ltd.

Somvanshi, S.P.S., Singh, H.P., Shaktwat, R.P.S. et al. (2017). Study of *sani* feeding techniques with supplementation of mineral mixture in dairy animals. *Plant Archives* 17 (2): 1219–1220.

Su, A.K. and Hsieh, R.C. (1999). Effect of feeding total mixed ration on the lactation performanceand blood parameters of dairy goats. *Journal of Taiwan Livestock Research* 32: 183–191.

Vagamashi, D.G., Murkute, V.D., Jangale, P.R., and Jotaniya, A.H. (2016). Impact of balanced feeding on milk production, milk fat and feeding cost in crossbred cows. *International Journal of Science, Environment and Technology* 5 (6): 3989–3992.

Zanton, G.I. and Heinrichs, A.J. (2007). Effect of controlled feeding of a high-forage or high-concentrate ration on heifer growth and first lactation milk production. *Journal of Dairy Science* 90: 3388–3396.

4

Multifaceted Impact of Lockdown During COVID-19 on Food Security and Smallholder Agricultural Systems

Aishwarya[1], Meenu Rani[2], Bhagwan Singh Chaudhary[3], Bharat Lal[1], Rajiv Nandan[1], and Pavan Kumar[4]

[1] *College of Agriculture, Rani Lakshmi Bai Central Agricultural University, Jhansi, Uttar Pradesh, India*
[2] *Department of Geography, Kumaun University, Nainital, Uttarakhand, India*
[3] *Department of Geophysics, Kurukshetra University, Kurukshetra, Haryana, India*
[4] *College of Horticulture and Forestry, Rani Lakshmi Bai Central Agricultural University, Jhansi, Uttar Pradesh, India*

CONTENTS

4.1 Introduction

With the emergence of coronavirus disease (COVID-19) as a public health emergency of international concern, India seems to be one of the top vulnerable countries. COVID-19 is the seventh member of the coronavirus family and in the past six other types have already been reported. The genetic and genome structure of COVID-19 is approximately 80% similar to that of the SARS virus detected in bats (Linton et al. 2020). The symptoms of its infection are almost identical to the seasonal flu where patients may have cold, cough, and fever.

Sustainable Agriculture Systems and Technologies, First Edition. Edited by Pavan Kumar, A.K. Pandey, Susheel Kumar Singh, S.S. Singh, and V.K. Singh.

However, the actual detection is only possible after the clinical test (Kumar et al. 2021; Singh et al. 2021), which is another big challenge for a populated country like India, with limited testing facilities. Looking at the pace of its spread and the severity of the situation, the WHO on 23 January 2020, declared COVID-19 outbreak as a public health emergency of international concern (Sohrabi et al. 2020; World Health Organization 2020). The earlier pandemic of Spanish flu in the year 1918 tolled 50 million deaths worldwide (Chatterjee et al. 2020) and remained active for more than the next two years until it was controlled (https://www.history.com/news/spanish-flu-second-wave-resurgence). Thus, it becomes essential to study the trajectories of infection, death, and recovery for the near future. This is expected to assist policymakers in formulating plans and adopt adequate measures to control any further damage caused by the COVID-19 (Raghunathan et al. 2019).

4.2 Predictive Model for Deflation of COVID-19 Spread in India

This result is based on the statistical model generated for the possible decline of COVID-19 cases and their duration in the country or impact based on parameters such as GDP, control rate, facilities, economic measures, and government policies. Here, the results are based partly on the possible recovery rates, which present a daily increase of 0.5, 1, 1.5, and 2%, and the timeline is set from 21 April 2020. The forecast predicts a cause for concern for the country. Even the precedented recovery rate 1 is 0.5%, recovery rate 2 is 1%, recovery rate 3 is 1.5%, and recovery rate 4 is 2% are assumed, it would take almost one year and nine months, 10 months, eight months, and five months respective recovery rate to completely recover from the pandemic. The decline in the number of the confirmed cases is the steepest and directly proportional to an increase in the recovery rates. Therefore, this hypothesis is totally based on the abovementioned parameters and how effectively they are executed in preventing the spread of this pandemic. It indicates the possible range of minimum and maximum duration required by the country to recover from this pandemic (Figure 4.1).

4.3 Impact on the National Economy

We are living in a historical time; the bad thing is that whenever this page of history is opened, it will open with sorrow and suffering. Everyone is talking about the impact of COVID-19,but only on national or on urban conditions. Unfortunately, not much is being said about the impact of COVID-19 on the rural sector, even though it is a large part of the economy and plays an important role in the product categories of the country. In the midst of the crisis of COVID-19, rural India is not becoming an issue of central discussion. Rather it should be in the hour of this crisis that rural economy should be discussed the most. At present, the agricultural sector is one such link which can make the Indian economy run smoothly again in the future. The agricultural sector has recorded a historic decline in the last few years. Annual growth rates of agriculture and allied sectors are 1.5, 5.6, −0.2, 0.6% in 2012–2013, 2013–2014, 2014–2015, 2015–2016, 2016–2017, 2017–2018, and 2018–2019 respectively 6.3, 5.0, and 2.9%. According to the report released by the National Statistics

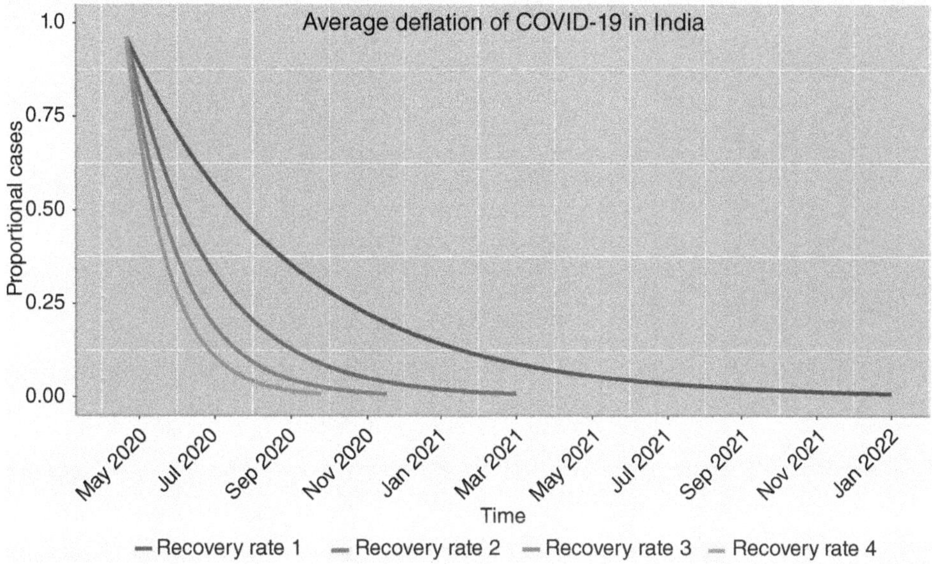

Figure 4.1 Predictive statistical model for the deflation of COVID-19 spread in India.

Office in January 2020, the estimate of agricultural growth rate in the year 2019–2020 was only 2.8%. If the crisis continues like this, agricultural development may once again be negative. Despite all the government announcements, due to the ongoing crisis of COVID, the market for agricultural products is ending rapidly. The agricultural sector is also witnessing a supply crisis along with demand. The important fact is that at the present time the economy cannot be divided between the rural and the city.

The situation caused by the COVID-19 crisis and extended lockdown period in India severely affects not only the supply but also the demand for agro-foods (Ceballos et al. 2020; Mahendra Dev 2020). This directly impacts the economy of 140 million Indian farmers and the share of the GDP associated with agriculture, which has similarly declined. In India, the agricultural yields from local farms for Rabi crops, cereals get stored in cold storages and/or are directly supplied to the Indian market. Food touch prices highly o urban market when mostly cereal production cannot get into market and this situation creates very difficult for both poor people as well as for local farmers (Kumar et al. 2020). To mitigate the immediate challenges related to lockdown, the Government of India (GoI) has provided economic aid or relief packages aimed at many different sectors. The economic aid in general has been critically needed but has also inversely impacted the national GDP (see Figure 4.2).

As shown in Figure 4.2, the national GDP was in March 2020 expected to be down about 0.9 trillion USD in value from FY2020-Q1 (March 2020) to FYP2020-Q3 (September 2020). A fair part of this drop is expected to be related to the agricultural sector. For comparison, a novel five-year scheme worth INR 15 000 crores introduced by the GoI aims to strengthen the state and national level systems in order to combat the health impacts of COVID-19 (Mohan et al. 2020). This fund was further circulated to intensive care units, supply of oxygen cylinders in medical centers, distribution of mask, sanitizer, oxygen generating plants,

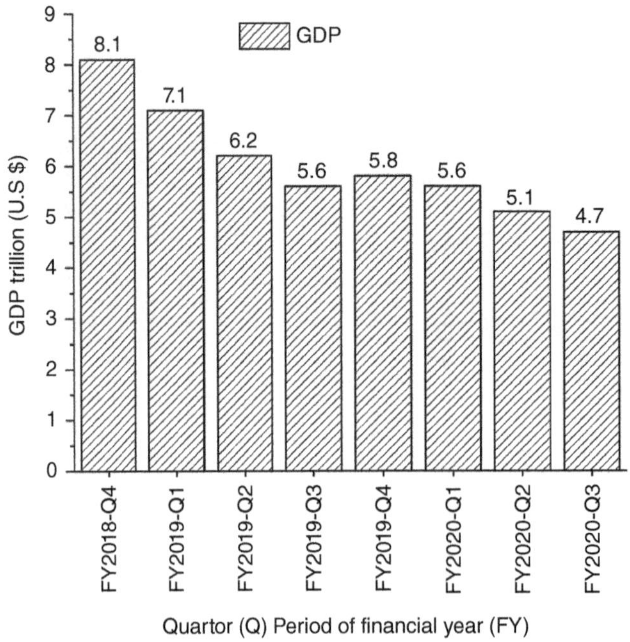

Figure 4.2 Recent and expected developments in the GDP of India since the fourth quarter (Q4) of the fiscal year 2018 (FY2018). The numbers shown are from the Reserve Bank of India (RBI), accessed 16 May 2020.

and many more. State government of India has planned for health system preparedness package into three phases from 2020 to 2024. The growth rate for FY2021 was initially rated by World Band and other rating agencies with the lowest figures in the last three decades since India's economic liberalization in the 1990s began. The economic packages announced by the GoI in the middle of May 2020 did not help and the country's GDP estimates were downgraded even more to negative figures, signaling a deep recession. Consequently, India was expected to lose over ₹32 000 crores (US$4.5 billion) every day during the first 21 days of complete lockdown declared following the COVID-19 outbreak (The Hindu Business Line 2020). It has been also estimated that lockdown may affect around 53% of enterprises across the country (The Indian Express 2020) severing almost all economic activity (Kumar et al. 2020). However, the GoI initiated several remedial measures beginning with food security and allocating extra funds for healthcare and states, sector related incentives and tax payment deadline extensions. In the very beginning, on 26 March 2020 itself, a number of economic relief measures for the poor were announced totaling over ₹170 000 crore. Reserve Bank of India also announced to make available ₹374 000 crore to the country's financial bounty. Global financial institutions like the World Bank and Asian Development Bank also supported India to tackle the coronavirus pandemic. Moving further, the Prime Minister announced an overall economic package worth ₹20 lakh crore (US$280 billion) constituting the 10% of India's GDP on 12 May 2020 which was further unfolded by the Indian Finance Minister in the next five days. As a final and booster dose to Indian economy, on 12 October 2020, the finance minister announced an economic stimulus package worth ₹46 675 crore which is 0.2% of the GDP.

4.4 Government of India and Local Government Initiatives

The migration of people back to their native rural communities combined with a decline in market availability for and sale of produce (and the social and economic issues relating to both) reinforces the importance of having strong policies in place to tackle these challenges. Timeline of Finance Minister announces measures to strengthen agriculture, infrastructure, logistics, food processing sectors, capacity building, governance, and administrative reforms for agriculture, and Fisheries during lockdown restrictions in India (Figure 4.2). The Indian Ministry of Finance help package under the existing Pradhan Mantri Garib Kalyan Yojana program specifically targets the poor to help them fight the battle against COVID-19, to cope with the nationwide lockdown and it's financial consequences. The rate of wages in India has also been revised and increased for all those working under the Mahatma Gandhi National Rural Employment Guarantee Act (MNREGA). This scheme is one of the world's largest wage schemes. Finally, the GoI has recently established a program called the Prime Minister's Citizen Assistance and Relief in Emergency Situations (PM-CARES) Fund, which was created on 27 March 2020. Hon'ble Narendra Modi is the chairman of PM-CARES funds and that trustees include Minister of Home Affairs, Amit Shah, the Minister of Defence, Rajnath Singh, and the Minister of Finance, Nirmala Sitharaman. PM-CARES aimed at building capacity to resolve national challenges caused by future pandemics for combating, and containment and relief efforts against the coronavirus outbreak and similar pandemic like situations in the future (Figure 4.3). Due to the completeness of the lockdown and primary focus on stopping the spread of the disease, consumers, agricultural marketers, farmers, daily wage earners, and other stakeholders suffered enormously.

4.5 The Economic Challenges of Local Farmers

India is an agricultural sector country and about 263 million people are directly engaged in farming in which only 45% have their own cropland. Rest 55% peoples rely on other nonagricultural occupation for their livelihood (Lowe and Roth 2020; Mahendra Dev and Sengupta 2020). Before COVID-19, the part of the GDP of India stemming from the agricultural sector experienced growth rates at around 3.2% per year for six continuous years from 2014 to 2019 and expected a similar growth of between 2.4 and 3.7% from the fiscal year 2019 (FY19) to the fiscal year 2020 (FY20) (National Statistical Office of India). This expected growth has been replaced by a sharp decline of about 5% in the first quarter (Q1) of the 2020 fiscal year (FY) due to the lockdown (FAO 2020). Trends in the agricultural part of the GDP of India can largely be attributed to variations in wages in the agricultural sector and hence to the operation of local farming systems. Figure 4.4 compares recent variations of the average wage related to agricultural activities ("agri wages," blue curves) vs. the average wage related to nonagricultural activities ("nonagri wages," brown curves) in India's rural areas (i.e. "rural wages"). The average wages are here differentiated into nominal and real wage, respectively. In general, progression in the real wage is linked to changes in the overall national GDP of India, whereas progression in nominal wages relate to the market conditions for agricultural products.

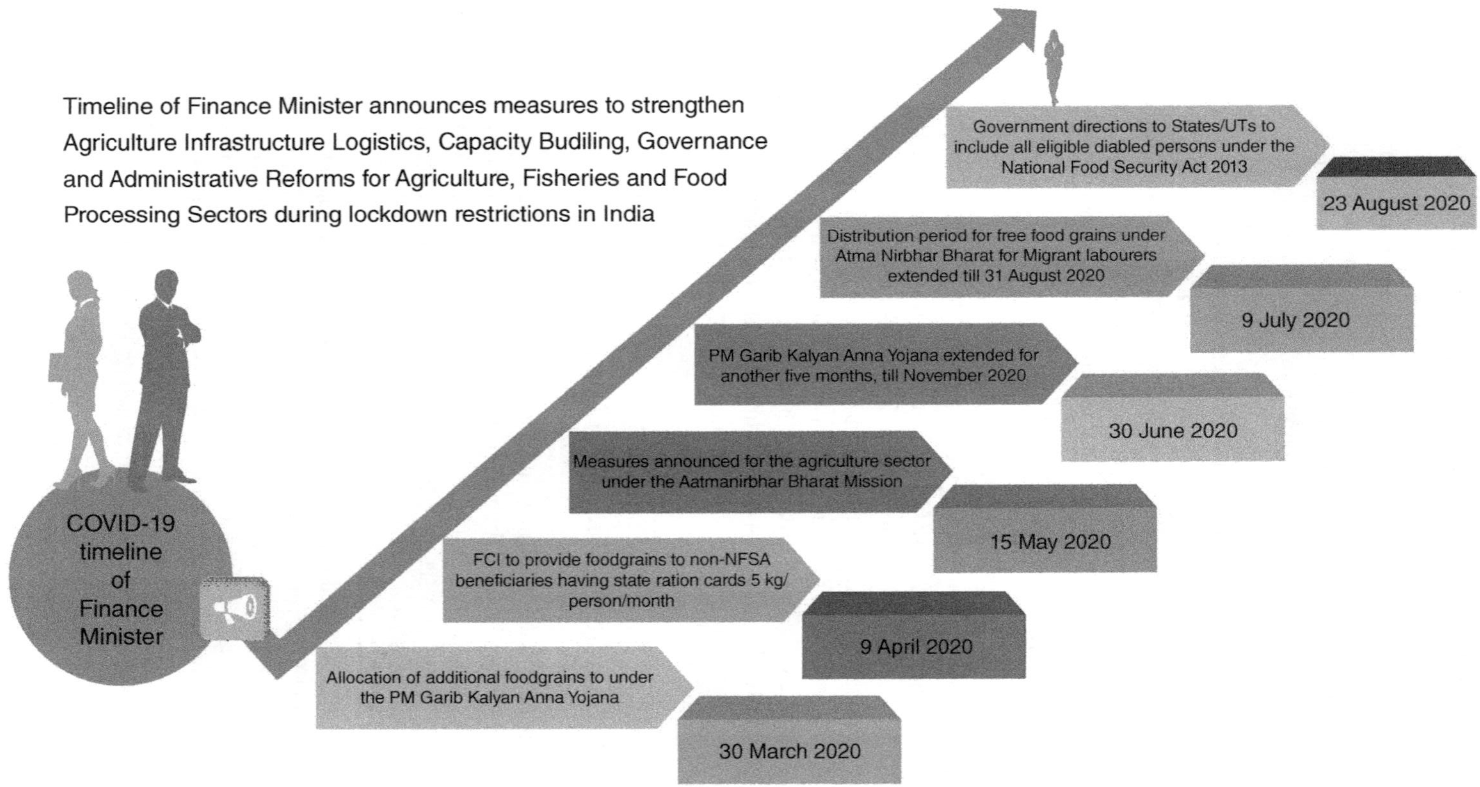

Figure 4.3 Timeline of Finance Minister announces lockdown restrictions in India. *Source*: https://prsindia.org.

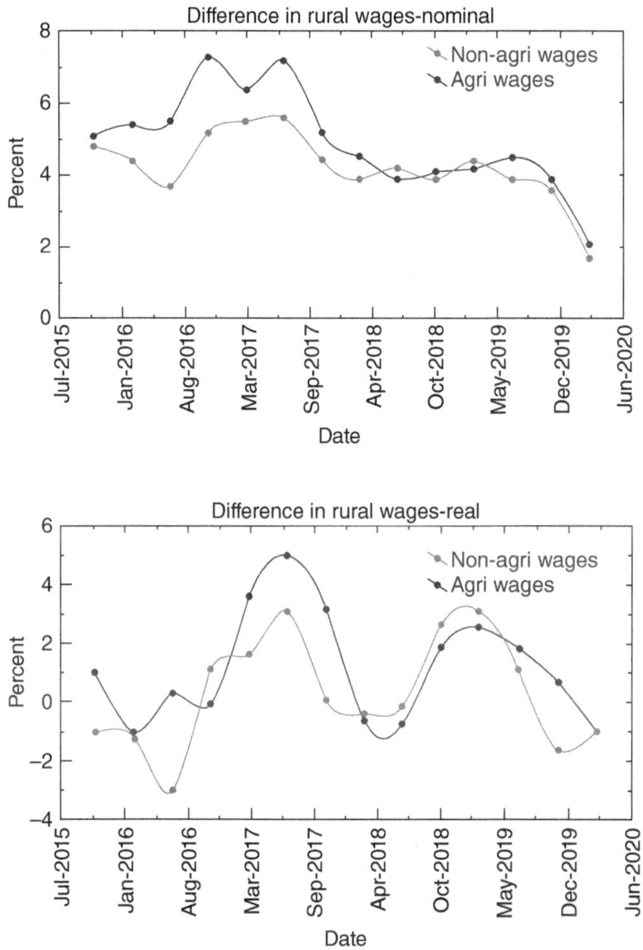

Figure 4.4 (Top panel) Variations in the average nominal wage (agriculturally related vs. nonagriculturally related in rural areas). (Bottom panel) Variations in the average real wage in rural areas based on data from RBI, accessed 16 May 2020. (*See insert for color representation of the figure.*)

From 2015 to 2020, the average nominal wage has decreased from 4.3 to 0.86% (Figure 4.2). Meanwhile, the real wage exhibited large fluctuations but as of December 2019 is negative for both the agricultural wage and nonagricultural wage. As expected this is synchronized with the drop in the national GDP of India seen from 2018 to 2019 (Figure 4.1). Meanwhile, the former seems to suggest that the current agricultural production in India is substantially higher than the demand. Combined, the data depict a situation, where despite an overall longtime positive growth of the Indian agriculture.

4.6 Impact on the Economy of Indian Farmers

The Ministry of Health and Family Welfare has confirmed a total of 15 500 active cases of COVID-19 with 592 cases of death in India by 21 April 2020 (www.mohfw.gov.in). The Prime Minister of India has already declared a nationwide "lockdown" on 24 March 2020 and now extended till 03 May 2020 (Phase 1: 25 March 2020–14 April 2020 [21 days] and Phase 2: 15 April 2020–3 May 2020 [19 days]). This lockdown has definitely impacted the farmers and farming operations because it coincides with the harvesting period of Rabi crops in the month of April/May 2020. The agricultural sector of India is encountering serious issues in hiring laborers. Since agriculture produce is an essential commodity, it is exempted from the directives of this lockdown, which is creating many problems to this sector, i.e. cutting and harvesting of rabi crops, other seeds and pulses, main associated issues storage and sale of crop in the mandi markets (Varshney et al. 2020b). The Indian agricultural sector, which suffered recently due to an uneven rainfall, is facing another hit due to disruption in activities as a consequence of the lockdown ensuing from the outbreak of COVID-19, because the harvesting season for Rabi crops is progressing rapidly, raising concerns regarding the management of matured crops by the farmers. However, the Ministry of Agriculture and Farmers' Welfare and Indian Council of Agricultural Research (ICAR) have released advisories involving crop, livestock, horticulture, and related enterprises in the wake of COVID-19 spread and, accordingly, practices specific to the state have been also suggested, considering the current situation. Experts of Krishi Vigran Kendras (KVKs) of the state have also circulated these advisories further among the farmers through mobile messaging, WhatsApp, and other interventions through the use of social sites. So far, in the state of Uttar Pradesh, a total of 679 related messages and advisories have been circulated by the KVK experts using Kisan Call Centers (Varshney et al. 2020a), mKISAN, and WhatsApp benefitting approximately 65 thousand, 4.20 lakh, and 49 thousand farmers, respectively.

With an increase in the severity of the global pandemic and subsequent rise in the cost of human life due to SARS-CoV-2, outbreak of the novel COVID-19 is adversely impacting the world economy (Kumar et al. 2020). India has been severely impacted by the SARS-CoV-2 outbreak (Singh et al. 2020), thus magnifying the preexisting risk to its outlook. However in the next kharif season agriculture practices are impacted due to nonavailability of agro-based economy, storage and prevailing state of COVID-19 disease (Kumar et al. 2020). The agricultural sector has been adversely affected due to the lockdown ensuing from the outbreak of COVID-19. This nationwide lockdown has impacted the economic system. The World Bank, in its South Asia Economic update report, while assessing the impact of the COVID-19 outbreak, has estimated the Indian economy to decelerate to 4.8% in 2020 and projected a sharp growth deceleration to 2.8% in a baseline scenario in fiscal 2021.

Although it is evident that farming is the only essential occupation, which can combat the food and livestock needs, some states in India, namely Punjab and Haryana, which are known as the "food bowls" of the country, are encountering major distress. The prime issues include appropriate operating of the machinery required for harvesting and threshing, winnowing, packaging, and transportation of the Rabi crop. Complete interruption of services has rendered the farmers more skeptical of their fate. Permission from the Government to harvest Rabi crops through the process of machine harvesting has encouraged the farmers. The implementation of the COVID-19 emergency response and health system preparedness package has been planned by the state government in Phase 1, 2, and 3. The center has funded a five-year scheme worth INR 15 000 crores to strengthen the nation and state level systems as the country is combating to control the infectious SARS-CoV-2.

4.6.1 Lack of Agricultural Labor Holds Up Harvesting

ICAR, which is the apex body of the government of India assessing the impact of the lockdown ensuing from the outbreak of the COVID-19 on agriculture and its allied sectors, is preparing the roadmap to minimize its demerits to prevent its impact on the food system (Chander 2020; Gupta et al. 2020a,b). While the government has exempted many agricultural operations including harvesting to the movement of produce to the markets according to the directives of the lockdown, ICAR has issued specific advisories to the farmers directing them to take general precautions and safety measures during crop harvesting, post-harvest operations, seed storage, and their marketing (Khan et al. 2016). However, considering the current situation and spread of the pandemic throughout the country, it is apparent that the farmers have little to hope for. Excluding the farmers of the Bundelkhand region, there can be no better assessment of the adverse impact of the lockdown ensuing from the outbreak of the COVID-19 than that was observed with other farmers of the country.

The nationwide lockdown has increased the social and economic burden of the farmers in the Bundelkhand region, shattering their hopes and sinking them into the debt trap (APEDA 2019). Subsequently, the farmers of Bundelkhand have employed all their resources to meet the challenges of unpredictable weather. However, these attempts have only been documented through official paperwork (Chaba and Damodara 2020). Due to the absence of any worthy support system, the farmers have put everything at stake to take loans and plan for the next crop season. All efforts to break the longstanding belief that Indian farmers are risk-averse have been made (Figure 4.5). They have experimented and invested for their betterment.

As per the government's data from Indian solicits, an average of 263 million people are directly engaged in the farming and agriculture sector. Of these, over 55% do not own the croplands (Chakraborty et al. 2017; Chander and Mohan 2020). A prime concern ensuing the announcement of the lockdown by the Prime Minister of India on 24 March 2020, due to the outbreak of COVID-19, lakhs of people have left the cities on account of unemployment and returned to their villages to avoid the pressures of maintaining a family in cities with a high cost of living. Moreover, they have returned unemployed.

4.6.2 Is Price Crash in the Post-Lockdown Phase a Reality?

Although it was predicted that the lockdown would be lifted after 21 days from 24 March 2020, now extended to extra 19 days 03 May 2020, farm experts are still burdened by the anticipation of a cash crash in the mandis, which might seriously impact their state of mental wellbeing. Everywhere, farmers are apprehensive and anticipating retraction of the lockdown. However, it is possible that, even if the lockdown ends, the mandis may not perform effectively in the post-lockdown period and the farmers may encounter a cash crash. This situation may potentially lead to the dumping of fruits, vegetables, and other farm produce, damaging the economic status of the farmers.

In the seven districts like Jhansi, Lalitpur, Banda, Jalaun, Hamirpur, Chitrakoot, and Mahoba of the Bundelkhand region of Uttar Pradesh, a change in the cropping pattern is being witnessed, which is not by choice but the only option available, which includes irrigation of the winter crop instead of the kharif crop. The failure to harvest in the past has

Figure 4.5 Wheat/rice crops are ready for harvesting at many places and farmers in India are worried that they would incur huge losses due to the lockdown ensuing from the outbreak of COVID-19.

led to a spate of suicides, which is continuing till date with the decadal numbers witnessing a spike. The failure of the existing government schemes in the Bundelkhand region is evident, in which, weather uncertainties are chronic. However, the situation is similar in the other regions of the country as well. Finally, quick policy measures by the government considering the specific need of time in relation to the specific regions throughout the country including direct procurement, landless laborers, direct cash transfers to the farmers and interest subvention through cash transfers and KCC modes are the ways to overcome this crisis and minimize the impact of this lockdown.

4.6.3 Steps Taken by the Government to Announce Packages, Do They Support?

The short-term lockdown resulting from outbreak of the COVID-19 pandemic has interrupted the state supply of food and stock and is mainly influencing the economic strategy of the country. The advisory board of the government for political and fiscal concerns has informed in a report from the Food Corporation of India (FCI) that food grains constitute 77 million tons of cereals against a buffer supply requirement of 21 million tons on 1 April 2020. Moreover, this lockdown is a humanity-related issue, in which, people are migrating back to the rural areas and reinforces the importance of having strong policies in place to tackle the decline in market availability for and sale of produce. The GoI has already announced that for the coming three months, 5 kg of free grains would be distributed to people registered under the National Food Security (NFS) Act. It is estimated that the FCI stockings may free up with help from the Government to help the migrants registered under the schemes from different states (Chakrabarti et al. 2019; Verma and Gustafsson 2020).

To overcome the current situation, the Indian Finance Minister has proclaimed a financial assistance of 1.7 trillion or INR 1.7 lakh crore rupees as subsidy to the people below the poverty line, which include 800 million people nationwide. Moreover, the GoI has announced a scheme called PM-KISAN for Pradhan Mantri Kisan Samman Nidhi (Varshney et al. 2020a). This scheme was offered amounts of INR 2000 thrice to the Indian farmer after every fourth month by April 2020. Despite provision of this financial assistance by the government, experts have noted that 87 million farmers in India alone have refrained from expressing their optimism. However, it stops farmer during long lockdown period to express their main concern to the government (Jat et al. 2019; Yadav et al. 2019). Mandis in the states are currently closed on account of this lockdown, preventing an influx in the income of the farmers and harvesters. Farmers require money, particularly now, because it is the season of harvest. However, the lockdown ensuing from the COVID-19 outbreak has impacted the mental stability and wellbeing of the farmers and is a precursor to a probable cash crash. Moreover, the government has not outlined a plan-B to deal with this situation, though "having life, hoping future" is the message to uphold currently. Some experts have claimed that the announced relief packages are pitiable and not fulfilling the requirement of the farmers, which is a serious concern.

4.7 ICAR Initiatives

In the state of Uttar Pradesh, after a complete lockdown Phase 1: 25 March 2020–14 April 2020 (21 days) of 17 days, the total cases of COVID-19 have risen above 500 with 35 deaths, although approximately 119 have recovered. Fortunately, this pandemic has not yet reached the stage of community spread (after stage 2: 15 April 2020–3 May 2020 [19 days]). Owing to complete lockdown, the daily wage earners, agricultural marketers, agricultural farmers, consumers, and related stakeholders are likely to suffer enormously. Therefore, the Government of Uttar Pradesh has issued two office orders permitting suppliers to continue the movement of honey, vegetables, mushrooms, fruits, milk, and related horticultural produce in the state. Secondly, permission to avail plant protection chemicals, seed fertilizers, agricultural commodities, and farm machineries has been granted. The effects of these efforts are apparent in the urban areas, as witnessed by the lack of panic in the purchase of fruits and vegetables in the cities. The ICAR has also issued advisories on specific farm practices applicable to this state, which have been circulated among farmers in every district by the KVKs.

4.8 Impact on State Agriculture

As the laborers from the nearby cities and Delhi have migrated to their respective villages, the availability of laborers for carrying out farm operations has increased. As a result, the daily wages at the village level have dropped by 15–20%. If available, their rates have increased by 30–40%. Due to reduced number of wholesalers in the mandis, the sale of vegetables has decreased, causing distress to the vegetable growers (Gaydon et al. 2017; Shyamsundar et al. 2019; Prasad 2020). However, the income of vegetable growers, and, in turn, the retailers has increased because of the higher prices quoted by them. The price of majority of the seasonal vegetables like cucurbits, brinjal, cabbage, potato, cauliflower, etc. has increased by 40–45%. Similarly, the fruit prices such as apple, grapes, banana, coconut, oranges, and many more have increased by 30–40%. Similarly, dairy farmers are experiencing the crisis of increased rates of feeds and concentrates (Gupta et al. 2020a,b). They need to sell milk at lower prices than the prevailing rates and even less quantities.

4.9 Conclusion

After lockdown, people below the poverty line are likely to overcome the impact slowly. However, farmers are expecting a recovery from the anticipated cash crash. A declaration of subvention for three to six months was previously assessed, while an initial step toward defeating the COVID-19 has received an overwhelming response. However, currently, harvesters are expecting clarification on the following questions: what shall be the source of income for the farmers? and will the money come from own sources? What is the solution to mitigate the issue of migrated labor? These issues are not addressed in the present plan. An extension of the lockdown shall allow generation of more factors, allowing further predictions of its impact on the economy of the farmers.

The government of all countries should take strict and significant actions to keep up with the control of the virus. Socioeconomic decisions of the respective countries could significantly impact the final outcome and an alarming situation in the future could be predicted, because we have also considered these factors in our report. This study may assist the planners in India to take stringent actions to prevent further spread of the virus by developing timely strategies and effectuating their implementation based on the different levels of severities of COVID-19 among the population of India presented here. It is highly desirable that other global regions take strict action to reduce exposure of their people for curtailing the spread of infection and provide timely support to facilitate the recovery of infected individuals.

References

Agricultural and Processed Food Products Export and Development Authority (APEDA) (2019). *Basmati Survey - Report-1., Basmati Acreage & Yield Estimation in Punjab, Haryana, Delhi, Uttarakhand, Himachal Pradesh. Western Uttar Pradesh and Parts of Jammu & Kashmir*, 52. New Delhi: Geotrans Technologies Pvt. Ltd., Basmati Export Development Foundation, APEDA.

Ceballos, F., Kannan, S., and Kramer, B. (2020). Impacts of a national lockdown on smallholder farmers' income and food security: empirical evidence from two states in India. *World Dev.* 136: 1050–1069.

Chaba, A.A. and Damodara, H. (2020). The Covid Nudge: Labour shortage makes Punjab, Haryana farmers switch from paddy to cotton. *The Indian Express*. https://indianexpress.com/article/india/covid-19-punjab-haryana-farmers-paddy-cotton6385600.

Chakrabarti, S., Khan, M.T., Kishore, A. et al. (2019). Risk of acute respiratory infection from crop burning in India: estimating disease burden and economic welfare from satellite and national health survey data for 250 000 persons. *Int. J. Epidemiol. 48* (4): 1113–1124.

Chakraborty, D., Ladha, J.K., Rana, D.S. et al. (2017). A global analysis of alternative tillage and crop establishment practices for economically and environmentally efficient rice production. *Nat. Sci. Rep.* 7: 9342.

Chander, M. (2020). COVID-19 crisis in India: How extension and advisory services can help. *Agrilinks, Feed the Future, USAID*. https://www.agrilinks.org/post/covid-19-crisis-india-how-extension-advisory-services-can-help (accessed 5 May 2020).

Chander, S. and Mohan, B. (2020). Effect of planting date of rice on brown planthopper. *Indian J. Entomol.* 82 (2): 277–279.

Chatterjee, P., Nagi, N., Agarwal, A. et al. (2020). The 2019 novel coronavirus disease (COVID-19) pandemic: a review of the current evidence. *Indian J. Med. Res.* https://doi.org/10.4103/ijmr.IJMR_519_20.

FAO (2020). *Covid-19 Pandemic: Impact on Food and Agriculture*. Rome: Food and Agricultural Organisation http://www.fao.org/2019-ncov/q-and-a/en.

Gaydon, D.S., Balwinder-Singh, P.L., Poulton, H. et al. (2017). Evaluation of the APSIM model in cropping systems of Asia. *Field Crop Res.* 204: 52–75.

Gupta, S., Kishore, A., Alvi, M., and Singh, V. (2020a). Designing better input support programs: lessons from zinc subsidies in Andhra Pradesh, India. *PLoS One* 15 (12) https://doi.org/10.1371/journal.pone.0242161.

Gupta, N., Tomar, A., and Kumar, V. (2020b). The effect of COVID-19 lockdown on the air environment in India. *Glob. J. Environ. Sci. Manag. 6* (Special Issue (Covid-19)): 31–40.

Jat, H.S., Sharma, P.C., Datta, A. et al. (2019). Re-designing irrigated intensive cereal systems through bundling precision agronomic innovations for transitioning towards agricultural sustainability in north-West India. *Nat. Sci. Rep.* 9: 17929.

Khan, T., Kishore, A., and Joshi, P.K. (2016). *Gender Dimensions on Farmers' Preferences for Direct-Seeded Rice with Drum Seeder in India*. New Delhi: International Food Policy Research Institute.

Kumar, A., Padhee, A.K., and Kumar, S. (2020). How Indian agriculture should change after COVID-19. *Food Secur.* 12 (4): 837–840.

Kumar, P., Singh, S.S., Pandey, A.K. et al. (2021). Multi-level impacts of the COVID-19 lockdown on agricultural systems in India: the case of Uttar Pradesh. *Agric. Syst.* 187: 103027.

Linton, N.M., Kobayashi, T., Yang, Y. et al. (2020). Incubation period and other epidemiological characteristics of 2019 novel coronavirus infections with right truncation: a statistical analysis of publicly available case data. *J. Clin. Med. 9* (2): 538.

Lowe, M. and Roth, B. (2020). *India's Supply Chain Unchained*. New Delhi, India. http://southasia.ifpr: International Food Policy Research Institute (IFPRI), South Asia Office http://i.info/2020/06/18/indias-supply-chains-unchained.

Mahendra Dev, S. (2020). Addressing COVID-19 impacts on agriculture, food security, and livelihoods in India, IFPRI Blog.

Mahendra Dev, S. and Sengupta, R. (2020). Covid-19: Impact on the Indian economy. Indira Gandhi Institute of Development Research, Mumbai Working Papers, Indira Gandhi Institute of Development Research, Mumbai, India.

Mohan, V., Dash, D.K., Dikshit, R. et al. (2020). How agriculture stayed resilient despite Covid shock, Times of India, Opinion Article, May 30, 2020. In: How agriculture stayed resilient despite Covid shock - Times of India. indiatimes.com.

Prasad, R. (2020). Effect of different dates of rice transplanting on incidence and abundance of major insect pests and the crop yield. *J. Eco-friend Agricult.* 15 (1): 67–69.

Raghunathan, K., Kannan, S., and Quisumbing, A.R. (2019). Can women's self-help groups improve access to information, decision-making, and agricultural practices? The Indian case. *Agric. Econ.* 50 (5): 567–580. https://doi.org/10.1111/agec.12510.

Shyamsundar, P., Springer, N.P., Tallis, H. et al. (2019). Fields on fire: alternatives to crop residue burning in India. *Science* 365 (6453): 536–538.

Singh, R.K., Drews, M., De la Sen, M. et al. (2020). Short-term statistical forecasts of COVID-19 infections in India. *Ieee Access 8*: 186932–186938.

Singh, R.K., Drews, M., De la Sen, M. et al. (2021). Highlighting the compound risk of COVID-19 and environmental pollutants using geospatial technology. *Sci. Rep.* 11 (1): 1–12.

Sohrabi, C., Alsafi, Z., O'Neill, N. et al. (2020). World Health Organization declares global emergency: a review of the 2019 novel coronavirus (COVID-19). *Int. J. Surg. 76*: 71–76.

The Hindu Business Line (2020). Covid-19 lockdown estimated to cost India $4.5 billion a day: acuité ratings. *The Hindu Business Line.* http://www.thehindubusinessline.com/economy/covid-19-lockdown-estimated-to-cost-india-45-billion-a-day-acuit-ratings/article31235264.ece.

The Indian Express (2020). Stock markets post worst losses in history; aensex crashes 3,935 points amid coronavirus lockdown. *The Indian Express.* https://indianexpress.com/article/business/market/bse-sensex-nse-nifty-stock-market-live-updates-coronavirus-global-markets6327415/.

Varshney, D., Joshi, P.K., Roy, D., and Kumar, A. (2020a). PM-KISAN and the adoption of modern agricultural technologies. *Econ. Polit. Wkly* 55 (23): 49.

Varshney, D., Roy, D., and Meenakshi, J.V. (2020b). Impact of COVID-19 on agricultural markets: assessing the roles of commodity characteristics, disease caseload and market reforms. *Indian Econ. Rev.* 55 (83–103): 1–21.

Verma, S. and Gustafsson, A. (2020). Investigating the emerging COVID-19 research trends in the field of business and management: a bibliometric analysis approach. *J. Bus. Res. 118*: 253–261.

World Health Organization (2020). Coronavirus disease 2019 (COVID-19) situation reports. https://www.who.int/emergencies/diseases/novel-coronavirus-2019/situation-reports (accessed 5 March 2020).

Yadav, O.P., Singh, D.V., Dhillon, B.S., and Mohapatra, T. (2019). India's evergreen revolution in cereals. *Curr. Sci.* 116: 1805–1808.

Section 2

Climate Change and Agriculture

5

Crop Diversification

An Approach for Productive and Climate-Resilient Production System

Rakesh Kumar[1], Bal Krishna[2], Prem K. Sundaram[3], Narendra Kumawat[4], Pawan Jeet[3], and Anil Kumar Singh[3]

[1] *Division of Crop Research, ICAR-Research Complex for Eastern Region, Patna, Bihar, India*
[2] *Department of Plant Breeding & Genetics, Bihar Agricultural University, Sabour, Bihar, India*
[3] *Division of Land & Water Management, ICAR-Research Complex for Eastern Region, Patna, Bihar, India*
[4] *AICRP for Dry Land Agriculture, College of Agriculture, Indore, Madhya Pradesh, India*

CONTENTS

Sustainable Agriculture Systems and Technologies, First Edition. Edited by Pavan Kumar, A.K. Pandey, Susheel Kumar Singh, S.S. Singh, and V.K. Singh.

5.1 Introduction

Various countries have adopted crop diversification for increasing production and growing high-value crop to meet the challenges of globalizing agriculture markets as well as emerging and changing requirement of population. Diversifying crop to high-value products such as medicinal, aromatic, vegetables, fruits, spices, and livestock. It increases chances of producing high-value crops in a given area/location, thus enhancing production-related activities while also reducing risk for farmers in the event of a specific crop failure. India has a population of 1.30 billion people. Vast population lives in rural area with agriculture occupation accounting for more than 70% of total population. Small farm holdings are prevalent in Indian agriculture. Average size of a farm is ~1.15 ha. A number of large and medium holdings account for less than 5%, they make up one-third of total area under cultivation. Marginal holdings account for 85% of total number of holdings and 45% of total operational area. Due to varying agroclimatic condition, large number of agricultural product is made in the country. Foodgrain crops (cereals/pulses) account for 66% of total cultivated land (Jeet et al. 2020). Crop diversification aims to provide a wide range of options for producing different crops in a given region, enhancing production-related activities on a variety of crops while also lowering the risk.

During the last 20 years, crop diversification has provided large-scale boosting to Indian agricultural economy in food security employment as well as income (Saha 2013). Since 1990, agriculture in India has experienced significant changes in production of paddy, wheat, horticultural crops, commercial crops, and plantation crops (Joshi et al. 2004). Indian agriculture faced problems like dominance of marginal/small farmers, poor rural infrastructure, seasonal and pseudo-employment, rural poverty, monsoon dependency, and nonmechanized farms (Saurabh et al. 2021). During recent years, agricultural sector has undergone a major change from traditional to modern agriculture and there is marvelous prospective to push overall agriculture growth above country targeted 4% level by using modernization, technology adoption, and changes in food habits of people (Bahadur 2010). Agriculture sector is directly related to food/nutritional security and poverty issues in urban/rural areas of emerging countries in the globe. Development of infrastructural facilities viz. motorable road, irrigation facilities, electricity used for agricultural purposes, market infrastructure greatly influence choices of cropping systems, and subsequently the extent and profitability of crop diversification.

5.2 What Is Diversification?

Diversification has evolved from the word "Diverge" which means to extend or move in the other direction from a common point (Dahiya et al. 2016). It is an important component for diversified agriculture production system. It should be adopted as a strategy of profit maximization by means of complementary and supplementary relationships for competitive products. It aims to provide more employment, achieve nutritional security, eco-friendly, poverty alleviation, and income generation (Samal et al. 2017).

5.3 Concept of Crop Diversification

In the present scenario of agriculture development, the word "diversification" is one of the most commonly used terms. Earlier, it was used more in subsistence farming, where subsistence means growing of many crops in farmer's field. But in the present condition, diversification means growing high-value crops by farmers in their area. Generally, it is used as risk management plan that include participation in more than one operation. In addition, it has beneficial effect on mitigating price risk and fluctuation in product. Advantage from such diversified system at farm level depends on the level of heterogeneity within farm with respect to biological/economic factors, soil and land resources, and optimization in utilizing resources in diversified production system (Jill and Erin 2005).

There are two approaches of crop diversification. Adding more crops to standing cropping systems is called horizontal diversification and this is the most widely used approach. This method of crop diversification, on other hand, broadens the base of system by simply adding more crops to an established cropping system using techniques, i.e. multiple cropping and other efficient management practices (Singh et al. 2014a). In vertical diversification, which involves performing diverse downstream activities? This may involve the value addition to manufactured product, i.e. fruits and vegetables that are canned or turned into juices or syrups, depending on situation. It represents the degree and stage of crop industrialization. It is also worth noting that diversification considers the viability of a variety of crops (Nanher et al. 2015).

5.4 Key Drivers of Crop Diversification

Important driving forces for diversification are:

- Enhancing the net profit of small land holding farmers
- Control price fluctuation
- Mitigate the ill-effects of abnormality of the weather
- Balanced food demands

- Enhancing fodders for livestock
- Conserve the natural resource
- Reducing the environment pollution
- Minimizing the reliance on off-farm inputs/products
- Lowering pest/disease problems

5.5 Urgent Need

Diversification of crops in Indian agriculture taking place horizontally or vertically or, it depends on market and sometimes on local/domestic requirements. With respect to utilization pattern of soil, land, and water, there are urgent needs to consider the following causes:

- Research and development of diversification should be in farmers participating modes as multidisciplinarians scientific teams including farmer will plan project until reaching valid conclusion as well as solution.
- Value addition of farm product offers employment under nonfarm areas.
- Needs to identify places which help in diversifying farming situations in different socioeconomic status, supply of inputs, domestic needs, and infrastructure of market.
- Farm enterprises diversification will generate more income possibilities and employment throughout the year.

5.6 Scope of Crop Diversification

Availability of natural resources viz., land, water, vegetation/plants, and sunlight are crucial for any diversification.

5.6.1 Land/Soil

It is scarce and practically no-extra land/soil is available for cultivation. Thus, new cropping system/crop planning need to adjust within existing land/area. So that crop diversification is only by crop substitution as there is hardly any scope for area expansion.

5.6.2 Climate

Climate is a key factor that affects crop production and productivity through change in rainfall intensity, variation in onset and end of monsoon, seasonal floods, and occurrence of cyclones and associated storm surges (Jeet et al. 2017). Due to climate change many negative consequences are seen in agricultural production system and there is a need to overcome this effect. Shifting to crop diversification may be a balanced and cost-effective solution. Crop diversification can help with resilience in a number of different ways:

- By enhancing potential to mitigate pest outbreaks and pathogen attacks, which are likely to intensify under future climate scenarios;
- By protecting crop production from the negative effects of increased climate variability and severe weather.

5.7 Key Elements for Diversification

Agronomical	Economical	Government policies
Soil and climate type (topography, fertility, irrigation, drainage)	Means for risk management	Nondiscriminatory policies that discriminate among crops
Required inputs (chemicals, fertilizer, tractors, credit, feed-grains)	Flow of markets signal and communications and information system	Demands driven efficient research and development extension programs, without any bias for major field crops or against high-value crops
Seed/plant and livestock of good genetic quality	Means for vertical diversification	Contracts farming opportunities
Management practices and quality manager	Venture capital and entrepreneurship	Present market planning to reduce production and marketing risk and costs
Appropriate scale and organizational form	Transparency of input and output prices	Rural credit and markets for other inputs
	Efficient marketing system	Marketing systems (included satisfactory quality standard) and partnership of the private sector

5.8 Plant Breeding Supports for Crop Diversification

Main goal of crop diversification is to enhance crop portfolio in order to reduce the dependence on single commodity and generate more income. If farmers grow only single crop, they are in high-risk zone in the event of fluctuating climatic. Agricultural system is extremely diverse, and diversification can occur in a number of ways, including genetic variety, organisms, and scales (within a field, within a crop, and at the landscape level), offering farmers a wide range of options and combinations for implementing this strategy (Lin 2011). Historically there are numerous examples of devastating effect relaying on narrow genetic base in food crops. Irish Potato Famine caused by devastating diseases, late blight (*Phytophthora infestans*) of potato because of widespread cultivation of genetically uniform clone caused 80% in yield loss (Zadoks and Schein 1979). In India, Bengal famine in 1943 was due to mass destruction of rice by a fungus, *Cochliobolus miyabean.*

There are many varieties/cultivars which can be included in diversified cropping system, ranging from resistance genes in monocropping (cultivar mixture, multilines) to species mixtures and very complex perennial polyculture (Finckh and Wolfe 2006). Genetics and plant breeding approach could be more efficient in diversifying genes in elite cultivars and broaden genetic base. Pre-breeding must be started to generate novel variability using elite landraces, wild-relative, and primitive cultivars. Prime objective of pre-breeding is to converge several divergent alleles in to elite genetic background and not be primarily focused on increasing yield as such. Practical aspect of pre-breeding is that there should be continuous supply of potential variability into the breeding pipeline to generate new high yield potential varieties/cultivars with a wide genetic base. The pre-bred traits could further be used for introgression of desirable genes for resistance to multiple stresses into genetic backgrounds of popular cultivars with minimum linkage drag. Breeding system greatly affects genetic diversity in population. In self-pollinated crops, use of multiline varieties having various genes for disease-resistance, may be expected to tolerate diseases and insect pest's attacks better than their pure components (Simmonds 1979).

Directional selection favors individuals with advantageous character, selecting genotypes of favors, and reducing genetic diversity by eliminating alternatives. Alternatively, disruptive selection favors different extremes phenotype and maintains variation in alleles. The recent advancement in biotechnological tools has great potential in facilitating gene reshuffling and broadening genetic bases of newly developed cultivars. Transgenic approaches allow to transfer, diverse range of genes of economic importance from related/unrelated taxa and probably resolve hurdle of conventional breeding. Gene diversification through integrated mode of conventional and innovative approaches will be crucial in enhancing overall crop diversification over space and time.

Evolutionary participatory plant breeding (EPD) forms a major area of research work geared toward enhancing crop diversification and is uniquely suited to improve crop varieties/cultivars for low cost input/natural conditions. Composite cross or multiple cross population made from distantly related parents possessing characteristics of interest are supposed to adapt to a wider range of environment as compared to mixed pure lines (Wolfe 1992). Additionally, plant breeders are contributing in diversification of existing cropping system by developing short duration, photo-thermal insensitive crop varieties which easily fit well in higher cropping intensity programmes, i.e. multiple and relay cropping.

In recent years, short-duration pulses are getting much attention because it fits well in cropping system and gives remunerative returns to the farmers. Development of short-duration varieties of black gram, pigeon pea, and green gram has paved a way for crop diversification and intensification in North India (Singh et al. 2009). In Indo-Gangetic plains, due to late onset of monsoon, generally it becomes difficult to grow rainfed *kharif* crop on time and under such circumstances popularization of short-duration *rabi* pulses is of great importance in order to increase cropping intensity (Mishra et al. 2019). Developments of short-duration gram varieties have played important role in crop diversification of cereal-based cropping system in Indo-Gangetic plains (Saurabh et al. 2021). The ICCV-96029 and ICCV-96030 short-duration super early desi gram varieties, which mature in 75–80 days and have early-growth vigor, semi-erect form, low temperature tolerance at vegetative phase, and can germinate and establish in high moisture and comparatively anaerobic soil, are excellent for late rabi planting. Summer/spring mung bean and urd bean varieties of 65–70 days duration are available for adoption for larger area. However, a majority of the recently released varieties has not different vegetative and reproductive phase so that these can fit well in narrow windows of cereals cropping system. Lentil is a climate smart legume having great potential for late sown conditions in paddy-fallow field of IGPs. Early-maturing cultivars have a higher dry matter production, disease resistance, and the ability to tolerate low temperatures during the vegetative phase and high temperatures during the reproductive phase. For large scale diversification of crops, there is a need to design ideotypes with a participatory approach in collaboration with farmers and breeders for adequate selection of varieties suitable for various cropping system.

5.9 Advantages of Agricultural Diversification

Climate change affects biotic/abiotic factors in cropping systems, threatening crop production, and sustainability. More diversification in agricultural production system will help crops in sustaining these climatic challenges. The following are some of the major ways that a more diverse system has been identified to boost crop production from climate change.

5.9.1 Soil Health

Lower soil fertility status and drought are major problems that limit foodgrain production. Poor cultivation practices, like continuous same cropping system with some inputs, might have gradually led to reduced soil fertility of small landholders. As a result, alternative and more sustainable production practices are critical for enhancing sustainable foodgrain production and preserving soil health. Microbial symbiosis is a key factor in plant nutrition and stress tolerance, both ecologically and intrinsically. For example, the phylum *Glomeromycota* belongs to vascular arbuscular mycorrhizal fungi (Smith and Read 2008). Furthermore, recent scientific studies have revealed that mycorrhizal symbiosis varies by crop cultivar (durum wheat), corn (An et al. 2010), and tomato. Cover crops with low input and organic farming practices are needed to maintain crop production. Cover crops have been shown to effectively minimize the occurrence of diseases (Ojaghian et al. 2012), weeds (Campiglia et al. 2010), and parasitic nematodes (Asmus et al. 2008).

5.9.2 Pest Suppression

Crop diversification is an essential component of ecosystem biodiversity. Diversified systems emulate more natural systems and are therefore able to sustain more diversity of animal species, many of which are natural enemies of crop insect-pests, due to spatial and temporal diversity in crop distribution and higher richness in crop species (Altieri 1999). Crop diversification has been shown in several studies to reduce insect infestations in agricultural systems (Cannon 1998). In one review examining pest control in agriculture, more degrees of complexity exist in increasing biodiversity for pest control (Gurr et al. 2003; Ratnadass et al. 2012). There are a variety of local conventional approaches used by rural dwellers to control pests and diseases that are both cost-effective and environmentally friendly. Among the most effective ways to combat pests and disease is to diversify crops by favoring species mixtures over sole culture (Tooker and Frank 2012). Crop mixtures discourage pests by interrupting disease cycles, increasing natural predators of insects and pests, volunteering crop plants and suppressing weeds, changing the microenvironment within the crop canopy, creating a dilution effect by decreasing resource concentration, or making pest and disease pathogen penetration extremely difficult (Yadav et al. 2018).

5.9.3 Disease Suppression

Plant diseases can greatly reduce crop yields, and climate change can have a direct impact on disease distribution and viability. Diseases are responsible for around 10% of global rice, maize, and wheat production losses (Oerke 2006). Pathogen load was nearly three times higher in sole culture fields, where host adequate was at a maximum, than in polyculture plots planted with 24 grassland plant spp., according to studies (Mitchell et al. 2002). At lower plant species richness, 11 diseases were more serious, with the majority of diseases correlating with host abundance. It means that disease transmission is aided by greater abundance of host species in plots with lower species diversity. With more varietal and species diversity in agricultural systems, disease transmission was reduced. Another choice for reducing disease risk is to include pulse, oilseed, and forage crops in cereal cropping (Krupinsky et al. 2002). While crop genetic diversity has been shown to achieve less predictable results in the control of plant diseases than in the control of pests, it has also been shown to produce more predictable results in the control of pests (Lin 2011). Crop diversification should therefore encourage soil/plant health while also incorporating other cost-effective and environmentally sustainable approaches, especially in the context of integrated pest management (Zeliang et al. 2020).

5.9.4 Yield Sustainability

The majority of small and marginal farmers depend on seasonal yields for both food and profit. Yield fluctuations mean less food for the family and less income for other necessities. Crop diversification ensures income stability by allowing them to fall back on another crop if one fails. According to several studies, it also protects small farmers from food insecurity by the yields, giving yield stability and an insurance impact (Singh et al. 2015). In future, change in local climatic condition and frequency of severe weather events are likely to worsen, potentially wreaking havoc on agricultural yield. As a result, techniques to make farming system more resilient to effects of climate change must be created (Singh et al. 2010a,b,c).

5.9.5 Food Security

According to food demand projections, South Asia would have a more diverse food basket. Vegetable, fruit, and milk demand will grow at a much faster pace than other food products. Due to a generational change in dietary patterns, crop diversification is projected to display a modest increase in consumer nutrition. It is expected that demand for cereals will decrease in South Asia, as calorie levels will be adequately met by evolving dietary patterns. This would strengthen food security by increasing the availability of micronutrients such as vitamin A and iron (Singh et al. 2010c, 2011a,b).

5.9.6 Poverty Alleviation and Employment Generation

Crop diversification through high-value commodity is projected to boost income and reduce poverty by creating jobs. Crop diversification is taking place in the crop, livestock, and fisheries sectors in India. Within crop market, non-food grain is steadily gaining popularity on food grains in terms of region allocation. Farmers also increased the amount of land dedicated to oilseeds, fruits, and vegetables. Increase in farm development activities to conventional cereal production by introducing more intensive crop rotations and crop–livestock combinations to provide jobs in the rural sector. Crop diversification can be an effective method for mitigating population pressure. In this context, India's crop diversification patterns are encouraging. There are several studies that show that diversifying agriculture in favor of commercial crops increases profits, creates job opportunities, and reduces rural poverty (Singh et al. 2017d; Zeliang et al. 2019).

Employment growth was sluggish from 1983 to 1988, but it accelerated from 1987 to 1994 as a result of economic reforms (Dev 2000). Agricultural diversification can also be related to trend in poverty reduction. Crop diversification should be promoted and maintained by closely monitoring the policies and institutions that are driving the change. According to Joshi (2005), Indian agriculture is increasingly diversifying toward high-value food commodities, with significant increases in the production of fruits, vegetables, milk, poultry, fish, and meat. Crop diversification that included high-value food crops increased income, empowered women farmers, created jobs, and protected natural resources (Singh et al. 2010c, 2017d,e, 2020).

5.10 Constraints in Crop Diversification

Crop diversification has increased areas under commercial crops like fruits and vegetables. Increased share allocation to commercial crops like cotton, sugarcane, and oilseeds has provided both income as well as yield sustainability. However, this has gained momentum in the last few years. Crop diversification adoption is hampered by the following factors, which have varying degrees of influence:

- Suboptimal and overexploitation of natural resources causing negative effect on the sustainability and environment in agriculture.
- Lack of supply of quality seeds and plant materials of improved varieties.
- Poor infrastructure such as power, transport, communications rural roads, etc.
- Improper post-harvest management technologies and lack of sufficient infrastructure for postharvest handling of perishable produce.
- Very poor agri-based industry.
- Poor linkage between research–extension–farmer.

- Improper trained people's resources with persistent and large-scale illiteracy among farmers.
- Various disease and insect-pests which affects the most crops.
- About 65% of the cropped area under rainfed conditions.
- Inspection of all enterprises is very difficult.
- Nonavailability and affordability of farm machines required for different unit operations.

5.11 Research and Development Support for Crop Diversification

Agriculture would have more opportunities for entrepreneurship growth in the age of globalization. Paradigm shifts in the research and technology development process, as well as technology transfer, are needed for effective crop diversification. The research system must not only resolve crop diversification problems, but also build a cadre of scientists through ongoing skill growth and human resource development. For technology transfer to farmers, scientists must create appropriate technologies, popularize them, and impart information and skills to extension workers.

5.12 Institutional and Infrastructure Development Toward Crop Diversification

To maintain and operationalize crop diversification projects, institutional support is needed. For developmental and technology transfer purposes, the Government of India has established crop directorates for each of the major crops and groups of crops such as pulses and oilseeds. These directorates serve as a liaison between research and development efforts. Also, Indian Government is promoting crop diversification through various schemes like *Pradhan Mantri Fasal Bima Yojana*, distribution of Soil Health Cards, *Paramparagat Krishi Vikas Yojana*, *Har Medh Par Ped*, Mission for Integrated Development of Horticulture, Micro Irrigation Fund, etc. (Anonymous 2020).

5.13 Strategies for Boosting Agricultural Production for Food Security

5.13.1 Farmers Are Benefitted Through Green Revolution

In India, the green revolution ensured food security. Although focusing on crop production research and development across the world, it is important to distinguish between areas that benefited from the green revolution and those that had a negligible effect. For example, Haryana, Punjab, Rajasthan, and Uttar Pradesh, reaped the benefits. Agriculture, on the other hand, is now at a standstill, and the soil environment is declining. A better strategy for changing the cropping pattern to rebuild soil productivity and raise crop yields while lowering production costs should be devised. This can be accomplished by taking the following steps:

5.13.1.1 Efficient Use of Water

The transformation from flood irrigation to micro-irrigation for more effective and judicious water usage, which saves water and nutrients for the plants.

5.13.1.2 Balanced Nutrition to Crops

To increase crop yields, different crops need balanced/judicious application of organic and inorganic fertilizers in appropriate proportions. Farmers nowadays use higher doses of micronutrients without using the proper amounts of farm yard waste, micronutrients, and bio-fertilizers. According to Chinese findings, only by applying higher doses of organic manure will the crop be able to use higher doses of chemical/synthetic fertilizers. This aids in increasing crop yield while preserving soil fertility. Simultaneously, applying micronutrients and soil amendments as required would boost crop yield much more. To reduce the cost of agrochemicals while reducing pollution, use of environmentally friendly inputs such as biofertilizers and biopesticides should be encouraged.

5.13.1.3 Crop Shifting

Crop rotation of paddy and wheat, which has been practiced for 35–40 years, is a major cause of declining soil fertility and crop yield reduction. As a result, an alternative cropping method must be implemented as soon as possible. While fully abandoning this rotation may be challenging, there are opportunities to introduce new farming systems on a small scale. Hi-tech dairy farming, which involves keeping 200–500 milking cows or buffaloes with intensive fodder production, is the most promising initiative among many. Dairy farms will not only save money on agrochemicals and labor, but they will also generate a large amount of FYM, which will aid in soil fertility and crop production (Yadav et al. 2013).

5.13.2 Management of Rainfed/Dryland Areas

More than 75% of the country's rural households are marginal farmer with land holdings of less than 2.0 ha. A majority of them relies on rainfed agriculture for livelihood and jobs. Crop yields are also poor due to depleted soil fertility, a lack of reliable irrigation, and a lack of vital inputs and technical know-how. It is important to increase agricultural production and reduce risk via crop diversification in order to ensure their food security, while also increasing their employment opportunities (Bhalla 2000; Singh et al. 2017f, 2017a,b,c; Chaudhary et al. 2020).

5.13.3 Tri-Dimensional Approach for Food Security

Any program's goal should be to boost food production by using all available resources. It should also include job opportunities for target groups, as well as effective monitoring and governance. Any programme should have as its goal the introduction of suitable and effective technologies for rainfed areas, wasteland management, and hilly regions. Water resource management should be a key component of any program's strategy. The above activities will generate income for both small landowners and landless people during the year. Development of essential agricultural inputs such as biofertilizers, biopesticides, mushrooms, vermicompost, silkworm larvae, cattle feed, spawn, and seeds of food and fodder crops can also generate jobs (Mauriya et al. 2013; Singh et al. 2014b).

Good governance is a necessary component for the program's effectiveness and long-term viability. This can be managed by project implementing agencies, committed civil society organizations, farmers' cooperatives, and farmer producer organization operating in project areas. These organizations should make it easier for committed organizations to form and allow them to design appropriate need-based activities and value chain growth. Engaging committed civil society organizations will help develop the capacity of members of participant families by providing preparation, organizing efficient food supply delivery in close collaboration with the public distribution system, and providing frequent mentoring to families for effective programme implementation. Although facilitating the aforementioned activities, the immediate emphasis should be on the production of rainfed agriculture, wastelands with effective water resource management, and livestock, which can provide both food security and income for the rural poor (Singh et al. 2013a,b, 2014a,b; Maurya et al. 2014; Prakash et al. 2017; Thirugnanavel et al. 2018).

5.13.4 Sectors for Achieving Food Security

5.13.4.1 Animal Husbandry Development

As a source of sustainable income, over 70–80% of rural households in India especially small farmer, own various species of livestock. However, due to lower milk production, over 80% of animals, especially cattle and buffalo are unprofitable. BAIF has demonstrated a unique approach to genetic improvement for generating more milk yield cattle and buffaloes at the doorsteps of small farmers in order to recognize this, and three such milking animals will provide a family with a sustainable livelihood. Goatery is another choice for supplementing income without raising herd size or putting a strain on natural resources. Breed enhancement, effective feeding by the use of available resources, health care, and direct marketing could increase a farmer's income by four to five times in a short period of time (12–18 months). Pigs and poultry may also have promising prospects in a few limited areas.

5.13.4.2 Promote to Agri-Horti-Forestry and Sustainable Development

There are numerous agro-forestry systems that are ideal for various agroclimatic conditions while meeting local needs. Denuded lands (receiving 800 mm rainfall) can be placed under agri-horti-forestry, whereas shallow depth land that is under moisture stress is suitable for a silvi-pasture scheme.

5.13.4.3 Capacity Building

When emphasizing on farmer capacity building in rainfed areas, it is critical to ensure that illiterate farmers and women face newly evolving challenges and handle them effectively. This can be accomplished by forming Farmers Produced Organizations, raising awareness through exposure visits, providing hands-on training on appropriate technology, and forging partnerships with development and financial institutions to provide free access to technologies and services. Members of local government, especially Gram Pahnchyats and Farmers' Cooperatives, as well as well-known Civil Society Organizations, must be made aware of the importance of actively supporting the programme. Such powerful grassroots organizations will be able to make the most of the government's numerous development programmes while creating productive value chains for various communities. These organizations may also take on the task of overseeing the public distribution system and controlling the Gram Pahnchyats' various development programmes.

5.14 Conclusion

Rice–wheat cropping system is widely cultivated in South Asia's IGPs; as energy and carbon-intensive, thus need to design effective agricultural production system that are more profitable, climate resilient, C- and energy-efficient. Rice and wheat may be diversified with adoption of suitable crops/cultivars like soybean, maize for cob/fodder, fodder jowar, and major millets like jowar/bajra during rainy season and maize, arhar, and mustard in winter in cropping system for different ecologies of eastern India for sustainable agricultural production system. It is high time to design a highly sustainable, energy- and C-efficient production program that encourages environmental safety for India's and South Asia's upland rainfed and irrigated ecosystems.

References

Altieri, M.A. (1999). The ecological role of biodiversity in agroecosystems agriculture. *Ecosystems and Environment* 74: 19–31.

An, G.H., Kobayashi, S., Enoki, H. et al. (2010). How does arbuscular mycorrhizal colonization vary with host plant genotype? An example based on maize (*Zea mays*) germplasms. *Plant and Soil* 327 (1–2): 441–453.

Anonymous (2020). Increasing farmers' income. Ministry of agriculture & farmers welfare. GoI. https://pib.gov.in/PressReleaseIframePage.aspx?PRID=1657220 (accessed 17 April 2021).

Asmus, G.L., Inomoto, M.M., and Cargnin, R.A. (2008). Cover crops for reniform nematode suppression in cotton: greenhouse and field evaluations. *Tropical Plant Pathology* 33 (2): 85–89. https://doi.org/10.1590/S1982-56762008000200001.

Bahadur, S. (2010). Horticulture: Key to India's agriculture growth. http://www.commodityonline.com/news/horticulture-key-to-indias-agriculture-growth-34627-3-34628.html (accessed 14 January 2021).

Bhalla, S. (2000). Behind poverty: the qualitative deterioration of employment prospects for rural Indians. Paper presented in the third Asian Conference of Agricultural Economists (18–20 October).

Campiglia, E., Mancinelli, E., Radicetti, E., and Caporali, F. (2010). Effect of cover crops and mulches on weed control and nitrogen fertilization in tomato (Lycopersicon esculentum Mill.). *Crop Protection* 29 (4): 354–363. https://doi.org/10.1016/j.cropro.2009.12.001.

Cannon, R.J.C. (1998). The implications of predicted climate change for insect pests in the UK, with emphasis on non-indigenous species. *Global Change Biology* 4: 785–796.

Chaudhary, S., Chaturvedi, S., and Dhyani, V.C. (2020). Effect of crop diversification and residue management techniques on yield attributes, yield and soil nutrient. *International Journal of Chemical Studies* 8: 1107–1111. https://doi.org/10.22271/chemi.2020.v8.i2q.8915.

Dahiya, M., Kundu, P., and Yadav, B. (2016). Preferential pattern of rural women for crop diversification in the villages of Hisar district. *IRA-International Journal of Management & Social Sciences* 4: 295–304.

Dev, M. (2000). Economic reforms, poverty, income distribution and employment. *Economic and Political Weekly* 35 (10): 823–835.

Finckh, M.R. and Wolfe, M.S. (2006). Diversification strategies. In: *The Epidemiology of Plant Disease* (eds. B.M. Cooke, D.G. Jones and K. Bernard), 269–308. Berlin: Springer.

Gurr, G.M., Wratten, S.D., and Luna, J.M. (2003). Multi-functional agricultural biodiversity: pest management and other benefits. *Basic and Applied Ecology* 4: 107–116.

Jeet, S., Singh, J.P., and Kumar, R. (2017). Production potential and nutrient uptake of QPM hybrids as influenced by nitrogen and sulphur fertilization. *Journal of AgriSearch* 4 (1): 27–30. http://dx.doi.org/10.21921/jas.v4i1.7415.

Jeet, P., Ghodki, B.M., Dolamani, A. et al. (2020). Enhancement of land and water productivity through participatory rural appraisal. *Journal of AgriSearch* 7 (4): 234–240.

Jill, L.C. and Erin, O.S. (2005). Land use and income diversification: comparing traditional and colonist population in the Brazilian Amazon. *Agricultural Economics* 32 (3): 221–237.

Joshi, P.K. (2005). *Crop Diversification in India: Nature, Pattern and Drivers.* New Delhi: International Food Policy Research Institute, IFPRA-South Asia.

Joshi, P.K., Gulati, A., Birthal, S.P., and Tewaru, L. (2004). Agriculture diversification in South Asia: patterns, determinates and policy implication. *Economic and Political Weekly* 39: 2457–2467.

Krupinsky, J.M., Bailey, K.L., McMullen, M.P. et al. (2002). Managing plant disease risk in diversified cropping systems. *Agronomy Journal* 94 (4): 198–209. https://doi.org/10.2134/agronj2002.955a.

Lin, B.B. (2011). Resilience in agriculture through crop diversification: adaptive management for environmental change. *Bioscience* 61 (3): 183–193. https://doi.org/10.1525/bio.2011.61.3.4.

Mauriya, A.K., Maurya, V.K., Kumar, S., and Kumar, R. (2013). Nutrient management in barley in water logged area. *Bioinfolet* 11 (3 B): 910–911.

Maurya, P., Kumar, V., Maurya, K.K. et al. (2014). Effect of potassium application on growth and yield of wheat varieties. *The Bioscan* 9 (4): 1371–1373.

Mishra, J.S., Poonia, S.P., Choudhary, J.S. et al. (2019). Rice mealybug (*Brevennia rehi*): a potential threat to rice in a long-term rice-based conservation agriculture system in middle Indo-Gangetic plains. *Current Science* 117 (4): 566–568.

Mitchell, C.E., Tilman, D., and Groth, J.V. (2002). Effects of grassland plant species diversity, abundance and composition on foliar fungal disease. *Ecology* 83: 1713–1726.

Nanher, A.H., Shamim, S.A., Tyagi, S. et al. (2015). Need of diversification of cropping system in India. *Rashtriya Krishi* 10 (1): 31–33.

Oerke, E.C. (2006). Crop losses to pests. *Journal of Agricultural Science* 144: 31–43. https://doi.org/10.1017/S0021859605005708.

Ojaghian, M.R., Cui, Z.Q., Xie, G.L. et al. (2012). Brassica green manure rotation crops reduce potato stem rot caused by *Sclerotinia sclerotium*. *Australasian Plant Pathology* 41 (4): 347–349. https://doi.org/10.1007/s13313-012-0142-6.

Prakash, V., Mishra, J.S., Kumar, R. et al. (2017). Thermal utilization and heat use efficiency of sorghum cultivars in middle Indo-Gangetic plains. *Journal of Agrometeorology 19* (1): 29–33.

Ratnadass, A., Fernandes, P., Avelino, J., and Habib, R. (2012). Plant species diversity for sustainable management of crop pests and diseases in agroecosystems: a review. *Agronomy for Sustainable Development* 32 (1): 273–303.

Saha, J. (2013). Crop diversification in Indian agriculture with special reference to emerging crops. *Transactions of the Institute of Indian Geographers 35* (1): 139–147.

Samal, S.K., Rao, K.K., Poonia, S.P. et al. (2017). Evaluation of long-term conservation agriculture and crop intensification in rice–wheat rotation of Indo–Gangetic Plains of South Asia: carbon dynamics and productivity. *European Journal of Agronomy* 90: 198–208. https://doi.org/10.1016/j.eja.2017.08.006.

Saurabh, K., Rao, K.K., Mishra, J.S. et al. (2021). Influence of tillage based crop establishment and residue management practices on soil quality indices and yield sustainability in rice-wheat cropping system of eastern Indo-Gangetic Plains. *Soil & Tillage Research* 206: 104841. https://doi.org/10.1016/j.still.2020.104841.

Simmonds, N.W. (1979). *Principles of Crop Improvement*. London; New York: Longman.

Singh, K.K., Ali, M., and Venkatesh, M.S. (2009). *Pulses in Cropping Systems*. Technical Bulletin. Kanpur: IIPR.

Singh, N.K., Rai, B., Kumar, R., and Singh, A.K. (2010a). Response of winter maize (*Zea mays* L.)-based intercropping system under irrigated condition of Uttar Pradesh. *Environment and Ecology* 28 (3): 1688–1690.

Singh, R.K., Singh, Y., Singh, A.K. et al. (2010b). Productivity and economics of mustard (*Brassica juncea*) varieties as influenced by different fertility levels under late sown condition. *Indian Journal of Soil Conservation* 38 (2): 121–124.

Singh, R.K., Singh, Y., Singh, A.K. et al. (2010c). Response of fertility levels on growth, yield and economics of Indian mustard [*Brassica juncea* (L.) Czern and Coss.] varieties under late sown condition. *International Journal of Tropical Agriculture* 28 (3–4): 379–383.

Singh, A.K., Kumar, R., Singh, A.K., and Kumari, A. (2011a). Bio–efficacy of sulfosulfuron on weed flora and irrigated wheat (*Triticum aestivum* L.) yield. *Environment and Ecology* 29 (2A): 834–838.

Singh, A.K., Kumar, R., Singh, A.K. et al. (2011b). Performance of sulfosulfuron against weeds in irrigated wheat (*Triticum aestivum* L.). *Environment and Ecology* 29 (2A): 831–833.

Singh, P., Rana, N.S., Shukla, U.N. et al. (2013a). Effect of genotypes and nitrogen levels on production potential of maize (*Zea mays* L.) under Indo–Gangetic Plain Zone of Western U.P. *The Bioscan* 8 (3): 777–781.

Singh, U.K., Kumar, R., Shukla, U.N., and Kumawat, N. (2013b). Effect of okra [*Abelmoschus esculentus* (L.) Moench] to fertilizers levels. *Bioinfolet* 10 (3B): 952–953.

Singh, S., Yadav, R.A., Pal, V. et al. (2014a). Influence of herbicides against weeds associated with wheat (*Triticum aestivum*). *Bioinfolet* 11 (3B): 855–857.

Singh, U.K., Kumawat, N., and Kumar, R. (2014b). Response of okra [*Abelmoschus esculentus* (L.) Moench] to levels of fertilizers. *Bioinfolet* 11 (1B): 165–167.

Singh, S.K., Lal, S.S., and Singh, R.K. (2015). Production potential of potato and maize inter-cropping as influenced by spacing, planting pattern and staggered sowing of maize. *Journal of AgriSearch* 2 (3): 167–174.

Singh, A.K., Singh, A.K., Choudhary, A.K. et al. (2017a). Towards oilseeds sufficiency in India: present status and way forward. *Journal of AgriSearch* 4 (2): 80–84.

Singh, A.K., Yasin, J.K., Kumar, R., and Sundaram, P.K. (2017b). Extent and pattern of seed setting in faba bean under self and open pollination arrangement and its influence on seed production. *Journal of AgriSearch* 4 (4): 226–230.

Singh, A.K., Singh, A.K., Kumar, R. et al. (2017c). Indian cereals saga: standpoint and way forward. *Journal of AgriSearch* 4 (1): 1–9.

Singh, A.K., Singh, R.S., Singh, S.P. et al. (2017d). Productivity, profitability and soil health of pigeon pea as influenced by phosphorus levels and bio-inoculants under eastern Uttar Pradesh. *International Journal of Current Microbiology and Applied Sciences* 6 (6): 1723–1732. https://doi.org/10.20546/ijcmas.2017.606.200.

Singh, D.P., Saxena, R.R., Sahu, Y.K. et al. (2017e). Statistical analysis of multi–environmental rice yield trial in Bastar district of Chhattisgarh. *Oryza* 54 (2): 239–243.

Singh, S.K., Abraham, T., Kumar, R., and Kumar, R. (2017f). Response of crop establishment methods and split application of nitrogen on productivity of rice under irrigated ecosystem. *Environment and Ecology* 35 (2A): 859–862.

Singh, A.K., Singh, R.S., Singh, A.K. et al. (2020). Effect of weed management on weed interference, nutrient depletion by weeds and production potential of long duration pigeonpea (*Cajanus cajan* L.) in irrigated ecosystem. *International Journal of Current Microbiology and Applied Sciences* 9 (1): 676–689.

Smith, S.E. and Read, D.J. (2008). *Mycorrhizal Symbiosis*, 3e. London: Academic Press.

Thirugnanavel, A., Deka, B.C., Kumar, R. et al. (2018). Evaluation of rajma bean landraces for growth, yield and quality under low altitude condition of Nagaland. *International Journal of Current Microbiology and Applied Sciences* 7 (10): 2641–2648.

Tooker, J.F. and Frank, S.D. (2012). Genotypically diverse cultivar mixtures for insect pest management and increased crop yields. *Journal of Applied Ecology* 49 (5): 974–985. https://doi.org/10.1111/j.1365-2664.2012.02173.x.

Yadav, L., Sharma, J., Singh, A.K. et al. (2013). Efficacy of 2, 4–D herbicide application on growth, yield and spike deformities in late sown wheat varieties in eastern Uttar Pradesh. *The Madras Agricultural Journal* 100 (1–3): 135–138.

Yadav, R.K., Kumawat, N., Singh, A. et al. (2018). Bio–efficacy of new herbicides on weed dynamics, productivity and nutrient uptake in maize (*Zea mays*) under rainfed condition of Jhabua hills. *Indian Journal of Agricultural Sciences* 88 (7): 1123–1128.

Zadoks, J.C. and Schein, R.D. (1979). *Epidemiology and Plant Disease Management*. New York: Oxford University Press.

Zeliang, P.K., Kumar, M., Kumar, R. et al. (2019). Varietal evaluation of frenchbean for higher productivity and nutritional security under the foot hill ecosystem of Nagaland. *Indian Journal of Hill Farming* (Special issue): 14–18.

Zeliang, P.K., Kumar, R., Kumar, M. et al. (2020). Diversity analysis of rice (*Oryza sativa*) genotypes for improving productivity in mid-hills of the eastern Himalaya. *Indian Journal of Agronomy* 65 (2): 161–165.

6

Impacts of Climate Variability on Food Security Dimensions in Indonesia

Reference from the Nusa Tenggara Timur Province

Boubacar Siddighi Balde[1], Martiwi Diah Setiawati[2], I. Wayan Nampa[3], and Mohamed Esham[4]

[1]*Integrated Research System for Sustainability Sciences (IR3S), The University of Tokyo, Kisohigashi, Tokyo, Japan*
[2]*Research Center for Oceanography, National Research and Innovation Agency (BRIN), Jakarta, Indonesia*
[3]*Department of Agribusiness, Faculty of Agriculture, Nusa Cendana University, Kupang, Indonesia*
[4]*Department of Agribusiness Management, Faculty of Agricultural Sciences, Sabaragamuwa University of Sri Lanka, Belihuloya, Sri Lanka*

CONTENTS

6.1 Introduction

Climate change continues to be a heated topic worldwide, particularly in areas where rain-fed agriculture is a source of income (Weldearegay and Tedla 2018). This variability may raise hunger risks by affecting all aspects of food security. Furthermore, climate variability

Sustainable Agriculture Systems and Technologies, First Edition. Edited by Pavan Kumar, A.K. Pandey, Susheel Kumar Singh, S.S. Singh, and V.K. Singh.

refers to change in the mean state and other statistics (such as standard deviations, the frequency of extremes, and so on) on both temporal and spatial scales beyond that of individual weather events. As a result, climate variability is therefore used to describe the statistical variations of individual weather events over a given period (e.g. a month, season, or year) from the long-term statistics relating to the corresponding calendar period (IPCC 2007).

6.1.1 Climate Variability and Food Security

Numerous studies have analyzed the relationship between climatic variability and food security dimensions (Badolo and Kinda 2015; Kinda 2016; Singh and Sharma 2018). In recent years, the debate on climate variability has led to a renewed interest in the effects of climate variability on agriculture. Lee et al. (2012) reveal that climate variability (such as high temperature and precipitation) raises agricultural production in Asian countries while high drop temperatures are detrimental. Christensen et al. (2007) reported that food production is highly susceptible to weather conditions, while another finding also said that rainfall shortage or excess hampers food production, causing food insecurity (Weldearegay and Tedla 2018).

Agriculture remains the most critical sector in many developing countries, where climate change can irreversibly harm the natural resource base, with severe implications for food security (Hoffmann 2011). Previous studies reported that climate change poses one of the most severe risks to food and nutrition security in Indonesia (Case et al. 2007; Measey 2010; Devendra 2012) and is projected to have an impact on all aspects of food security including availability, accessibility, stability, and food utilization (Schmidhuber and Tubiello 2007; Krishnamurthy et al. 2012). Climate change may have a constant, discontinuous, or permanent effect (Boer and Kartikasari 2014) and continues to cause a major threat to food and nutrition security, especially for households whose livelihoods depend on food production. As the climate becomes increasingly erratic, rainfall deviation, increased frequency and intensity of climate-related events, and increases in pest and crop disease will have negative impacts on farmers, ranging from difficulties in timing agricultural activities to effects on the yields and productivity of crops, undermining farmers' overall resilience (BKP and WFP 2015).

6.1.2 Food Insecurity in NTT

Nusa Tenggara Timur (NTT) was identified as one of the most food-insecure provinces in eastern Indonesia, with 23.36% of people living below the national poverty line (BKP and WFP 2015). About 74.65 and 27.62% lacked appropriate sanitation and safe drinking water (BPS 2016). Furthermore, the literacy rate is low among other provinces (BPS 2016). According to National Basic Health Research (RISKESDAS), in 2013, among young children under five years of age, 33.1% were underweight, and 51.7% report stunting (Riskesdas 2013). These conditions are alarming public health concerns as per WHO standards (WHO 2000). In addition, the prevalence rate of stunting and underweight children is high among low-income families living in rural areas (Riskesdas 2013). Food insecurity becomes significant as more than 90% of the NTT population lives in rural areas (BPS 2016).

6.1.3 Climate Variability and Food Security in NTT

Climate variability and anomaly and its potential impact on the agricultural sector in the NTT have received heightened attention from the Indonesian government. According to the Intergovernmental Panel on Climate Change (IPCC) fifth assessment report, the temperature in NTT province is estimated to increase by more than 4 °C, as well as the rainfall variability is projected to increase by the end of the twenty-first century (Hijioka et al. 2014). Moreover, NTT is already experiencing climate extremes in 7 drought events and 112 flood events from 1990 to 2009 (BNPB 2017). These climate extremes have caused a severe impact on all dimensions of food security: availability, accessibility, utilization, and stability. Food availability relates to the physical properties of food from agriculture or allied sectors, mainly focused on food production (ADB 2012). In NTT, rice, maize, and cassava are the main food crops, of which about 70% of annual requirements are locally produced, mainly in rain-fed cropland (66%) (NTT Regional Government 2010). Most of the sub-district in NTT province are food self-sufficient. However, 15% of NTT's sub-districts are nutrition-deficient due to drought, low soil productivity, land conversion from food crops to industrial crops, and shifting cultivation systems (NTT Regional Government 2010).

The accessibility of food refers to affordability, allocation, and preference, and it is directly related to the financial ability of the population to fulfill nutritional requirements. Agriculture is the main livelihood of NTT communities (BPS 2016). It contributed 40% to the gross regional domestic product (GRDP) in the past four years (NTT Regional Government 2010). Even though the agricultural sector has the highest contribution to NTT GRDP, the growth rate is still lower than other sectors (i.e. trade and service sectors). Infrastructure occupies a significant role in uplifting the livelihoods of the NTT communities. However, more than 14% of villages in NTT cannot be reached by car during the rainy season, and almost 60% of the households are without electricity (NTT Regional Government 2010).

Food stability refers to the condition where food is regularly available, which is directly linked to extreme weather events such as drought, flood, landslide, forest fire, and hurricane. These climate extremes have caused a severe impact on food production and rural livelihood. In particular, droughts have damaged more than 70 000 ha of agricultural land (BNPB 2017). NTT rainfall variance was influenced by four main factors, namely, global phenomenon (i.e. El-Nino/Southern Oscillation [ENSO], Indian Ocean Dipole Mode [IOD]), regional phenomenon (i.e. Monsoon circulation, Madden Julian Oscillation, sea-surface temperature), synoptic factor (i.e. low-pressure area, tropical cyclone), and local factor (i.e. altitude, island position, land cover, wind circulation). In particular, the ENSO and IOD events caused extreme weather conditions, resulting in an increased incidence of flood and drought in the study area (As-syakur et al. 2013).

Food utilization is mainly linked with the nutritional value of food and food safety. Among the other provinces in Indonesia, NTT has the highest rate of undernutrition, with more than one-third of children under five years old considered underweight (Riskesdas 2013). According to the Ministry of Health, NTT is ranked the lowest among the provinces regarding access to sanitation (i.e. good sanitation only can be accessed by 25.35% of the NTT region) (Riskesdas 2013). Furthermore, climate change and the rapid population growth in NTT, which has expanded by 81% over the last 35 years, have severe implications for regional food security (BPS 2016). According to Handoko et al. (2008), climate change loss of agricultural production is valued at 382 million US dollars by 2050. In addition, climate change also affects human health by aggravating dengue fever and malaria incidence in the region (Hect 2016).

In this research, we evaluate the possible impacts of climate variability on food security dimensions based on climate data obtained from 10 weather stations scattered in different districts of the province. Thus, the main objective of this research is to analyze the effects of climate variability on food security over the period 2002–2014 in NTT province. The study undertook district-wise analysis based on secondary data.

6.2 Method

6.2.1 Study Area

NTT is located around 8°–12° South latitude and 118°–125° East longitude, with a land area of roughly 47 000 km^2 and 1192 small islands, four of which are considerably larger: Flores, Sumba, Timor, and Alor as shown in Figure 6.1. Furthermore, the landscape is primarily rugged and steep (Figure 6.1). The capital is Kupang, with a population of 0.3 million in 2011 (BPS 2012). The province's dry and rainy seasons are longer and shorter than in other country areas, with the wet season occurring just four months from December to

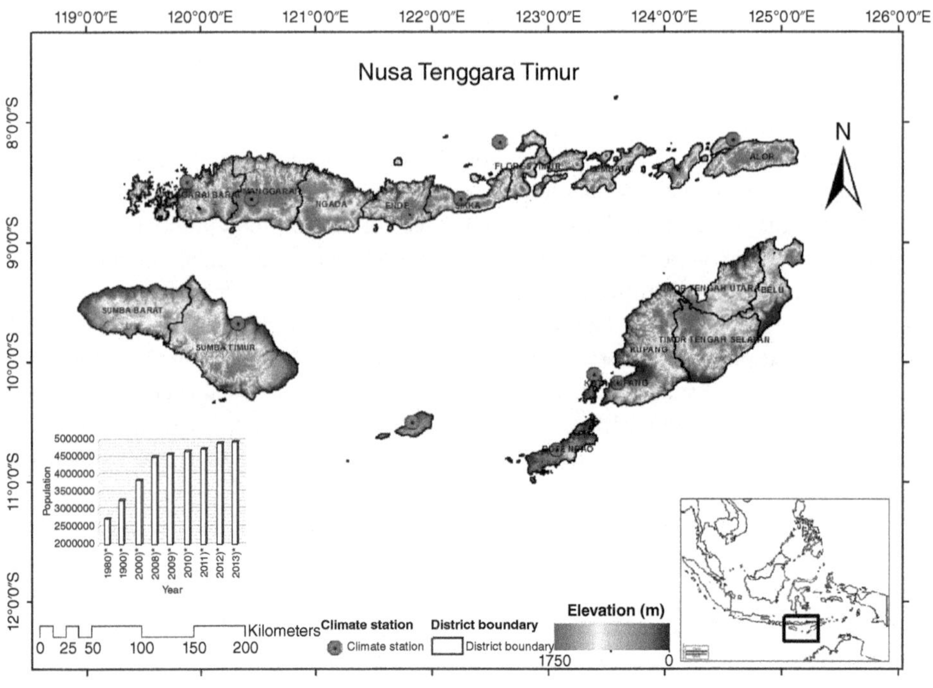

Figure 6.1 Study area. *Source:* Boubacar Siddighi Balde, Martiwi Diah Setiawati, I Wayan Nampa, Mohamed Esham. (*See insert for color representation of the figure.*)

March. The study area has a population of 4.7 million with a poverty rate of 21%, much higher than the national average of 12% (BPS 2011). Rice and maize are the two most important crops, and rice is considered the staple food of households (Salim 2010) and also enhances the status of those who consume it (Kawanishi 2014).

Of the poor people in Indonesia, 62.65% live in rural areas and more than 47.02% (8 167 890 people) in densely populated Java (BPS 2016). Papua, Papua Barat, NTT, Maluku, Gorontalo, and Bengkulu were the provinces with the highest poverty rate, which ranged from 17.09% in Bengkulu to 27.80% in Papua in 2014 (BKP and WFP 2015).

Agriculture is NTT's primary source of revenue (80%), where the rural payments are around six times lower than urban (BPS 2016). It primarily takes the form of subsistence-based rain-fed crop production where maize, rice, and tuber are included as the essential crops. Households also have some livestock; however, this is not the case with the poorest of families. Coastal communities also participate in marine activities such as fishing (mostly near the coast) and, in some areas, seaweed production. Other sources of income may include handicrafts by women and seasonal labor in urban areas, mainly by men (UNDP 2013).

NTT has a tropical climate that ranges from semi-humid to semi-arid. In the highlands, the monsoon begins in October, but in the lowlands occurs in late November. The main factor of climate-related to agriculture in NTT is the annual rainfall variability, and especially between years at the start and end of the rainy season, which influences when farmers can plant their crops and their selection of cultivars (UNDP 2013).

6.2.2 Data Collection and Description of Variables

The analysis used two main categories: climate and food security-related data, as shown in Table 6.1. The climate data variables consist of maximum temperature, minimum temperature, precipitation, relative humidity, sunshine duration, zero rainfall frequencies, and extreme rainfall frequencies. Zero rainfall was defined as less than 1 mm/day, while excessive rainfall was above 58 mm/day (Iwasaki 2014). The daily observed climate data from 10 stations were retrieved from Indonesia Meteorological Agency (BMKG) during 1980 and 2015. The distribution of climate stations over the region was shown in Figure 6.1, and the data availability can be accessed online at http://dataonline.bmkg.go.id (BMKG 2016). The food security-related data was obtained from NTT statistical agency from 2002 until 2014 at the district level. The data are also available online on their publication homepage (http://ntt.bps.go.id). In addition, the total livestock unit (TLU) calculation was referred to Chilonda and Otte (2006). Since the climate data and food security-related data have different sample numbers, we assumed that the non-climate data district used the nearest neighbor climate data station.

Annual historical data of natural disasters and ENSO index were used to support the primary analysis. The historical data of natural disasters was obtained from the online database of Indonesia National Agency for Disaster Management (BNPB) (http://dibi.bnpb.go.id), and the ENSO index was obtained from NOAA climate prediction center (http://www.cpc.ncep.noaa.gov). The ENSO index was computed by sea surface temperature anomaly, and it was necessary to determine the anomaly climate event in the region. In addition, the threshold value was 0.5, where +0.5 or higher is El Nino (warm period), and −0.5 or lower is La Nina (cold period) (Figure 6.5 in supplementary material).

Table 6.1 Coxidered variables in the study of effects of climate variability on food security dimexiox.

Category	Dimexiox	Indicators	Unit	Abbreviation
Food security variables (respoxes)	Food availability	Paddy production	Tons	Paddy
		Maize production	Tons	Maize
		Tuber crops production	Tons	Tubers
		Leguminous production	Tons	Leguminous
		Coffee production	Tons	Coffee
		Cacao production	Tons	Cacao
		Coconut production	Tons	Coconut
		Other plantation production	Tons	Others plantation
		Fish production	Tons	Fish
		Total livestock holding	TLU	Livestock
		Average yield food grain production	q/ha	Yield of food grain
	Food accessibility	GDP per capita in PPP (purchasing power parity)	Rupiah (IDR)	Income per capita
		Population below national poverty line	%	Household poverty
		Expenditures	Rupiah (IDR)	Expenditures
	Food utilization	Population without access to safe water	%	No safe water
		Prevalence of waterborne diseases	Number	Risk of diarrhea
		Underweight children under five	Number	Underweight
		Low birth weight (<2500 g)	Number	Undernutrition
		Pregnant women with anemia	%	Anemia
Climate variables (predictors)		Maximum temperature	°C	T. max
		Minimum temperature	°C	T. min
		Average temperature	°C	T. average
		Precipitation	mm per year	Annual rainfall
		Relative humidity	%	RH
		Suxhine duration	Hours per day	Suxhine
		Frequency of no rainfall (less 1 mm per day)	%	Zero rainfall days
		Extreme rainfall (>58 mm per day)	%	Extreme rainfall

Source: Boubacar Siddighi Balde, Martiwi Diah Setiawati, I Wayan Nampa, Mohamed Esham.

6.2.3 Analytical Framework and Methods

Figure 6.2 describes the analytical framework adopted to analyze the impact of climate variability on food security. It comprises two parts; first, we use the climate data (1980–2015), natural disaster data, and ENSO index to materialize the annual trend of the climate variability. In addition, we use the climate data (2002–2014) of 16 districts and food security variables to model the impact of climate variability on food security. In order to determine

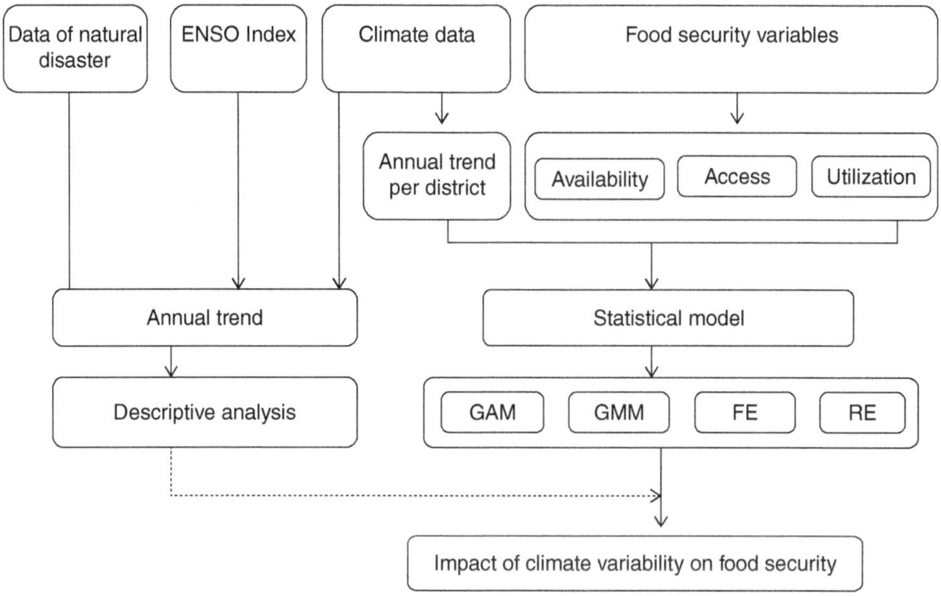

Figure 6.2 Analytical framework. *Source:* Boubacar Siddighi Balde, Martiwi Diah Setiawati, I Wayan Nampa, Mohamed Esham.

the effects of climate variability on food security, some empirical models have been employed. The study applies the generalized additive method (GAM), a dynamic panel data estimation (i.e. generalized method of moments [GMM]), and static panel estimation techniques: fixed effects (FE) and random effects (RE). These estimations were carried out for an alternative comparison. The models are described below.

6.2.3.1 Generalized Additive Method

GAM was used to investigate the link between climate variables and food security components. In this study, we analyzed the data independently to minimize data reduction due to too many missing values of each predictor variable. GAM is a semi-parametric enhancement of a generalized linear model, which has a smooth element in the predictor variables (Wood 2006). The further advantage of GAM is its ability to consider highly nonlinear and non-monotonic relationships between the response and predictor variables (Wood 2006). GAM was assigned in this paper because GAM goes along with the shape of the response curved data (Setiawati et al. 2015). This statistical model gives high flexibility to show the relationship between response and predictor variables. GAM model was applied in this study to assess the influence of climate factors on the 19 food security indicators with 192 samples of each food security indicators (Table 6.1). This method has been frequently applied for ecological studies (Guisan et al. 2002; Setiawati et al. 2015) but has rarely been applied for food security assessment. GAM model was built in R version 3.3.0 software, using the gam function of the *mgcv* package (Wood 2006), with the food security indicators as a response variable and climate data as predictor variables. GAM model is explained in Eq. (6.1). In addition, the GAM model also was employed to assess the trend of climate variability for the annual scale, where time was used as predictor variables and climate data as response variables.

$$g(\mu) = \alpha + f_i X_i \tag{6.1}$$

where $g(\mu)$ is response variable, f_i is smoothing function, n is number of predictor variables, and X_i is predictor variable. The model was evaluated based on the significance level of predictors (P-value), deviance explained (DE). In addition, DE has an identical role as determination coefficient (R^2) in linear regression analysis.

Furthermore, we also used this method to plot the trend of climate variability in the study area where eight variables were evaluated independently. The annual trend of climate variables was shown in Figure 6.3, where year is a predictor variable, and climate data is response variable. Y-axis was defined as residual (i.e. the difference between average value and the estimated values), and x-axis was defined as years.

6.2.3.2 Fixed Effects (FE) and Random Effects (RE)

FE and RE represent mixed-effect regression models and are commonly used techniques in panel data (Djennas 2016). In most cases, these models were chosen to provide better control for the influence of missing or unobserved variables (Verbeek 2008). Moreover, these models can account for intertemporal and individual differences among our units of analysis. In panel estimations, some important issues are considered (Ulusoy and Demiralay 2017). FE and RE are static panel estimation techniques to account for unobserved heterogeneity in panel data.

The equation for the FE model is represented as follows:

$$Y_{it} = \beta_1 X_{it} + \alpha_i + U_{it} \tag{6.2}$$

where Y_{it} is the dependent variable, where represents the districts and <t> the time; β_1 is the coefficient for the considered independent variables; X_{it} represents one independent variable; α_i ($i = 1 \ldots n$) is the unknown intercept for each district (n district-specific intercepts); U_{it} is the error term.

The random effects model is:

$$Y_{it} = \beta_1 X_{it} + \alpha_i + U_{it} + \varepsilon_{it} \tag{6.3}$$

U_{it} is the error term between-entity (district) also called between-entity error.
ε_{it} is the error term within-entity (district) also called within-entity error.

The appropriateness of these models was checked using the Hausman test. To decide between FE or RE, we run a Hausman test where the null hypothesis is that the preferred model is REs versus the alternative FEs (Torres-Reyna 2007). It basically tests whether the unique errors (U_i) are correlated with the regressors, the null hypothesis is they are not. We run a fixed effects model and save the estimates, then run a random model and save the estimates, then perform the test (see supplementary materials Tables S1–S4).

The appropriateness of these models was checked using the Hausman test.

6.2.3.3 Generalized Method of Moments

Effects of climatic variability on food security (availability, accessibility, and utilization) were also estimated using the generalized method of moments (GMM) from the

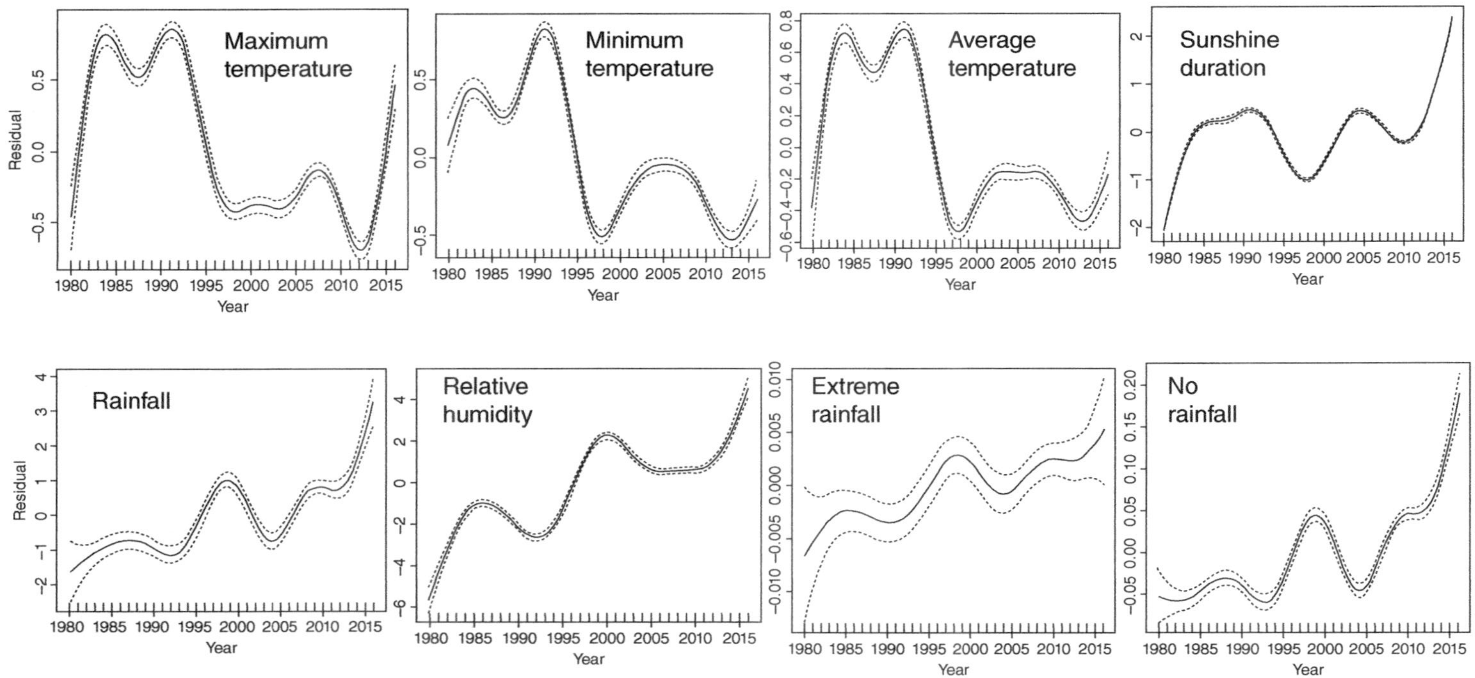

Figure 6.3 Annual trend of climate variability. *Source:* Boubacar Siddighi Balde, Martiwi Diah Setiawati, I Wayan Nampa, Mohamed Esham.

Arellano-Bond dynamic panel data estimation. We estimate the dynamic panel data model in which random effects and fixed effects estimators are inconsistent and biased due to autocorrelated lagged dependent variable resulting correlation with the error term. The bias gets larger when the period gets large. The two estimators remain inconsistent and biased in the large cross-section with a short period (Ulusoy and Demiralay 2017).

$$FS_{i,t} = c_0 + c_1 FS_{i,t-1} + c_2 CV_{i,t} + \alpha_i + \epsilon_{i,t} \tag{6.4}$$

where $FS_{i,\,t}$ represents the food security indicators (Table 6.1) for a specific district across time. "i" is a specific district from the 16 districts of the NTT province. "t" stands for the time period (2001–2014). $CV_{i,\,t}$ stands for the predictors defined here as the climatic variables as listed in Table 6.1. $FS_{i,\,t-1}$ is the lag value of different dependent variables, food security indicators. $\epsilon_{i,\,t}$ is a random disturbance or error term. α_i indicates an unobservable district-specific effect that is constant across time.

6.3 Results

6.3.1 Annual Trend of Climate Variability

Figure 6.3 shows the annual trend of climate data between 1980 and 2015 from 10 rain gauge stations where the distribution spread evenly in the study area (Figure 6.1). The result shows that the temperature trend after 1995 tends to decrease. In contrast, sunshine duration, rainfall, relative humidity, extreme rainfall, and zero rainfall frequency increase. However, excessive rain and zero rainfall frequencies have increased over time in this region. Consequently, NTT has more extreme events that cause increasing natural disaster incidents such as drought, hurricane, landslide, and flood.

6.3.2 Extreme Events and Their Impacts on Livelihood and Food Insecurity in NTT

Figure 6.4 documented the historical data of NTT natural disasters from 1951 to 2014. The result stated that the frequency of these natural disasters has gradually increased since 2003 in the region. Furthermore, Table 6.2 summarizes some direct and negative impacts of these natural disasters on the local communities. The data revealed that flood-related disasters caused severe causality in the area. Also, even though drought is causing fewer casualties among other natural disasters, it has a significant impact at the household level (i.e. affecting production and reducing farmers' income) (WFP 2016). According to Indonesian National Board for Disaster Management, drought had damaged more than 70 000 ha of agricultural land in the study area in 2009 (BNPB 2017).

Figure 6.5 shows the three months running mean ENSO index in December, January, and February. It shows that in 2008, 2009, 2011, and 2012, there were weak to moderate La Nina events, while in 2003, 2005, 2007, 2010, and 2016, there were weak to very strong El Nino events. Also, in 2008 and 2009, there was a high annual frequency of floods, landslides, and storms during the moderate La Nina event, which is shown in Figure 6.4 However, in 2010, there was a moderate El Nino event, which caused drought (Figure 6.4).

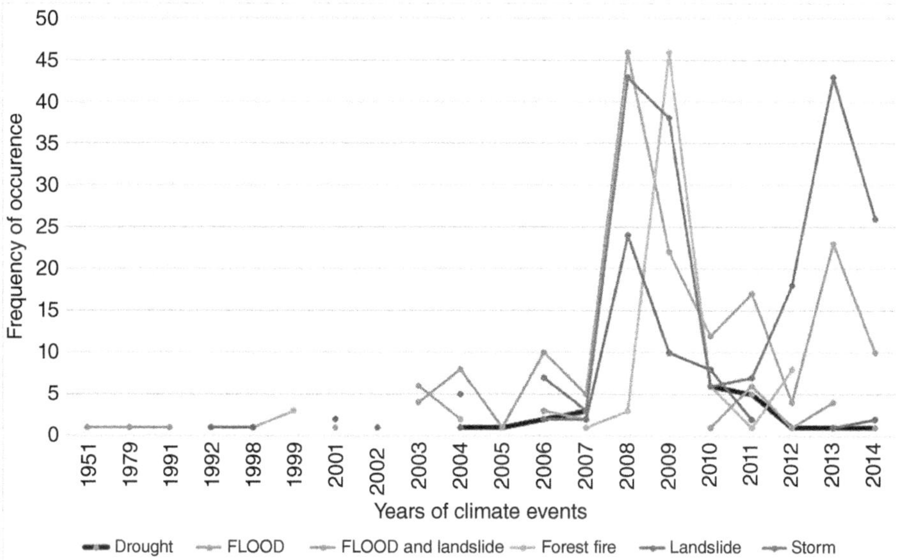

Figure 6.4 Types of natural disasters and frequency of occurrence in NTT. *Source:* Boubacar Siddighi Balde, Martiwi Diah Setiawati, I Wayan Nampa, Mohamed Esham.

Table 6.2 Natural disasters in NTT from 1951 to 2014.

Climate events	Death	Disappeared	Injured	Evacuate
Drought	2	0	0	0
Flood	198	72	2930	24586
Flood and landslide	126	78	140	5964
Forest fire	7	0	3	271
Landslide	16	0	14	190
Storm	38	0	36	1699
Total	387	150	3123	32710

Computed by authors using data from BNPB (2017).
Source: Boubacar Siddighi Balde, Martiwi Diah Setiawati, I Wayan Nampa, Mohamed Esham.

In addition, from 2015 to 2016, there were very strong El Nino events, which caused severe drought in NTT (WFP 2016), confirming our findings in Figure 6.5. Following Figures 6.4 and 6.5, we noticed a high annual frequency of flood, landslide, and storm during the moderate La Nina event (2008 and 2009).

ENSO influences climate variability over a large part of the earth's surface (McPhaden et al. 2006), and it strongly influences hydrological processes and societal risk, including drought, flood, and storm (As-syakur et al. 2013, Ward et al. 2014). In particular, floods and storms are related to rainfall intensity, and the high-rainfall events in Indonesia during the December–January–February season (As-syakur et al. 2013). Also, the previous study

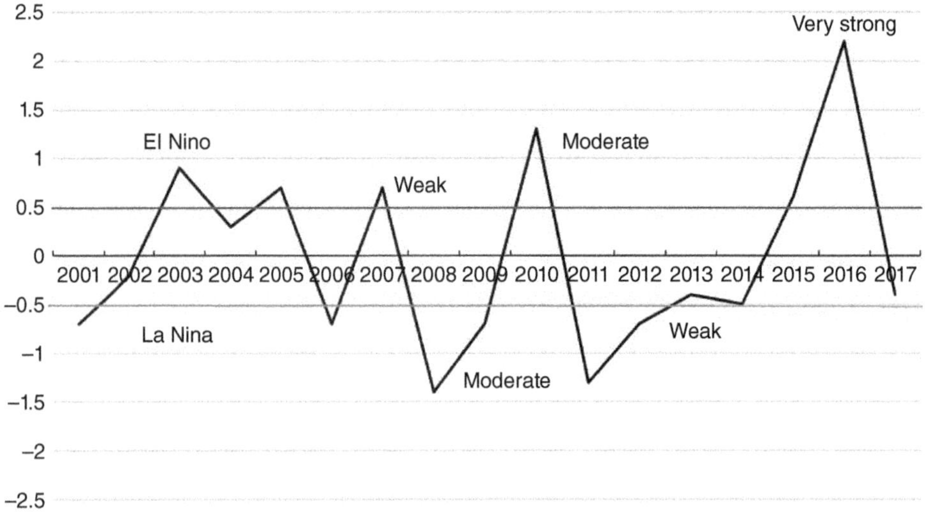

Figure 6.5 Three months running mean of El Niño Southern Oscillation (ENSO) index in December, January, and February. *Source:* Boubacar Siddighi Balde, Martiwi Diah Setiawati, I Wayan Nampa, Mohamed Esham.

conducted by Hidayat (1999) found that some areas in Indonesia, particularly Eastern Island, have experienced heavy rainfalls leading to severe floods during the La Nina event. Also, during El Nino, some parts of Indonesia were hit by drought and forest fire (Hidayat 1999). Consequently, natural disasters, including storms, floods, and droughts, have the potential to reduce farm productivity, damage farm inputs and facilities and/or infrastructure, and limit farm planting options (Israel and Briones 2013). Furthermore, individually storms and floods can damage farm supply routes and cause death or injury.

Consequently, these factors can further lead to indirect and negative impacts on agriculture and the economy as a whole. Especially as a result of the storm, floods, and droughts, the overall cost of agricultural production increases; agricultural production output declines; food supply fall, and, as a result, food prices rise. Altogether, the direct and indirect negative impacts on agriculture threaten food security in the affected districts. Furthermore, changes in climate variability and extreme events may substantially impact the prevalence and distribution of pests and disease and the proliferation of weeds (Thornton et al. 2014). Thus, climate change and the rapid population growth in NTT, which has expanded by 81% over the last 35 years (BPS 2016), have serious implications for regional food security.

6.3.3 Impacts of Climate Variability on Food Security Dimensions

This section documents the impacts of climate variability on food security dimensions using the analytical methods discussed in the Section 6.2.3.

6.3.3.1 Generalized Additive Method

Table 6.3 presents the summary of GAM analysis. How to read the results was described as follows, paddy vs. Tmax has DE 35.2%, which means that Tmax influenced 35.2% of paddy

Table 6.3 GAM analysis result of deviance explained, significance level, and trend indication.

Food security variables	T max.	T ave.	T min.	Annual rainfall	RH	Sunshine duration (SD)	Extreme rainfall	Zero rainfall events
Paddy	(35.2%)*** (−)	(37.3%)*** (−)	(28.7%)*** (−)	(32.5%)*** (+)	(13.7%)*** (+)	X	(10.1%)** (+)	(20.7%)*** (−)
Maize	X	X	(2.29%)* (+)	X	(3.1%)* (−)	X	X	X
Tubers	X	X	X	X	X	X	X	X
Leguminous	X	X	(4.32%)*** (+)	X	(16.6%)*** (−)	X	X	X
Coffee	(52%)*** (−)	(33%)*** (−)	(26.5%)*** (−)	(24%)*** (+)	(9.24%)*** (+)	X	(14.4%)** (+)	(17.5%)*** (−)
Cacao	X	X	(17.7%)*** 23 < T min < 25 (+)	X	(2.55%)* 75 < RH < 84	X	(2.2%)** (−)	X
Coconut	X	X	X	X	X	X	X	X
Other plantation	X	X	X	(6.82%)* (−)	(8.62%)* 75.8 < RH < 84 (+)	(13.9%)** 7 < SD < 8 (+)	(6.27%)** (−)	X
Fish production	X	X	X	X	X	X	X	X
TLU	X	X	(14%)*** (+)	X	(5.41%)* RH < 77 & RH > 84 (+)	(12.2%)** (+)	X	(12%)* 38 < No rain < 55 (+)
Grain	X	X	X	(9.71%)*** (+)	X	(43.6%)*** (+)	X	X

(Continued)

Table 6.3 (Continued)

Food security variables	T max.	T ave.	T min.	Annual rainfall	RH	Sunshine duration (SD)	Extreme rainfall	Zero rainfall events
GDP/capita	X	X	(13.2%)*** (+)	X	X	(16.4%)*** (6 < opt<7)	X	(16.8%)*** (−)
Expenditure	X	X	(15.2%)*** (+)	(10.08%)* (−)	X	(26.8%)*** (6 < SD < 7) (+)	X	(35.7%)*** (+)
Poor family	X	X	X	X	X	(9.83%)* (5.8 < SD < 7.5) (+)	X	X
Risk of diarrhea	(17.7%)** (+)	X	X	X	X	(33.4%)*** 6.5 < SD < 7.5 (+)	X	X
Access of unclean water	X	X	X	X	X	X	(3.17%)* (−)	X
Underweight	X	X	X	X	X	(46.4%)** (−)	X	X
Malnutrition	X	X	(30.7%)** (+)	X	X	X	X	X
Anemia	(17.2%)** (−)	X	–	X	X	X	X	X

Note: x (not significance), (−, +,opt; trend indication).
* (α: 5%).
** (α: 1%).
*** (α: 0.1%).
Source: Boubacar Siddighi Balde, Martiwi Diah Setiawati, I Wayan Nampa, Mohamed Esham.

production, and other variables influenced the rest. The negative sign indicates a negative relationship between the variable (i.e. paddy vs. Tmax; high temperature negatively impacts paddy production).

The model results indicate that high temperature, low RH, and low precipitation negatively influence the production of paddy and coffee in the region. A previous study by Masutomi et al. (2009) found increasing temperature during the growing period as the leading cause for low rice yield. In addition, high temperatures during the grain-filling period result in an increased incidence of chalky rice, which has become a significant problem for rice production (Masutomi et al. 2015). In the case of coffee production, a study by Fain et al. (2018) revealed that high temperature and low precipitation levels resulted in reduced quality and yield and increased exposure and susceptibility to some insects and diseases. Moreover, six out of eight climate variables had a statistically significant (i.e. $a = 0.1\%$) impact on rice and coffee production.

In contrast, livestock (TLU) production responded differently to climate variables, as shown in Table 6.2. The model results show that increasing minimum temperature and sunshine duration had a positive impact on livestock production. It is probably because the increasing minimum temperature can reduce the extreme weather condition where it can cause losses of livestock production by increased animal mortality and reduced feeding efficiency (Belasco et al. 2015). There was no significant relationship between climate variables and tuber, coconut, and fish production.

Figure 6.6 shows the spatial distribution of the NTT food production between 2002 and 2014, where red color indicates high production and white indicates low production. The result demonstrates that paddy and coffee production has increased evenly in all NTT,

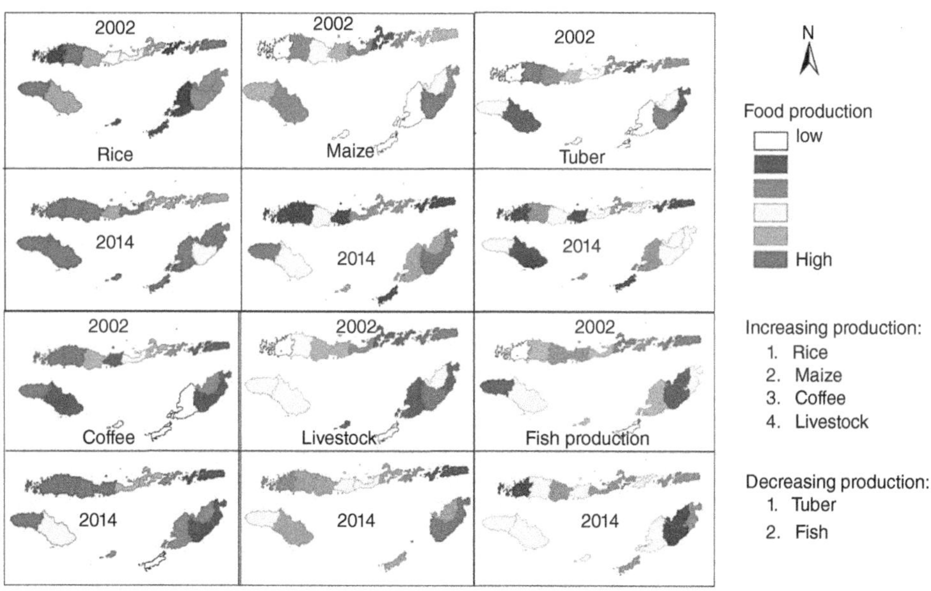

Figure 6.6 Food production between 2002 and 2014 in NTT provinces. *Source:* Boubacar Siddighi Balde, Martiwi Diah Setiawati, I Wayan Nampa, Mohamed Esham. (*See insert for color representation of the figure.*)

followed by maize and livestock. However, tubers and fish production decreased almost in all districts of NTT. Growing those commodities was part of the NTT medium-term development plan (RPJMD) (NTT Regional Government 2013), which stated that those commodities could increase the farmer's income.

As shown in Table 6.3, we examined three primary parameters: GDP/capita, expenditure, and low-income family in terms of food access. The result indicated that minimum temperature, zero rain events, and sunshine duration influence GDP/capita. Also, the increase of zero rain events negatively affected GDP/capita in the region, where low rainfall negatively affected agricultural output and rural income in developing countries (Dell et al. 2014). In addition, when the sunshine duration was higher than 7 h/day or lower than 6 h/day, it negatively affected GDP per capita. It is because sunshine duration is highly related to solar radiation (Watanabe et al. 2011) as one of the main essential factors for crop yields (Masutomi et al. 2009; Masutomi et al. 2015). Moreover, climate variability, particularly minimum temperature, annual rainfall, sunshine duration, and zero rain events, also influenced the expenditure. The analysis indicated that the spending would be positively affected by increasing annual rainfall and negatively affected by zero rain events.

Concerning food utilization, risk of diarrhea, access to clean water, underweight, malnutrition, and anemia were evaluated, as shown in Table 6.3. This model indicates that increasing maximum temperature influenced the risk of diarrhea by 17%. This model almost has the same agreement with the previous study by Philipsborn et al. (2016), which found that a 1 °C increase in mean temperature is associated with an increase in the incidence of diarrhea genic *Escherichia coli* by 8%. Moreover, increased temperatures can increase replication rates and survival of bacteria in the environment and alter *E. coli* gene expression (Vanelsas et al. 2011).

Furthermore, access to clean water, underweight, malnutrition, and anemia also were influenced by climate variability. In particular, underweight and malnutrition were decreased by increasing sunshine duration and lowering the minimum temperature. Lloyd et al. 2011 found that climate change would affect malnutrition and underweight in Southeast Asia around 21 and 18%, respectively. Overall, the GAM model results show that climate variability influenced four main components of food security.

6.3.3.2 Random Effects and Fixed Effects Models

Table 6.4 documents the results of the effects of climate variability on food security (accessibility, utilization, and availability) with different econometric methods: FEs and REs. The appropriateness of these models was checked using the Hausman test. We considered three food access indicators to explain the effects of climate variability on food accessibility. Table 6.4 shows that FE is more appropriated for explaining this climate variability effect on poor family and expenditures. The RE model is appropriated for the GDP per capita. In both FE and RE models, income per capita and expenditures would be negatively affected by higher temperature and no rainfall events (supplementary material Table S1). Meanwhile, both indicators are positively affected by average temperature and precipitation under the estimation of both empirical models. In addition, income per capita is positively influenced by relative humidity and sunshine duration. FE model indicated that population below the national poverty line is negatively associated with relative humidity and sunshine duration.

Table 6.4 Fixed and random effect models.

Food security indicators		Fixed effect	Random effect	Hausman test (Prob > chi^2)
Access	GDP/capita	×	✓	0.9817
	Poor family	✓	×	0.0000
	Expenditure	✓	×	0.0000
Utilization	No safe water	×	✓	0.7547
	WB diseases	✓	✓	0.0001
	Underweight	×	✓	0.3190
	Malnutrition	✓	×	0.0014
	Anemia	✓	×	0.0394
Availability	TLU	×	✓	0.7896
	Yield grain	✓	×	0.0000
	Paddy	×	✓	0.7660
	Tubers	×	✓	0.7988
	Leguminous	×	✓	0.9282
	Other plants	×	✓	0.8199
	Fish	✓	×	0.0000
	Coffee	×	✓	0.8929
	Cacao	×	✓	0.9807
	Coconut	×	✓	0.8863

Note: x is not significant, ✓ is significant. *Source:* Boubacar Siddighi Balde, Martiwi Diah Setiawati, I Wayan Nampa, Mohamed Esham.

Five indicators are considered for the food utilization as shown in Table 6.4. The FE model is more appropriated for explaining this climate variability effect on waterborne diseases, malnutrition, and anemia. The RE model is appropriated for the no safe water and underweight. Results indicated a lack of safe water, and waterborne diseases increase as a response to increasing temperature. Therefore, the prevalence of waterborne diseases declines in response to minimum temperature, relative humidity, and average temperature.

Eleven indicators are considered for the food availability as shown in Table 6.4. The FE model is more appropriate for explaining this climate variability effect on yield grain and fish. The RE is more appropriated for the remaining indicators except for maize. Tables S2 and S4 in the supplementary material also display the continued food availability. These tables document the effects of climate variability on food availability. Results generated from static models (FE and RE) show that paddy production would be positively and negatively affected as a response to sunshine duration and zero rainfall events, respectively. As for tubers and legumes, their production is negatively influenced by extreme rainfall (Table S2). For leguminous production, both static models revealed that it would be positively affected by sunshine duration and precipitation. Tuber production is also found to be positive and significantly affected by average temperature. Only the RE appears to explain

Table S1 Impacts of climate variability on food access and utilization.

	Food accessibility						Food utilization			
	GDP/capita		**Poor family**		**Expenditure**		**No safe water**		**Waterborne diseases**	
	Fixed-effects	Random-effects	Fixed-effects	Random-effects	Fixed-effects	Random-effects	Fixed-effects	Random-effects	Fixed-effects	Random-effects
	Coef. (Std. Err.)	Coef. (Std. Err.)	Coef. (Std. Err.)	Coef. (Std. Err.)	Coef. (Std. Err.)	Coef. (Std. Err.)	Coef. (Std. Err.)	Coef. (Std. Err.)	Coef. (Std. Err.)	Coef. (Std. Err.)
T. max	−2383257*** (481863)	−2280895*** (443 751.3)	7.24 (4.89)	6.28* (3.78)	−188 918.9*** (59 591.54)	−131 567.8*** (42 512.17)	6.30** (3.06)	3.74 (2.54)	303136*** (63 232.32)	114065*** (37910.26)
T. min	432 960.8*** (162 673.6)	448 838.8*** (154 905)	0.68 (1.64)	−0.25 (1.44)	10 084.69 (19 377.74)	13 648.22 (16 645.69)	0.38 (0.99)	−0.04 (0.92)	−56028*** (19 641.93)	−44116*** (15 376.61)
T. average	2157824*** (505 193.1)	1 967 691*** (466 555.5)	−0.84 (4.79)	−6.57* (3.72)	249 303.4*** (59 462.84)	137 664.9*** (41 922.96)	−1.84 (3.06)	−3.47 (2.58)	−178 661*** (61 676.51)	−63 186* (36 464.61)
Annual Rainfall	623.44* (334.36)	595.72* (325.92)	−0.01* (0.003)	−0.01* (0.003)	119.12*** (39.34)	68.63* (37.75)	−0.003 (0.002)	−0.002 (0.002)	−53.17 (39.21)	−7.63 (36.49)
RH	1989.02** (917.42)	1836.89** (836.16)	−0.02** (0.01)	−0.01 (0.01)	206.14** (105.11)	102.33 (68.17)	−0.01 (0.01)	−0.01** (0.004)	51.92 (132.27)	−136.05** (59.99)
Sunshine	390 285.8* (206 268.3)	376 413.8* (199 487.7)	−3.26* (1.89)	−0.71 (1.76)	11 749.7 (23 624.01)	717.11 (21 823.97)	−0.94 (1.22)	−0.53 (1.16)	−15 387.61 (23 815.81)	12 287.96 (20 796.7)
Zero rainfall events	−34 550.49** (7284.77)	−34 880.82*** (6810.528)	0.29*** (0.08)	0.19*** (0.07)	−4722.59*** (871.27)	−3907.49*** (759.40)	0.09** (0.04)	0.08** (0.04)	96.02 (884.49)	−14.76 (718.63)
Extreme Rainfall	106 045.9 (167 879.5)	113 833.3 (165 229.5)	1.54 (1.56)	0.65 (1.53)	13 850.22 (19 296.79)	9283.99 (19 218.47)	0.30 (0.99)	0.14 (0.97)	−1687.25 (19 225.14)	2483.07 (18 858.04)
constant	1.04e+07 (1.05e+07)	1.21e+07* (6692089)	−182.16 (113.48)	11.19 (47.41)	−541 987.4 (1 329 414)	663 862.6 (526 595.7)	−142.05** (68.51)	−10.56 (32.20)	−3241214** (1 370 167)	−887069* (490 382.3)

sigma_u	1 976 863.6	2 298 616.7	23.89	6.60	154 861.81	48 636.35	14.08	6.81	212 306.12	0
sigma_e	1 642 875.9	1 642 875.9	14.14	14.14	185 233.9	185 233.9	9.55	9.55	183 652.61	183 652.61
rho	0.59	0.66	0.74	0.18	0.41	0.06	0.68	0.34	0.57	0
Number of obs	201	201	160	160	184	184	185	185	172	172
Number of districts	16	16	16	16	16	16	16	16	16	16
Obs per district (min)	9	9	8	8	8	8	8	8	8	8
Obs per district (avg)	12.6	12.6	10.0	10.0	11.5	11.5	11.6	11.6	10.8	10.8
Obs per district (max)	14	14	11	11	13	13	13	13	13	13
R^2	0.3526	0.3521	0.1507	0.1011	0.3231	0.3071	0.0773	0.0536	0.2256	0.1642
Hausman test (chi square)	1.98		96.42		83.62		5.03		31.39	
Hausman test (Prob>chi2)	0.9817		0.0000		0.0000		0.7547		0.0001	

*, **, *** significance level at 10, 5, and 1%, respectively. Values in parentheses are standard errors.

Table S2 Impacts of climate variability on food availability.

	Food availability									
	Paddy		**Tubers**		**Leguminous**		**Other plants**		**Fish**	
	Fixed-effects	Random-effects	Fixed-effects	Random-effects	Fixed-effects	Random-effects	Fixed-effects	Random-effects	Fixed-effects	Random-effects
	Coef. (SE)	Coef. (SE)	Coef. (SE)	Coef. (SE)	Coef. (SE)	Coef. (SE)	Coef. (SE)	Coef. (SE)	Coef. (SE)	Coef. (SE)
T. max	−2714.47 (5674.08)	−4736.76 (5307.32)	−6027.10 (9451.51)	−7071.62 (9009.33)	289.14 (323.85)	309.50 (308.64)	−957.44* (529.38)	−953.88* (511.29)	−39 297.32 (25 120.54)	−36727** (15 770)
T. min	581.52 (1925.21)	191.18 (1863.11)	−3905.74 (3413.85)	−4219.74 (3298.84)	72.28 (115.41)	93.99 (111.33)	9.72 (188.65)	7.69 (184.82)	6186.53 (8131.12)	4084.15 (6415.79)
T. average	−2257.96 (5892.08)	−3895.19 (5531.60)	17 038.8* (10366)	11 692.49 (9725.84)	−386.55 (355.14)	−295.25 (333.23)	1399.62** (580.54)	1193.69** (551.16)	−1887.54 (25 035.49)	29818** (15 120)
Annual rainfall	−1.34 (3.90)	−1.42 (3.85)	−0.45 (6.93)	0.79 (6.86)	0.65*** (0.23)	0.64*** (0.23)	0.53 (0.38)	0.46 (0.39)	12.34 (16.49)	3.96 (15.43)
RH	10.59 (10.63)	7.60 (9.91)	−2.72 (18.57)	−9.38 (17.46)	0.48 (0.64)	0.37 (0.60)	5.19*** (1.04)	4.60*** (0.99)	0.74 (44.36)	20.01 (25.51)
Sunshine	9180.77*** (2381.57)	9698.45*** (2331.63)	−1979.74 (4258.46)	−1482.61 (4191.15)	314.88** (146.48)	297.56** (143.52)	92.67 (239.44)	53.48 (238.61)	−5884.30 (9977.44)	−3316.99 (8805.04)
Zero rainfall events	−211.11** (85.76)	−215.87*** (81.54)	284.33* (151.37)	222.95 (143.80)	0.77 (5.17)	1.89 (4.91)	−1.998 (8.45)	−0.22 (8.14)	44.78 (368.16)	447.28 (301.65)
Extreme Rainfall	−670.50 (1953.53)	−779.78 (1940.24)	−7208.19** (3481.07)	−6775** (3482.24)	−350.89*** (119.30)	−334*** (118.63)	−11.38 (195.01)	−16.26 (197.59)	−12 503.96 (8136.49)	−5689.37 (8029.02)
Constant	129 309.7 (123 401.6)	241 980.7*** (83 514.06)	−109 797.4 (208 937.1)	68 994.75 (139 602)	−489.07 (7170.12)	−4061.73 (4933.11)	−5161.49 (11 720.82)	713.45 (7901.32)	1194981** (559 573.3)	275 096.7 (205 662.9)
sigma_u	32 012.512	30 203.79	57 841.67	49 336.04	1885.93	1867.16	3472.79	2768.09	98 672.85	0

sigma_e	18938.59	18938.59	33914.97	33914.97	1167.66	1167.66	1908.75	1908.75	78282.81	78282.81
rho	0.74	0.72	0.74	0.68	0.72	0.72	0.77	0.68	0.61	0
Number of obs	195	195	199	199	203	203	203	203	186	186
Number of districts	16	16	16	16	16	16	16	16	16	16
Obs per district (min)	9	9	9	9	9	9	9	9	8	8
Obs per district (avg)	12.2	12.2	12.4	12.4	12.7	12.7	12.7	12.7	11.6	11.6
Obs per district (max)	14	14	14	14	14	14	14	14	13	13
R^2.	0.1370	0.1326	0.0656	0.0595	0.1096	0.1077	0.1663	0.1645	0.0648	0.0264
Hausman test (chi2)	4.92		4.61		3.10		4.39		30.17	
Hausman test (Prob>chi2)	0.7660		0.7988		0.9282		0.8199		0.0002	

*, **, *** significance level at 10%, 5% and 1% respectively. Values in parentheses are standard errors.

Table S3 Impacts of climate variability on food utilization and availability.

	Food utilization						Food availability			
	Underweight		Malnutrition		Anemia		TLU		Yield of food grain	
	Fixed-effects Coef. (SE)	Random-effects Coef. (SE)	Fixed-effects Coef. (SE)	Random-effects Coef. (SE)	Fixed-effects Coef. (SE)	Random-effects Coef. (SE)	Fixed-effects Coef. (SE)	Random-effects Coef. (SE)	Fixed-effects Coef. (SE)	Random-effects Coef. (SE)
T. max	−40.01 (61.64)	−33.08 (44.64)	863.63 (603.49)	80.11 (328.53)	−22.79 (16.20)	−13.14* (7.59)	−20 801.1*** (8226.53)	−20 093*** (7857.10)	−4.43 (4.15)	−0.97 (2.80)
T. min	−31.40* (16.64)	−14.36 (14.07)	316.31* (162.93)	224.18** (101.55)	−7.18 (4.37)	−2.03 (2.21)	7426.94*** (2931.59)	7951*** (2840.17)	3.22** (1.41)	3.10*** (1.15)
T. average	51.09 (50.15)	32.18 (40.50)	−113.70 (491.03)	−21.36 (294.29)	14.41 (13.18)	13.30** (6.55)	10 719.64 (9021.44)	9688.38 (8469.79)	−11.69*** (4.36)	−4.96* (2.74)
Annual rainfall	−0.05* (0.03)	−0.06** (0.02)	0.37 (0.28)	0.63*** (0.24)	0.002 (0.01)	−0.01 (0.01)	6.21 (6.03)	6.07 (5.98)	−0.005* (0.003)	−0.007** (0.002)
RH	−0.08 (0.07)	−0.09* (0.06)	0.73 (0.69)	0.52 (0.41)	−0.03 (0.02)	−0.01 (0.01)	46.17*** (16.21)	38.63*** (15.27)	0.008 (0.008)	−0.001 (0.01)
Sunshine	−13.21 (12.85)	−18.74 (11.97)	−128.74 (125.84)	40.71 (106.95)	−4.08 (3.38)	−1.88 (2.67)	10 404.53*** (3720.86)	10140*** (3666.75)	13.60*** (1.74)	9.82*** (1.57)
Zero rainfall events	−0.49 (0.74)	−0.31 (0.62)	5.13 (7.25)	−2.21 (5.10)	−0.24 (0.19)	0.08 (0.12)	−308.23** (131.39)	−297.04** (125.03)	−0.31*** (0.06)	−0.18*** (0.05)
Extreme Rainfall	25.66*** (9.66)	27.23*** (9.54)	130.69 (94.63)	83.31 (91.40)	0.46 (2.54)	0.19 (2.43)	−2925.72 (3030.47)	−2290.95 (3036.36)	3.36** (1.43)	2.99** (1.43)
Constant	974.95 (921.73)	914.57* (522.33)	−30375*** (9024.61)	−7333.37* (4056.94)	568.60** (242.29)	161.48 (102.15)	210 280.7 (182 139.3)	203 998.8* (121 442.3)	348.10*** (90.32)	82.99** (36.65)

sigma_u	125.43	101.48	2216.47	354.90	48.47	0	47 280.28	43 037.43	28.68	1.69
sigma_e	78.28	78.28	766.48	766.48	20.58	20.58	29 661.56	29 661.56	13.79	13.79
rho	0.72	0.63	0.89	0.18	0.85	0	0.72	0.68	0.81	0.015
Number of obs	78	78	78	78	78	78	203	203	193	193
Number of districts	16	16	16	16	16	16	16	16	16	16
Obs per district (min)	4	4	4	4	4	4	9	9	8	8
Obs per district (avg)	4.9	4.9	4.9	4.9	4.9	4.9	12.7	12.7	12.1	12.1
Obs per district (max)	5	5	5	5	5	5	14	14	14	14
R^2	0.2889	0.2692	0.3088	0.1872	0.1952	0.0466	0.1499	0.1484	0.4267	0.3755
Hausman test (chi2)	8.16		25.32		16.22		4.70		84.12	
Hausman test (Prob>chi2)	0.3190		0.0014		0.0394		0.7896		0.0000	

*, **, *** significance level at 10, 5, and 1%, respectively. Values in parentheses are standard errors.

Table S4 Impacts of climate variability on food availability.

	Food availability					
	Coffee		**Cacao**		**Coconut**	
	Fixed-effects	Random-effects	Fixed-effects	Random-effects	Fixed-effects	Random-effects
	Coef. (SE)	Coef. (SE)	Coef. (SE)	Coef. (SE)	Coef. (SE)	Coef. (SE)
T. max	−161.20 (299.28)	−286.98 (283.75)	57.92 (189.07)	0.50 (190.82)	−519.62 (327.52)	−449.26 (316.76)
T. min	−147.34 (97.89)	−160.63* (95.50)	−7.18 (61.31)	44.20 (59.39)	−142.64 (107.59)	−126.31 (106.33)
T. average	48.16 (302.26)	−8.45 (288.65)	−20.49 (186.89)	−1.23 (192.19)	378.74 (324.70)	420.06 (317.39)
Annual rainfall	−0.22 (0.19)	−0.22 (0.19)	0.06 (0.12)	0.002 (0.12)	0.08 (0.22)	0.06 (0.21)
RH	−0.15 (1.14)	−0.40 (1.08)	32.98 (20.73)	−0.67 (1.36)	0.14 (0.58)	0.15 (0.57)
Sunshine	30.86 (119.31)	51.79 (117.28)	125.75* (75.46)	80.80 (75.36)	100.62 (132.37)	78.86 (131.09)
Zero rainfall events	0.40 (4.40)	0.50 (4.30)	−0.09 (2.85)	−0.27 (2.93)	−3.42 (5.08)	−2.46 (5.00)
Extreme rainfall	37.81 (97.86)	24.46 (96.81)	22.74 (59.57)	44.12 (60.94)	−81.02 (106.70)	−76.80 (106.5)
Constant	8707.39 (6665.78)	14 215*** (4946)	−3956.78 (4564.48)	−569.30 (3240.13)	12 808.89* (7293.96)	9246.99 (6149.22)
sigma_u	2040.88	2036.59	9824.06	1303.05	3259.92	3207.18
sigma_e	895.35	895.35	507.47	507.47	1015.66	1015.66
Rho	0.84	0.84	0.997	0.87	0.91	0.91
Number of obs	171	171	164	164	183	183
Number of districts	16	16	16	16	16	16
Obs per district (min)	2	2	1	1	8	8
Obs per district (avg)	10.7	10.7	10.3	10.3	11.4	11.4
Obs per district (max)	13	13	13	13	13	13
R^2	0.0332	0.0301	0.0399	0.0209	0.0323	0.0307
Hausman test (chi square)	3.58		2.01		2.99	
Hausman test (Prob>chi2)	0.8929		0.9807		0.8863	

*, **, *** significance level at 10, 5, and 1%, respectively. Values in parentheses are standard errors.

climate variability on the fish catch (Table S2). The impact of climate variables differs from one response variable to another.

Supplementary material Table S3 presents the continued indicators of food utilization (i.e. underweight, malnutrition, and anemia) and some indicators of food availability (total livestock holding an average yield food grain production). Results revealed that underweight decreases as a response to precipitation, minimum temperature, and relative humidity. However, it is positively affected by extreme rainfall. The analysis indicated that climate variability (minimum temperature and precipitation) is positive and significantly associated with malnutrition. The ratio of pregnant women with anemia would be negatively and positively affected as a response to maximum temperature and average temperature, respectively. The empirical findings revealed that climate variabilities (maximum temperature and no rain events) are negatively associated with livestock. Therefore, livestock would positively respond to minimum temperature, relative humidity, and sunshine duration. As for the average yield of food grain production, the analysis suggests that it is positively affected as a response to minimum temperature, sunshine duration, and extreme rainfall. However, this food availability indicator is negatively affected due to zero rain events, precipitation, and average temperature.

Results indicate that fish catch is positively and negatively affected by average temperature and maximum temperature, respectively. The effect of climate variability on food availability related to industrial crops is presented in supplementary material (Table S4). Results from the RE model revealed that the minimum temperature is negatively associated with coffee production. Meanwhile, the FE model shows that cacao production would be positively affected by sunshine duration.

6.3.3.3 Generalized Method of Moments

Tables 6.5 and 6.6 present the empirical results from the Arellano-Bond dynamic panel data estimation. Results in Table 6.5 show the effects of climate variability on food access (GDP per capita, population below national poverty and expenditures) and utilization (population without access to safe water, the prevalence of waterborne diseases, underweight, malnutrition [low birth weight], and anemia). The results indicated the significance of the initial or lagged variable of food access, which has a positive coefficient value.

Table 6.6 presents the effects of climate variability on food availability. The model results indicated that the number of livestock holding increases positively in response to minimum temperature, relative humidity, and sunshine duration. However, it declines in response to the extension (lengthiness) of days without rain. As for the average yield of food grain production, the empirical findings suggest that it would decline in response to average temperature and no rains. Meanwhile, sunshine duration and extreme rainfall positively affect the average yield of food grains (Table 6.6). The previous study (Porter and Semenov 2005) indicated that climate variability and extreme events could also be important for yield quality. Sunshine duration has a positive and significant relationship with paddy production, while no rains have a negative and significant association with paddy production. This implies that paddy production increases as a response to the sunshine duration and declines as a response to zero rainfall events (Table 6.6). The empirical findings revealed that tubers and leguminous production decrease due to extreme rainfall, while they increase differently as a response to no rainfall events. In addition, production of tubers also declines as a response to sunshine duration. Furthermore, the analysis

Table 6.5 Impacts of climate variability on food access and utilization.

	Food accessibility			Food utilization				
	GDP/capita	Poor family	Expenditure	No safe water	Water borne	Underweight	Malnutrition	Anemia
	Coef. (Std. Err.)	Coef. (Std. Err.)	Coef. (Std. Err.)	Coef. (Std. Err.)	Coef. (Std. Err.)	Coef. (Std. Err.)	Coef. (Std. Err.)	Coef. (Std. Err.)
Lag (dependent)	1.03*** (0.02)	0.33** (0.14)	0.93*** (0.06)	−0.03 (0.09)	0 0.002 (0.01)	−0.56 (0.42)	−0.02 (0.06)	−0.42* (0 0.23)
T. max	−307 610.9*** (122 637.9)	−4.27 (3.90)	40 079.18 (49 617.19)	11.58*** (4.64)	17 800.63*** (5369.54)	45.91 (99.99)	296.43 (329.83)	−28.76 (18.68)
T. min	−38 175.43 (38 826.4)	0.99 (1.11)	25 312.18 (16 301.66)	3.18** (1.53)	−22.72 (1653.56)	−25.27 (22.20)	61.21 (77.86)	−9.01** (4.54)
T. average	246 543.6** (120 341.1)	5.10 (3.21)	55 474.94 (44 503.95)	−6.43 (3.97)	−17 290.9*** (4390.25)	29.22 (66.05)	−311.22 (248.16)	30.94** (14.24)
Annual rainfall	73.82 (74.09)	0.001 (0.002)	101.34*** (25.44)	−0.003 (0.002)	−3.52 (2.47)	0.02 (0.03)	−0.10 (0.14)	−0.002 (0.008)
RH	551.67*** (200.06)	−0.005 (0.008)	169.64*** (64.32)	−0.01 (0.01)	−2.07 (11.71)	0.19 (0.19)	0.33 (0.78)	0.02 (0.04)
Sunshine	−55 989.93 (44 957.08)	−2.32** (0.96)	−5049.76 (14 842.19)	−1.52 (1.43)	−2153.92 (1398.97)	−11.74 (23.71)	34.99 (64.11)	−8.40** (3.68)
Zero rainfall days	−196.91 (1975.74)	0.099 (0.06)	−2325.87*** (710.51)	0.09 (0.06)	218.66*** (70.93)	−2.97** (1.54)	−7.21 (4.94)	0.09 (0.27)
Extreme rainfall	−58 120.59* (33 461.21)	0.81 (0.69)	−718.09 (11 184.79)	0.16 (1.04)	334.59 (1000.23)	18.86* (10.59)	126.49*** (39.85)	2.66 (2.22)
Constant	4468364** (2273007)	−4.49 (61.21)	−3195361*** (848636)	−242.72*** (85.09)	−76 806.86 (87 593.73)	−1105.44 (1541.99)	−1763.80 (5489.40)	379.88 (305.88)

Number of obs	164	92	153	153	120	44	44	44
Number of districts	16	16	16	16	16	16	16	16
Obs per district (min)	6	4	6	6	5	1	1	1
Obs per group (avg)	10.25	5.75	9.56	9.56	7.5	2.75	2.75	2.75
Obs per district (max)	12	6	11	11	11	3	3	3
Wald chi2	5219.70	27.64	468.94	23.47	56.26	21.15	29.20	18.25
Prob > chi2	0.0000	0.0011	0.0000	0.0052	0.0000	0.0120	0.0006	0.0324

*, **, *** significance level at 10%, 5%, and 1% respectively.
Source: Boubacar Siddighi Balde, Martiwi Diah Setiawati, I Wayan Nampa, Mohamed Esham.

Table 6.6 Impacts of climate variability on food availability.

	Food availability						
	TLU	Grain	Paddy	Tubers	Leguminous	Other plants	Fish
	Coef. (Std. Err.)	Coef. (Std. Err.)	Coef. (Std. Err.)	Coef. (Std. Err.)	Coef. (Std. Err.)	Coef. (Std. Err.)	Coef. (Std. Err.)
Lag	0.05 (0 0.11)	0 0.64** (0 0.32)	0 0.57*** (0 0.14)	0 0.08 (0 0.08)	0 0.28*** (0 0.08)	0 0.37*** (0 0.10)	0 0.07 (0.09)
T. max	−10 523.66 (7895.21)	−0.09 (5.11)	1928.59 (6376.97)	−8611.67 (11 939.33)	−57.15 (413.03)	389.02 (570.73)	−58600* (35 779.33)
T. min	5355.42* (2833.61)	1.45 (1.81)	−2552.83 (2419.66)	−5368.18 (4443.14)	94.50 (143.28)	268.55 (216.70)	2436.06 (11 606.41)
T. average	2886.20 (7891.67)	−20.36*** (−3.87)	−6084.11 (6242.12)	12 839.92 (11 861.98)	−588.02 (393.58)	47.72 (545.30)	10 738.18 (31 995.07)
Annual rainfall	4.49 (4.77)	−0.004 (0.003)	−0.82 (3.78)	−7.44 (7.01)	0 0.15 (0.23)	−0.04 (0.35)	5.23 (19.57)
RH	37.56*** (13.82)	0.01 (0.01)	16.39 (10.89)	−7.97 (24.13)	0.45 (0.80)	4.87*** (1.09)	−9.54 (65.30)
Sunshine	9432.61*** (2949.92)	15.59*** (1.86)	8260.35*** (2335.41)	−7107.16* (4303.07)	179.41 (150.60)	83.47 (206.49)	−13 506.08 (11 561.66)
Zero rainfall events	−292.51** (126.16)	−0.42*** (0.08)	−337.75*** (99.62)	633.12*** (190.85)	12.97** (6.18)	5.12 (8.96)	437.23 (520.96)
Extreme rainfall	−607.80 (2095.34)	2.37* (1.39)	−317.19 (1730.02)	−5756.94* (3209.28)	−251.59** (106.76)	80.16 (155.92)	−14 897.7* (9105.17)
Constant	151 552.3 (147 675.7)	460.51*** (110.67)	147 800 (124 122.5)	127 588.7 (227 161)	15 143.26** (7770.02)	−18 011.23* (10 936.3)	15 79 920** (651 926)

Number of obs	165	156	159	159	165	165	154
Number of districts	16	16	16	16	16	16	16
Obs per district (min)	6	5	6	6	6	6	6
Obs per district (avg)	10.31	9.75	9.94	9.94	10.31	10.31	9.63
Obs per district (max)	12	12	12	12	12	12	11
Wald chi2	32.59	163.85	63.66	26.39	51.87	32.30	13.35
Prob > chi2	0.0002	0.0000	0.0000	0.0018	0.0000	0.0002	0.1472

*, **, *** significance level at 10%, 5% and 1% respectively. Values in parentheses are standard errors.
Source: Boubacar Siddighi Balde, Martiwi Diah Setiawati, I Wayan Nampa, Mohamed Esham.

included some local crops, and their production revealed a positive and significant association as a response to the relative humidity. Results revealed that fish catch declines as a response to maximum temperature and extreme rainfall.

A study by Cheung et al. (2009) found that marine fisheries productivity may be affected by changes in ocean conditions resulting from climate change, including changes in food web structure and species distribution. Studies show that marine fish and invertebrates tend to shift their distributions toward higher latitudes and deeper waters in response to climate change. As some ecosystems become less suitable for some species, their relative abundance may change.

Results indicated that the GDP per capita based on purchasing power parity (PPP) would be negatively affected with a decreasing value of 307 611 Indonesian Rupiah (IDR) for every 1% of temperature increase (Table 6.3). Similarly, it would be negatively affected with a decrease of 58 121 IDR for every 1% increase in extreme rainfall. However, average temperature and relative humidity have a significant and positive relationship with the GDP per capita based on PPP. With 1% increases in average temperature and relative humidity, the GDP per capita increases by 246 544 and 552 IDR, respectively. Results show that sunshine duration has a significant and negative relationship with the share of the population below national poverty. The analysis indicated that expenditure would be positively affected by an increase of 101 and 170 IDR (Table 6.3) for every 1% increase of precipitation and relative humidity, respectively. Therefore, the expenditure would be negatively affected (−2326 IDR) as a result of a −1% increase in days without rains.

Both maximum and minimum temperatures have a significant and positive relationship with the share of the population without access to safe water. Results indicated that 1% increases in maximum and minimum temperatures lead to an increase in the population's share without access to safe water by 12 and 3%, respectively. The prevalence of waterborne diseases would increase under high temperatures and no rainfall. Therefore, it decreases under the average temperature. The analysis showed that the ratio of underweight children under five years would be negatively affected and decreases by 3% for every 1% increase of the number of days without rains (Table 6.3). As for extreme rainfall, it has a significant and positive relationship with malnutrition (low birth weight). The model results demonstrate that the ratio of pregnant women with anemia declined by 9 and 8.4% due to minimum temperature and sunshine, respectively. However, it increases by 30.94% as a response to average temperature (Table 6.3).

6.4 Discussion

Some empirical models were assessed to evaluate climate variability on all components of food security. The impact of climate variables differs from one response variable to another, as shown in Table 6.4, where shows the summary of climate variability's impact on food access, food utilization, and food availability.

6.4.1 Food Accessibility

The climate variables influenced GDP/capita, poor family, and expenditure using four different models. Moreover, almost all models indicate the same trend for each response variables corresponding to climate variability. The results revealed that GDP/capita was the most affected by climate variables, followed by expenditure and household poverty. Zero rainfall days and annual rainfall significantly influenced food accessibility. A study by Jury (2001) found that high rainfall and drought significantly influenced GDP/capita in South Africa. Lindsay (1990) stated that GDP in the dry season is 27% lower than a rainy season in South Africa, and Glantz et al. (1997) estimated that the US$60 million in foreign exchange was lost in Zimbabwe during a drought in 1992. As shown in Figure 6.3, zero rainfall days (no rainfall intensity) and annual rainfall in NTT show an increasing trend. This condition implies that NTT is more likely to experience extreme events due to the ENSO phenomenon. Based on Table 6.4, a 1% increase in no rainfall intensity reduced GDP/capita by IDR 34 550/month.

6.4.2 Food Utilization

According to Table 6.4, climate variables affected access to clean water, risk of diarrhea, underweight, malnutrition, and anemia for pregnant women. The result indicated that the risk of diarrhea is greatly affected by climate variability, particularly maximum temperature. There is a positive association between maximum temperature and risk of diarrhea. Research on *E. coli* O157 (i.e. pathogen which caused diarrhea) has shown that there is greater transmission at warmer temperatures and decreased survival of bacteria under conditions of temperature fluctuation (Semenov et al. 2007). In addition, Dowel (2001) stated, "higher temperatures may alter human susceptibility due to host physiology."

6.4.3 Food Availability

Table 6.4 shows the summary of all empirical models for food availability, which consist of 11 variables. The models show that the impact of variables differs from one to the others. The result revealed that rice and livestock were significantly affected by climate variability, followed by paddy and coffee production. In contrast, maize and coconut were mainly not affected by climate variability. Moreover, high temperature, high zero rainfall days, and deficient rainfall have positively impacted production of tubers. Thus, this commodity was a vital food under water-stressed conditions.

6.5 Conclusion

This study explores the impacts of climate variability on food security in Indonesia with reference to NTT province, which has the highest rank of undernutrition, poor access to sanitation, poverty level, infant mortality rate, and lack of infrastructure among the 34 provinces in Indonesia. The study uses longitudinal panel data from 2002 to 2014 targeting 16 districts. Results from four different models confirm the impacts of climate variability were significant on food security. However, except annual rainfall on food availability (yield of food grains) and food accessibility (expenditure) as models show different impacts. The frequency of natural disasters (e.g. storms, floods, landslides, drought, etc.) in the province increased gradually from 2003 to 2014.

Results indicated that despite the increasing amount of annual rainfall, more extreme events were prevalent. Our findings revealed both increase (rice, maize, coffee, and livestock) and decrease (tubers and fish) in food production. It was important to point out that the decrease in tubers and fish production was also attributed to the provincial policy orientation (excluding them from the medium-term development plan) rather than the effect of climate variability alone. Understanding the impact of these types of natural disasters is crucial to ensure that the most appropriate policies and practices are implemented. The FAO emphasizes that disaster aid should better represent the effect of disasters on agriculture. Investments in disaster response and recovery should also build resilience to future shocks through risk reduction and management measures, particularly in countries experiencing climate extremes and where agriculture is a crucial source of livelihoods, food, nutrition security, and a key driver of the economy.

Acknowledgments

The authors wish to acknowledge the Indonesia Meteorological Agency (BMKG), NTT Statistical Agency (BPS-NTT), National Agency for Disaster Management of Indonesia (BNPB), and National Oceanic and Atmospheric Administration (NOAA) for providing the data.

Author Contribution

Boubacar Siddighi Balde conducted data analysis and writing the draft manuscript. Martiwi Diah Setiawati conducted meteorological data collection, data analysis, and writing the manuscript. Boubacar Siddighi Balde, Martiwi Diah Setiawati, and Mohammad Esham reviewed and edited the manuscript. Nampa Wayan collected statistical data of NTT province from four dimension. Boubacar Siddighi Balde and Martiwi Diah Setiawati have equal contribution.

References

ADB (2012). Food security and poverty in Asia and the pacific: key challenge and policy issues. Mandaluyong. Philiphines: s.n. https://www.adb.org/publications/food-security-and-poverty-asia-and-pacific-key-challenges-and-policy-issues (Accessed 4 December 2016).

As-syakur, A.R., Tanaka, T., Osawa, T., and Mahendra, M.S. (2013). Indonesian rainfall variability observation using TRMM multi-satellite data. *International Journal of Remote Sensing* 34 (21): 7723–7738. https://doi.org/10.1080/01431161.2013.826837.

Badolo, F. and Kinda, S.R. (2015). Climatic variability and food security in developing countries. https://halshs.archives-ouvertes.fr/halshs-00939247v2/document (Accessed 20 May 2021).

Belasco, E., Cheng, Y., and Schroeder, T. (2015). The impact of extreme weather on cattle feeding profits. *Journal of Agricultural and Resource Economics* 40 (2): 285–305.

BKP and WFP (2015). Food security and vulnerability atlas of Indonesia 2015. https://documents.wfp.org/stellent/groups/public/documents/ena/wfp285131.pdf?_ga=2.52783976.1786513060.1621582991-444357105.1621582991 (Accessed 1 December 2016).

BMKG (2016). Online data base of Indonesia Meteorological Agency. http://dataonline.bmkg.go.id/webfaq (Accessed 1 December 2016).

BNPB (2017). Statistic of natural disaster in Indonesia. http://dibi.bnpb.go.id (Accessed 2 December 2017).

Boer, R. and Kartikasari, K. (2014). Climate Change Impact on Food Security in Southeast Asia in RSIS Centre for Non-Traditional Security (NTS) Studies, 'Climate change impact on food security in Southeast Asia' (RSIS Special Policy Report). Singapore: S. Rajaratnam School of International Studies [RSIS]. https://www.zef.de/fileadmin/user_upload/1jlassa_download_RSIS_ERIA.pdf (Accessed 20 January 2017).

BPS (2011). Nusa Tenggara Timur dalam angka [East Nusa Tenggara in figures]. Kupang, East Nusa Tenggara, Indonesia. https://ntt.bps.go.id/publication.html (Accessed 1 December 2016).

BPS (2012). Kota Kupang dalam angka [Kupang in figures]. Kupang, NTT, Indonesia. https://ntt.bps.go.id/publication.html (Accessed 1 December 2016).

BPS (2016). NTT in Figures 2015, Kupang: NTT Statistical agency. https://ntt.bps.go.id/publication.html (Accessed 1 December 2016).

Case, M., Ardiansyah, F., and Spector, E. (2007). Climate change in Indonesia. Implications for Humans and Nature (Gland, Switzerland: World Wide Fund for Nature (WWF)). http://assets.wwf.org.uk/downloads/indonesian_climate_ch.pdf (Accessed 21 May 2021).

Cheung, W.W.L., Lam, V.W.Y., Sarmiento, J.L. et al. (2009). Large-scale redistribution of maximum fisheries catch potential in the global ocean under climate change. *Global Change Biology.* 16 (1): 24–35. https://doi.org/10.1111/j.1365-2486.2009.01995.x.

Chilonda, P. and Otte, J.(2006). Indicators to monitor trends in livestock production at national, regional and international levels. LivestockResearch for Rural Development, 18(8): 117.

Christensen, J., Hewitson, B., Busuioc, A. et al. (2007). Regional climate projections. In: *Climate Change 2007: The Physical Science Basis.* Contribution of Working Group 1 to the Fourth Assessment Report of the Intergovernmental Panel on Climate Change (eds. S. Solomon, D. Qin, M. Manning, et al.), 848–926. Cambridge, United Kingdom and New York, NY: Cambridge University Press; IPCC.

Dell, M., Jones, B. F., & Olken, B. A. (2008). Climate change andeconomic growth: Evidence from the last half century (No. w14132). National Bureau of Economic Research.

Devendra, C. (2012). *Climate Change Threats and Effects: Challenges for Agriculture and Food Security.* Malaysia: ASM Series on Climate Change Academy of Sciences. https://www.agrilinks.org/sites/default/files/resource/files/climate_change_threats_effects_strategies.pdf (Accessed 20 May 2021).

Djennas, M. (2016). Business cycle volatility, growth and financial openness: Does Islamic finance make any difference?. Borsa Istanbul Review, 16(3):121–145.

Dowel, S. (2001). Seasonal variation in host susceptibility and cycles of certain infectious diseases. *Emerging Infectious Diseases.* 7: 369–374.

Fain, S. J., Quiñones, M., Álvarez-Berríos, N. L. et al. (2018). Climate changeand coffee: assessing vulnerability by modeling future climate suitability inthe Caribbean island of Puerto Rico. ClimaticChange, 146(1): 175–186.

Glantz, M., Betsill, M., and Crandall, K. (1997). Food security in Southern Africa: assessing the use and value of ENSO information. Technical Report. NCAR NOAA. https://agris.fao.org/agris-search/search.do?recordID=XF2015043526 (Accessed 12 May 2017).

Guisan, A., Edwards, T., and Hastie, T. (2002). Generalized linear and generalized additive models in studies of species distributions: setting the scene. *Ecological Modelling* 157: 89–100.

Handoko, Sugiarto, Y., and Yusman, S. (2008). *Keterkaitan perubahan iklim dan produksi pangan strategis: telaah kebijakan independen dalam bidang perdagangan dan pembangunan*. Bogor: SEAMEO-BIOTROP and Kemitraan. https://opac.perpusnas.go.id/DetailOpac.aspx?id=609631 (Accessed 1 December 2016).

Hect, J.E. (2016). *Indonesia Cost of Climate Change 2050 Policy Brief*. Washington: United stated Agency for International Development. https://www.climatelinks.org/resources/indonesia-costs-climate-change-2050-policy-brief (Accessed 20 May 2017).

Hidayat, P. (1999). *ENSO Impact on Indonesia Seasonal Rainfall*. Hanoi: Asian Pacific FRIEND and GAME Joint workshop on ENSO. https://www.researchgate.net/publication/286239271_ENSO_IMPACT_ON_INDONESIA_SEASONAL_RAINFALL_1 (Accessed 1 December 2016).

Hijioka, Y., Lin, E., Pereira, J.J. et al. (2014). Asia. In: *Climate Change 2014: Impacts, Adaptation, and Vulnerability. Part B: Regional Aspects*. Contribution of Working Group II to the Fifth Assessment Report of the Intergovernmental Panel on Climate Change (eds. V.R. Barros, C.B. Field, D.J. Dokken, et al.), 1327–1370. Cambridge, United Kingdom; New York, NY: Cambridge University Press.

Hoffmann, U. (2011). *Assuring Food Security in Developing Countries under the Challenges of Climate Change: Key Trade and Development Issues of a Fundamental Transformation of Agriculture*. Geneva: UNCTAD.

IPCC (2007). Appendix 1 glossary. In: *Climate Change 2007: Impacts, Adaptation and Vulnerability. Contribution of Working Group II to the Fourth Assessment Report of IPCC* (eds. M.L. Parry, O.F. Canziani, J.P. Palutikof, et al.). Cambridge, UK: Cambridge University Press https://www.ipcc.ch/site/assets/uploads/2018/02/ar4-wg2-app-1.pdf.

Israel, D.C. and Briones, R.M. (2013). Impacts of natural disasters on agriculture, food security, and natural resources and environmental in the Phillipines. ERIA Discussion paper series.

Iwasaki, H. (2014). Increasing trends in heavy rain during the warm season in eastern Japan and its relation to moisture variation and topographic convergence. *International Journal of Climatology*. 35: 2154–2163.

Jury, M.R. (2001). Economic impact of climate variability in South Africa and development of resource prediction models. *Journal of applied meteorology*. 41: 46–55.

Kawanishi, M. (2014). Assessment of climate and socio-economic impacts and responses for rice production and distribution in Indonesia.

Kinda, S.R. (2016). Climatic shocks and food security: the role of foreign aid. https://hal.archives-ouvertes.fr/hal-01260846/document (Accessed 20 May 2021).

Krishnamurthy, P.K., Lewis, K., and Choularton, R.J. (2012). Climate impacts on food security and nutrition: a review of existing knowledge. United Nations World Food Programme 2012. https://documents.wfp.org/stellent/groups/public/documents/communications/wfp258981.pdf (Accessed 20 May 2021).

Lee, J., Nadolnyak, D., and Hartarska, V. (2012). Impact of climate change on agricultural production in Asian countries: evidence from panel study. *2012 Annual Meeting* (4–7 February 2012), Birmingham, Alabama 119808. Southern Agricultural Economics Association. https://core.ac.uk/download/pdf/6718824.pdf (Accessed 21 May 2021).

Lindsay, C. (1990). *Economical and Socio-Political Impact of Rainfall Variation*. Johannesburg: University of Witwatersrand.

Lloyd, S., Kovats, R., and Chalabi, Z. (2011). Climate change, crop yield, and undernutrition: development of a model to quantify the impact of climate scenarios on child undernutrition. *Environmental Health Perspective* 119 (12): 1817–1823.

Masutomi, Y., Takahashi, K., Harasawa, H., and Yuzuru, M. (2009). Impact assessment of climate change on rice production in Asia in comprehensive consideration of process/parameter uncertainty in general circulation models. *Agriculture, Ecosystem and Environment* 131: 281–291.

Masutomi, Y., Arakawa, M., Minoda, T. et al. (2015). Critical air temperature and sensitivity of the incidence of chalky rice kernels for the rice cultivar "Sai-no-kagayaki". *Agricultural and Forest Meteorology* 203: 11–16. https://doi.org/10.1016/j.agrformet.2014.11.016.

McPhaden, M., Zebiak, S., and Glantz, M. (2006). ENSO as an integrating concept in earth science. *Science* 314: 1740–1745.

Measey, M. (2010). Indonesia: a vulnerable country in the face of climate change. *Global Majority E-Journal* 1 (1): 31–45.

NTT Regional Government (2010). *Food Security Map of NTT Provice*. Kupang: NTT Regional Government, WFP. http://irgsc.org/files/FSVA-NTT15Aug11.pdf (Accessed 7 January 2017).

NTT Regional Government (2013). *Rencana Pembangunan Jangka Menengah Daerah (RPJMD) 2013-2018*. Kupang: NTT Regional Governmnet.

Philipsborn, R., Ahmed, S., Brosi, B., and Levy, K. (2016). Climatic drivers of diarrheagenic *Escherichia coli* Incjdence: a systematic review and meta-analysis. *Journal of Infectious Diseases.* 214: 6–15.

Porter, J.R. and Semenov, M.A. (2005). Crop responses to climatic variation. *Philosophical Transactions of the Royal Society B* 360: 2021–2035.

Riskesdas (2013). *Riset Kesehatan Dasar*. Jakarta: Indonesia Health Minister. http://labdata.litbang.kemkes.go.id/images/download/laporan/RKD/2013/Laporan_riskesdas_2013_final.pdf (Accessed 1 December 2016).

Salim, Z. (2010). *Food Security Policies in Maritime Southeast Asia: The Case of Indonesia*, Series on Trade and the Food Security-Policy Report 1. International Institute for Sustainable Development: Winnipeg, Manitoba, Canada.

Schmidhuber, J. and Tubiello, F.N. (2007). Global food security under climate change. *Proceedings of the National Academy of Sciences of the United States of America.* 104 (50): 19703–19708. https://doi.org/10.1073/pnas.0701976104.

Semenov, A.V., van Bruggen, A.H.C., van Overbeek, L. et al. (2007). Influence of temperature fluctuations on *Escherichia coli* O157:H7 and *Salmonella enterica* serovar Typhimurium in cow manure. *FEMS Microbiology Ecology* 60 (3): 419–428. https://doi.org/10.1111/j.1574-6941.2007.00306.x.

Setiawati, M.D., Sambah, A.B., Miura, F. et al. (2015). Characterization of bigeye tuna habitat in the Southern Waters off Java–Bali using remote sensing data. *Advances in Space Research* 55 (2): 732–746. https://doi.org/10.1016/j.asr.2014.10.007.

Singh, A.K. and Sharma, P. (2018). Implications of climatic and non-climatic variables on food security in developing economies: a conceptual review. *MOJ Food Process Technology* 6 (1): 00138. https://doi.org/10.15406/mojfpt.2017.06.00138.

Torres-Reyna, O. (2007). Linear regression using Stata. Data andstatistical services, Princeton University. Retrieved from http://dss.princeton. edu/training/, accessed on, 28(07), 2015.

Thornton, P.K., Ericksen, P.J., Herrero, M., and Challinor, A.J. (2014). Climate variability and vulnerability to climate change: a review. *Global Change Biology* 20 (11): 3313–3328. https://doi.org/10.1111/gcb.12581.

Ulusoy, V. and Demiralay, S. (2017). Energy demand and stock market development in OECD countries: a panel data analysis. *Renewable and Sustainable Energy Reviews* 71: 141–149.

UNDP (2013). Strategic planning and action to strengthen climate resilience of rural communities in Nusa Tenggara Timur (SPARC) Project.

Van Elsas, J. D., Semenov, A. V., Costa, R., & Trevors, J. T. (2011).Survival of Escherichia coli in the environment: fundamental and public healthaspects. The ISME Journal, 5(2): 173–183.

Verbeek, M. (2008). *A Guide to Modern Econometrics*. Wiley.

Ward, P.J., Jongman, B., Kummu, M. et al. (2014). Strong influence of El Niño southern oscillation on flood risk around the world. *Proceedings of the National Academy of Sciences* 111 (44): 15659–15664. https://doi.org/10.1073/pnas.1409822111.

Watanabe, S., Komori, D., Aoki, M. et al. (2011). Estimation of daily solar radiation from sunshine duration in Thailand. *Journal of Meteorology Society of Japan* 89A: 355–364.

Weldearegay, S.K. and Tedla, D.G. (2018). Impact of climate variability on household food availability in Tigray, Ethiopia. *Agriculture & Food Security* 7 (6): 1–9. https://doi.org/10.1186/s40066-017-0154-0.

WFP, 2016. The Impact of Drought on Households in Four Provinces in Eastern Indonesia, Jakarta: World Food Programe. https://www.wfp.org/publications/indonesia-impact-drought-households-four-provinces-eastern-indonesia-february-201 (Accessed 7 January 2017).

WHO (2000). Classification of severity of malnutrition in Community for childern under 5 years of age from the management of nutrition in Major emergencies. Geneva: s.n.

Wood, S. (2006). *Generalized Additive Models, an Introduction with R*. London: Chapman Hall.

7

Knowledge-Intensive Livestock Resource Management in a Changing Environment

Avijit Haldar[1], Indranil Samanta[2], and Amlan Kumar Patra[3]

[1] *Department of Animal Reproduction, ICAR-Agricultural Technology Application Research Institute (ATARI), Kolkata, West Bengal, India*
[2] *Department of Veterinary Microbiology, West Bengal University of Animal and Fishery Sciences, Kolkata, West Bengal, India*
[3] *Department of Animal Nutrition, West Bengal University of Animal and Fishery Sciences, Kolkata, West Bengal, India*

CONTENTS

Sustainable Agriculture Systems and Technologies, First Edition. Edited by Pavan Kumar, A.K. Pandey, Susheel Kumar Singh, S.S. Singh, and V.K. Singh.

7.1 Introduction

The climate change poses unprecedented challenges to the food industry, creating the major hurdle in formation of the hunger-free world due to the sensitivity of agriculture-livestock production system to changing climate conditions. The climate change issue became relevant since 1980 with the formation of the Intergovernmental Panel on Climate Change (IPCC) by the United Nation General Assembly (Resolution 43/53, 1988). The global scientific community makes a consensus that some climate change is already occurring and that further change is unavoidable (IPCC 2013; Quere et al. 2014). According to World Meteorological Organization (WMO 2014), continued greenhouse gas (GHGs) emissions at or above current rates will lead to further increase in global warming in this century. Carbon dioxide (CO_2), methane (CH_4), and nitrous oxide (N_2O) are the three major GHGs. CH_4 also has dangerous effect on high atmosphere ozone formation. The IPCC in fifth assessment report identified 0.3–4.8 °C increase in global average surface temperature by 2100 (IPCC 2014). Most climatologists agree that the rise in GHGs in the atmosphere results an increase in air temperature, as well as an alteration in rainfall pattern with increased frequency and severity of extreme weather events, such as frosts, heat waves, droughts, and floods that pose a threat to agriculture-livestock food production system worldwide (IPCC 2001; Thompson 2010). For example, Bangladesh is one of the disaster-prone areas and worst sufferers for global warming and climate change due to rising sea level (Chen et al. 2012). According to Global Climate Risk Index, 2018 report, Bangladesh is ranked eighth as a high-risk country for suffering extreme weather events (Eckstein et al. 2018).

As there is a rising trend of human population growth, as well as improvement in global standard of living, especially in the tropical and subtropical region of the globe, the demand for agricultural products will increase by about 70% by the year 2050 (FAO 2009). The rising demand is dominated by animal origin food due to urbanization and coming up of affluent "middle class" consumers who preferably move to animal protein-based diet from the grain-based diet (Rae and Nayga 2010). Livestock origin products are becoming important agricultural commodity in the global foods security arena because they provide 17% of global kilocalorie consumption and 33% of global protein consumption (Rosegrant et al. 2009). Thus, the animal origin food demand is expected to rise by 62% for milk and by 77% for meat by the year 2050 as compared with 2005–2007 (Alexandratos and Bruinsma 2012). The global farm animal production is expected to be doubled by 2050 from the present levels.

The livelihoods of one billion of the poorest population in the world depend on livestock sector (Hurst et al. 2005). At least 90% of them live in Asia and sub-Saharan Africa. They are small farm households, working on land plots smaller than 2 ha (Lowder et al. 2014). Smallholder farmers constitute the largest proportion of the 570 million farms worldwide (Lowder et al. 2016). India is a land of small farmers, constituting about 86.21% of the country's farmers (Agriculture Census 2015–2016). In China, nearly 98% of farmers cultivate farms smaller than 2 ha. In Ethiopia and Egypt, farms smaller than 2 ha constitute nearly 90% of the total number of farms. The half of the farmer population is small in Mexico (Rapsomanikis 2015). Smallholder farmers produce and supply a substantial proportion of food to the world's population. Climate change is a real threat to such 570 million smallholder farmers who depend on livestock for their livelihoods.

The livestock sector shares 14.5% of global GHG emissions (Gerber et al. 2013). The major proportion of total agricultural emission of methane comes from ruminant livestock such as cattle, buffalo, sheep, and goats (Steinfeld et al. 2006). Direct and indirect sources of GHG emissions in animal production systems include physiological processes of the animal (enteric fermentation and respiration), rearing systems (intensive or extensive), animal feeds, manure storage, manure management (compost and anaerobic treatment), use of chemical fertilizers in grazing land or fodder cultivation (Casey et al. 2006). At the same time, climate change has serious effect on livestock. There are direct effects of air temperature, humidity, wind speed, and other climate factors on animal growth, milk production, wool production, and reproduction (Hounghton et al. 2001). Climate change has indirect effects on the quantity and quality of feedstuffs such as pasture, forage, grain, and the severity and distribution of livestock diseases and parasites (Seo and Mendelsohn 2006). However, the available information on effect of climate change on livestock is still fragmented (Herrero and Thornton 2013). Collecting and synthesizing livestock data is always a challenge because of different production systems, varied agro-ecological zones, and heterogeneous production objectives (Pica-Ciamarra et al. 2014; Robinson et al. 2014). Besides, there are a variety of practices across production systems depending on cultural, socioeconomic, and institutional conditions (Steinfeld et al. 2006; Thornton et al. 2009). Thus, generating research evidence of climate change impacts on livestock is an important implication for the development of the sector and the people depending on it.

The challenge is to maintain a balance between livestock productivity and environmental preservation (Wright et al. 2012). Responding to the challenge of climate change requires designing of suitable adaptation and mitigation measures for the livestock sector (Gerber et al. 2013). Reducing the emissions of GHGs from livestock production system is therefore a top priority, because it can curb warming fairly rapidly. There are many adaptation and mitigation options, which can reduce climatic impacts on livestock, but the rational use of such technologies is crucial for the survival and profitability of the livestock enterprise. Therefore, the objectives are to (i) realize the responsibility of livestock sector to climate change; (ii) illustrate the influences of climate change on livestock production; (iii) synthesize adaptation response and mitigation strategies to combat climate change effects on livestock; and (iv) make the stakeholders aware about the effects of climate change on livestock and mitigation measures.

7.2 Sources GHGs from Livestock Sector

7.2.1 Different Livestock Activities

Worldwide food production systems (livestock and vegetable-origin foods) contribute 18 Gt GHG (CO_2, CH_4, N_2O, and fluorinated gases) emissions in CO_2 equivalent (CO_2e) (non-CO_2 gases are expressed as CO_2e based on the warming potential of the gases) accounting one-third (34%) of total global GHG based on detailed life cycle assessment (LCA) analysis

(Crippa et al. 2021). Different livestock activities such as livestock rearing, feed production, land use and land use change (LULUC), manure management, transport, slaughtering, processing, and storage contribute significantly to the total anthropogenic GHG emissions and are considered an important driver of global climatic change in the food-system emissions. Fossil fuel burning and LULUC result in most of the CO_2 emission with the greatest share of global GHG emissions. But non-CO_2 GHG such as CH_4 and N_2O also add significantly to global anthropogenic GHG emissions, which predominantly arise due to livestock production. From the livestock sector, total non-CO_2 GHG emissions were estimated to be 2.45 Gt CO_2e in 2000 (Herrero et al. 2013). It has been estimated that GHG emissions from livestock share about 70% of total emissions from the agriculture, forestry, and other land use (Tubiello et al. 2013). In livestock production, direct CH_4 emissions from enteric fermentation and manure, and N_2O emission during the process of nitrification and denitrification of the manure nitrogen comprise for about 9% of total GHG emissions (IPCC 2007). Along with GHG emissions from enteric and manure sources, emissions occurring in the livestock products supply chains such as transport and feed production account for half of food-origin GHG emissions (Jan Kramer et al. 1999; European Commission 2006) and 18% of global anthropogenic GHG emissions (Steinfeld et al. 2006). Considering emissions from animals, manure, feed production, and land use change, livestock production is responsible for 14.5% of total anthropogenic GHG emissions globally (Gerber et al. 2013). Increasing demand for livestock products could further increase enteric CH_4 production by 31% between 1990 and 2030 and manure N_2O emission by 20% (EPA 2011). In LCA analysis, the major GHG emissions occur at the farm stage (e.g. rearing of animals, feeds, grazing, and manure management), and subsequent activities such as slaughtering, processing, supply chain activities, retailing, and transport play minor roles in the GHG shares (Berlin 2002).

Total direct non-CO_2 GHG emissions of enteric and manure sources globally increased from 1.77 Gt CO_2e in 1961 to 2.77 Gt CO_2e in 2010 at an annual growth rate of 0.92% estimated following IPCC (2006) tier I methodology (Patra 2014a). This increased rate of GHG emissions has mostly arisen from the developing countries (Caro et al. 2014). As per FAO estimated by Steinfeld et al. (2006), worldwide GHG emissions was 2.32 Gt CO_2e from enteric fermentation and manure management in 2004. In another estimate, livestock globally contributed 3.1 Gt CO_2e in 2010 from enteric fermentation and manure (Caro et al. 2014). In India, total non-CO_2 GHG emissions was 0.33 Gt CO_2e from enteric fermentation and manure in 2007 estimated following IPCC (2006) tier II methodology (Patra 2012a).

7.2.2 Different Livestock Species

Major contributions to GHG emissions were resulted from cattle (79.3%), followed by swine (9.57%) and sheep (7.40%) production globally, and from cattle (68.3%), buffalo (13.7%), and goat (5.4%) production for developing countries (Patra 2014a). In other estimate, beef cattle production contributed the most (54%) of total livestock GHG, subsequently dairy cattle, sheep, buffaloes, pigs, and goats production (17, 9, 7, 5, and 4%, respectively) (Caro et al. 2014). In India, the contributions of GHG were 55, 37, 4, 2, and 1% for cattle, buffalo, sheep, goat, and pig, respectively (Patra 2017). Global GHG emissions are predicted to become 3.52 Gt CO_2e by 2050 due to increased animal populations driven by greater meat and dairy product demands (Patra 2014a). In another study, total cattle population accounted for 77% of GHG emissions (Herrero et al. 2013). The monogastric livestock contributed about 10% of total livestock emissions, which was mostly due to CH_4 production from manure (56% of total emissions by monogastric animals) (Herrero et al. 2013). The developing countries shared 75% of world GHG emissions from ruminants and 56% of emissions from monogastric animals (Herrero et al. 2013). Mixed crop–livestock systems from ruminants contributed greatest to the total emissions (61%), whereas grazing systems produced 12% of emissions, and urban and other systems contributed the rest (Herrero et al. 2013). Estimates of GHG emissions differ largely based on the number of livestock activities included in the LCA methodology. In sub-Saharan Africa, GHG emission intensities are very high due to lower animal productivity in large arid land areas, scarcity of feeds along with low nutritive values, and low genetic potential animals (Herrero et al. 2013). Ruminants in sub-Saharan Africa are mostly raised for meat, which is associated with lower feed conversion efficiency and higher emission intensities in comparison with milk (Herrero et al. 2013).

7.2.3 Livestock Feeds

In livestock, GHG emissions are mainly contributed by CH_4 and N_2O emissions. Direct CO_2 emission arising primarily from field machinery use, processing, and transport is less important for ruminants although their contribution is greater for intensive poultry and pig production (Garnett 2009). Northern European studies suggested that CH_4 and N_2O emissions predominated, while direct CO_2 emission was of less significant in livestock production (Schils et al. 2005; Olesen et al. 2006). However, the indirect contribution of CO_2 is substantial, which is mainly associated with feed production or feeding of animals (Garnett 2009). A broader LCA approach in livestock production system needs to quantify GHG impacts associated with fodders, pasture, and concentrates considering the GHG emissions attributable to feed production activities including fertilizer production, machinery, soil N_2O emission, CO_2 releasing impacts of pasture, and fodder crop production or LULUC (Garnett 2009; Kløverpris et al. 2008; Searchinger et al. 2008). But LULUC impacts in LCA methodology of livestock GHG have not been attempted adequately, especially in the developing countries. The land for livestock rearing is directly or indirectly used for mainly pasture, fodder crops, cereals, and oilseeds.

Cereals are a major part in the diets of poultry, pigs, dairy cows, and intensive beef cattle systems. Globally, one-third (FAO 2002) or more (37%) of total cereal production (WRI 2004) is used for livestock feeding. It has been argued that GHG emissions per unit of food would be more efficient if cereals are consumed directly by humans because a significant fraction of nutrient contents in cereals is lost during conversion from cereals to animal products and GHG emissions arising from plant-based foods are lower (Gerbens-Leenes and Nonhebel 2002). For instance, beef is associated with GHG emission of about 16 kg CO_2e/kg versus 0.8 kg CO_2e/kg for wheat (Williams et al. 2006). Among livestock product types, LCA finds that GHG burden for poultry and pigs is lower as they efficiently convert plant nutrients into animal products and thus consuming chickens and pork in preference to ruminant meat would be more environment efficient (McMichael et al. 2007). In contrast, it can be argued that the diets of pigs and poultry are mainly based on cereals and oilseed meal compared to the diets of ruminants. Considering the feed conversion efficiency of cereal grains and soybean to animal products and CO_2 emissions arising from LULUC for extra production of these feeds (pigs and poultry consume around 60% of soybean meal in the European Union), substitution of white meat for red may not be greatly environment friendly (Garnett 2009). Another major component of diets is oilseed cakes or meals, which are main protein ingredients with high-quality protein, and they account for about two thirds of the economic value of the crops (FAO 2008). Livestock GHG LCA analysis sometimes includes the emissions arising from oilseed production (for example soybean meal) and its associated inputs (Williams et al. 2006). However, LCA analysis of livestock GHG emissions does not usually include LULUC arising from oilseed production and the associated CO_2 release of stored soil carbon (Garnett 2009). Animals are also fed several byproducts from other agricultural sectors such as cereal and legume straws, molasses, brewers grains, vegetable and fruit residues, and bran and husks. In contrast to cereals and oilseeds, the use of these wastes in animal diets would be resource efficient because these otherwise wastes are converted to meat and milk for human consumption, which leads to lower net GHG emissions.

A significant portion of ruminant production system depends upon grazing pasture lands. Livestock production in natural grasslands gives economic value to grasslands acting as carbon sinks (Garnett 2009). Any land use changes that disturb the soil such as tillage practice and urbanization cause releases of stored soil carbon into the atmosphere. The carbon sequestration outweighs the CH_4 and N_2O emissions by cattle reared in unfertilized grasslands without the use of feed inputs or additional fertilizer (Allard et al. 2007). However, many grasslands are fertilized with nitrogen applications leading to N_2O and CO_2 emissions (Garnett 2009). Moreover, when grassland is overgrazed by livestock, the vegetative loss and soil degradation can result in soil carbon losses and CO_2 release (Zhou et al. 2020). An estimate of about 20% of land globally and up to 73% in drylands has been degraded, and overgrazing is a major problem in the developing world (Steinfeld et al. 2006).

7.2.4 Enteric Fermentation

In livestock, CH_4 production from enteric fermentations, especially from the rumen of ruminants predominantly, contributes to the GHG emissions. Methanogenic archaea inhabit in the rumen and the lower digestive tract of animals, where they reduce carbon dioxide, formic acid, or methylamines to produce CH_4 using hydrogen derived from feed fermentation (Patra et al. 2017). Out of 2.45 Gt CO_2e of non-CO_2 GHG produced from different livestock activities, enteric CH_4 from ruminants was the greatest source contributing 1.6 Gt CO_2e (Herrero et al. 2013). Of the total GHG emissions from enteric fermentation and manure of livestock, enteric CH_4 emission was the largest source accounting 85.6% of total emissions from world livestock in 2010 (Patra 2014a). Global enteric CH_4 emission was estimated to increase by 54.3% from 1961 to 2010 (61.5 to 94.9×10^9 kg annually) with the greatest annual growth rate for goat (2.0%), followed by buffalo and swine (1.57 and 1.53%, respectively) (Patra 2014a). Global enteric CH_4 is predicted to become 120×10^9 kg per year in 2050 (Patra 2014a) due to increased ruminant populations unless proper mitigation strategies are adopted.

7.2.5 Livestock Manure

Livestock manure comprising of feces and urine (excreta) is responsible for CH_4 and N_2O (both directly and indirectly) emissions. Emission of CH_4 from manure results from anaerobic and facultative bacterial fermentation of organic matter present in manure (Bouwman 1996). The CH_4 emission potential of manure is influenced by the composition of the manure, ambient temperature, and management conditions. Manure management produces direct N_2O via both nitrification (oxidation of ammonium to nitrate by aerobic microbes) and denitrification (reduction of nitrate to nitrogen gas by anaerobic microbes) of nitrogen present in animal wastes (Barton and Atwater 2002; Miller et al. 2009). Direct N_2O emission depends upon total amount of N excretion by livestock and manure management conditions. Indirect N_2O emission occurs through volatilization of nitrogen mainly as ammonia and NOx. Nitrogen loss starts at the time of excretion and continues in storage management and treatment systems (Oenema and Tamminga 2005). Indirect N_2O emission also results from leaching/runoff of nitrogen applied in managed soils (Meyer et al. 2002).

Similar to managed manure, direct and indirect N_2O emission in soil occurs from manure deposited by grazing ruminants on pasture and paddock (Bateman and Baggs 2005). The GHG emissions owing to manure applied to soil are accounted for the fertilized pasture and crops, not directly from livestock. FAO estimate suggests that manure contributes to about 5% of total anthropogenic GHG emissions, mostly N_2O (Steinfeld et al. 2006). Global CH_4 emission from manure increased by 67.6% (6.81×10^9 kg in 1961 to 11.4×10^9 kg in 2010) and is expected to further increase to 15×10^9 kg by 2050 (Patra 2014a). In India, enteric CH_4 production was the greatest source of GHG amounting 89.7% of total GHG, but manure CH_4 (9.2%) and N_2O (1.2%) contributed to a less extent because feces from cattle and buffaloes was used for fuel after drying them (Patra 2017). In other estimate, CH_4 and N_2O emissions from manure were 0.25 and 0.21 Gt CO_2e, and N_2O from manure applied to soil was 0.49 Gt CO_2e (Herrero et al. 2013).

Livestock manure contributes vital nutrients, especially nitrogen and phosphorus to the soil improving soil quality and productivity, reducing the use of more GHG-producing mineral fertilizer (Fließbach et al. 2007; Leip et al. 2019). Manure fertilizer applied to soil is more biologically active than mineral fertilizers and can also increase soil carbon storage potential (Fließbach et al. 2007). Globally, around 38% of total phosphate and 22% of total nitrogen applied in soil come from animal origin as pointed out by Garnett (2009). Therefore, GHG emissions from manure should be carefully allocated to livestock contribution (Leip et al. 2019).

7.2.6 Livestock Products

Expression of GHG emissions (usually in CO_2e) per unit of livestock products or nutrients (i.e. GHG emission intensity or carbon footprint) is an important indicator to compare different products for their environmental impacts. Globally, mean non-CO_2 emission intensity for all livestock products was 41 kg CO_2e/kg edible animal protein

(Herrero et al. 2013), which had great variability in different production systems, regions, and type of products. Moderate emission intensities are estimated throughout the developing world, in Amazonian regions (beef cattle production), arid regions, and South Asia (Herrero et al. 2013). In most of the developed countries, carbon footprints of livestock products are low due to improved and intensive feeding practices and inherently higher feed quality in temperate conditions (Herrero et al. 2013). The carbon footprint of meat and eggs from monogastrics is substantially low than milk and meat from ruminants (Patra 2017). Global GHG intensity was estimated to be 24 kg CO_2e/kg protein for pork production and 3.7 kg CO_2e/kg edible protein for poultry meat and eggs (Herrero et al. 2013). The emission intensities of chicken, pork, and beef were 3.7–6.9, 3.9–10, and 14–32 kg CO_2e/kg, respectively (de Vries and de Boer 2010).

In India based on partial LCA analysis, the average carbon footprint values of fresh milk production were highest for indigenous cattle (2.96 kg CO_2e/kg) and lowest for crossbred cows (1.21 kg CO_2e/kg), followed by buffaloes (1.85 kg CO_2e/kg) and goats (2.54 kg CO_2e/kg) with wider variations among the states (Patra 2017). Carbon footprint for milk protein was also lesser for crossbred cattle (34.7 kg CO_2e/kg), followed by buffalo (45.8 kg CO_2e/kg), and higher for indigenous cattle (85.1 kg CO_2e/kg) and goat (76.4 kg CO_2e/kg). However, carbon footprints for milk energy output were similar for buffaloes and crossbred cattle (0.42 and 0.41 kg CO_2e/MJ), but were greater for goats (0.83 kg CO_2e/MJ) and indigenous cattle (1.0 kg CO_2e/MJ) (Patra 2017). In New Zealand and Sweden, the average carbon footprint values at the farm gate were 1.00 and 1.16 kg CO_2e/kg milk, respectively (Flysj€o et al. 2011).

The carbon footprints for animal protein ranged between 24 and 38 kg CO_2e/kg for milk, 21 and 53 kg CO_2e/kg for pork, 18 and 36 kg CO_2e/kg for chicken, 30 and 38 kg CO_2e/kg for eggs, and 75 and 170 kg CO_2e/kg for beef (Flysj€o et al. 2011). Also, Williams et al. (2006) analyzed that carbon footprint for protein was lower for chicken (30–36 kg CO_2e) than for pork (47–49 kg CO_2e) or egg (32–38 kg CO_2e) and was comparable to milk (28–31 kg CO_2e). In India, average carbon footprint per unit of livestock-derived food energy was lower for duck eggs (0.12 kg CO_2e/MJ) and hen eggs (0.31 kg CO_2e/MJ) and highest for sheep (3.08 kg CO_2e/MJ) production (Patra 2017). Average emission intensities for protein was also lower for duck eggs (6.8 kg CO_2e/kg), followed by chicken meat (14.7 kg CO_2e/kg) and hen eggs (17.8 kg CO_2e/kg) and highest in ruminants animals (45.5 kg CO_2e/kg for goats and 90.7 kg CO_2e/kg for cattle) (Patra 2017). The lower emission intensities for chicken and pigs are attributable to the industrial pig and poultry production, which uses high-quality balanced concentrate diets and high genetic potential animals (Herrero et al. 2013). In India, GHG intensity for protein or energy from the duck and chickens was considerably lower because they are mostly reared in backyard farming with provision of less supplementary feeds, mostly of byproducts feeds. Although livestock products have higher emission intensities compared with other vegetable food groups, livestock products are of high quality and serve great nutritional security in human diets, especially for the developing countries, which needs to be judged appropriately when comparing the GHG intensity of non-livestock-derived foods (Herrero et al. 2013; Adesogan et al. 2020).

7.3 Effect of Climate Change on Livestock Production System

Climate change has multidimensional direct and indirect negative effects on livestock production farming system. Collecting and synthesizing research evidence of climate change impacts on livestock is an important implication for the development of the sector.

7.3.1 Nutritional Stress

The climate-change-associated nutritional stress in livestock occurs due to lack of access to the pastures and quality feeds and fodders (Wheeler and Von Braun 2013). Elevated carbon dioxide, fluctuation in temperature, and depletion of ozone layer are the major climate change related issues, which affect the quality of fodders. Elevated temperature altered rainfall pattern and the scarcity of water increased evatransporation of the plants (grasses) and consequently changed the productivity and composition of grassland, the major pasture covering almost 70% of the agricultural land throughout the world (Soussana and Lüscher 2007). In contrary, elevated CO_2 increased the photosynthesis and water use efficiency of the plants, especially in the water-limited systems, which increased the biomass of the forage with elevated growth rate, especially in C4 (tropical) plants than the C3 (temperate) counterpart (Long et al. 2004). The availability of the feed will be increased in the intensive livestock rearing system, but elevated CO_2 decreases the quality of the feed with reduced protein content and digestibility, especially in the C3 plants (Stokes and Howden 2010). The digestibility of the plants diminished due to increased lignin content and associated reduced degradability and nutrient availability to the animals (Polley et al. 2013).

The effect of climate change on fodder quality and availability varies with the location. Increased pasture growth rate in winter and early spring was predicted in temperate region with increased surface temperature (Cullen et al. 2009). Elevation of temperature was also correlated with increased grazing time by 2.5 months in European countries (Phelan et al. 2016). In subtropical region, decreased pasture production was expected due to lower rainfall and increased intensity of drought, which was compensated a little by the elevated carbon dioxide related increased plant biomass production (Howden et al. 2008). The climate change will negatively affect the quality and availability of fodder more in tropical and subtropical region and the extensive animal farming depending on the pasture (Sautier et al. 2013; Rust 2019). Livestock rearing in tropical/subtropical region (South Asia, Middle East and North Africa [MENA]) is characterized by extensive farming of the ruminants mostly, within frequent intensive rearing of monogastric species (20–25%). The ruminants of the region prefer to consume fibrous feed consists of grass (73%), crop residues (95%), and occasional feeds (90%) (Herrero et al. 2013). The estimate detected reduction of crop production (wheat, maize, sorghum, millet) by 8% by the year 2050 in South-Asia and Africa due to climate change (Knox et al. 2012). Reduction in crop yield would produce severe negative effect on livestock production system in the countries like India where 40–70% of the livestock feed comes from the crop residue, not from the pasture that shares only 3.4% of the total geographical area (Sirohi and Michaelowa 2007). Different studies explored the negative correlation of temperature humidity index (THI) with the milk yield and dry matter intake (DMI). The reduction of feed consumption occurs by 3–5% for each 1 °C rise of temperature above the optimum level (10–30 °C) (Davis-Reddy and Vincent 2017).

The climate change is also associated with depletion of ozone layer and elevation of environmental ozone level, the devastating air pollutant of the century. Elevated ozone can reduce plant growth and biomass yield. Reduction of grassland biomass by 23% was detected in a five-year span field study exposed to the elevated ozone (Volk et al. 2006). Prolonged exposure to ozone may also reduce the digestibility of fodder due to increased lignin content, decreased leaf/stem ratio, increased levels of phenolic acids and flavonoids in the plants, which affect the rumen microbial growth in a negative way (Booker and Miller 1998).

7.3.2 Water Stress

Drought is a kind of environmental stress when the human, animals, and plants are exposed to limited water or no water. Prediction says that 64% of the world population is going to suffer with the water scarcity by the year 2025 (Rojas-Downing et al. 2017). The livestock sector is responsible for 8% of global water use, and higher temperature will increase the thrust of the animals with increasing water use (Nardone et al. 2010).

The water scarcity during drought reduces forage yield and the quality too. The plants can conserve water by adapting morphological changes such as deep penetrating root, leaf orientation, waxing of leaf surface, and leaf senescence to reduce evatransporation-mediated water loss (Jones and Corlett 1992). Few plants employ osmotic adjustment to protect the water level of plant tissues such as apical meristem (Baumgard et al. 2012). The composition of pasture changes with more growth of deeper rooting legumes, which intensifies with degree of water scarcity (Gauly and Ammer 2020). The adaptation strategy taken by the plants to preserve water such as closed leaf stomata reduces carbon assimilation per unit of leaf surface and consequently, the photosynthesis (Bruce et al. 2002). The fodder cultivated under water stress is mostly found unsuitable for livestock feeding.

Other than the quantity, the quality of water is also compromised due to climate change. Increased salinity of water will reduce seed germination and seedling emergence in grasses and legumes depending on the plant species (Shannon and Noble 1995). The polluted water contaminated with heavy metals, coliform bacteria, virus, parasites increases the risk of toxicity or pathogen transmission into the livestock and affect the production quantity and quality in a negative way (Silanikove and Koluman 2015). The research gap exists in data related to direct effect of water stress in livestock production.

7.3.3 Heat Stress

Heat stress is a major economic burden producing $900 million extra cost per year for dairy industry in the United States alone (Pollmann 2010). Heat stress can be measured through THI, an indicator which combines the effects of environmental temperature and relative humidity on animal performances. Currently, black globe temperature humidity index

(BGTHI), equivalent temperature index (ETI), and heat load index (HLI) are also used as indicator considering the wind velocity and solar radiations other than the temperature and relative humidity (Silva and Passini 2017).

The mammals and birds always try to maintain the "homeostasis"/"euthermia" by the neural regulatory process known as "thermoregulation." To maintain the body temperature little above the environmental temperature, the mammals prefer to dissipate the excess heat through gradient dependent and sensible routes of heat loss (conduction, convection, and radiation). The average surface temperature above upper critical limit (>25–37 °C in general) blocks all the sensible routes of heat loss, and the dissipation can take place through evaporation (panting and sweating) only (Kumar et al. 2011). The evaporation route of heat loss is also gradient dependent (vapor pressure), and relative humidity of the surrounding environment controls the rate. The optimum temperature (thermal neutral zone) for the best performance of the livestock is 10–30 °C (Davis-Reddy and Vincent 2017). When the environmental temperature crosses the upper or lower limit of the neutral temperature, the animal has to expend the energy to maintain the euthermia, which produces thermal stress. The upper and lower critical temperature varies with the age, physiological state, species, and breeds of the animals (Collier et al. 1982). Thermoregulation to maintain homeostasis is the neural regulatory process, which leads to physiological changes in livestock such as panting/sweating, rise in body temperature (>102.5 °F), respiration rate (>70–80/minute), blood flow, and increase in maintenance energy requirement by 20–30% depending on the environmental temperature (Pereira et al. 2008). The neural sensors can collect the information about the external temperature and accordingly the efferent responses are produced within the body to maintain the internal homeostasis.

The response to thermal stress will result "acclimation" (physiological response to a single stressor) and "acclimatization" (coordinated response against multiple stressors such as heat, humidity, photoperiod) characterized by reduced feed intake and utilization, disturbances in enzyme activity, reduced weight gain, low availability of energy for milk production, and metabolic acidosis due to increased sodium and potassium ion losses (Marai et al. 2008; Nardone et al. 2010). Recent study estimated the increased body temperature, pulse rate, and respiration rate with significant drop in milk production in the lactating Holstein Friesian cows managed on automated robotic dairy exposed to variable THI (Osei-Amponsah et al. 2020).

The thermal stress response can be sub-divided into two phases, i.e. acute and chronic. The acute response as the name suggests can take place within a few minutes and lasts for few days. The sensory thermal receptors present in skin and hypothalamus are associated with the acute response. Exposure to high temperature alters neuro-endocrine function and activates the sympathetic-adrenal medullar axis (SAM), including the hypothalamic–pituitary–adrenal (HPA) axis to release catecholamines and glucocorticoids (Shini et al. 2008). Cortisol and corticosterone are the primary glucocorticoids detected in cattle, pigs, sheep and fish (cortisol), and poultry and rodents (corticosterone). The released hormones can alter the metabolic process and induce the transcription factors, which take active participation in the process of the physiological changes such as tachycardia and others (Collier and Gebremedhin 2015). The glucocorticoids also increase the glucose synthesis required for survival of mammals and poultry during stress (Ognik and Sembratowicz 2012). Consequently, reduced feed intake occurred with increased THI. The

THI between 73–78, 79–88, and 89–98 are considered to generate mild, moderate, and severe heat stress, respectively (McDowell 1972).

Poultry are the worst affected birds to the thermal stress due to absence of sweat glands. The feed conversion ratio (FCR) increased more during heat stress in fast-growing broilers than the layers, which also increased the production cost sharply (Loyau et al. 2013). On the other hand, the commercial broilers faced reduced feed intake and weight gain, immune-suppression, endocrine disorder, respiratory alkalosis (due to air sac mediated panting), and electrolyte imbalance, which reduced the profit (Lara and Rostagno 2013). The reduced feed intake also adversely affected the appearance, texture, juiciness, flavor, and functionality of commercial poultry meat (Feng et al. 2008). The respiratory alkalosis reduced free calcium availability in the blood circulation and the formation of egg shell was compromised (El-Tarabany 2016). The research gap exists regarding the direct effects of heat stress on backyard or rural poultry reared by extensive or semi-intensive system, especially in low- and-middle-income group of countries (LMICs).

7.3.4 Productive Stress

Decreased milk production was recorded by 0.2 kg (36%) per unit increase in THI after crossing the cut-off value (THI = 72) (Ravagnolo and Misztal 2000). Later observed that the milk production started to decline actually at lower cut-off value of THI (THI = 68) (Zimbleman et al. 2009). Another study detected reduction of DMI by 9.6% and milk production by 21% with increased THI value from 68 to 78 (Spiers et al. 2004). Reduced milk production in cattle due to increased THI occurs due to direct effect of heat stress and reduced feed intake (indirect effect) and the drop in production is more pronounced in high yielding cattle (Zimbelman et al. 2010; Baumgard and Rhoads 2012). Direct effect occurs through increased basal insulin level, which affects milk synthesis due to shift in glucose utilization in non-udder tissues and reduced udder cell proliferation during dry period (Rhoads et al. 2013). Reduction (6%) in milk production in Alpine goats in Brazil was comparatively lower than detected in Holstein cattle (de Albuquerque Brasil et al. 2000).

Other than yield, the heat stress changes the milk composition such as reduced milk protein (whey protein and casein), modified triacylglycerol profile, and reduced phospholipid with no effect on milk sugar (lactose) (Cowley et al. 2015; Liu et al. 2017). Altered milk casein content (especially α and κ fraction) was detected as direct effect of exposure to high temperature (Bernabucci et al. 2015). Effect of heat stress on milk fat quantity is debatable, but majority of the studies detected reduced milk fat in cow milk during summer than winter (Bernabucci et al. 2015). Reduced milk phosphorus, increased milk pH, somatic cell count, and milk coagulation time were also detected as direct effects of heat stress in cattle (Summer et al. 2019). Modification in the composition of milk negatively effects the preparation of downstream products such as cheese, butter, and so on. The beef cattle are less affected with heat stress than the milch cattle due to lower metabolic rate and body heat production, which shifts the THI cutoff value toward the higher side (Nardone et al. 2010).

The swine industry suffers with lowered profit associated with heat stress due to reduced growth of the animals, decreased carcass value owing to enhanced lipid and decreased protein content, and increased disease prevalence and treatment-related costs (Collin et al. 2001). At the ambient temperature, the pigs experience the "lipid accretion," i.e. decreased lipid-to-protein ratio in the tissues, and consequently, less amount of lipids are deposited in the carcasses, and the lean carcasses are produced (Oresanya et al. 2008). The heat stress also induces skeletal muscle catabolism indicated by increased circulating plasma urea nitrogen (PUN) level (Rhoads et al. 2009).

The buffaloes are most productive within the ambient temperature (13–18 °C), relative humidity (55–65%), and wind velocity (5–8 km/hour). Like the cattle, the THI crossing the cutoff value (THI = 72) produces heat stress in buffaloes (Payne 1990). Reduced feed intake (9–13% or up to 40% in severe condition) was detected in buffaloes as the most prominent effect of heat stress, which was associated with decreased cud chewing, reduced feed digestion, and decreased availability of buffer and water in the rumen (Hooda and Singh 2010; Savsani et al. 2015). Reduced wheat straw intake in the buffaloes was detected in hot-dry (29%) and hot-humid climate (30%) (Korde et al. 2006). Reduced feed intake in buffaloes occurred due to depression of right hypothalamus during exposure to high temperature (Mishra 2021), and it was found inversely proportional to the body weight gain (8–18% reduction) (Habeeb et al. 2012). Use of foggers, fans within the shed, or wallowing of the buffaloes can increase the feed intake during thermal stress (Ahmad et al. 2017). The thermal stress produced negative impact on milk yield, milk composition (lower milk fat, protein, and SNF), lactation length, dry period and calving interval in Murrah and Egyptian buffaloes (Aggarwal and Singh 2006; Marai et al. 2009; Singh et al. 2013).

Similarly, the poultry production experiences reduction in body weight (32.6%), high FCR (25.6%), reduced carcass weight, and egg production (28%) with increased ambient temperature (Sohail et al. 2012; Zhang et al. 2017). The quality of the eggs is compromised with high environmental temperature, and reduced egg weight (3.2%), egg shell weight (9.9%), egg shell quantity (0.66%), and shell thickness(1.2%) are observed (Nardone et al. 2010; Mack et al. 2013). The poultry meat quality is also compromised with thermal stress due to reduced muscle growth and protein content of meat, electrolyte imbalance, and lipid peroxidation causing increased fat deposit (Dai et al. 2012; Sokołowicz et al. 2016).

7.3.5 Reproductive Stress

Impaired reproductive performance was detected in cattle due to heat stress associated with anoestrus, reduced conception rate (20–30%), alterations in the follicle development, polyspermy, altered capacity of zygote for further development, and many more (Kadokawa et al. 2012; Campen et al. 2018). The conception rate was decreased by 3.2% for each unit increase of THI above 70 in *Bas taurus* cattle (Amundson et al. 2006). Presence of infection such as mastitis along with the heat stress produces additional negative effect on the fertility (Roth and Wolfenson 2016).

In cattle, reduced pulse amplitude of luteinizing hormone (LH) was detected due to impaired feed intake associated with heat stress. Reduced LH caused decreased production of estradiol and progesterone, which consequently hampered the follicular growth and ovulation (Szymanski et al. 2007). High temperature can also induce the secretion of adrenocorticotropic hormone (ACTH), which can prevent the estradiol-induced estrus in cattle (Hansen and Areéchiga 1999). Decreased reproductive activity with reduced concentration of circulating triiodothyronine (T3) and thyroxin (T4) was also associated with heat stress (Farghaly 1984). Other than the direct damage of oocytes to the elevated temperature, the indirect effect of heat stress in cattle produces prolonged estrus cycle. The prolonged estrus cycle may ovulate an "aged" oocyte with reduced potential for development (Baumgard et al. 2012). The studies explored that the swine oocytes exposed to heat during the first 21 hours of the development could not reach the metaphase II stage and consequently the development of the oocytes got arrested (Baumgard et al. 2012).

Experimental study in sows revealed that the heat stress during late phase of gestation period (102–110 days) reduced the numbers of live offspring (Omtvedt et al. 1971). The number of offspring viability was not affected during exposure of heat stress in sows in early phase of gestation (3–30 days) (Liao and Veum 1994). Similarly in cattle, embryonic death due to exposure of cows to the thermal stress was detected due to reduced protein synthesis and increased concentration of free radicals (Edwards and Hansen 1997). Redistribution of blood flow to facilitate the heat dissipation from the body also reduced blood supply to the placenta, which acted as additional cause for retardation of fetal growth and embryonic death (Collier et al. 1982). Moreover, maternal heat stress reduced uterine blood flow and consequently decreased birth weight of lambs (Dreiling et al. 1991). Reduced lactation performance was also attributed to heat stress in sows due to impaired postabsorptive metabolism (Johnston et al. 1999).

Exposure to higher temperature also generated more numbers of abnormal sperms with reduced sperm motility due to increased generation of reactive oxygen species (ROS) (Nichi et al. 2006). Testicular damage due to heat stress may be counteracted with gherlin, the hormone produced by enteroendocrine cells of the gastrointestinal tract.

The buffaloes are the worst affected livestock species with thermal stress due to less numbers of sweat glands and nonavailability for places of wallowing (Vaidya et al. 2010). THI crossing the cut-off value of 75 produces severe negative impact on reproductive efficiency of buffaloes, especially in tropical countries (Vale 2007). The poor expression of estrus (silent estrus) and low conception rate are the consistent reproductive complications in buffaloes during summer. The highest proportion of studied buffaloes exhibited the estrus during early morning and evening and the lowest during the noon when the environmental temperature was at the peak among the day (El-Wardani and El-Asheeri 2000). The sexual activity of buffaloes was correlated with decreased day length and temperature. The follicular development and ovulation occur normally, but the animals cannot express the obvious signs of estrus and lack of "progesterone priming" was identified as a causative factor in buffaloes exposed to high temperature (Singh et al. 2013). Low level of circulating reproductive hormones such as progesterone, LH, and specially estradiol during the day of estrus in buffaloes during summer was also found to be associated with poor expression of estrus (Upadhyay et al. 2009). The altered endocrine activity of pineal hypothalamo–hypophyseal–gonadal axis was detected in buffaloes. Reduction in estradiol concentration occurs either due to lack of optimum follicular growth or high rate of metabolism of the estradiol (Awasthi et al. 2007). Similarly, the prolonged service period and reduced conception rate in Murrah buffaloes (18% reduction) were detected with increased THI after crossing the value of 75 (El-Wishy 2007; Das et al. 2016).

The poultry birds also experience reduced surge of LH and follicle-stimulating hormone (FSH), and infertility due to exposure to high temperature (Ayo et al. 2011). Delayed ovulation associated with smaller follicles, and reduced concentration of estradiol was detected in heat exposed birds (Kala et al. 2017). The egg yolk quality and maturation rate were compromised due to lipid peroxidation, increased ROS, and altered fatty acid profile in birds (Papadopoulou et al. 2017). The testicular function was adversely affected in male birds due to blockage of intracellular ion exchange and the sperm motility was reduced (Tosti and Gallo 2015). The embryonic growth was adversely affected in birds continuously exposed to heat stress (Noiva et al. 2014).

7.3.6 Livestock Diseases

The effect of climate change on livestock diseases is difficult to predict due to multivariate nature of the climate change. Few studies established the linkage between climate and livestock disease distribution, timing, and intensity of outbreaks (Baylis and Githeko 2006). Climate influences microbial density and distribution, distribution of vector-borne diseases, food and water shortages, or foodborne diseases (Lacetera et al. 2013). Increased temperature and humidity favor the multiplication of pathogens, as well as their vectors such as flies, mosquitoes, and ticks, which accelerate disease transmission into the host (Thornton et al. 2009). Increased temperature at night favors the flight activity of vector (mosquito) and replication or transmission of viral pathogens. Similarly, alteration in rainfall pattern with high velocity wind helps in intra-continental spread of vector population and severe humidity after drought causes enhanced replication of vector and pathogens (Trape et al. 1996). Change in distribution of vector population was detected to modify the distribution of viral pathogens such as Rift valley fever (Rolin et al. 2013). A model simulated by Wittmann et al. (2001) demonstrated that an increase of 2 °C of environmental temperature can extensively spread *Culicoides imicola*, which is responsible for the transmission of bluetongue virus in sheep cattle, goats, and also wild ruminants. This virus has been spreading globally since 1990, and global warming is the cause for spreading this virus (Lacetera 2019). Hemoprotozoan infections such as *Babesia spp.*, *Anaplasma spp.*, and *Theileria spp.* are important protozoa health hazards of livestock in tropical countries that can be transmitted by ticks during summer season (Bary et al. 2018; Caminade et al. 2019). The developmental rate of free-living larval stage of the *Haemonchus contortus* has been reported to increase with rising environmental temperature in the tropical regions (Fox et al. 2015). Survivability of the pathogens and vectors with increased transmission rate was detected in higher temperature. Higher prevalence of vector-borne, parasitic, and zoonotic infections such as leishmaniasis, cryptosporidiosis, giardiasis, trypanosomiasis, schistosomiasis, filariasis, onchocerciasis, loiasis, theileriosis, African horse sickness, African swine fever, blue tongue, Japanese encephalitis, lumpy skin disease, Rift valley fever, equine encephalitis, West Nile fever, tularemia, piroplasmosis was noted throughout the world, especially in Asia, Africa, and South America (Patz et al. 2000; Rose et al. 2015; Omazic et al. 2019).

Other than infectious diseases, a number of studies indicated a greater risk of mortality of farm animals during the hottest summer months (Dechow and Goodling 2008; Purusothaman et al. 2008; Vitali et al. 2015). High temperatures might be the cause of heat stroke, heat exhaustion, heat syncope, heat cramps, and ultimately organ dysfunction leading to death. Elevated temperatures may cause increased intake of concentrate-based diet with less roughage during heat stress leading to development of acidosis and lameness in ruminants. Reduced feed intake also produced clinical or sub-clinical ketosis, especially in high-yielding dairy cattle during exposure to high temperature (Lacetera et al. 1996). Heat stress induces oxidative stress with increase of plasma reactive oxygen metabolite substances in farm animals (Akbarian et al. 2016; Mirzad et al. 2018). The heat stress also caused increased sloughing and damage to the intestinal epithelium with shortened villi or crypts. Increased intestinal permeability, especially for bacterial endotoxin, was detected due to phosphorylation of type 2 myosin light chain, which opens the tight junction present in the epithelium (Lambert 2009). The bacterial endotoxin (LPS) is a potent immune system stimulator, which can cause inflammation, septicemia, and death in severe cases. Activation of immune system partitions the energy and nutrients in livestock and reduces the growth and production (Gabler and Spurlock 2008). Due to global climate change, natural disasters such as cyclones, tidal surges, floods, salinity intrusions, and droughts may have severe impact on livestock health and production with many negative consequences such as nutritional deficiency, lack of fresh water, increased incidence of diarrhea, skin diseases, liver fluke, loss of bodyweight, and breakdown of the immune system (Dastagir 2015; Alam et al. 2017).

Exposure to high temperature compromises gut wall integrity in poultry birds due to occluding-mediated disruption of tight junction, the protein complex which regulates the transmembrane transport of materials and maintains the barrier functions (Tellez et al. 2017). Disruption of gut wall barrier produces reduced nutrient absorption, immune system disorders, increased susceptibility to the infection, and mortality of the birds (Varasteh et al. 2015). The immunosuppression of the thermal stressed birds was associated with decreased weight of the primary and secondary lymphoid organs, reduced concentration of circulatory antibodies (IgG and IgM), reduced numbers of intraepithelial lymphocytes and Ig-A secreting lymphocytes in the intestine, and reduced phagocytic capacity of the macrophages (Nawab et al. 2018). Activation of toll-like receptor (TLR 4) and nuclear transcription factors (NF-κB)-mediated release of pro-inflammatory cytokines also caused immunosuppression of the birds when exposed to high temperature (Karnati et al. 2015). The immunosuppressed birds suffer more with bacterial, viral, protozoal and parasitic infections and infestations. Lecchi et al. (2016) reported that high temperatures impaired significantly the functionality of neutrophils, which have a vital role in the protection of the mammary gland against infections. As an immune response to bacterial invasion of the teat canal to the cow's udder, there is a greater risk of the occurrence of mastitis during the summer months (Vitali et al. 2016).

Global climate change affects the seasonal temperature variation of migratory birds traveling from one country to another country. Changing climate is considered as a major determinant for increased pathogenicity and transmission potential of bird origin zoonotic infections such as avian influenza (H5N1). Prolonged summer, high-mean air pressure, and low-mean specific humidity changed the migration pattern of wild birds and increased the risk for H5N1 outbreaks in migratory birds in Western countries (Tian et al. 2015), while the migratory birds were identified as the potential source of avian influenza spread in domestic ducks in Bangladesh during the winter season (Haider et al. 2017; Sarker et al. 2017). Changing migratory pattern of wild birds also generated novel species congregation in a specific locality with increased possibility of mixing avian influenza viral strains in the hosts (Van Hemert et al. 2014). More genetic reassortment in segmented RNA viruses is associated with not only increased virulence but also adaptation into more numbers of hosts. Moreover, high temperature, change in rainfall pattern, and more paddy cultivation to cater the rising demand for food in South-East Asia identified central and coastal China, the western Korean Peninsula, Japan, North India, and the Nile Delta as the potential hotspot for avian influenza in the coming days (Fuller et al. 2013).

7.4 Adaptation and Mitigation Strategies to Combat Climate Change Effects on Livestock

In livestock sector, GHG emission depends on climatic regions (tropical, dry, temperate, etc.), type of animal species (ruminants or nonruminants), quality of animals (high/medium/low productive or nonproductive), production systems (extensive, intensive, or semi-intensive), and production practices (type and quality of feeds and fodders, herd management, manure management, energy use, etc.). Several mitigation recommendations and adaptation strategies are available to reduce the effect of livestock on climate change and vice versa. Depending on the geographical regions, emission sources, livestock species, production systems, and production practices, mitigation solutions and adaptation strategies vary. The impact of livestock on climate change or vice versa is reduced by applying mitigation measures (Dickie et al. 2014), and livestock productivity is made resilience to climate change by following adaptation strategies (USDA 2013). Thornton and Herrero (2014) indicate that three dimensions of sustainability (economic, environmental, and social) need to be considered before identifying appropriate mitigation options. Kates et al. (2012) outline that three classes of adaptations may bring change at a much larger scale or be new to a particular area or shift locations. Mitigation measures and adaptation strategies may be successful for bringing about beneficial changes on climate, as well as livestock, if they are included in national and regional policies and actions (FAO 2009).

7.4.1 Livestock Production System Approaches

GHG emissions vary with different livestock production systems. Extensive livestock production system has relatively high pressure compared with intensive production systems. For example, grain-fed beef cattle maintained in intensive systems consume more energy, but produce less GHGs per unit of meat produced than the emission recorded under extensive grass-fed systems (Williams et al. 2006; Clark and Tilman 2017). Dairy cattle reared under intensive and mixed farming systems in industrialized countries generate less GHGs per kg of fat and protein corrected milk (FPCM) than GHGs emitted under extensive grazing systems in developing countries (Gerber and Vellinga 2009). Moreover, crop byproducts feeding based mixed system is always better than rangeland-based grazing system (Herrero et al. 2008). Thus, a move from extensive to intensive production system may be beneficial for reducing GHG emissions in developing countries. However, intensive livestock farming will involve high production cost and very-high-quality skill set to justify the system (Rust 2019).

There are several livestock species-level models suitable for different agro-climatic situations. Monogastric species like pigs and chickens, because their unique characteristics, produce less GHGs than ruminants and thus pigs and poultry birds are considered as more environmentally friendly for promotion in developing countries (Kantanen et al. 2015; Escarcha et al. 2018). Thus, the shift from ruminant animal farming to monogastric animal farming may provide co-benefits for lessening GHGs emissions and bringing down the health risks associated with overconsumption of ruminant animal products, particularly in developed countries (Davis et al. 2016; Springmann et al. 2018; Willett et al. 2019).

Seo and Mendelsohn (2008) suggest a model in which the farmers can shift to goat and sheep farming from cattle and poultry farming when environmental temperature rises. In drought-prone areas, subsistence form of farming with small ruminants (goats and sheep) is recommended for livelihood purposes and income generation (Maiti et al. 2014). Because of diverse species portfolios, small farms are contemplated as more climate change resilient farms in developing countries and the local breed portfolios are found to be more adapted to the local systems (Seo and Mendelsohn 2007, 2008).

Jones and Thornton (2009) described another model where the ecologically vulnerable and socially backward marginal farmers can adopt crop-livestock systems for sustainability and economical viability. Mixed crop-livestock farming systems, practiced in two-thirds of the world, produce more than half of the crops, milk, and meat (Herrero et al. 2012). Judicial changes like proper integration of crop and livestock, efficient use of available resources, crop rotation, crop diversification, intercropping, precision livestock feeding in mixed crop-livestock farming systems can enhance food production and farmer income, as well as reduce GHGs emissions (Steinfeld et al. 2006; Herrero et al. 2010; Nepstad et al. 2019). Hence, the traditional livestock farmers prefer to maintain high diversity with multispecies to guard against climatic and economic crunches (FAO 2009).

7.4.2 Species and Breed Selection

Generally, the farmers first use adaptation technologies that can adapt their animal's genetics under changing environment, and when it becomes unavoidable, the producers only then change their genetic resources portfolio. Cattle and buffaloes have the major share in GHG emission in the world. Thus, the dairy sector offers enormous scope to mitigate climate change (FAO 2010), particularly in India, which ranks first in total population of cattle and buffalo in the world (FAO 2011). As mitigation measures, Gerber et al. (2013) suggest reducing the overhead stock of the breeding herd and selecting the animals based on production performances for lowering GHGs emissions per unit of product. Therefore, a strategic breeding program needs to be designed and implemented with the aim of decreasing the inferior quality of cattle and buffalo populations and maintaining the quality animals at the herd in India.

High-producing animals of temperate climate show lower acclimatization ability to the high temperature, but higher acclimatization ability to the low temperature (Hansen 2004). High-yielding cattle are the most susceptible to heat stress than the low-yielding counterpart. Among the zebu cattle reared mostly in tropical/subtropical region (Brazil), Gir is the breed found least adapted to the climate changes and the Sindhi and Girolando are the breeds with higher acclimatization ability to the high temperature (Cardoso et al. 2015). In India, Tharparkar is the most preferred cattle breed due to high degree of thermotolerance, tick, and disease resistance. The transcriptome analysis of Tharparkar cattle reveals the presence of ubiquitin, protector of naïve unfolded protein, and antioxidants to counter the heat stress (Khan et al. 2021). However, the goats, camels, and buffaloes are much more resilient and adaptive to the heat stress, especially in the tropical and subtropical region than the cattle. Low body mass, low metabolic requirements, proficient grazing behavior, and useful digestive system of the goats help them in maximum feed utilization even in scarcity. The striking feature of goat digestive tract to modify the anatomy and function with the season and diet such as changes in the volume of foregut, salivary glands, and surface area of mucosa make them the most adaptive livestock species with the heat stress (Silanikove and Koluman 2015).

Thermal tolerance is an important trait of animals for consideration under climate change scenario. The consideration of a THI and genotype-by-environment interactions (G × E) in breeding indices is the need of the hour. Carabaño et al. (2019) explore the possibility of selecting farm animals for thermal tolerance. Finocchiaro et al. (2005) propose to choose heat-resistant sheep for breeding purpose in hot climates. Though various commercial breeding programs have already made a significant genetic improvement on worm resistance in livestock species in Australia, New Zealand, and South Africa (de Greef 2009), it is arduous to combine both heat tolerance and production potential traits for selection purposes in high-output breeds like dairy animals, because heat tolerance, milk yield, and reproductive performance are regulated by different physiological and metabolic processes (Ravagnolo and Misztal 2002; Bohmanova et al. 2007, 2008).

Planned cross breeding programs can be useful to address the issue of production, reproduction, heat tolerance, and disease resistance at a time (Rowlinson 2008; Henry et al. 2012). In tropical grasslands of northern Australia, composite cattle breeds have shown greater productive and reproductive potential with better heat tolerance and disease resistance ability than the pure shorthorn native breeds (Bentley and Hegarty 2008). As the locally adapted breeds are not only tolerant to heat, parasites, and diseases but also thrive under poor nutrition, crossbreeding programs among the native breeds may be more useful and relevant with climate change than the imported breeds for production purpose (Hoffmann and Vogel 2008; Cardoso et al. 2015). However, such kind of strategic crossbreeding programs requires structured phenotypes and performance records of targeted native breeds and reference populations for developing very high-density SNP panels, which are largely missing in developing countries. Hayes et al. (2009) suggest for marker-assisted selection of sires whose daughters seem to manage milk production with higher heat tolerance ability under low feeding schedules.

7.4.3 Livestock Production Management Technologies

Intensive livestock production mitigation management measures have greater capacity to adapt than mitigation measures in extensive system of livestock management. In intensive livestock system, the microenvironment of the animal house can be modified by suitable management and technologies to make the housed animals comparatively insensitive to climate change and reduce the effects of climate change on high-output breeds for retaining optimum production (Adams et al. 1998). During short-term heat waves, shading or sprinkling or fan systems may be effective to reduce excessive heat loads (Marcillac-Embertson et al. 2009). Arrangements of fans and showering systems are shown in Figures 7.1 and 7.2. Some high-efficiency cooling technologies include conductive cooling (Silva et al. 2009; Cabezon et al. 2017), tunnel ventilation (Zhang and Bjerg 2017), and radiative cooling (Pang et al. 2010). It is crucial to give an adequate access to water by increasing 1.2–2 times during heat stress time. Exposure to high heat load although increased water intake from 32 litre to 82 litre per steer per day as observed in a feedlot study to make up the evaporation (Gaughan et al. 2010). Provision of cooled drinking water (Jeon et al. 2006; Renaudeau et al. 2012), water

Figure 7.1 Fan system in cow shed at the farm of ICAR Research Complex for NEH Region, West Tripura, India.

Figure 7.2 Overhead showering system for buffaloes at the farm of ICAR Research Complex for NEH Region, West Tripura, India.

baths (Huynh et al. 2006), or wallows (Bracke 2011; de Mello et al. 2017) are effective to protect cattle and buffaloes from heat stress.

The extensive system through pastoral grazing is common in developed countries like USA, Denmark, Netherland, Germany, Australia, New Zealand, etc. But, it is limited in developing countries, such as India, Pakistan, Bangladesh, the Philippines, Malaysia, Thailand, and Indonesia. Pastoral system for grazing of cattle and sheep in New Zealand are presented in Figures 7.3 and 7.4.

In extensive pastoral systems, local breed portfolios are more adaptive and resilient to the local systems than high-output breeds. Improved pasture management has a positive influence on livestock productivity (Smith et al. 2007). Shade trees in the pastoral lands

Figure 7.3 Pastoral system for grazing of cattle in Dunedin, New Zealand.

Figure 7.4 Pastoral system for grazing of sheep in Dunedin, New Zealand.

can protect animals from heat stress (Thornton and Herrero 2010). Improving grazing management by reducing grazing pressure can lower down the enteric emission intensities, because the animals under lower grazing pressures can get a broad choice of forage and choose more nutritious forage leading to rapid live weight gain (Rolfe 2010). Rotational grazing can be an efficient option of grazing management for quality pasture availability, which can help in reducing methane emissions per unit of area, and at the same time, enhancing digestibility and augmenting productivity of the animals (Eagle et al. 2012). The low-cost, bamboo made and/or iron mesh net fencings for rotational grazing of goats are shown in Figures 7.5 and 7.6.

Figure 7.5 Low-cost, bamboo-made fencing for rotational grazing of Black Bengal goat in West Tripura, India.

Figure 7.6 Bamboo- and iron-mesh-net-made fencing for rotational grazing of goats in Banka, Bihar, India.

7.4.4 Grazing and/or Fodder Land Management and Carbon Sequestration

Globally, grasslands occupying about 70% of the global agricultural area are the largest terrestrial carbon sink, which accounts 343 billion tonnes of carbon (FAO 2010). Thus, a little change in soil carbon stocks in grassland can cause a significant change in global carbon balance (Sacks et al. 2014). Mitigation measures (such as rebuilding soil organic matter, lowering soil erosion, diminishing biomass losses due to burning and overgrazing) for the improvement of grazing land conditions can help in soil C-sequestration by around 0.15 gigatonnes CO_2-eq per year globally, boost biodiversity, and enhance livestock productivity as well (Smith et al. 2007; Henderson et al. 2015). Up to 25% reduction of GHG emissions along with improvement of individual beef cattle and herd performance due to grazing land management and change in land use has been reported in Brazil (Gerber et al. 2013).

Pasture land management measures may be carried out by replacing native grasses with higher yielding and more digestible grasses, improved varieties of pasture, perennial fodders, and legumes (Bentley and Hegarty 2008). Access to more digestible and nutritious forages can lead to higher animal growth rates, reduction of age at first calving, improvement in fertility rate, and minimization in mortality rate (FAO 2013). Grazing land mitigation measures also include conservation tillage operations, erosion prevention techniques, soil acidity management, multi-cropping practices, crop rotations, incorporation of shade trees in the grazing land, regular soil testing, use of organic fertilizers and/ or technologically advanced fertilizers, introduction of earthworms, and more, which can lead to carbon sequestration in pasture land and decrease GHG emissions (Steinfeld et al. 2006; Denef et al. 2011; Dickie et al. 2014). Reduction in grazing pressure can bring back soil carbon losses and revive soil carbon stocks by sequestering up to 148 Mt. CO_2-eq per year globally, while a major share of this potential (81%) is represented by developing countries (Henderson et al. 2015).

7.4.5 Shelter Management

It is necessary to provide proper housing or shelter for protecting livestock from any kind of climate-related stress. The physiological, physical, and behavioral studies coupled with meteorological record and economic data are to be considered for designing a proper animal shelter. Animal house or shelter should be designed considering optimization of space requirements, proper flooring, suitable roofing, thermal stress alleviation arrangement, and methane mitigation system (Ambazamkandi et al. 2015). Though climate change-related modifications in animal house differ from livestock to livestock, but the general rules are the same.

Dairy sector is the most important subsidiary business in livestock sector. The dairy house needs to be designed in a scientific way to provide comfort to the animals for optimizing production. In cattle house, some high-efficiency cooling technologies like thermal conductive cooling of floor, tunnel ventilation system, suitable roofing materials, THI-based wetting, and other important thermal stress alleviation techniques and measures including proper waste management are relevant to reduce methane emission (Vanegas et al. 2006; Vijayakumar et al. 2009). Among species, sheep and goats have more heat tolerance ability than cattle (Silanikove 2000). Goats are well adapted under varied agro-climatic situations, including extreme climatic conditions, particularly in hot, dry areas, and so goats can be reared in almost all agro-ecological zones of the world. However, sheep and goats are slightly susceptible to wet and moist conditions, and hence, the floor height of goat and sheep house should be maintained at 1–1.5 m above the ground. Proper designing of goat and sheep units not only protects animals from climatic stress but also avoids diseases and prevents production loss and guarantees maximum profit (Peacock and Sherman 2010). In south-eastern Asian countries, the small, marginal, and even landless families rear goats for subsidiary income and they provide low-cost housing structures for night shelter of the goats. In a study in rural West Bengal, India, Nandi et al. (2011) described low-cost goat houses where paddy straw or earthen tiles or tin sheets were used as the roofing materials. A model of low-cost, raised, bamboo-made goat house is shown in Figure 7.7. Pigs are non-sweating species and thus very vulnerable to inclement weather. Hence, proper pig house with waste management facility is essential to protect pigs from inclement weather and prevent parasitic and other diseases like swine flu under climate change scenario. The floor of the pig house must be hard and impervious to water and easy to clean, and so the concrete floor with a slope toward the drainage is always designed. In pig rearing and management, waste management is a very important aspect to prevent the spread of diseases (Ambazamkandi et al. 2015). In a smallholder farming system, the pigs are maintained in either semipermanent, brick-cement, half-walled pig house or locally available bamboo-made pig shelter (Haldar et al. 2017). A model of low-cost, half-walled pig shelter for resource poor farmers is shown in Figure 7.8. Among livestock, climate change can cause maximum impact on poultry birds. Poultry birds are highly sensitive to stress, especially due to climatic variations, because poultry birds can only tolerate narrow temperature ranges, and beyond these ranges, the birds suffer from stress. Hence, the ameliorating measures need to be undertaken to reduce the stress and provide congenial macro- and microenvironments, which determines the success of poultry farming (Ambazamkandi

Figure 7.7 Low-cost, raised, bamboo-made goat house in South Tripura, India.

Figure 7.8 Low cost, half-walled pig shelter for smallholder pig farming in West Tripura, India.

et al. 2015). The possible measures against climatic variations include cooling of poultry house and reduction of house humidity in summer, lighting arrangement, or heating devices inside the poultry house in winter and optimal ventilation. In poultry house, suitable mitigation facilities need to be available so that optimal seasonal temperatures are maintained inside the house and the birds can feel comfort and the risk of climate change including more dramatic events, such as storms, heat waves, and cold waves, are reduced.

7.4.6 Enteric Methane Mitigation

Livestock production at certain level can deal with GHG emissions by contributing to soil carbon sequestration, proper utilizing byproducts, and residues that are inedible to humans (Garnett 2009). Nevertheless, at current levels of livestock production, the disbenefits of livestock in terms of GHG emissions far surpass the benefits, which will intensify more with greater livestock production due to their demands (Garnett 2009). Clearly, mitigation of GHG emissions from livestock is urgently needed.

Methane production from enteric fermentation contributes to the greatest extent to global livestock GHG emissions, and also it is a loss of feed energy for ruminants. Enteric CH_4 from ruminants share about 6% of anthropogenic GHG emissions and 40% of total livestock emissions (Gerber et al. 2013). Therefore, CH_4 mitigation in ruminants could serve both economic and environmental benefits. Several mitigation options have been explored in ruminants. Some mitigation strategies such as chemical inhibitors (CH_4 analogs), ionophores (monensin), and defaunation reduce methane production directly or indirectly in the rumen, but their consistent effects for practical use have not been proved (Patra 2012b; Patra 2016). A variety of nutritional options such as increased use of concentrate in diets (Wang et al. 2016), inclusion of some condensed tannins containing leguminous forages (Patra and Saxena 2010; Beauchemin et al. 2020), supplementing low-quality forages with protein and readily fermentable carbohydrates (Wang et al. 2016), addition of fats (Patra 2013, 2014b), and utilization of tree fodders rich in several plant bioactive compounds (Pal et al. 2015) show promise for CH_4 mitigation. These nutritional strategies can also enhance the efficiency of feed utilization and productivity, which will be useful for practical mitigation. The addition of nitrate, which acts as both direct inhibitors of methanogens and alternative hydrogen sinks, have two benefits – reduction of methanogenesis and nitrogen supplements (Patra and Yu 2013; Beauchemin et al. 2020). Plant secondary metabolites including essential oils, tannins, and saponins have shown promise in different studies (Patra and Saxena 2010; Patra and Yu 2012), but still their use in practical diets is not consistent. Several new potential technologies including probiotics, alternative electron acceptors, CH_4 oxidation by methylotrophs, stimulation of acetogens, immunization, genetic selection of low CH_4-producing animals, and early life programming have emerged to mitigate enteric CH_4 emission, but extensive research is needed before these technologies can be recommended to livestock farmers (Patra 2012b). It has been suggested that individually, some of these options might have moderate effects (<20% decrease) with the exception of the 3-nitrooxypropanol, a direct methanogens inhibitor, which have been shown to decrease CH_4 consistently by 20–40% (Beauchemin et al. 2020). Therefore, different CH_4 mitigation options should be adopted together to decrease CH_4 emission from ruminants to a large extent at the farm levels (Patra and Yu 2015a, b). Most of the anti-methanogenic compounds often show inconsistent results due to variations in rumen microbiome, adaptation to anti-methanogenic inhibitors, and diets (Patra et al. 2017; Beauchemin et al. 2020). Because methanogenesis in the rumen is a complex process relating methanogenic archaea and many different groups of H_2-producing microbiota, such challenge is anticipated. Feed intake, nutrient digestion, and rumen fermentation are adversely affected when ruminants are fed at high doses to obtain effective mitigation. Ideally, CH_4 mitigation options should be cost effective showing consistent CH_4 mitigation response without adverse effect on animal performance and health and without presence of residues in animal-derived foods, but this model technology is still needed to be explored for the practical adoption by the farmers.

7.4.7 Precision Livestock Farming

The "high-input high-output system" in livestock farming has recently been shifted to "precision livestock farming" (PLF) through the use of modern monitoring and control systems. The term "precision livestock farming" coined in 2004 delineate "a management system based on continuous automatic real-time monitoring and control of production, reproduction, animal health and welfare, and the environmental impact of livestock production" (Berckmans 2014; Halachmi and Guarino 2016). PLF includes some technologies that ensure the sustainability of production, based on the concept of "producing more with less." PLF has been come to light from the demand to inform the farmers on precise and effective use of available resources for taking real-time and evidence-based decisions on the animals' needs to achieve sound health and optimum production (Norton and Berckmans 2018). Livestock farmers in high-income countries have adapted PLF technologies. However, the rate of adaptation is not much encouraging at this moment. PLF is IT-based management systems that include big data analytics, real-time data monitoring, and advanced robotics for the use of optimum amount of inputs for each animal and achieving maximum yields and reducing farm waste including GHG emissions. PLF offers a platform to monitor nutrition status, health status, animal behavior, animal growth, milk, meat and egg production, and aspects related to the physical microenvironment and GHG emissions. PLF allows the farmers to monitor resources' use, as well as the health, performance, and welfare of the animals (Berckmans 2014; Bell and Tzimiropoulos 2018). Basically, the function of the animal's own intrinsic (e.g. genetics, health, nutritional status) and extrinsic (environmental and social stressors) factors can be monitored in real-time through PLF techniques (Hauschild et al. 2015). In high-input high-output pig production systems, PLF techniques have been applied successfully through the use of exact amount of high-quality resources for realizing maximum production potential, minimizing losses, and waste (Rauw et al. 2020). PLF technologies have already been implemented successfully to grazing systems through the use of bioacoustics (virtual fences, sensors) or drones for capturing real-time images of herds and flocks and thus real-time monitoring on health and productivity of both animals and the land they graze (Rutter 2014).

In livestock production systems, feed accounts for 60–70% of the overall production costs. Hence, precision livestock feeding is a major component of PLF, which allows to provide the right amount of feed, in the right nutrient composition, at the right time to each animal individually that maximizes nutrient utilization for bringing about the optimum performances (White and Hall 2017; Pomar et al. 2019). Precision livestock feeding allows automatic data collection (e.g. feed intake, body weight, analytes), data processing, and computational estimation of the nutrient requirements based on these inputs, and then formulation of precise diet that will be offered to the individual animal for achieving the desired production trajectory (Pomar et al. 2019).

7.4.8 Livestock Reproduction Management

Subfertility, infertility, pubertal anoestrus, long postpartum anoestrus, poor conception rate, and low pregnancy rate in various livestock species during the summer season lead to a substantial economic loss in many tropical and subtropical countries like Australia, Brazil, India, the Philippines, Vietnam, Malaysia, Indonesia, etc. Livestock reproduction mitigation measures under climate change scenario include planned feeding program, favorable microenvironment through artificial modifications, adaptation of synchronized and controlled breeding program at particular time, improved reproductive disease management, application of embryo transfer technique, and other improved reproductive technologies.

The primary factor for curbing fertility in ruminant animals is inadequate nutrition in developing countries (FAO 2013); thus, the adequate, balanced, quality feeding regimen can improve fertility. There is a positive relationship between the digestibility of feed and the growth rate of animals (Manninen et al. 2011; Keady et al. 2012), and thus, the age at first calving can be reduced by increasing digestibility in feeding program. At the same time, stress minimizing measures and preventive health care can lessen stress, as well as disease infection rate that further decrease mortality rate, improve health, early maturity, and enhance fertility rate leading to decrease GHG emissions per unit of output.

The two most serious reproductive problems in dairy animals are anoestrus and repeat breeding leading to a huge economic loss in terms of losing milk production. The farmers are very much concerned with anoestrus period (non-occurrence of first estrus/late puberty) in heifer; however, in many cases, they overlook prolonged postpartum anoestrus (even after three- or four-month postpartum), as the cow is in milking stage. In repeat breeding case, the farmers experience a loss of milk production of 21 days, which is basically delayed due to missing of each heat, in addition to bearing the maintenance cost of an animal. Thus, anoestrus and repeat breeding contribute "nonproductive days" and reduce profitability tremendously in livestock farming. The application of various mitigation strategies under the traditional farming system in a big way will help to lessen summer subfertility and/or infertility and enhance the economic return from livestock species to the farmers. Of late, developments focus on CL and follicle control to synchronize ovulation using some suitable and economical protocols. Because of the two major benefits, the farmers agree to use new generation of ovulation-synchronization protocols. Two benefits are diminishment of the number and frequency of handling animal and removal of estrus detection by applying Fixed- time/Timed AI (TAI) protocols. Since 1995 (Pursley et al. 1995), many TAI protocols have been developed. These protocols include Ovsynch Protocol (Pursley et al. 1997), Co-Synch (Geary et al. 2001), Pre Synch-Ovsynch (Moreira et al. 2001; Wiltbank and Pursley 2014), Heat Synch (Geary et al. 2000; Stevenson et al. 2004), etc. Many of these protocols are very much effective for improving fertility during the summer.

Embryo transfer has been found to be effective technology for revitalizing pregnancy rate during the summer months (Drost et al. 1999). However, embryo transfer is not a widely adopted technique, because of some limitations. Embryo transfer technology may be popularized and implemented successfully through huge out-reach program, lowering the cost of commercially available embryos, development of skill on handling frozen embryo and timed embryo transfer.

7.4.9 Livestock Disease Surveillance and Health Management

Livestock disease patterns are changing day by day under changing climatic scenario, and there is continuous occurrence of new and emergence type of diseases. A quick data analysis and generation of information is required for a timely and systematic delivery of control and eradication activities. Livestock disease surveillance is a key for generating data and use in risk analysis, detection of disease or infection, monitoring disease trends, early warning, and prevention of the spread of disease or infection (Merianos 2007; Kshirsagar et al. 2013). The importance of surveillance of the spread of vector-borne bluetongue virus in ruminants and camelid species, because of the appropriate climate conditions for vector maintenance, has been reported in endemic areas of South America, Central America, the Caribbean, Middle East, and Asia (Legisa et al. 2014; Saminathan et al. 2020; Rizzo et al. 2021). Surveillance data also permit evaluation of extreme weather-related casualties due to cyclones, floods, heat waves or cold snaps, and forecasting mitigation measures based on the analysis of extreme weather events. Better surveillance can build better adaptive capacity, if the resulting data are made available in a timely fashion and appropriate manner. One of the most critical hurdles to develop our understanding of climate-disease linkages is the lack of high-quality epidemiological data on disease incidence for many locations. Much attention should be given to collect sufficient epidemiological data from the areas where climate-sensitive or weather-related diseases or outbreaks are often appeared.

Modern computing tool is being used to improve surveillance through geographical information systems (GIS) and spatial statistical analyses. GIS is a computer-aided spatial data information management system. GIS enables the spatial data management (collection, storage maintenance), implements the analysis (statistics spatial modeling), and graphics display (figures, satellite images, remote sensing data and maps) for knowing disease incidence, prevalence, mortality, morbidity, transmission pattern of diseases or infections on farm, region, or national levels (Yong et al. 2006). GIS helps in better monitoring and surveillance through mapping of different locations of farms and animals on a single map and thus permits in diseases forecasting, prediction of outbreaks, identification of disease cluster or hotspot, and taking different mitigations strategies to prevent the spread of diseases or infections during sudden disease outbreaks of livestock that might occur as a result of climate change (Siddiqui et al. 2018). Basically, GIS can be used in a real-time outbreak notification and mitigation measures.

Information systems are the crucial tool for making information available for risk analysis. At the international level, the World Animal Health Information System (WAHIS) of the World Organization for Animal Health (OIE) is the reference for all national or regional systems. Through its web interface and e-alert system, WAHIS disseminates a large quantity of information on animal disease outbreaks worldwide (Ben Jebara 2007). At the national level, several countries have Web-GIS based on a variety of disease specific and non-disease specific information systems by which the veterinary service can visualize and analyze geographic distribution of diseases on the Internet, make plan for the inspections of farm where outbreak occurs, and so expedite the process of disease control and eradication (Colangeli et al. 2011; Di Lorenzo et al. 2019). More and more countries are now using simple, low-cost GIS for mapping and management of outbreak of livestock diseases. For examples, GIS offers the capability to demonstrate vector–environment relationships and potentially forecast the risk of disease outbreaks and eradication plan of theileriosis (Lessard et al. 1990), trypanosomiosis (Rogers 1991), East Coast fever (Lawrence 1991), avian influenza, and Rift valley fever (Kshirsagar et al. 2013). Thus, preventive health measures such as vaccinations, deworming, and stress reduction management may be followed in right time in right location to control the outbreak of diseases or infections.

7.4.10 Manure Management

Different manure management options are available to alleviate emissions of CH_4 and N_2O during handling, storage, and spreading of manures. Dispersion of manure on pasture is a common practice in the world's livestock production system and thus emissions of CH_4 and N_2O occur in the field. But it is often difficult to take mitigation measures when manure is dispersed on pasture (Dickie et al. 2014). The method of application and timing, selection of better match plant for their requirements, and avoid of heavy rains are the important considerations for reducing nitrogen losses to the environment (Smith and Conen 2004; Van Groenigen et al. 2010). The emissions of N_2O in crop and grazing lands can also be diminished by using nitrification inhibitors (Clough et al. 2009; Snyder et al. 2009).

The emissions of N_2O from the open fields can be prevented by using manure collection and storage practices that reduce nitrogen losses to the environment from volatilization and leaching (Smith et al. 2008). The collection, storing, covering, and compacting of manure are the simple techniques to prevent emissions of N_2O from the open fields (Chadwick 2005; Chadwick et al. 2011). The anaerobic digestion of manure slurries or liquid manure using anaerobic digester prior to apply to soils is one of the sophisticated and promising practices to mitigate GHG emissions from manure (Masse et al. 2003a, b; Yang et al. 2019). This technology is convenient for relatively confined production systems available in cattle, pig and poultry farm, and highly profitable in warm climates (Gerber et al. 2008). Anaerobic digestion of manure is also useful for producing biogas containing 60–80% $CH_{4,}$ depending on the substrate and operational conditions and meeting on-farm energy demands and thus bringing down both GHG emissions and costs (Roos et al. 2004; Gerber et al. 2013). Anaerobic digesters are the tanks where manure is digested under anaerobic conditions to produce biogas and combust it for generating renewable energy. This process converts CH_4 into CO_2 and reduces the potential of GHG emissions (ICF International 2013). As the anaerobic digester is costly, alternatively manure may be collected in some digging place or tanks and then covered tightly for anaerobic digestion of manure to mitigate GHG emissions, particularly CH_4 (ICF International 2013).

In smallholder farming system, the livestock owners do not utilize farmyard manure properly, particularly in low-income countries where manure may be a cause of threat for both human health and the environment. Composting and vermicomposting processes may be one option for the biological stabilization of solid waste (Lazcano et al. 2008). The limitations of composting include reduction of agronomic value of compost, GHG emission, loss of nutrients during compost making, and energy loss due to labor involvement or fuel required for turning the compost heap to ensure aeration (Hao et al. 2001). Vermicomposting is a mesophilic bio-oxidation and stabilization process of organic materials performed by the joint action of earthworm and microorganisms (Nagavallemma et al. 2004; Chew et al. 2019). It is a bio-ecologically effective, economically sustainable, and eco-friendly waste management technology that helps to reduce environmental impacts and associated health risks (Mahaly et al. 2018; Yadav and Garg 2019; Raza et al. 2021). Vermicomposts are excellent sources of biofertilizer and earthworms are the good source of protein to fishes and monogastric animals (Kopec et al. 2018; Hussain and Abbasi 2018). Vermicomposting has a higher potential in conserving nutrients compared to composting, stockpiling, and anaerobic digestion and besides, GHG emission during vermicomposting is lesser than the emission during composting and stockpiling (Jjagwe et al. 2019). Nigussie et al. (2016) reported that vermicomposting could lower the emissions of N_2O and CH_4 by 25–36 and 22–26%, respectively. Figures 7.9 and 7.10 show stockpiling of cow dung and vermicomposting in Bhojpur, Bihar, India, respectively.

The solids separator may be used for farm waste management purposes (Dickie et al. 2014). The solids separator is useful in confinement systems. This separator removes solids like bedding materials from manure streams, which then enter in the treatment or storage systems. CH_4 emissions are decreased and crust formation is prevented by removing the solids from manure streams (ICF International 2013). Though this practice requires more time and more effort, it is generally low-cost technology.

Figure 7.9 Stockpiling of cow dung in Bhojpur, Bihar, India.

Figure 7.10 Vermicomposting in Bhojpur, Bihar, India.

7.5 Awareness and Capacity Development of the Stakeholders

Climate change is a global issue. Livestock and livestock products supply chains are an international business. Hence, livestock farming-related mitigation actions need to be addressed at local, region, and global perspectives. The fact is that the general population is conscious about some popular environmental issues like air or water pollution, industrial pollution, deforestation, or wildlife conservation (Dunlap and Jorgenson 2012). But, the people are hardly aware of the environmental damages caused by the animal food production and supplying sector and thus they continue to consume animal products (Krystallis et al. 2009). Escarcha et al. (2018) point out that there is a lack of information on impacts of climate change on livestock and impact-specific adaptation options across the countries in the world, particularly in Asia and South America and thus suggest to put forward more research and extension work on climate change issues and mitigation strategies in livestock systems in developing countries.

To understand and deal with climate change, the awareness and the capacity of the livestock producers on mitigation strategies need to be improved on priority basis. All stakeholders (private and public sector, civil society, research and academia, and international organizations) should be involved for their participation in campaigning toward the diversity and complexity of agriculture system, animal husbandry, and climate change issues. Further, various programs on multi-stakeholder round-table meeting, consumers' awareness, farmers' awareness, farmers' networks, farmers' training and capacity building, farmer field schools, etc. need to be organized for the adoption of better technologies and mitigation practices.

Mitigation technologies may not be adopted by the majority of the farmers for a number of reasons. The farmers may consider risk factors, labor and cash constraints, costs of technologies, uncertainty of benefits, sociocultural norms, and unstable markets. While the farmers are more concerned about the agricultural production and marketing of agriculture produces, climate change and its effect on agriculture and livestock production system may not be considered serious issue in the farmers' agendas in many cases. Education, family farm succession planning, and social interaction among farmers and farming communities may improve farmers' risk perception of climate change and farmer decision-making ability (Barnes 2013). It is thus need of the hour to develop capacity of the farmers and farmers' family members toward identification of the problems and adopt climate change adaptation and mitigation measures (Jones et al. 2013).

Multi-stakeholder platforms would have a significant role to play in the current issue. Public and private sector stakeholders, pressure groups, farmers associations, supermarkets and other retailers, support services, and consumers need to work together to develop a viable, credible strategy and to implement an appropriate agenda of action. Support structures need to be strengthened for enabling the producers toward adoption of new technologies. On one hand, public investment in basic infrastructures like roads, storage, broadband, connectivity, etc. is necessary; and on other hand, subsidies and provision of microfinance schemes on new technology adaption and practices should be envisaged.

7.6 Conclusions

Livestock production system has been marked as a major contributor to GHGs for global climate change. At the same time, the negative effect of climate change on livestock production has been figured out. However, there is still limited information regarding the impacts of climate change on livestock production system across different global agro-climatic zones. Most of the studies were country-specific, and most of those countries were from North America, Europe, and Australia continents. More research on climate change and its effects on livestock are required from the countries under Asia, Africa, and South America continents. There is a demand of country-wise regional or local studies under varied agro-climatic and socioeconomic conditions. Most of the research concerning the effect of climate change on livestock production focuses on cattle; more studies on nonruminants are required.

There is a dearth of information related to the current use of adaptation and mitigation measures to combat effects of climate change on different livestock species and livestock farming systems defined for different locations or regions under varied agro-climatic and socioeconomic situations. Technologies and practices are available for reduction of GHG emissions, but are not widely used. Their adoption and use by the farmers can reduce GHG emissions by livestock significantly. There is a range of mitigation options that have high potential, but require further testing and refinement at the farmers' field before they can be considered technically feasible and economically viable and culturally fit. Based on evidences, adaptation strategies will be planned in such a manner that livestock can ascertain its role in meeting the demands of the growing population for animal-based products under climate change. More detailed adaptation research is needed for recommending more adaptation options for local, national, and regional-level policy making. Public and private sector involvements also have a vital role to play in supporting research and development to provide benefits of mitigation options and advance the applicability of the existing technologies and practices in livestock farming.

The present information is an important first step toward understanding the challenges the livestock farmers will face due to climate change. This understanding will enable the farmers to plan the location-specific, farm-specific, livestock species-specific mitigation measures for the future. However, constant education and sensitization are essential in order to aware and make the farmers ready to adapt mitigation measures to combat the possible effects of climate change. The success of mitigation strategies depends upon how efficiently those strategies are being transferred to the farming communities. There is a need to enhance the capacity of livestock producers through organizing suitable capacity development program to understand climate change science, vulnerability, and deal with the impacts of climate change on livestock and ecosystems. Hence, various extension activities on climate change scenario and mitigation strategies will help the farmers to adapt to climate change and alleviate its negative impacts on livestock production system.

References

Adams, R.M., Hurd, B.H., Lenhart, S., and Leary, N. (1998). Effects of global climate change on agriculture: an interpretative review. *Climate Research* 11: 19–30.

Adesogan, A.T., Havelaar, A.H., McKune, S.L. et al. (2020). Animal source foods: sustainability problem or malnutrition and sustainability solution? Perspective matters. *Global Food Security* https://doi.org/10.1016/j.gfs.2019.100325.

Aggarwal, A. and Singh, M. (2006). Effect of water cooling on physiological responses, milk production and composition of Murrah buffaloes during hot-humid season. *Indian Journal of Dairy Science* 59 (6): 386–389.

Agriculture Census (2015–2016). All India Report on Agriculture Census 2015-16. Department of Agriculture, Cooperation and Farmers Welfare, Ministry of Agriculture and Farmers Welfare, Government of India, New Delhi, India.

Ahmad, M., Bhatti, J.A., Abdullah, M. et al. (2017). Effect of different ambient management interventions on milk production and physiological performance of lactating Nili-Ravi buffaloes during hot humid summer. *Livestock Research for Rural Development* 29: 230.

Akbarian, A., Michiels, J., Degroote, J. et al. (2016). Association between heat stress and oxidative stress in poultry; mitochondrial dysfunction and dietary interventions with phytochemicals. *Journal of Animal Science and Biotechnology* 7: 37. https://doi.org/10.1186/s40104-016-0097-5.

Alam, M.Z., Carpenter-Boggs, L., Mitra, S. et al. (2017). Effect of salinity intrusion on food crops, livestock, and fish species at Kalapara Coastal Belt in Bangladesh. *Journal of Food Quality* 2017: 1–23.

de Albuquerque Brasil, L.H., Wechesler, F.S., Baccari Júnior, F. et al. (2000). Thermal stress effects on milk yield and chemical composition and thermoregulatory responses of lactating alpines goats. *Revista Brasileira de Zootecnia* 29 (6): 1632–1641.

Alexandratos, N. and Bruinsma, J. (2012). *World Agriculture towards 2030/2050: The 2012 Revision*, ESA Working Paper No. 12-03, 1–154. Rome, Italy: FAO.

Allard, V., Soussana, J.-F., Falcimagne, R. et al. (2007). The role of grazing management for the net biome productivity and greenhouse gas budget (CO_2, N_2O and CH_4) of semi-natural grassland. *Agriculture, Ecosystems and Environment* 121: 47–58.

Ambazamkandi, P., Thyagarajan, G., Sambasivan, S. et al. (2015). Shelter design for different livestock from a climate change perspective. In: *Climate Change Impact on Livestock: Adaptation and Mitigation* (eds. V. Sejian, J. Gaughan, L. Baumgard and C. Prasad), 399–424. India: Springer https://doi.org/10.1007/978-81-322-2265-1_23.

Amundson, J.L., Mader, T.L., Rasby, R.J., and Hu, Q.S. (2006). Environmental effects on pregnancy rate in beef cattle. *Journal of Animal Science* 84 (12): 3415–3420.

Awasthi, M.K., Kavani, F.S., Siddiquee, G.M. et al. (2007). Is slow follicular growth the cause of silent estrus in water buffaloes? *Animal Reproduction Science* 99 (3–4): 258–268.

Ayo, J.O., Obidi, J.A., Rekwot, P.I. (2011). Effects of heat stress on the well-being, fertility, and hatchability of chickens in the northern Guinea savannah zone of Nigeria: a review. *International Scholarly Research Notices 2011*.

Barnes, A.P. (2013). Heterogeneity in climate change risk perception amongst dairy farmers: a latent class clustering analysis. *Applied Geography* 41: 105–115.

Barton, P.K. and Atwater, J.W. (2002). Nitrous oxide emissions and the anthropogenic nitrogen in wastewater and solid waste. *Journal of Environmental Engineering* 128: 137–150.

Bary, M.A., Ali, M.Z., Chowdhury, S. et al. (2018). Prevalence and molecular identification of haemoprotozoan diseases of cattle in Bangladesh. *Advances in Animal and Veterinary Sciences* 6: 176–182.

Bateman, E.J. and Baggs, E.M. (2005). Contributions of nitrification and denitrification to N2O emissions from soils at different water-filled pore space. *Biology and Fertility of Soils* 41: 379–388.

Baumgard, L. and Rhoads, R.P. (2012). Effects of environment on metabolism. In: *Environmental Physiology of Livestock* (eds. R.J. Collier and J.L. Collier), 81–100. United States: Wiley Blackwell.

Baumgard, L.H., Rhoads, R.P., Rhoads, M.L. et al. (2012). Impact of climate change on livestock production. In: *Environmental Stress and Amelioration in Livestock Production* (eds. L.H. Baumgard, R.P. Rhoads, M.L. Rhoads, et al.), 413–468. Berlin, Heidelberg: Springer.

Baylis, M. and Githeko, A.K. (2006). The Effects of Climate Change on Infectious Diseases of Animals. *Report for the foresight project on detection of infectious diseases, Department of trade and industry, UK Government 35.*

Beauchemin, K., Ungerfeld, E., Eckard, R., and Wang, M. (2020). Review: fifty years of research on rumen methanogenesis: lessons learned and future challenges for mitigation. *Animal* 14 (S1): S2–S16.

Bell, M. and Tzimiropoulos, G. (2018). Novel monitoring systems to obtain dairy cattle phenotypes associated with sustainable production. *Frontiers in Sustainable Food Systems* 2: 31. https://doi.org/10.3389/fsufs.2018.00031.

Ben Jebara, K. (2007). WAHIS and the role of the OIE's reference laboratories and collaborating centres. *Developmental Biology (Basel)* 128: 69–72.

Bentley, D. and Hegarty, R. (2008). Managing livestock enterprises in Australia's extensive rangelands for greenhouse gas and environmental outcomes: a pastoral company perspective. *Australian Journal of Experimental Agriculture* 48: 60–64.

Berckmans, D. (2014). Precision livestock farming technologies for welfare management in intensive livestock systems. *Revue Scientifique et Technique (Paris)* 33: 189–196.

Berlin, J. (2002). Environmental life cycle assessment (LCA) of Swedish semi-hard cheese. *International Dairy Journal* 12: 939–953.

Bernabucci, U., Basiricò, L., Morera, P. et al. (2015). Effect of summer season on milk protein fractions in Holstein cows. *Journal of Dairy Science* 98 (3): 1815–1827.

Biggar, S., Man, D., Moffroid, K. et al. (2013). *Greenhouse Gas Mitigation Options and Costs for Agricultural Land and Animal Production Within the United States.* Washington, DC: ICF International, Department of Agriculture Climate Change Program Office http://www.usda.gov/oce/climate_change/mitigation_technologies/GHG_Mitigation_Options.pdf.

Bohmanova, J., Misztal, I., and Cole, J.B. (2007). Temperature-humidity indices as indicators of milk production losses due to heat stress. *Journal of Dairy Science* 90: 1947–1956.

Bohmanova, J., Misztal, I., Tsuruta, S. et al. (2008). Short communication: genotype by environment interaction due to heat stress. *Journal of Dairy Science* 91 (2): 840–846.

Booker, F.L. and Miller, J.E. (1998). Phenylpropanoid metabolism and phenolic composition of soybean [*Glycine max* (L.) Merr.] leaves following exposure to ozone. *Journal of Experimental Botany* 49 (324): 1191–1202.

Bouwman, T. (1996). Direct emission of nitrous oxide from agricultural soils. *Nutrient Cycling in Agroecosystems* 46: 53–70.

Bracke, M.B.M. (2011). Review of wallowing in pigs: description of the behaviour and its motivational basis. *Applied Animal Behaviour Science* 132: 1–13.

Bruce, W.B., Edmeades, G.O., and Barker, T.C. (2002). Molecular and physiological approaches to maize improvement for drought tolerance. *Journal of Experimental Botany* 53 (366): 13–25.

Cabezon, F.A., Schinckel, A.P., and Stwalley, R.M. III (2017). Thermal capacity of hog-cooling pad. *Applied Engineering in Agriculture* 33: 891.

Caminade, C., McIntyre, K.M., and Jones, A.E. (2019). Impact of recent and future climate change on vector-borne diseases. *Annals of the New York Academy of Sciences* 137: 119–129.

Campen, K.A., Abbott, C.R., Rispoli, L.A. et al. (2018). Heat stress impairs gap junction communication and cumulus function of bovine oocytes. *Journal of Reproduction and Development* 64: 385–392.

Carabaño, M.J., Ramón, M., Menéndez-Buxadera, A. et al. (2019). Selecting for heat tolerance. *Animal Frontiers* 9 (1): 62–68.

Cardoso, C.C., Peripolli, V., Amador, S.A. et al. (2015). Physiological and thermographic response to heat stress in zebu cattle. *Livestock Science* 182: 83–92.

Caro, D., Davis, S.J., Bastianoni, S., and Kaldeira, K. (2014). Global and regional trends in greenhouse gas emissions from livestock. *Climatic Change* 126: 203–216.

Casey, K.D., Bicudo, J.R., Schmidt, D.R. et al. (2006). Air quality and emissions from livestock and poultry production/waste management systems. In: *Animal Agriculture and the Environment* (eds. J.M. Rice, D.F. Caldwell and F.J. Humenik). St. Joseph, Mich: National Center for Manure and Waste Management White Papers, American Society of Agricultural and Biological Engineers, 40.

Chadwick, D.R. (2005). Emissions of ammonia, nitrous oxide and methane from cattle manure heaps: effect of compaction and covering. *Atmospheric Environment* 392005: 787–799.

Chadwick, D., Sommer, S., Thorman, R. et al. (2011). Manure management: implications for greenhouse gas emissions. *Animal Feed Science and Technology* 166-67: 514–531.

Chen, C.C., McCarl, B., and Chang, C.C. (2012). Climate change, sea level rise and rice: global market implications. *Climate Change* 110: 543–560.

Chew, K.W., Chia, S.R., Yen, H.W. et al. (2019). Transformation of biomass waste into sustainable organic fertilizers. *Sustainability* 11: 2266.

Clark, M. and Tilman, D. (2017). Comparative analysis of environmental impacts of agricultural production systems, agricultural input efficiency, and food choice. *Environmental Research Letters* 12: 064016.

Clough, T.J., Ray, J.L., Buckthought, L.E. et al. (2009). The mitigation potential of hippuric acid on N2O emissions from urine patches: an in situ determination of its effect. *Soil Biology and Biochemistry* 41 (10): 2222–2229.

Colangeli, P., Iannetti, S., Cerella, A. et al. (2011). The national information system for the notification of animal diseases in Italy. *Veterinaria Italiana* 47 (3): 303–312.

Collier, R.J. and Gebremedhin, K.G. (2015). Thermal biology of domestic animals. *Annual Review of Animal Biosciences* 3: 513–532.

Collier, R.J., Beede, D.K., Thatcher, W.W. et al. (1982). Influences of environment and its modification on dairy animal health and production. *Journal of Dairy Science* 65 (11): 2213–2227.

Collin, A., van Milgen, J., Dubois, S., and Noblet, J. (2001). Effect of high temperature and feeding level on energy utilization in piglets. *Journal of Animal Science* 79 (7): 1849–1857.

Cowley, F.C., Barber, D.G., Houlihan, A.V., and Poppi, D.P. (2015). Immediate and residual effects of heat stress and restricted intake on milk protein and casein composition and energy metabolism. *Journal of Dairy Science* 98 (4): 2356–2368.

Crippa, M., Solazzo, E., Guizzardi, D. et al. (2021). Food systems are responsible for a third of global anthropogenic GHG emissions. *Nature Food* 2: 198–209.

Cullen, B.R., Johnson, I.R., Eckard, R.J. et al. (2009). Climate change effects on pasture systems in South-Eastern Australia. *Crop & Pasture Science* 60 (10): 933–942.

Dai, S.F., Gao, F., Xu, X.L. et al. (2012). Effects of dietary glutamine and gamma-aminobutyric acid on meat colour, pH, composition, and water-holding characteristic in broilers under cyclic heat stress. *British Poultry Science* 53 (4): 471–481.

Das, K.S., Singh, J.K., Singh, G. et al. (2016). Heat stress amelioration measures in lactating Nili-Ravi buffaloes: effect on body weight changes, dry matter intake, milk production and economics. *Indian Journal of Animal Research* 50 (2): 242–249.

Dastagir, M.R. (2015). Modeling recent climate change induced extreme events in Bangladesh: a review. *Weather and Climate Extremes* 7: 49–60.

Davis, K.F., Gephart, J.A., Emery, K. et al. (2016). Meeting future food demand with current agricultural resources. *Global Environmental Change* 39: 125–132.

Davis-Reddy, C.L. and Vincent, K. (2017). *Climate Risk and Vulnerability: A Handbook for Southern Africa*, 2e. Pretoria, South Africa: CSIR.

Dechow, C.D. and Goodling, R.C. (2008). Mortality, culling by sixty days in milk, and production profiles in high- and low-survival Pennsylvania herds. *Journal of Dairy Science* 91: 4630–4639. https://doi.org/10.3168/jds.2008-1337.

Denef, K., Archibeque, S., and Paustian, K. (2011). Greenhouse Gas Emissions from U.S. Agriculture and Forestry: A Review of Emission Sources, Controlling Factors, and Mitigation Potential. *Interim report to USDA under Contract #GS-23F-8182H.*

Di Lorenzo, A., Savini, L., Candeloro, L. et al. (2019). Web-GIS and livestock trace tools for epidemiological surveillance, control and management. *Frontiers in Veterinary Science Conference Abstract: GeoVet 2019. Novel Spatio-temporal Approaches in the Era of Big Data.* doi: https://doi.org/10.3389/conf.fvets.2019.05.00041.

Dickie, A., Streck, C., Roe, S. et al. (2014). Strategies for mitigating climate change in agriculture: Abridged report. Climate focus and california environmental associates, prifadred with the support of the climate and land use Alliance. Report and Supplementary Materials. www.agriculturalmitigation.org.

Dreiling, C.E., Carman, F.S. III, and Brown, D.E. (1991). Maternal endocrine and fetal metabolic responses to heat stress. *Journal of Dairy Science* 74 (1): 312–327.

Drost, M., Ambrose, J.D., Thatcher, M.J. et al. (1999). Conception rates after artificial insemination or embryo transfer in lactating dairy cows during summer in Florida. *Theriogenology* 52: 1161–1167.

Dunlap, R.E. and Jorgenson, A.K. (2012). *Environmental Problems. The Wiley-Blackwell Encyclopedia of Globalization*. Hoboken, NJ: Blackwell Publishing Ltd.

Eagle, A.J., Olander, L.P., Henry, L.R. et al. (2012). *Greenhouse Gas Mitigation Potential of Agricultural Land Management in the United States. A Synthesis of Literature*, Report NI R 10-04, 3e. Durham, NC: Nicholas Institute for Environmental Policy Solutions, Duke University.

Eckstein, D., Hutfils, M.L., and Winges, M. (2018). *Global Climate Risk Index 2019*. Bonn, Germany: Germanwatch eV Available at http://www.germanwatch.org/en/cri.

Edwards, J.L. and Hansen, P.J. (1997). Differential responses of bovine oocytes and preimplantation embryos to heat shock. *Molecular Reproduction and Development: Incorporating Gamete Research* 46 (2): 138–145.

El-Tarabany, M.S. (2016). Effect of thermal stress on fertility and egg quality of Japanese quail. *Journal of Thermal Biology* 61: 38–43.

El-Wardani, M.A. and El-Asheeri, K. (2000). Influence of season and number of heat checks on detecting of ovulatory estrus in Egyptian buffaloes. *Egyptian Journal of Animal Production* 37 (1): 1–8.

El-Wishy, A.B. (2007). The postpartum buffalo: I. Endocrinological changes and uterine involution. *Animal Reproduction Science* 97 (3–4): 201–215.

EPA (2011). Global Anthropogenic Non-CO_2 Greenhouse Gas Emissions: 1990–2030. *Final Report, 182*. http://www.epa.gov/climatechange/EPAactivities/economics/nonco2projections.html.

Escarcha, J.F., Lassa, J.A., and Zander, K.K. (2018). Livestock under climate change: a systematic review of impacts and adaptation. *Climate* 6: 54. https://doi.org/10.3390/cli6030054.

European Commission (2006). Environmental Impact of Products (EIPRO): Analysis of the Life Cycle Environmental Impacts Related to the Total Final Consumption of the EU25. *European Commission Technical Report EUR 22284 EN.*

FAO (2002). *World Agriculture: Towards 2015/2030. Summary Report*. Rome, Italy: Food and Agriculture Organization of the United Nations.

FAO (2008). *Food Outlook: Global Market Analysis. Statistical Appendix Table A24- Selected International Prices for Oilcrop Products and Price Indices*. Rome, Italy: Food and Agriculture Organization of the United Nations.

FAO (2009). *The State of Food and Agriculture: Livestock in the Balance*. Rome, Italy: Food and Agriculture Organization of the United Nations.

FAO (2010). *Greenhouse Gas Emissions From the Dairy Sector: A Life Cycle Assessment*. Rome, Italy: Animal Production and Health Division, 10–11. Food and Agriculture Organization of the United Nations.

FAO (2011). *FAO Statistical Database, FAOSTAT*. Rome, Italy: Food and Agriculture Organization of the United Nations http://faostat.fao.org/site/567/default.aspx.

FAO (2013). Interactions among mitigation practices. In: *Mitigation of Greenhouse Gas Emissions in Livestock Production – A Review of Technical Options for Non-CO2 Emissions*, Animal Production and Health Paper No. 177 (eds. P.J. Gerber, B. Henderson and H. Makkar). Rome, Italy: Food and Agriculture Organization of the United Nations.

Farghaly, H.A. (1984). Thyroid and some ovarian hormones and their relation to reproduction in Egyptian water buffalo. MSc. Thesis. Faculty of Agriculture Cairo, University, Cairo.

Feng, J., Zhang, M., Zheng, S. et al. (2008). Effects of high temperature on multiple parameters of broilers in vitro and in vivo. *Poultry Science* 87 (10): 2133–2139.

Finocchiaro, R., van Kaam, J.B.C.H.M., Portolano, B., and Misztal, I. (2005). Effect of heat stress on production of Mediterranean dairy sheep. *Journal of Dairy Science* 88: 1855–1864.

Fließbach, A., Oberholzer, H.-R., Gunst, L., and Mader, P. (2007). Soil organic matter and biological soil quality indicators after 21 years of organic and conventional farming. *Agriculture, Ecosystems and Environment* 118: 273–284.

Flysj€o, A., Henriksson, M., Cederberg, C. et al. (2011). The impact of various parameters on the carbon footprint of milk production in New Zealand and Sweden. *Agricultural Systems* 104: 459–469.

Fox, N.J., Marion, G., Davidson, R.S. et al. (2015). Climate-driven tipping points could lead to sudden, high-intensity parasite outbreaks. *Royal Society Open Science* 2: 140296.

Fuller, T.L., Gilbert, M., Martin, V. et al. (2013). Predicting hotspots for influenza virus reassortment. *Emerging Infectious Diseases* 19 (4): 581–588.

Gabler, N.K. and Spurlock, M.E. (2008). Integrating the immune system with the regulation of growth and efficiency. *Journal of Animal Science* 86 (Suppl 14): E64–E74.

Garnett, T. (2009). Livestock-related greenhouse gas emissions: impacts and options for policy makers. *Environmental Science and Policy* 12: 491–503.

Gaughan, J.B., Bonner, S., Loxton, I. et al. (2010). Effect of shade on body temperature and performance of feedlot steers. *Journal of Animal Science* 88 (12): 4056–4067.

Gauly, M. and Ammer, S. (2020). Review: challenges for dairy cow production systems arising from climate changes. *Animal* 14 (S1): S196–S203.

Geary, T.W., Downing, E.R., Bruemmer, J.E., and Whittier, J.C. (2000). Ovarian and estrous response of suckled beef cows to the select synch estrous synchronization protocol. *The Professional Animal Scientist* 16: 1–5.

Geary, T.W., Salverson, R.R., and Whittier, J.C. (2001). Synchronization of ovulation using GnRH or hCG with the CO-synch protocol in suckled beef cows. *Journal of Animal Science* 79 (10): 2536–2541.

Gerbens-Leenes, P.W. and Nonhebel, S. (2002). Consumption patterns and their effects on land required for food. *Ecological Economics* 42 (S): 185–199.

Gerber, P. and Vellinga, T. (2009). GHG emissions from animal food chains. Development of a quantification model using life cycle analysis approach. 23 September 2009, IDS, Berlin.

Gerber, P.J., Carsjens, G.J., Pak-uthai, T., and Robinson, T.P. (2008). Decision support for spatially targeted livestock policies: diverse examples from Uganda and Thailand. *Agricultural Systems* 96: 37–51.

Gerber, P.J., Steinfeld, H., Henderson, B. et al. (2013). *Tackling Climate Change Through Livestock: A Global Assessment of Emissions and Mitigation Opportunities*. Rome, Italy: Food and Agriculture Organization of the United Nations.

de Greef, J. (2009). Actual standing and perspectives for the sustainable use and development of parasite resistant or tolerant breeds in developed regions: Australia and NZ as an example. Presentation, Joint FAO/INRA Workshop on animal genetic resources and their resistance/tolerance to diseases, with special focus on parasitic diseases in ruminants, Paris (10 July 2009).

Habeeb, A.A.M., Gad, A.E., and El-Tarabany, A.A. (2012). Effect of hot climatic conditions with different types of housing on productive efficiency and physiological changes in buffalo calves. *Isotope and Radiation Research* 44 (1): 109–126.

Haider, N., Sturm-Ramirez, K., Khan, S.U. et al. (2017). Unusually high mortality in waterfowl caused by highly pathogenic avian influenza A (H5N1) in Bangladesh. *Transboundary and Emerging Diseases* 64: 144–156.

Halachmi, I. and Guarino, M. (2016). Editorial: precision livestock farming: a per animal approach using advanced monitoring technologies. *Animal* 10: 1482–1483.

Haldar, A., Das, D., Saha, B. et al. (2017). Smallholder pig farming for rural livelihoods and food security in North East India. *Journal of Animal Research* 7 (3): 471–481.

Hansen, P.J. (2004). Physiological and cellular adaptations of zebu cattle to thermal stress. *Animal Reproduction Science* 82: 349–360.

Hansen, P.J. and Areéchiga, C.F. (1999). Strategies for managing reproduction in the heat-stressed dairy cow. *Journal of Animal Science* 77 (Suppl 2): 36–50.

Hao, X., Chang, C., Larney, F.J., and Travis, G.R. (2001). Greenhouse gas emissions during cattle feedlot manure composting. *Journal of Environmental Quality* 30 (2): 376–386.

Hauschild, L., Lovatto, P.A., Pomar, J., and Pomar, C. (2015). Development of sustainable precision farming systems for swine: estimating real-time individual amino acid requirements in growing-finishing pigs. *Journal of Animal Science* 90: 2255–2263.

Hayes, B.J., Bowman, P.J., Chamberlain, A.J. et al. (2009). A validated genome wide association study to breed cattle adapted to an environment altered by climate change. *PLoS One* 4: e6676. https://doi.org/10.1371/journal.pone.0006676.

Henderson, B., Gerber, P., Hilinski, T. et al. (2015). Greenhouse gas mitigation potential of the world's grazing lands: modelling soil carbon and nitrogen fluxes of mitigation practices. *Agriculture, Ecosystems and Environment* 207: 91–100.

Henry, B., Charmley, E., Eckard, R. et al. (2012). Livestock production in a changing climate: adaptation and mitigation research in Australia. *Crop & Pasture Science* 63: 191–202.

Herrero, M. and Thornton, P.K. (2013). Livestock and global change: emerging issues for sustainable food systems. *The Proceedings of the National Academy of Sciences United States of America* 110: 20878–20881.

Herrero, M., Thornton, P.K., Kruska, R., and Reid, R.S. (2008). Systems dynamics and the spatial distribution of methane emissions from African domestic ruminants to 2030. *Agriculture, Ecosystems, and Environment* 126: 122–137.

Herrero, M., Thornton, P.K., Notenbaert, A.M. et al. (2010). Smart investments in sustainable food production: revisiting mixed croplivestock systems. *Science* 327: 822–825.

Herrero, M., Thornton, P.K., Notenbaert, A. et al. (2012). *Drivers of Change in Crop-Livestock Systems and Their Potential Impacts on Agro-Ecosystems Services and Human Wellbeing to 2030: A Study Commissioned by the CGIAR System Wide Livestock Programme*. Nairobi, Kenya: International Livestock Research Institute.

Herrero, M., Havlík, P., Valin, H. et al. (2013). Biomass use, production, feed efficiencies, and greenhouse gas emissions from global livestock systems. *The Proceedings of the National Academy of Sciences United States of America* 110 (52): 20888–20893.

Hoffmann, M.T. and Vogel, C. (2008). Climate change impacts on African rangelands. *Rangelands* 30: 12–17.

Hooda, O.K. and Singh, G. (2010). Effect of thermal stress on feed intake, plasma enzymes and blood biochemicals in buffalo heifers. *Indian Journal of Animal Nutrition* 27 (2): 122–127.

Hounghton, J.T., Ding, Y., Griggs, D.J. et al. (2001). *Climate Change: The Scientific Basis. Contribution of Working Group I to the Third Assessment Report of the Intergovernmental Panel on Climate Change*. New York: Cambridge University Press.

Howden, S.M., Crimp, S.J., and Stokes, C.J. (2008). Climate change and Australian livestock systems: impacts, research and policy issues. *Australian Journal of Experimental Agriculture* 48 (7): 780–788.

Hurst, P., Termine, P., and Karl, M. (2005). *Agricultural Workers and Their Contribution to Sustainable Agriculture and Rural Development*. Rome: FAO.

Hussain, N. and Abbasi, S.J.S. (2018). Efficacy of the vermicomposts of different organic wastes as clean fertilizers: state-of-the-art. *Sustainability* 10: 1205.

Huynh, T.T.T., Aarnink, A.J.A., Truong, C.T. et al. (2006). Effects of tropical climate and water cooling methods on growing pigs' responses. *Livestock Science* 104: 278–291.

IPCC (2001). Technical Summary: Contribution of Working Group I to the Third Assessment Report. *Climate Change 2001: Synthesis Report*. Intergovernmental Panel on Climate Change.

IPCC (2006). Emissions from livestock and manure management. In: *2006 IPCC Guidelines for National Greenhouse Gas Inventories* (eds. H.S. Eggleston, L. Buendia, K. Miwa, et al.). Japan: IGES.

IPCC (2007). Summary for policymakers. In: climate change 2007: mitigation. In: *Contribution of Working Group III to the Fourth Assessment Report of the Intergovernmental Panel on Climate Change* (eds. B. Metz, O.R. Davidson, P.R. Bosch, et al.). Cambridge, UK: Cambridge University Press.

IPCC (2013). Impacts, adaptation and vulnerability. Intergovernmental Panel on Climate Change. http://ipcc-wg2.gov/AR5/images/uploads/WGIIAR5-TS_FGDall.pdf.

IPCC (2014). Climate change 2014: impacts, adaptation, and vulnerability. In: *Part A: Global and Sectoral Aspects* (eds. C.B. Field, V.R. Barros, D.J. Dokken, et al.). Contribution of working group II to the fifth assessment report of the Intergovernmental Panel on Climate Change. Cambridge, UK: Cambridge University Press.

Jan Kramer, K., Moll, H.C., Nonhebel, S., and Wilting, H.C. (1999). Greenhouse gas emissions related to Dutch food consumption. *Energy Policy* 27: 203–216.

Jeon, J., Yeon, S., Choi, Y. et al. (2006). Effects of chilled drinking water on the performance of lactating sows and their litters during high ambient temperatures under farm conditions. *Livestock Science* 105: 86–93.

Jjagwe, J., Komakech, A.J., Karungi, J. et al. (2019). Assessment of a cattle manure vermicomposting system using material flow analysis: a case study from Uganda. *Sustainability* 11: 5173. https://doi.org/10.3390/su11195173.

Johnston, L.J., Ellis, M., Libal, G.W. et al. (1999). Effect of room temperature and dietary amino acid concentration on performance of lactating sows. NCR-89 committee on swine management. *Journal of Animal Science* 77: 1638–1644.

Jones, H.G. and Corlett, J.E. (1992). Current topics in drought physiology. *The Journal of Agricultural Science* 119 (3): 291–296.

Jones, P.G. and Thornton, P.K. (2009). Croppers to livestock keepers: livelihood transitions to 2050 in Africa due to climate change. *Environmental Science and Policy* 12: 427–437.

Jones, A.K., Jones, D.L., Edwards-Jones, G., and Cross, P. (2013). Informing decision making in agricultural greenhouse gas mitigation policy: a best-worst scaling survey of expert and farmer opinion in the sheep industry. *Environmental Science and Policy* 29: 46–56.

Kadokawa, H., Sakatani, M., and Hansen, P.J. (2012). Perspectives on improvement of reproduction in cattle during heat stress in a future Japan. *Animal Science Journal* 83 (6): 439–445.

Kala, M., Shaikh, M.V., and Nivsarkar, M. (2017). Equilibrium between anti-oxidants and reactive oxygen species: a requisite for oocyte development and maturation. *Reproductive Medicine and Biology* 16 (1): 28–35.

Kantanen, J., Løvendahl, P., Strandberg, E. et al. (2015). Utilization of farm animal genetic resources in a changing agro-ecological environment in the Nordic countries. *Frontiers in Genetics* 6: 52.

Karnati, H.K., Pasupuleti, S.R., Kandi, R. et al. (2015). TLR-4 signalling pathway: MyD88 independent pathway up-regulation in chicken breeds upon LPS treatment. *Veterinary Research Communications* 39 (1): 73–78.

Kates, R.W., Travis, W.R., and Wilbanks, T.J. (2012). Transformational adaptation when incremental adaptations to climate change are insufficient. *The Proceedings of the National Academy of Sciencesin the United States of America* 109 (19): 7156–7161.

Keady, T.W.J., Marley, C.M., and Scollan, N.D. (2012). Grass and alternative forage silages for beef cattle and sheep: Effects on animal performance. *Proceedings of the XVI International Silage Conference*, Hämeenlinna, Finland (2–4 July 2012).

Khan, R.I.N., Sahu, A.R., Malla, W.A. et al. (2021). Systems biology under heat stress in Indian cattle. *bioRxiv* https://doi.org/10.1101/2020.04.09.031153.

Kløverpris, J., Wenzel, H., and Nielsen, P.H. (2008). Life cycle inventory modelling of land use induced by crop consumption part 1: conceptual analysis and methodological proposal. *International Journal of Life Cycle Assessment* 13 (1): 13–21.

Knox, J., Hess, T., Daccache, A., and Wheeler, T. (2012). Climate change impacts on crop productivity in Africa and South Asia. *Environmental Research Letters* 7 (3): 034032.

Kopec, M., Gondek, K., Mierzwa-Hersztek, M., and Antonkiewicz, J. (2018). Factors influencing chemical quality of composted poultry waste. *Saudi Journal of Biological Science* 25: 1678–1686.

Korde, J.P., Jadhao, S.V., Varshney, V.P. et al. (2006). Long term effects of heat exposure on nutrient digestibility and digesta flow rate in buffalo calves. *Buffalo Bulletin* 22 (1): 25–32.

Krystallis, A., de Barcellos, M.D., Kügler, J.O. et al. (2009). Attitudes of European citizens towards pig production systems. *Livestock Science* 126: 46–56.

Kshirsagar, D.P., Savalia, C.V., Kalyani, I.H. et al. (2013). Disease alerts and forecasting of zoonotic diseases: an overview. *Veterinary World* 6 (11): 889–896.

Kumar, S.B.V., Ajeet, K., and Meena, K. (2011). Effect of heat stress in tropical livestock and different strategies for its amelioration. *Journal of Stress Physiology and Biochemistry* 7 (1): 45–54.

Lacetera, N. (2019). Impact of climate change on animal health and welfare. *Animal Frontiers* 9: 26–31.

Lacetera, N., Bernabucci, U., Ronchi, B., and Nardone, A. (1996). Body condition score, metabolic status and milk production of early lactating dairy cows exposed to warm environment. *Rivista Di Agricoltura Subtroppicale Troppicale* 90 (1): 43–55.

Lacetera, N., Segnalini, M., Bernabucci, U. et al. (2013). Climate induced effects on livestock population and productivity in the Mediterranean area. In: *Regional Assessment of Climate Change in the Mediterranean. Advances in Global Change Research 51* (eds. A. Navarra and L. Tubiana), 135–156. Dordrecht, The Netherlands: Springer Science + Business Media.

Lambert, G.P. (2009). Stress-induced gastrointestinal barrier dysfunction and its inflammatory effects. *Journal of Animal Science* 87 (Suppl 14): E101–E108.

Lara, L.J. and Rostagno, M.H. (2013). Impact of heat stress on poultry production. *Animals* 3 (2): 356–369.

Lawrence, J.A. (1991). Retrospective observations on the geographical relationship between Rhipicephalus appendiculatus and East Coast fever in southern Africa. *Veterinary World* 128 (8): 180–183.

Lazcano, C., Gomez-Brand, M., and Dominguez, J. (2008). Comparison of the effective of composting and vermicomposting for the biological stabilization of cattle manure. *Chemosphere* 72 (7): 1013–1019.

Lecchi, C., Rota, N., Vitali, A. et al. (2016). in vitro assessment of the effects of temperature on phagocytosis, reactive oxygen species production and apoptosis in bovine polymorphonuclear cells. *Veterinary Immunology and Immunopathology* 182: 89–94.

Legisa, D.M., Gonzalez, F.N., and Dus Santos, M.J. (2014). Bluetongue virus in South America, Central America and the Caribbean. *Virus Research* 182: 87–94.

Leip, A., Ledgard, S., Uwizeye, A. et al. (2019). The value of manure: manure as co-product in life cycle assessment. *Journal of Environmental Management* 241: 293–304.

Lessard, P., L'Eplattenier, R.L., Norval, R.A.I. et al. (1990). Geographical information systems for studying the epidemiology of cattle diseases caused by *Theileria parva*. *Veterinary Record* 126: 255–262.

Liao, C.W. and Veum, T.L. (1994). Effects of dietary energy intake by gilts and heat stress from days 3 to 24 or 30 after mating on embryo survival and nitrogen and energy balance. *Journal of Animal Science* 72 (9): 2369–2377.

Liu, Z., Ezernieks, V., Wang, J. et al. (2017). Heat stress in dairy cattle alters lipid composition of milk. *Scientific Reports* 7 (1): 1–10.

Long, S.P., Ainsworth, E.A., Rogers, A., and Ort, D.R. (2004). Rising atmospheric carbon dioxide: plants FACE the future. *Annual Review of Plant Biology* 55: 591–628.

Lowder, S.K., Skoet, J., and Singh, S. (2014). *What Do We Really Know About the Number of Farms in the World?* ESA Working Paper No. 14-02. Rome, Italy: Agricultural Development Economics Division: FAO.

Lowder, S.K., Skoet, J., and Raney, T. (2016). The number, size, and distribution of farms, smallholder farms, and family farms worldwide. *World Development* 87: 16–29.

Loyau, T., Berri, C., Bedrani, L. et al. (2013). Thermal manipulation of the embryo modifies the physiology and body composition of broiler chickens reared in floor pens without affecting breast meat processing quality. *Journal of Animal Science* 91 (8): 3674–3685.

Mack, L.A., Felver-Gant, J.N., Dennis, R.L., and Cheng, H.W. (2013). Genetic variations alter production and behavioral responses following heat stress in 2 strains of laying hens. *Poultry Science* 92 (2): 285–294.

Mahaly, M., Senthilkumar, A.K., Arumugam, S. et al. (2018). Vermicomposting of distillery sludge waste with tea leaf residues. *Sustainable Environment Research* 28: 223–227.

Maiti, S.J., Jha, S.K., Garai, S. et al. (2014). Adapting to climate change: traditional coping mechanism followed by the Brokpa pastoral nomads of Arunachal Pradesh, India. *Indian Journal of Traditional Knowledge* 13: 752–761.

Manninen, M., Honkavaara, M., Jauhiainen, L. et al. (2011). Effects of grass-red clover silage digestibility and concentrate protein concentration on performance, carcass value, eating quality and economy of finishing Hereford bulls reared in cold conditions. *Agricultural and Food Science* 20: 151–168.

Marai, I.F.M., El-Darawany, A.A., Fadiel, A., and Abdel-Hafez, M.A.M. (2008). Reproductive performance traits as affected by heat stress and its alleviation in sheep. *Tropical and Subtropical Agroecosystems* 8 (3): 209–234.

Marai, I., Daader, A., Soliman, A., and El-Menshawy, S. (2009). Non-genetic factors affecting growth and reproduction traits of buffaloes under dry management housing (in sub-tropical environment) in Egypt. *Livestock Research for Rural Development* 4: 6.

Marcillac-Embertson, N.M., Robinson, P.H., Fadel, J.G., and Mitloehner, F.M. (2009). Effects of shade and sprinklers on performance, behavior, physiology, and the environment of heifers. *Journal of Dairy Science* 92: 506–517.

Masse, D.I., Croteau, F., Patni, N.K., and Masse, L. (2003a). Methane emissions from dairy cow and swine manure slurries stored at 10°C and 15°C. *Canadian Biosystems Engineering* 45 (6): 1–6.

Masse, D.I., Masse, L., and Croteau, F. (2003b). The effect of temperature fluctuations on psychrophilic anaerobic sequencing batch reactors treating swine manure. *Bioresource Technology* 89: 57–62.

McDowell, R.E. (1972). *Improvement of Livestock Production in Warm Climates*, vol. 711. San Francisco, CA: Freeman.

McMichael, A.J., Powles, J.W., Butler, C.D., and Uaua, R. (2007). Food, livestock production, energy, climate change, and health. *The Lancet* 370 (9594): 1253–1263.

de Mello, J.L.M., Berton, M.P., de Cassia Dourado, R. et al. (2017). Physical and chemical characteristics of the longissimus dorsi from swine reared in climate controlled and uncontrolled environments. *International Journal of Biometeorology* 61 (10): 1723–1731.

Merianos, A. (2007). Surveillance and response to disease emergence. *Current Topics in Microbiology and Immunology* 315: 477L509.

Meyer, R.L., Kjær, T., and Revsbech, N.P. (2002). Nitrification and denitrification near a soil-manure interface studied with a nitrate-nitrite biosensor. *Soil Science Society of America Journal* 66: 498–506.

Miller, M.N., Zebarth, B.J., Dandie, C.E. et al. (2009). Influence of liquid manure on soil denitrifier abundance, denitrification and nitrous oxide emissions. *Soil Science Society of America Journal* 73 (3): 760–768.

Mirzad, A.N., Tada, T., Ano, H. et al. (2018). Seasonal changes in serum oxidative stress biomarkers in dairy and beef cows in a daytime grazing system. *Journal of Veterinary Medical Science* 80: 20–27.

Mishra, S.R. (2021). Thermoregulatory responses in riverine buffaloes against heat stress: an updated review. *Journal of Thermal Biology* 96: 102844.

Moreira, F., Orlandi, C., Risco, C.A. et al. (2001). Effects of presynchronization and bovine somatotropin on pregnancy rates to a timed artificial insemination protocol in lactating dairy cows. *Journal of Dairy Science* 84: 1646–1659.

Nagavallemma, K., Wani, S., Lacroix, S. et al. (2004). Vermicomposting: Recycling Wastes into Valuable Organic Fertilizer. *Global Theme on Agrecosystems Report no. 8*. pp. 20. International Crops Research Institute for the Semi-Arid Tropics. Patancheru, Andhra Pradesh, India.

Nandi, D., Roy, S., Bera, S. et al. (2011). The rearing system of black Bengal goat and their farmers in West Bengal, India. *Veterinary World* 4 (6): 254–257.

Nardone, A., Ronchi, B., Lacetera, N. et al. (2010). Effects of climate changes on animal production and sustainability of livestock systems. *Livestock Science* 130 (1–3): 57–69.

Nawab, A., Ibtisham, F., Li, G. et al. (2018). Heat stress in poultry production: mitigation strategies to overcome the future challenges facing the global poultry industry. *Journal of Thermal Biology* 78: 131–139.

Nepstad, L.S., Gerber, J.S., Hill, J.D. et al. (2019). Pathways for recent Cerrado soybean expansion: extending the soy moratorium and implementing integrated crop livestock systems with soybeans. *Environmental Research Letters* 14: 044029.

Nichi, M., Bols, P.E.J., Züge, R.M. et al. (2006). Seasonal variation in semen quality in *Bos indicus* and *Bos taurus* bulls raised under tropical conditions. *Theriogenology* 66 (4): 822–828.

Nigussie, A., Kuyper, T.W., Bruun, S., and de Neergaard, A. (2016). Vermicomposting as a technology for reducing nitrogen losses and greenhouse gas emissions from small-scale composting. *Journal of Cleaner Production* 139 https://doi.org/10.1016/j.jclepro.2016.08.058.

Noiva, R.M., Menezes, A.C., and Peleteiro, M.C. (2014). Influence of temperature and humidity manipulation on chicken embryonic development. *BMC Veterinary Research* 10 (1): 1–10.

Norton, T. and Berckmans, D. (2018). Precision livestock farming: the future of livestock welfare monitoring and management? In: *Animal Welfare in a Changing World* (ed. A. Butterworth). Wallingford: CABI.

Oenema, O. and Tamminga, S. (2005). Nitrogen in global animal production and management options for improving nitrogen use efficiency. *Science in China Series C: Life Sciences* 48: 871–887.

Ognik, K. and Sembratowicz, I. (2012). Stress as a factor modifying the metabolism in poultry. A review. *Annales UMCS, Zootechnica* 30 (2).

Olesen, J.E., Schelde, K., Weiske, A. et al. (2006). Modelling greenhouse gas emissions from European conventional and organic dairy farms. *Agriculture, Ecosystems and Environment* 112: 207–220.

Omazic, A., Bylund, H., Boqvist, S. et al. (2019). Identifying climate-sensitive infectious diseases in animals and humans in Northern regions. *Acta Veterinaria Scandinavica* 61: 53.

Omtvedt, I.T., Nelson, R.E., Edwards, R.L. et al. (1971). Influence of heat stress during early, mid and late pregnancy of gilts. *Journal of Animal Science* 32 (2): 312–317.

Oresanya, T.F., Beaulieu, A.D., and Patience, J.F. (2008). Investigations of energy metabolism in weanling barrows: the interaction of dietary energy concentration and daily feed (energy) intake. *Journal of Animal Science* 86 (2): 348–363.

Osei-Amponsah, R., Dunshea, F.R., Leury, B.J. et al. (2020). Heat stress impacts on lactating cows grazing Australian summer pastures on an automatic robotic dairy. *Animals* 10 (5): 869.

Pal, K., Patra, A.K., Sahoo, A., and Kumawat, P.K. (2015). Evaluation of several tropical tree foliages for methane production potential, degradability and rumen fermentation in vitro. *Livestock Science* 180: 98–105.

Pang, Z., Li, B., Xin, H. et al. (2010). Characterisation of an experimental water-cooled cover for sows. *Biosystems Engineering* 105: 439–447.

Papadopoulou, A., Petrotos, K., Stagos, D. et al. (2017). Enhancement of antioxidant mechanisms and reduction of oxidative stress in chickens after the administration of drinking water enriched with polyphenolic powder from olive mill waste waters. *Oxidative Medicine and Cellular Longevity*: 1–11.

Patra, A.K. (2012a). Estimation of methane and nitrous oxide emissions from Indian livestock. *Journal of Environmental Monitoring* 14: 2673–2684.

Patra, A.K. (2012b). Enteric methane mitigation technologies for ruminant livestock: a synthesis of current research and future directions. *Environmental Monitoring and Assessment* 184: 1929–1952.

Patra, A.K. (2013). The effect of dietary fats on methane emissions, and its other effects on digestibility, rumen fermentation and lactation performance in cattle: a meta-analysis. *Livestock Science* 155: 244–254.

Patra, A.K. (2014a). Trends and projected estimates of GHG emissions from Indian livestock in comparisons with GHG emissions from world and developing countries. *Asian-Australasian Journal of Animal Sciences* 27: 592–599.

Patra, A.K. (2014b). A meta-analysis of the effect of dietary fat on enteric methane production, digestibility and rumen fermentation in sheep, and a comparison of these responses between cattle and sheep. *Livestock Science* 162: 97–103.

Patra, A.K. (2016). Recent advances in measurement and dietary mitigation of enteric methane emissions in ruminants. *Frontiers in Veterinary Science* https://doi.org/10.3389/fvets.2016.00039.

Patra, A.K. (2017). Accounting methane and nitrous oxide emissions, and carbon footprints of livestock food products in different states of India. *Journal of Cleaner Production* 162: 678–686.

Patra, A.K. and Saxena, J. (2010). A new perspective on the use of plant secondary metabolites to inhibit methanogenesis in the rumen. *Phytochemistry* 71: 1198–1222.

Patra, A.K. and Yu, Z. (2012). Effects of essential oils on methane production and fermentation by, and abundance and diversity of, rumen microbial populations. *Applied and Environmental Microbiology* 78: 4271–4280.

Patra, A.K. and Yu, Z. (2013). Effective reduction of enteric methane production by a combination of nitrate and saponin without adverse effect on feed degradability, fermentation, or bacterial and archaeal communities of the rumen. *Bioresource Technology* 148: 352–360.

Patra, A.K. and Yu, Z. (2015a). Effects of adaptation of in vitro rumen culture to garlic oil, nitrate and saponin and their combinations on methanogenesis, fermentation, and abundances and diversity of microbial populations. *Frontiers in Microbiology* https://doi.org/10.3389/fmicb.2015.01434.

Patra, A.K. and Yu, Z. (2015b). Effects of garlic oil, nitrate, saponin and their combinations supplemented to different substrates on in vitro fermentation, ruminal methanogenesis, and abundance and diversity of microbial populations. *Journal of Applied Microbiology* 119: 127–139.

Patra, A.K., Park, T., Kim, M., and Yu, Z. (2017). Rumen methanogens and mitigation of methane emission by anti-methanogenic compounds and substances. *Journal of Animal Science and Biotechnology* 8 https://doi.org/10.1186/s40104-017-0145-9.

Patz, J.A., Graczyk, T.K., Geller, N., and Vittor, A.Y. (2000). Effects of environmental change on emerging parasitic diseases. *International Journal for Parasitology* 30 (12–13): 1395–1405.

Payne, W.J.A. (1990). *An Introduction to Animal Husbandry in the Tropics*, Tropical Agricultural Series, 4e. Harlow, England; New York: Longman Scientific & Technical; Wiley.

Peacock, C. and Sherman, D.M. (2010). Sustainable goat production: some global perspectives. *Small Ruminant Research* 89: 70–80.

Pereira, A.M., Baccari, F., Titto, E.A., and Almeida, J.A. (2008). Effect of thermal stress on physiological parameters, feed intake and plasma thyroid hormones concentration in Alentejana, Mertolenga, Frisian and Limousine cattle breeds. *International Journal of Biometeorology* 52 (3): 199–208.

Phelan, P., Morgan, E.R., Rose, H. et al. (2016). Predictions of future grazing season length for European dairy, beef and sheep farms based on regression with bioclimatic variables. *The Journal of Agricultural Science* 154 (5): 765.

Pica-Ciamarra, U., Baker, D., Morgan, N. et al. (2014). *Investing in the Livestock Sector: Why Good Numbers Matter, a Sourcebook for Decision Makers on how to Improve Livestock Data*. Rome, Italy: FAO.

Polley, H.W., Briske, D.D., Morgan, J.A. et al. (2013). Climate change and North American rangelands: trends, projections, and implications. *Rangeland Ecology & Management* 66 (5): 493–511.

Pollmann, D.S. (2010). Seasonal effects on sow herds: industry experience and management strategies. *Journal of Animal Science* 88 (Suppl 3): 9.

Pomar, C., Van Milgen, J., and Remus, A. (2019). Precision livestock feeding, principle and practice. In: *Poultry and Pig Nutrition: Challenges of the 21st Century* (eds. W.H. Hendriks, V. Mwa and L. Babinszky), 397–418. Wageningen: Wageningen Academic Publishers.

Pursley, J.R., Mee, M.O., and Wiltbank, M.C. (1995). Synchronization of ovulation in dairy cows using PGF2alpha and GnRH. *Theriogenology* 44: 915–923.

Pursley, J.R., Kosorok, M.R., and Wiltbank, M.C. (1997). Reproductive management of lactating dairy cows using synchronization of ovulation. *Journal of Dairy Science* 80: 301–306.

Purusothaman, M.R., Thiruvenkadan, A.K., and Karunanithi, K. (2008). Seasonal variation in body weight and mortality rate in Mecheri adult sheep. *Livestock Research for Rural Development* 20: 150. http://www.lrrd.org/lrrd20/9/thir20150.htm.

Quere, C.L., Peters, G.P., Andrew, R.M. et al. (2014). Global carbon budget 2013. *Earth System Science Data* 6: 235–263.

Rae, A. and Nayga, R. (2010). *Trends in Consumption, Production and Trade in Livestock and Livestock Products*, 11–34. Washington, DC: Island Press.

Rapsomanikis, G. (2015). *The Economic Lives of Smallholder Farmers: An Analysis Based on Household Data from Nine Countries*, 1–39. Rome, Italy: FAO.

Rauw, W.M., Rydhmer, L., Kyriazakis, I. et al. (2020). Prospects for sustainability of pig production in relation to climate change and novel feed resources. *Journal of the Science of Food and Agriculture* 100 (9): 3575–3586.

Ravagnolo, O. and Misztal, I. (2000). Genetic component of heat stress in dairy cattle, parameter estimation. *Journal of Dairy Science* 83 (9): 2126–2130.

Ravagnolo, O. and Misztal, I. (2002). Effect of heat stress on nonreturn rate in Holsteins: fixed-model analyses. *Journal of Dairy Science* 85: 3101–3106.

Raza, S.T., Tang, J.L., Ali, Z. et al. (2021). Ammonia volatilization and greenhouse gases emissions during vermicomposting with animal manures and biochar to enhance sustainability. *International Journal of Environmental Research and Public Health* 18 (1): 178. https://doi.org/10.3390/ijerph18010178.

Renaudeau, D., Collin, A., Yahav, S. et al. (2012). Adaptation to hot climate and strategies to alleviate heat stress in livestock production. *Animal* 6: 707–728.

Rhoads, M.L., Rhoads, R.P., VanBaale, M.J. et al. (2009). Effects of heat stress and plane of nutrition on lactating Holstein cows: I. Production, metabolism, and aspects of circulating somatotropin. *Journal of Dairy Science* 92 (5): 1986–1997.

Rhoads, R.P., Baumgard, L.H., Suagee, J.K., and Sanders, S.R. (2013). Nutritional interventions to alleviate the negative consequences of heat stress. *Advances in Nutrition* 4 (3): 267–276.

Rizzo, H., Balaro, M.F.A., Matos, A.C.D. et al. (2021). Is bluetongue virus a risk factor for reproductive failure in tropical hair sheep in Brazil? *Acta Scientiae Veterinariae* 49: 1812. https://doi.org/10.22456/1679-9216.112591.

Robinson, T.P., Wint, G.R.W., Conchedda, G. et al. (2014). Mapping the global distribution of livestock. *PLoS One* 9: e96084.

Rogers, D.J. (1991). Satellite imagery tsetse and trypanosomiosis in Africa. *Preventive Veterinary Medicine* 11: 201–220.

Rojas-Downing, M.M., Nejadhashemi, A.P., Harrigan, T., and Woznicki, S.A. (2017). Climate change and livestock: impacts, adaptation, and mitigation. *Climate Risk Management* 16: 145–163.

Rolfe, J. (2010). Economics of reducing grazing emissions from beef cattle in extensive grazing systems in Queensland. *The Rangeland Journal* 32: 197–204.

Rolin, A.I., Berrang-Ford, L., and Kulkarni, M.A. (2013). The risk of Rift Valley fever virus introduction and establishment in the United States and European Union. *Emerging Microbes and Infections* 2 (1): 1–8.

Roos, K.F., Martin, J.H., and Moser, M.A. (2004). *AgSTAR Handbook: A Manual for Developing Biogas Systems at Commercial Farms in the United States*, 2e. US Environmental Protection Agency. EPA-430-B-97-015.

Rose, H., Wang, T., van Dijk, J., and Morgan, E.R. (2015). GLOWORM-FL: A simulation model of the effects of climate and climate change on the free-living stages of gastro-intestinal nematode parasites of ruminants. *Ecological Modelling* 297: 232–245.

Rosegrant, M.W., Fernandez, M., and Sinha, A. (2009). Looking into the future for agriculture and AKST. In: *International Assessment of Agricultural Knowledge, Science and Technology for Development (IAASTD): Agriculture at a Crossroads* (eds. B.D. McIntyre, H.R. Herren, J. Wakhungu and R.T. Watson), 307–376. Washington, DC: Island Press.

Roth, Z. and Wolfenson, D. (2016). Comparing the effects of heat stress and mastitis on ovarian function in lactating cows: basic and applied aspects. *Domestic Animal Endocrinology* 56: S218–S227.

Rowlinson, P. (2008). Adapting livestock production systems to climate change: temperate zones. In: *Livestock and Global Climate Change Conference Proceeding* (eds. P. Rowlinson, M. Steel and A. Nefzaoui), 61–63. Tunisia: Cambridge University Press.

Rust, J.M. (2019). The impact of climate change on extensive and intensive livestock production systems. *Animal Frontiers* 9 (1): 20–25.

Rutter, S.M. (2014). Smart technologies for detecting animal welfare status and delivering health remedies for rangeland systems. *Scientific and Technical Review of the Office International des Epizooties (Paris)* 33 (1): 181–187.

Sacks, A.D., Teague, R., Provenza, F. et al. (2014). Restoring atmospheric carbon dioxide to pre-industrial levels: re-establishing the evolutionary grassland-grazer relationshi. In: *Geotherapy* (eds. T. Goreau, R.W. Larson and J. Campe), 155–194. Boca Raton, FL: CRC Press.

Saminathan, M., Singh, K.P., Vineetha, S. et al. (2020). Virological, immunological and pathological findings of transplacentally transmitted bluetongue virus serotype 1 in IFNAR1-blocked mice during early and mid-gestation. *Scientific Reports* 10 (1): 2164.

Sarker, R.D., Giasuddin, M., Chowdhury, E.H., and Islam, M.R. (2017). Serological and virological surveillance of avian influenza virus in domestic ducks of the north-east region of Bangladesh. *BMC Veterinary Research* 13: 180.

Sautier, M., Duru, M., and Martin-Clouaire, R. (2013). Use of productivity-defined indicators to assess exposure of grassland-based livestock systems to climate change and variability. *Crop & Pasture Science* 64 (7): 641–651.

Savsani, H.H., Padodara, R.J., Bhadaniya, A.R. et al. (2015). Impact of climate on feeding, production and reproduction of animals – a review. *Agricultural Reviews* 36 (1): 26–36.

Schils, R.L.M., Verhagen, A., Aarts, H.F.M., and Šebek, L.B.J. (2005). A farm level approach to define successful mitigation strategies for GHG emissions from ruminant livestock systems. *Nutrient Cycling in Agroecosystems* 71: 163–175.

Searchinger, T., Heimlich, R., Houghton, R.A. et al. (2008). Greenhouse gases through emissions from use of U.S. croplands for biofuels increases land-use change. *Science* 319: 1238.

Seo, S.N. and Mendelsohn, R. (2006). *The Impact of Climate Change on Livestock Management in Africa: A Structural Ricardian Analysis*, CEEPA Discussion Paper No. 23. Centre for Environmental Economics and Policy in Africa, University of Pretoria.

Seo, S.N. and Mendelsohn, R. (2007). *An Analysis of Livestock Choice: Adapting to Climate Change in Latin American Farms*, Policy Research Working Paper 4164, 18. World Bank.

Seo, S.N. and Mendelsohn, R. (2008). Measuring impacts and adaptations to climate change: a structural Ricardian model of African livestock management. *Agricultural Economics* 38: 151–165.

Shannon, M.C. and Noble, C.L. (1995). Variation in salt tolerance and ion accumulation among subterranean clover cultivars. *Crop Science* 35 (3): 798–804.

Shini, S., Kaiser, P., Shini, A., and Bryden, W.L. (2008). Biological response of chickens (Gallus gallusdomesticus) induced by corticosterone and a bacterial endotoxin. *Comparative Biochemistry and Physiology Part B: Biochemistry and Molecular Biology* 149 (2): 324–333.

Siddiqui, M.F.M.F., Digraskar, S.U., Shaikh, S.R. et al. (2018). Application of GIS in animal disease monitoring and surveillance. *International Journal of Chemical Studies* 6 (4): 2886–2888.

Silanikove, N. (2000). The physiological basis of adaptation in goats to harsh environments. *Small Ruminant Research* 35: 181–193.

Silanikove, N. and Koluman, N. (2015). Impact of climate change on the dairy industry in temperate zones: predications on the overall negative impact and on the positive role of dairy goats in adaptation to earth warming. *Small Ruminant Research* 123 (1): 27–34.

Silva, D.C. and Passini, R. (2017). Physiological responses of dairy cows as a function of environment in holding pen. *Engenharia Agrícola* 37 (2): 206–214.

Silva, B.A.N., Oliveira, R.F.M., Donzele, J.L. et al. (2009). Effect of floor cooling and dietary amino acids content on performance and behaviour of lactating primiparous sows during summer. *Livestock Science* 120: 25–34.

Singh, M., Chaudhari, B.K., Singh, J.K. et al. (2013). Effects of thermal load on buffalo reproductive performance during summer season. *Journal of Biological Sciences* 1 (1): 1–8.

Sirohi, S. and Michaelowa, A. (2007). Sufferer and cause: Indian livestock and climate change. *Climatic Change* 85 (3): 285–298.

Smith, K.A. and Conen, F. (2004). Impacts of land management on fluxes of trace greenhouse gases. *Soil Use and Management* 20: 255–263.

Smith, P., Martino, D., Cai, Z. et al. (2007). Policy and technological constraints to implementation of greenhouse gas mitigation options in agriculture. *Agriculture, Ecosystems and Environment* 118: 6–28.

Smith, P., Martino, D., Cai, Z. et al. (2008). Greenhouse gas mitigation in agriculture. *Philosophical Transactions of the Royal Society B* 363: 789–813.

Snyder, C.S., Bruulsema, T.W., Jensen, T.L., and Fixen, P.E. (2009). Review of greenhouse gas emissions from crop production systems and fertilizer management effects. *Agriculture, Ecosystems and Environment* 133 (3–4): 247–266.

Sohail, M.U., Hume, M.E., Byrd, J.A. et al. (2012). Effect of supplementation of prebiotic mannan-oligosaccharides and probiotic mixture on growth performance of broilers subjected to chronic heat stress. *Poultry Science* 91 (9): 2235–2240.

Sokołowicz, Z., Krawczyk, J., and Świątkiewicz, S. (2016). Quality of poultry meat from native chicken breeds: a review. *Annals of Animal Science* 16 (2): 347–368.

Soussana, J.F. and Lüscher, A. (2007). Temperate grasslands and global atmospheric change: a review. *Grass and Forage Science* 62 (2): 127–134.

Spiers, D.E., Spain, J.N., Sampson, J.D., and Rhoads, R.P. (2004). Use of physiological parameters to predict milk yield and feed intake in heat-stressed dairy cows. *Journal of Thermal Biology* 29 (7–8): 759–764.

Springmann, M., Clark, M., Mason-D'Croz, D. et al. (2018). Options for keeping the food system within environmental limits. *Nature* 562: 519–525.

Steinfeld, H., Wassenaar, T., and Jutzi, S. (2006). Livestock production systems in developing countries: status, drivers, trends. *Revue Scientifique et Technique (Paris)* 25: 505–516.

Stevenson, J.S., Tiffany, S.M., and Lucy, M.C. (2004). Use of Estradiol cypionate as a substitute for GnRH in protocols for synchronizing ovulation in dairy cattle. *Journal of Dairy Science* 87: 3298–3305.

Stokes, C. and Howden, M. (2010). *Adapting Agriculture to Climate Change: Preparing Australian Agriculture, Forestry and Fisheries for the Future*. CSIRO Publishing.

Summer, A., Lora, I., Formaggioni, P., and Gottardo, F. (2019). Impact of heat stress on milk and meat production. *Animal Frontiers* 9 (1): 39–46.

Szymanski, L.A., Schneider, J.E., Friedman, M.I. et al. (2007). Changes in insulin, glucose and ketone bodies, but not leptin or body fat content precede restoration of luteinising hormone secretion in ewes. *Journal of Neuroendocrinology* 19 (6): 449–460.

Tellez, G., Tellez-Isaias, G., and Dridi, S. (2017). Heat stress and gut health in broilers: role of tight junction proteins. *Advances in Food Technology and Nutritional Sciences – Open Journal* 3: 1–4.

Thompson, L.G. (2010). Climate change: the evidence and our options. *The Behavior Analyst* 33: s153–s170.

Thornton, P.K. and Herrero, M. (2010). Potential for reduced methane and carbon dioxide emissions from livestock and pasture management in the tropics. *The Proceedings of the National Academy of Sciences in the United States of America* 107 (46): 19667–19672.

Thornton, P.K. and Herrero, M. (2014). Climate change adaptation in mixed crop-livestock systems in developing countries. *Global Food Security* 3 (2): 99–107.

Thornton, P.K., van de Steeg, J., Notenbaert, A., and Herrero, M. (2009). The impacts of climate change on livestock and livestock systems in developing countries: a review of what we know and what we need to know. *Agricultural Systems* 101: 113–127.

Tian, H., Zhou, S., Dong, L. et al. (2015). Climate change suggests a shift of H5N1 risk in migratory birds. *Ecological Modelling* 306: 6–15.

Tosti, E. and Gallo, A. (2015). Ion currents involved in gamete physiology. *International Journal of Developmental Biology* 59 (7–8–9): 261–270.

Trape, J.F., Godeluck, B., Diatta, G. et al. (1996). The spread of tick-borne borreliosis in West Africa and its relationship to sub-Saharan drought. *The American Journal of Tropical Medicine and Hygiene* 54 (3): 289–293.

Tubiello, F.N., Salvatore, M., Rossi, S. et al. (2013). The FAOSTAT database of greenhouse gas emissions from agriculture. *Environmental Research Letters* 8: 1–10.

Upadhyay, R.C., Ashutosh, R.V., Singh, S.V., and Aggarwal, P. (2009). Impact of climate change on reproductive functions of cattle and buffaloes. In: *Global Climate Change and Indian Agriculture* (ed. P.K. Aggarwal), 107–110. New Delhi, India: ICAR.

USDA (2013). *Climate Change and Agriculture in the United States: Effects and Adaptation.* Washington, DC: USDA technical bulletin. United States Department of Agriculture http://www.usda.gov/oce/climate_change/effects_2012/CC%20and%20Agriculture%20Report%20%2802-04-2013%29b.pdf.

Vaidya, M., Kumar, P., and Singh, V.S. (2010). Circadian variations in heat production and heat loss in Murrah buffaloes during different season. *Revista Veterinaria* 21 (1): 777–779.

Vale, W.G. (2007). Effects of environment on buffalo reproduction. *Italian Journal of Animal Science* 6 (Supl 2): 130–142.

Van Groenigen, J.W., Velthof, G.L., Oenema, O. et al. (2010). Towards an agronomic assessment of N2O emissions: a case study for arable crops. *Europian Journal of Soil Science* 61 (6): 903–913.

Van Hemert, C., Pearce, J.M., and Handel, C.M. (2014). Wildlife health in a rapidly changing north: focus on avian disease. *Frontiers in Ecology and the Environment* 12 (10): 548–556.

Vanegas, J., Overton, M., Berry, S.L., and Sischo, W.M. (2006). Effect of rubber flooring on claw health in lactating dairy cows housed in free-stall barns. *Journal of Dairy Science* 89 (11): 4251–4258.

Varasteh, S., Braber, S., Akbari, P. et al. (2015). Differences in susceptibility to heat stress along the chicken intestine and the protective effects of galacto-oligosaccharides. *PLoS One* 10 (9): e0138975.

Vijayakumar, P., Pandey, H.N., Singh, M., and Dutt, T. (2009). Effect of heat amelioration measures on growth and physiological response of buffalo heifers during summer. *The Indian Journal of Animal Sciences* 79 (4): 437–441.

Vitali, A., Felici, A., Esposito, S. et al. (2015). The impact of heat waves on dairy cow mortality. *Journal of Dairy Science* 98: 4572–4579.

Vitali, A., Bernabucci, U., Nardone, A., and Laceter, N. (2016). Effect of season, month and temperature humidity index on the occurrence of clinical mastitis in dairy heifers. *Advances in Animal Biosciences* 7: 250–252.

Volk, M., Bungener, P., Contat, F. et al. (2006). Grassland yield declined by a quarter in 5 years of free-air ozone fumigation. *Global Change Biology* 12 (1): 74–83.

de Vries, M. and de Boer, I.J.M. (2010). Comparing environmental impacts for livestock products: a review of life cycle assessments. *Livestock Science* 128: 1–11.

Wang, M., Wang, R., Xie, T.Y. et al. (2016). Shifts in rumen fermentation and microbiota are associated with dissolved ruminal hydrogen concentrations in lactating dairy cows fed different types of carbohydrates. *Journal of Nutrition* 146: 1714–1721.

Wheeler, T. and Von Braun, J. (2013). Climate change impacts on global food security. *Science* 341 (6145): 508–513.

White, R.R. and Hall, M.B. (2017). Nutritional and greenhouse gas impacts of removing animals from US agriculture. *The Proceedings of the National Academy of Sciences in the United States of America* 114: E10301–E10308.

Willett, W., Rockström, J., Loken, B. et al. (2019). Food in the Anthropocene: the EAT-Lancet Commission on healthy diets from sustainable food systems. *The Lancet Commissions* 393: 447–492.

Williams, A.G., Audsley, E., and Sandars, D.L. (2006). Determining the Environmental Burdens and Resource Use in the Production of Agricultural and Horticultural Commodities. *Main Report Defra Research Project ISO205*. Cranfield University and Defra, Bedford. http://randd.defra.gov.uk/Document.aspx?Document¼IS0205_3959_FRP.doc.

Wiltbank, M.C. and Pursley, J.R. (2014). The cow as an induced ovulator: timed AI after synchronization of ovulation. *Theriogenology* 81 (1): 170–185.

Wittmann, E.J., Mellor, P.S., and Baylis, M. (2001). Using climate data to map the potential distribution of Culicoides imicola (Diptera: Ceratopogonidae) in Europe. *Revue Scientifique et Technique* 20: 731–740.

WMO (2014). Weather and climate: engaging youth. *World Meteorological Organization WMO Bulletin* 63 (1): 1–44.

WRI (2004). *Food Security: Grain Fed to Livestock as a Percent of Total Grain Consumed 2003/4 Data*. World Resources Institute http://earthtrends.wri.org.

Wright, I.A., Tarawali, S., Blummel, M. et al. (2012). Integrating crops and livestock in subtropical agricultural systems. *Journal of the Science of Food and Agriculture* 92: 1010–1015.

Yadav, A. and Garg, V. (2019). Biotransformation of bakery industry sludge into valuable product using vermicomposting. *Bioresource Technology* 274: 512–517.

Yang, B., Ma, Y., and Xiong, Z. (2019). Effects of different composting strategies on methane, nitrous oxide, and carbon dioxide emissions and nutrient loss during small-scale anaerobic composting. *Environmental Science and Pollution Research* 26: 446–455.

Yong, X., Siqing, Z., and Yang, Y. (2006). Research and application of geographic information systems for prophylaxis and control disease. *Disease Monitor* 21 (1): 45–47.

Zhang, G. and Bjerg, B.S. (2017). Developments of thermal environment techniques of animal housing in hot climate–a review. In: *Animal Environment and Welfare - Proceedings of International Symposium* (eds. J.Q. Ni, T.T. Lim, C. Wang and L. Zhao), 375–384. Beijing: China Agriculture Press.

Zhang, P., Yan, T., Wang, X. et al. (2017). Probiotic mixture ameliorates heat stress of laying hens by enhancing intestinal barrier function and improving gut microbiota. *Italian Journal of Animal Science* 16 (2): 292–300.

Zhou, Y., Ding, Y., Li, H. et al. (2020). The effects of short-term grazing on plant and soil carbon and nitrogen isotope composition in a temperate grassland. *Journal of Arid Environments* https://doi.org/10.1016/j.jaridenv.2020.104198.

Zimbelman, R.B., Baumgard, L.H., and Collier, R.J. (2010). Effects of encapsulated niacin on evaporative heat loss and body temperature in moderately heat-stressed lactating Holstein cows. *Journal of Dairy Science* 93 (6): 2387–2394.

Zimbleman, R.B., Rhoads, R.P., Baumgard, L.H., and Collier, R.J. (2009). Revised temperature humidity index (THI) for high producing dairy cows. *Journal of Dairy Science* 92 (E-Suppl. 1): 347.

8

Aquaculture Resources and Practices in a Changing Environment

Shib Kinkar Das[1], Amit Mandal[2], and Sachin Onkar Khairnar[2]

[1]*Department of Aquaculture, Faculty of Fishery Sciences, West Bengal University of Animal and Fishery Sciences (WBUAFS), Kolkata, India*
[2]*Department of Aquaculture, College of Fisheries, Guru Angad Dev Veterinary and Animal Sciences University (GADVASU), Ludhiana, Punjab, India*

CONTENTS

8.1 Introduction

Toward providing nutrition to the 3.3 billion populations globally, aquaculture has established itself as the fastest evolving natural food producing and entrepreneurial sector as well with annual growth rate of 5.3%. This is more so in the developing countries in the tropics and sub-tropics where remarkable contribution of aquaculture toward food security,

Sustainable Agriculture Systems and Technologies, First Edition. Edited by Pavan Kumar, A.K. Pandey, Susheel Kumar Singh, S.S. Singh, and V.K. Singh.

improving gender equity, employment, and income generation is highly conspicuous. Among the aquaculture produces, fish alone provides 16–22% of the animal protein, essential minerals, and $n-3$ fatty acid viz. among others (Pradeepkiran 2019). Global fish production of 179 million tonnes, valued around $401 billion were recorded during 2018 of which share of aquaculture production was 82 million tonnes ($250 billion). Out of the total aquaculture production, contributions of fish, aquatic algae, and, ornamental seashells and pearl were 82.1, 32.4 and 0.026 million tonnes (FAO 2020). Around 156 million tonnes of fish were used for direct human consumption as the per capita fish consumption was 20.5 kg per annum (FAO 2020). On the other hand, global yield from capture fisheries was only 96.4 million tonnes with a reducing growth rate of 5.4%.

In India, total fish production in 2019–2020 was 14.16 million tonnes including both capture and culture fishery resources with an average annual growth rate of 4.35% in which Andhra Pradesh contributed maximally (3.61 million tonnes) followed by West Bengal (1.782 million tonnes) and Orissa (0.0.818 million tonnes) in 2019–2020 (DFAHD 2020). The major production basically comes from freshwater aquaculture sector which supplies 12.8% of total animal protein (Anand 2019). Inland aquaculture contributed 73.67% of total fish production with an average annual growth rate of 7.37% in which Andhra Pradesh contributed maximally (3.61 million tonnes) followed by West Bengal (1.62 million tonnes), and Uttar Pradesh (0.69 million tonnes) in 2019–2020 (DFAHD 2020).

India as the carp culture country (Dhawan and Kaur 2002) is a major producer of fish through aquaculture (Goswami and Zade 2015). Different commercially important freshwater species (carps, catfish, cichlids, prawn, etc.) and brackish water species (mullet, milk fish, sea bass, shrimp, mud crab, etc.) are being cultured in most of the Asian and European countries. Coastal aquaculture is mainly for Penaeid shrimp culture like tiger shrimp (*Penaeus monodon*), white leg shrimp (*Litopenaeus vannamei*). Diversification in terms of farming practices, resource use, and species of culture is of utmost necessity toward enhancement of productivity and sustainability of this sector.

The aquaculture sector as a whole is facing lots of challenges and opportunities that warranted sustainable ecosystem-based approaches. Though technological advancement is felt obvious for augmenting sustainable aquaculture production as well as its expansion, unregulated and excessive uses of drugs, antibiotics, and chemicals are cause of concern both in the culture systems and in the postharvest phases of processing and marketing of the produce. Therefore, global aquaculture industry needs scientific innovations and adoptions in improving environmental, social, and financial stability toward fulfilling food and nutrition demand (Lu et al. 2019). Multidimensional intra- and inter-husbandry integration as well as dynamicity in selection and adoption of modern technologies viz. pond and reservoir cage farming system, riverine pen system, waste water aquaculture system, flow through and raceway system, recirculatory aquaculture system (RAS), biofloc technology (BFT), aquaponics and hydroponics, and integrated multi-trophic aquaculture system (IMTA) is crucial in mitigating challenges and opening up of newer frontiers in future aquaculture practices. Effective utilization of untapped resources, optimum use of water and land resources as well as application of modern scientific technological interventions are the key factors in that direction. Instead of any generalized package of practice, site specific micro-planning of a particular culture system taking into account the agro-eco-climatic zones holds the key in sustainability in pond-based aquaculture practices.

8.2 Aquaculture Resources and Production

Globally, freshwater and coastal brackish water contributed 50–60% and 40–50% respectively in total aquaculture production in 2018 though constituting only 3% of planet's water and 3% of surface water. Most of the inland aquaculture is usually pond-based systems though tanks, cages, pens, and raceway tanks are used sporadically. In spite of the diversity and species richness in aquaculture, production is primarily being contributed by a small number of staple species (FAO 2020). The culture systems are also very diverse in terms of culture methods, practices, and integration approaches with agriculture and live stocks. China continued to be the top aquaculture production country contributing 47.559 million tonnes distantly followed by India (7.066 million tonnes), and Indonesia (5.427 million tonnes) (DFAHD 2020). Carps and cyprinids are mostly cultured in ponds employing modified extensive to semi-intensive level of intensification and contributed 36.90% in 2018. During the period, tilapia and catfish are mostly cultured in ponds, tanks, and cage using semi-intensive and intensive methods, through monoculture system, contributed 10.20% and 5.20%, respectively; prawn is usually cultured in ponds contributed 5%. It is important to note that inland aquaculture production of finfish reduced from 97.2% in 2000 to 91.5% in 2018.

Water resources are diverse and in the tropical south and south-east Asian countries including India, inland water bodies viz. ponds, rivers, canals, streams, lakes, and reservoirs abound for aquaculture purpose. India is having reservoirs (2.926 million ha), ponds and tanks (2.432 million ha), beels, jheels, and derelict water bodies (1.07 million ha), upland lakes (0.72 million ha), and, rivers and canals (0.19 million ha) (Figure 8.1). Besides, in recent years, nonconventional water resources like abandoned open cast coal mines, irrigation canals, jute retting ditches and small impoundments, household backyard

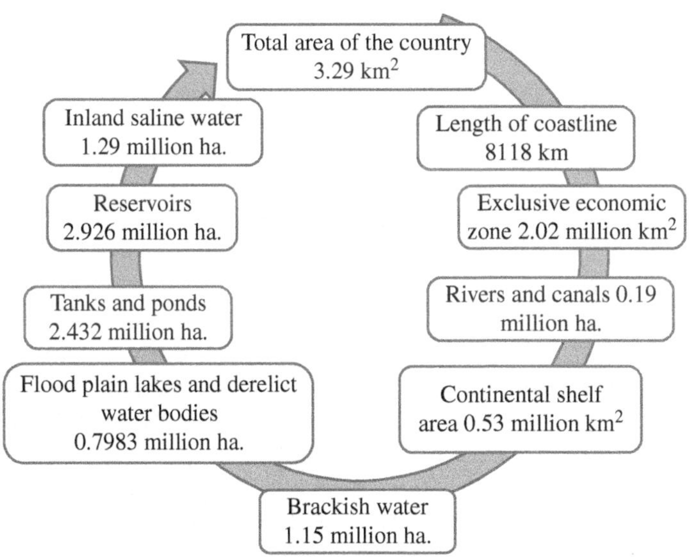

Figure 8.1 Fisheries and aquaculture resources of India. *Source*: DAHDF (2017) and DFAHD (2020). © John Wiley & Sons.

impoundments, and constructed cement cisterns are also being popular as water resources for fish farming in the rural areas. During the last two decades, inland saline land mostly found in the arid and semi-arid regions of the non-coastal Indo-Gangetic planes in northern India (Ansal et al. 2016) became attractive resource for farming of brackish water finfish and shrimp species as such lands are not suitable for agriculture crop production. Moreover, India is naturally endowed with 11.6 million ha of water logged area (Chowdhury et al. 2018) which have huge potential for aquaculture. The community village ponds are also untapped resource for fish production in India.

Reservoirs are the most potential resources for cage and pen culture high-valued fish species. Maharashtra, Jharkhand, and Uttar Pradesh are pioneers in producing carps, catfishes, and tilapia in cage system in reservoirs, whereas, Assam, West Bengal, Bihar, and Odisha are utilizing ox-bow lakes, wet lands, marshy water logged areas for fish production. The major river systems like the Ganges and the Brahmaputra along with the associated floodplains are contributing immensely as fish seed resource and conserving viable fish germplasm in India. However, vulnerability of such resources has already been badly manifested because of increasing anthropogenic activities, salinization of water, urbanization, industrial development, natural calamities. The floodplain wetlands are challenged due to heavy siltation, transition of the ox-bow lakes from live to dead loosing connection from the mother rivers, weed infestation, increased pollution loads, and decreased water retention capacity (NFDB 2020). The estuarine lakes such as Chilika, Pulicat, Vembanad, and backwater resources in Kerala continued to serve as valuable resource in providing protein sources through community-based cluster aquaculture approaches. The importance of coastal aquaculture resources in low lying estuarine areas in West Bengal (bheries), Kerala (pokkali), Karnataka (gazani), and Goa (khazan) has been well established. Besides, the coastal saline soil area known as kharland in Maharashtra is also considered for aquaculture production (Raut and Gawande 2019).

Coastal aquaculture is significantly contributing to marine aquaculture production, constituting the Penaeid shrimp species like *P. monodon*, *L. vannamei*, *Fenneropenaeus indicus* in coastal ponds and bheries in semi-intensive and intensive system in coastal states and Union Territories along the 8118 km coastline in India. The vast resources with varied biodiversity have big potential to enhance the fish and shellfish production and also can increase export in global market (ICAR 2000). Mariculture is mainly confined in South-West coast of India for culture of seaweeds, pearly oysters, edible oysters, and mussels through rack, ren, and ring system.

The potential coastal area for shrimp culture in India is about 1.24 million ha. Traditional brackish water aquaculture in India used to be dominated by culture of penaeid shrimp species for long. However, the first boom with commercial culture of tiger prawn (*P. monodon*) was short-lived and faced sudden burst during the mid-1980s due to injudicious approaches and incidence of diseases. The second cycle toward booming started with culture of white leg shrimp (*L. vannamei*) in the mid-2000s (Vijayan and Balasubramanian 2019) and yet to face significant reverses because of cautious and planned approaches. Two-thirds of this area belongs to West Bengal (47.51%), Andhra Pradesh (17.60%), Maharashtra (9.38%), Kerala (7.62%), Tamil Nadu (6.57%), and Gujarat (4.41%). As of now, only 17% of the available area has been developed for aquaculture. Andhra Pradesh is holding the highest brackish water culture area for shrimp and scampi about 0.64 thousand million ha, followed by West Bengal (0.55 million ha) and Orissa (0.115 million ha) (DFAHD 2018).

Although, Central Marine Fisheries Research Institute (CMFRI) has developed many culture technologies for important candidate species, but many technologies are not achieving the target due to lack of public awareness and entrepreneurship approaches (Devaraj and Appukuttan 2000). Besides, several technologies have also been developed by the Central Institute of Brackish water Aquaculture (CIBA) for medium to large scale commercial culture of shrimps and finfishes. Mariculture and coastal aquaculture contributed 30.8 million tonnes in 2018 of which finfish (7.3 million tonnes) and crustaceans (5.7 million tonnes) contributed 42.50%. Because of the crush in penaeid shrimp production during the mid-1980s and associated environmental issues raised, coastal aquaculture practices in India have been brought under the jurisdiction of Coastal Aquaculture Authority Act, 2005 aimed toward ensuring sustainability and environmental stability.

8.3 Aquaculture-Environmental Interaction and Conservation

Environmental issues are intricately involved with aquaculture development as the relationship between environment and aquaculture is complex because of overdependence on resources and limited capacity of biological productivity, interdependence among the culture units, and associated environmental factors (Bailly and Paquotte 1996). The sustainability of biological functions largely depends upon the physicochemical factors of water that are continuously being impacted by environmental factors. Several other factors like landscaping, quality and volume of water, primary productivity, and protected species resources emerged as potential contributors in aquaculture systems. Increasing growth of population, anthropogenic activities, pollution of water bodies due to domestic and industrial waste, urbanization and industrialization are the possible threat for environmental sustainability and conservation which greatly affected inland freshwater and coastal aquaculture systems. Both endogenous and exogenous factors resulted in several environmental issues that lower production efficiency by increasing the cost of aquaculture production. The over stocking density of cultured species beyond the carrying capacity is another factor affecting the aquatic system which leads to sudden mortality and undesirable species growth in culture system. Conflict remains among aquaculture practices, conservation of nature and natural resources through potential effects like genetic interaction and disturbance of wildlife ecommunities (Gowen and Rosenthal 1993). Modification in aquaculture systems with respect to intensification has also great impact on environment. Freshwater salinization, ground water salinization, decrease of coastal water salinity, dependence on wild seed, in situ organic waste generation and release at surrounding water bodies, unregulated and very often excessive use of drugs, antibiotics, and aqua-chemicals are also considered as major constrains/barriers for maintaining homeostasis and healthy interaction in the environment and ecology of the culture units.

The economic growth in aquaculture sector is at the cross roads because of the sustainable issue (Tisdell and Leung 1999). Aquaculture conservation needs to support the use of natural resources in a sustainable manner through strategies like mitigation, prevention, and restoration measures. The conservation of aquaculture will enable in commodity protection and habitat restoration toward improving wild population, development of aquaculture-based ecosystem approaches in lessening environmental footprints (Froehlich et al. 2017). Species level conservation can reduce the overdependence and pressure on wild stocks and restoration of their habitats. The demand of food security for increasing population is badly affected aquaculture as the exploitation of species resources are increasing day-by-day. Protection and conservation of indigenous fish species by conserving and developing suitable breeding environment, seed production and culture practices are logical means to restore the species richness. Therefore, conservation of hatcheries should be ensured for enhancement of fish stocks (Costa-Pierce 2008; Taylor et al. 2017). Hatcheries exert positive impacts on improvement of culture system and ecological stability (Araki and Schmid 2010). Moreover, biogenic restoration of habitat is crucial in conserving aquaculture at ecosystem level though the major challenge is habitat restoration (Wortley et al. 2013). Culture of different economically important aqua-crops in sustainable way has the potential to improve water quality (Humphries et al. 2016), carbon sequestering (Greiner et al. 2013), and habitat of available wild species (Seitz et al. 2014) like culture of bivalves and sea weeds in coastal areas (Dumbauld et al. 2009; Kim et al. 2015). The effect of aquaculture on culture of wild species may have positive impact on the system and environment as well (Hehre and Meeuwig 2016).

Although the policy makers are diverse in formulating policies for maintaining healthy environment and ecosystems, lapses in public awareness and several social, political, and economic attributes are responsible for degradation of environment stability. Therefore, responsible participatory management will be the best approach for good management practices.

8.4 Climate Change and Aquaculture

The impact of climate change on aquaculture warrants number of alterations in culture practices, species level competition, conservation issues, etc. Global warming-associated severe weather phenomena like cyclones and heat waves are the most conspicuous manifestations of anthropogenic climate change (Seneviratne et al. 2012; Field et al. 2014; Lin 2019). Inland aquatic system is one of the most stressed ecosystems (Katiha et al. 2014; Sarkar et al. 2019) with respect to climatic extremes (Dudgeon et al. 2006). Such extremes adversely impact inland fisheries (Easterling et al. 2000; Dudgeon et al. 2006; Mills et al. 2012), with reduced natural productivity and biodiversity (Jones et al. 2018; Ruthrof et al. 2018; Smale et al. 2019) and affect the livelihood of the fishing communities (Barange et al. 2018; Sarkar et al. 2019).

Different aspects of variability viz. region, culture practices, time, space, size, and changeability significantly impact aquaculture production (De Silva and Soto 2009). The direct and indirect impact of climate change on natural resources like land and water, seed availability, and energy distribution are the major concerns in augmenting productivity and profitability (Oguntuga et al. 2009). The exposure to people vis-a-vis property and vulnerability of socioeconomic and environmental factors are considered as major risks in inland aquaculture sector (Westlund et al. 2007). Inland aquatic systems, being the major service provider of all the four ecosystem services incur huge loss (especially in terms of intangible and tangible goods and services) in the context of climate changing scenario (Poff et al. 2002). The possible impacts of climate changes related to aquaculture are droughts, floods, tropical cyclones, scorching heat and heat waves, changes on seasonality and precipitation patterns, and, surface and groundwater salinization. Climate extremes such as heat waves affect the hydrobiological processes of inland aquatic systems and alter the functions and services of inland ecological systems (Nelson et al. 2013).

Increasing trends of intensification in aquaculture coupled with global warming resulted in eutrophication along with stratification of water bodies and circulation of water masses emerged as major factors of alteration of species heterogeneity in aquaculture (Ficke et al. 2007). Upwelling events can cause oxygen depletion in water bodies due to changes in rainfall pattern and wind action. The major river systems like Ganges and Brahmaputra in India are facing challenges due to irregular water shortage and excessive flood conditions (Vass et al. 2009). Declining of fish stocks is also a big issue due to failure of reproductive and breeding capability.

Impact of climate change in brackish water aquaculture can be direct or indirect on culture both of finfish and shellfish. A change in water salinity reduces the species viability (Ahmed and Diana 2015) which in turn results in biodiversity loss in the coastal ecosystem. Irregular rainfall pattern causes flooding that physically damage aquafarms and escaping of species to the wild. Increase in mortality of larvae and postlarvae occurs due to acidification of water (Allan and Maguire 1992a,b). Increase in physical turbidity causes decline in natural productivity through cascading effects in the photosynthesis process which cause oxygen depletion in water (Ahmed and Diana 2015) and increased mortality rate of culture species resulting the poor production and low economic output. Heavy rainfall decreases water temperature and sudden drop in surface water salinity which hampers physiological process and growth of fish and shellfish (Kumlu and Kir 2005). Prolonged drought conditions can pose irreversible alternation of farm activities by enhancing the load of waste metabolites and pollution (Ahmed and Diana 2015).

8.5 COVID-19 and Aquaculture

The ongoing pandemic has impacted the farming sectors including aquaculture and related activities around the globe. In a suddenly changed perspective, particularly during the lockdown period, fish farmers in general had been left with no experience to cope with the situation and tide over the crisis. In the freshwater sector, small scale aqua farming was least impacted and the recovery period was also comparatively less to that of the coastal shrimp farming sectors (Hait 2021). As the latter is largely dependent upon export market and huge capital involvement toward input and management, financial loss was significantly higher and recovery to normalcy was very slow (Vignesh 2021). Therefore, importance of low-input sustainable aquaculture (LISA) in inland freshwater small scale fish farming has been tested for the first time during the pandemic and it successfully overcame the sudden crisis because of formation of localized micro level marketing channels, involvement of family laborers, low-cost input use among others. Likewise, for climate change also aquaculture sector should emphasize not only climate smart but also climate adjusted reasonable practice (CARP) focusing the small scale farmers and farming practices with low risk and high shock absorption capacity.

8.6 Adaptive Measures

India is holding the second largest farmed fish producing country in the world, producing 7.7 mmt of farmed fish during 2018–2019. In spite of that, the freshwater aquaculture sector is primarily traditional to the culturable species and system diversification, production efficiency in inland aquaculture sector is facing lots of challenges due to scarcity of healthy brood stocks, nonavailability of quality seeds and varied agroclimatic conditions with changes in seasonal patterns and irregular distribution patterns in precipitations. The major lacuna in inland aquaculture sector is insignificant diversification in terms of species of culture and cropping patterns. Further, logical intensification in pond culture systems and enclosure-based culture system in large water bodies are also not properly addressed. Therefore, interventions are necessary for the horizontal and vertical expansion approaches like:

a) Diversification in species and culture spectrum
b) Adoption of eco-friendly viable technologies
c) Setting up of brood bank
d) Setting up of fish seed supplying centers in different agroclimatic regions
e) Reviving of aquaculture through culture of indigenous minor carps and freshwater prawn (*Macrobrachium rosenbergii*)
f) Development of rainfed aquaculture system
g) Promotion of coldwater species like rainbow trout in higher altitudes
h) Use of renewable resources for environmental sustainability
i) Setting up of business incubation center for capturing domestic and international markets
j) Use of village community ponds for better utilization of resources to enhance production.

8.6.1 Horizontal Expansion: Use of Untapped Resources

Reclamation and restoration of untapped aquaculture resources is considered as the most pragmatic approach of diversification in fish culture, horizontal and vertical expansion of culture system, and increase of per unit fish production. The multispecies culture through rational use of such untapped resources can enhance the fish production and productivity in aquaculture.

8.6.1.1 Aquaculture Potential in Abundant Open Cast Coal Mines

The abundant mine pits are deserted barren land which is unfit for agriculture activities. Such area can be used for aquaculture of hardy species in enclosures which can tolerate the perturbed environmental conditions because of mining activities. Such closed water aquaculture system is stated to be ecologically viable and it can have the capacity to enhance fish production through reclamation and restoration that could support the ecological and economic diversification of livelihood of the neighboring populations. However, risk factors in such practice include irregular horizontal and vertical dimensions of the pits, physical ability of pit walls, flood-ability, water availability and quality, climate change related effects, and availability of good candidate fish species for culture (Otchere et al. 2004).

The common impact of coal mine activities includes reduced oxygen level in water because of huge depth and, high level of sulfate and turbidity (Mylliemngap and Ramanujam 2011). Excessive accumulation of acid mine drainage which degrade water quality and causes loss of biodiversity in the coal mining areas was reported by Pentreath (1994). Besides, highly uneven bottom surface along with excessive depth cause uneven distribution of nutrients resulting in stratification and anoxic horizon at the bottom. Therefore, floating cage culture of hardy species is the best option in this particular type of water bodies. In India, farming of fish in abandoned coal mine pits has been experimentally practiced under cages in West Bengal, Chhattisgarh, and Meghalaya. Joint venture approach between the Department of Fisheries, Govt. of West Bengal, and Eastern Coalfields Ltd. has been initiated in open-cast mines to create employment opportunity and to enhance fish production. It has been observed that species like common carp, air breathing catfishes, and murrels are good candidate species in such culture system with seven to nine months culture period following multiple stocking and multiple harvesting (MSMH) practice protocols.

However, aquaculture practice in such water bodies is at its extremely juvenile stage and package of practice must be standardized further. Moreover, approaches should be cautious and judicious because of chances of heavy metals contamination of the produce and eventually bioaccumulation in human body through consumption of the produce.

8.6.1.2 Potential of Aquaculture in Undrainable Village Pond

Roadside transitory ditches in semi-urban and rural areas have huge socioeconomic and ecological potentials (Ghosh and Biswas 2015). Several factors that usually degrade such water bodies are agricultural run-offs, jute retting, macrophyte infestation, over exploitation and climate changes that results in loss of biodiversity and ecological efficiency. Homestead village ponds that hold huge potential for aquaculture and livelihood generation (Ansal et al. 2010) are very often targeted as dumping ground of human and animal wastes. As a result, they are perturbed and most of them ends up through eutrophication and macrophyte infestation. Such untapped resource through adoption of reclamation strategies like duckweed-based auto-bioremediation can be used as low-input small scale aquafarming (LISSA) to mitigate climate change. The trapped nutrients in such water bodies can fulfill the nutritional requirement of cultured fish species and can be economically harvested. The favourable candidate species are likely to be carps, tilapia, catfishes, and snakeheads. Several strategies should be taken before taking up viable aquaculture practices in such village ponds like: (i) proper lay out of the culture system and development of suitable drainage facility, (ii) development of sanitary facility for household practices for raising livestock animals, and (iii) development of duck weed-based phytoremediation rural model.

The effect of jute retting on aquatic environment is really a major concern as it causes huge organic load, decrease in pH and dissolved oxygen, discoloration of water and anoxic condition out of decomposition. Restoration of such pond can be done after the complete loss of toxicity of jute retting which usually takes two to three months based upon the complete water drainage facility and rainfall. Aquaculture practice can only be possible when water toxicity will completely lose. However, with liming and application of potassium permanganate, restoration is possible within a short time after which hardy species like catfish, tilapia, and common carps can be cultured as short duration cropping for four to five months.

8.6.1.3 Potential of Aquaculture in Inland Saline Water

The inland saline aquaculture system is known as "desert aquaculture" in Israel which commercially started during the late 1980s (Allan et al. 2009). It is the land-based aquaculture system which uses mainly the subsoil saline water and has been popular in Israel, the United States, India, and Australia. However, inland saline water may create major environmental problem by inhibiting the productivity of agriculture lands and in extreme cases turned into barren land with fragile ecosystem and vulnerable infrastructure (Allan et al. 2001; Anon 2001). Aquaculture in such system allows culture of euryhaline fin fishes like seabass (*Lates calcarifer*, *Dicentrarchus labrax*), shrimp (*P. monodon*, *L. vannamei*, *Marsupenaeus japonicus*), mollusks (*Saccostrea glomerata*), salmons (*Oreochromis mykiss*), tilapia (*Oreochromis niloticus*), prawn (*Macrobrachium rosenbergii*), *Mugil cephalus*, *Etroplus suratensis*, and common carp.

In India, out of a total area of 6.74 million ha that are affected by salt (including coastal saline soils), about 1.20 million ha are located in the non-coastal Indo-Gangetic plains in northern India covering seven states, viz. Punjab (0.15 million ha), Haryana (0.23 million ha), Rajasthan (0.38 million ha), Bihar (0.15 million ha), Uttar Pradesh (0.14 million ha), Madhya Pradesh (0.14 million ha), and Jammu and Kashmir (0.17 million ha) (Ansal and Singh 2019). Central Institute of Fisheries Education (Indian Council of Agricultural Research [ICAR]), Mumbai has done experimental trial on culture of giant freshwater prawn (*M. rosenbergii*) and white leg shrimp (*L. vannamei*) in Haryana. After successful trials, the technology was promoted to fish farmers in Punjab also. Guru Angad Dev Veterinary and Animal Science University, Punjab standardized the culture of Pacific white leg shrimp in Fazilka district of Punjab in 2014. This led to the start-up of financial assistance to the poor farming communities to adopt the technology and reclaim the lands. The technology has been carried out in 15 ha of salt affected areas of Punjab and expected to increase over 200 ha in 2019. The reported productivity ranged from 8 to 10 tonnes/ha per crop (cycles of four months) and a corresponding net profit of US$14 345–17 216 (Indian Rupee [INR] 1–1.2 million). Such shrimp farming has converted wastelands into gold mines, especially for large farmers with relatively stronger economic backgrounds and higher investment and risk-bearing capacity (Ansal and Singh 2019). Though such culture technology has less negative impact on environment compared to coastal aquaculture system, release of waste water from the culture units to the surrounding area may create environmental problems in the long run. Therefore, package of practice should be standardized and refined further, incorporating ecological issues into account, so as to make such resource use sustainable.

8.6.2 Vertical Expansion: Species and Culture Diversification

Vertical expansion through diversification of species and culture systems is considered as the principal determinant for "aquaplosion" in the country aimed toward up-scaling fish production, reducing the dependence of less species availability and increasing the viability of culture system. The diversification in production process should be carried out in an organized manner to maintain the sustainability and socioecological stability. The adaptation of culturable finfish, shellfish, and sea weed species to the environment will help to increase the production efficiency through available resource utilization and reduction of environmental deterioration (Table 8.1). The use of hydrobiological resources in aquaculture sector can be the new dimension for commercial aquaculture practices.

Table 8.1 Economically important culturable species in aquaculture in India.

Culture system	Species name	Scientific name
Freshwater aquaculture	Catla	*Labeo catla*
	Rohu	*Labeo rohita*
	Mrigal	*Cirrhinus mrigala*
	Silver carp	*Hypophthalmichthys molitrix*
	Grass carp	*Ctenopharyngodon idella*
	Common carp (Leather carp)	*Cyprinus carpio nudus*
	Common carp (Mirror carp)	*Cyprinus carpio specularis*
	Common carp (Scale carp)	*Cyprinus carpio communis*
	Amur carp	*Cyprinus carpio haematopterus*
	Bata	*Labeo bata*
	Calbasu	*Labeo calbasu*
	Deshi magur	*Clarias batrachus*
	Singhi	*Heteropneustes fossilis*
	Pangas	*Pangasius pangasius*
		Pangasianodon hypophthalmus
	Climbing perch	*Anabas testudineus*
	Mozambique tilapia	*Oreochromis mossambicus*
	Nile tilapia	*Oreochromis niloticus*
	GIFT tilapia	*Oreochromis niloticus*
	Giant snakehead	*Channa marulius*
	Striped snakehead	*Channa striata*
	Rainbow trout	*Oncorhynchus mykiss*
	Mahseer	*Tor putitora*
	Freshwater Prawn	*Macrobrachium rosenbergii*
	Freshwater mussel	*Lamellidens marginalis*

(Continued)

Table 8.1 (Continued)

Culture system	Species name	Scientific name
Brackish water aquaculture	Gray mullet	*Mugil cephalus*
	Parsia	*Liza parsia*
	Milk fish	*Chanos chanos*
	Asian Seabass	*Lates calcarifer*
	Cobia	*Rachycentron canadum*
	Snubnose Pompano	*Trachinotus blochii*
	Tiger prawn	*Penaeus monodon*
	White leg shrimp	*Litopenaeus vannamei*
	Kuruma shrimp	*Marsupenaeus japonicus*
	Mud crab	*Scylla serrata*
	Sand lobster	*Thenus orientalis*
	Spiny lobster	*Panulirus homarus*
	Indian backwater oyster	*Crassostrea madrasensis*
	Indian rock oyster	*Saccostrea cucullata*
	Sea weed	*Gracilaria edulis* *Gelidiella acerosa* *Kappaphycus alvarezii* *Hypnea valentiae* *Hypnea musciformis* *Acanthophora spicifera*

Breeding and culture techniques of diverse groups of finfish and shellfish species have been developed and standardized with respect to agroclimatic conditions. In freshwater aquaculture, besides Indian major carps and exotic carps, the commercial culture of different high valued fish species have been promoted for culture like pangas (*Pangasianodon hypophthalmus*), magur (*Clarias batrachus*), singhi (*Heteropneustes fossilis*), pabda (*Ompok pabda*), chital (*Notopterus chitala*), climbing perch and Vietnam koi (*Anabas testudineus*), and Java punti (*Puntius javanicus*), etc. Effective and time-tested strategies have been developed viz. polyculture and composite culture of carps and prawns, mono culture of catfish and tilapia, fish seed rearing-based cluster system, production of stunted fingerlings in shallow or rain fed ponds, inter-cropping system, adoption of MSMH, waste fed and canal fed aquaculture, integration with agriculture and live stocks and cage and pen culture in open water bodies like reservoirs and rivers, etc. to enhance fish production and productivity.

The scientifically managed fish culture system enhances productivity by increasing the fish production per unit area through stocking of commercially important, fast growing, and compatible fish species which can have the capability to utilize all ecological niches of the water body (Goswami and Samajdar 2011). Over the decades, composite culture system is contributing toward inland aquaculture production in sustainable manner (Biswas et al. 2018a) and also it is considered as the efficient proven technology for obtaining higher yield and return from unit area (Hussein 2012). During the last five decades of its inception, composite fish farming technology has been changed in many ways so far the management practices are concerned. In spite of having lack of education and low financial support from Govt. organization, experienced farmers are continuously modifying the management practices to enhance growth and production of fish thereby the farmers have been able to attain productivity of 5 tonnes/ha/year (Biswas et al. 2018b). MSMH is the modern approach for augmentation of carp production with stocking density of 12 000–14 000 fingerlings with culture duration of three to four months have become popular depending upon fish seed availability. Such adoption of approach resulted in enhanced production efficiency up to two to three times higher than traditional polyculture system. Harvesting of fish species is usually done after the fish attains a size of 0.5 kg or more. The intensity of the system requires professional management practices through scientific interventions.

Wastewater aquaculture plays an important role in India. Sewage fed aquaculture is mainly popular in East Kolkata wetlands of West Bengal where the fish culture is practiced in bheries. MSMH are usually followed by the farmers with high stocking density (20 000–50 000 nos/ha). The use of untreated urban waste water to fertilize the pond is really unique where large number of communities encompassing heterogeneous stake holders derive their livelihood from sewage fed aquaculture integrate with horticultural crops maintaining ecological stability (Jana 1998). The most suitable fish species in such systems are bottom grazing omnivores belonging to carps, tilapia, and cat fishes that can tolerate considerably low dissolved oxygen and high ammonia level. The average fish production in this culture system ranged from 1500 to 2000 kg/ha (ICAR 2011).

Periphyton-based aquaculture practices have great potential in the eastern states of India. The practice allows coculture of natural food-based culture system in which the racked bamboo sticks act as good substrate for colonization of complex fish food organisms consisting of protozoans, planktons, microbes, invertebrates, etc. collectively known as periphyton. The system is advantageous for rearing of herbivorous species like carps which effectively graze upon periphyton (Das and Sharma 2010).

Cage and pen culture in large open water bodies are of great potential in inland aquaculture development in the changed environmental scenario. The potentiality has been successfully demonstrated in India where 50 kg/m^3has been recorded within a short span of culture duration. The expansion and intensification of such culture systems in reservoirs which are basically untapped resource can contribute huge production along with enhanced livelihood generation and food security. Globally many countries are practicing cage culture wherein the contribution of China is around 68.4% followed by Vietnam (12.2%), Indonesia (6.6%), and Philippines (5.9%) to the global freshwater cage culture production. Presently many species are being cultured in cage system like the dominant species is Pangasius (41.10%), Nile tilapia (26.7%), common carp (6.6%), Mozambique tilapia (5.1%), rainbow trout (4.1%), and salmon (3.7%).

Recently, Govt. of India has taken initiatives to promote cage culture in reservoirs under various schemes like National Mission for Protein Supplementation that promoted cage culture to increase fish production in Tamil Nadu and Jharkhand. The semi-intensive cage culture allows low requirement of protein feed (<30%) which reduces the cost of production and can demonstrate high production performance (Radhakrishnan et al. 2019). Commercial culture of fish in cages has received great importance during last five years, especially with Pangas catfish, *P. hypophthalmus*. Production levels of 3–4 tonnes in a culture period of six months in cages of 6 m × 4 m × 4 m are being realized in several reservoirs, with the states of Chhattisgarh and Jharkhand taking the lead. However, cage culture system often faces the problem of poor water circulation through cages and seems problematic in private impoundments because of disease outbreaks, poor water quality, and modest growth, whereas raceway system has fewer issues than most of the other culture systems. Pen culture also has huge dimension in open water bodies for raising fish seed and fish production with a great emphasis on culture diversification as well as species diversification for conservation, sustainability and profitability point of view (Sarkar et al. 2020). The recommended stocking size of fish in pen system is about >100 mm long to achieve higher survival and production (Selvaraj et al. 1990). Pen culture can be considered as cheaper alternative system if the pen system is improved in terms of lowering of construction and management cost (Murugesan et al. 2005).

Integrated aquaculture system also has great potential for enhancement of fish production in India as the system requires low inputs like less land and water resource, less amount of feed, and no fertilizers. In India, integrated paddy-fish farming is practiced in various states like Bihar, West Bengal, Orissa, and Assam, where deep water paddy fields are available, whereas, livestock-fish farming is practiced in states like Andhra Pradesh, Bihar, Haryana, Kerala, West Bengal, Uttar Pradesh, Maharashtra, Orissa, and Tamil Nadu (Majumdar et al. 2018). Integrated farming system is considered as the most suitable and ecosystem-based approach to utilize the agriculture and livestock in such a way that the culture environment maintains its ecological balance as well as sustainability besides proven potentiality in enhancement of livelihood generation, high economic profitability in terms less production cost, and environmental amelioration through use of animal and agricultural wastes. Production levels of 3000–5000 kg/ha/year of fish have been demonstrated from such integrated farming practices.

Diversification of fish/shellfish species in brackishwater culture system is also another strategy to uphold the production efficiency as well as enhancement of fish production to develop sustainable production system. Coastal aquaculture has long been practiced as shrimp-based production system although the commercial production potential for other finfishes and bivalves has recently been recognized. After the tiger shrimp (*P. monodon*), the booming of culture of Pacific white leg shrimp has recorded phenomenal success in coastal aquaculture in India. With the continuously changing environmental perspectives, Coastal Aquaculture Authority in India has implemented rules and regulation under Coastal Aquaculture Authority Act, 2005, to ensure sustainable farming practices without any adverse impact on environment and ecosystem. Moreover, the policy makers are also concerned with the improvement of biosecurity and health management of culture species, additional quarantine facilities, and adoption of best management practices.

8.7 Application of Modern Technologies

8.7.1 Cluster Aquaculture

Small scale aquaculture practices play an important role in contributing toward country's fish production as well as economic growth. Usually, this type of producer are mainly having limited land, water, input material holding capacity, and financial support which lead to low productive performance in aquaculture in many developing countries in Asia and Africa. The livelihood generation is mainly based on integrated aquaculture approaches and on- or off-farm activities through traditional household management practices (Subasinghe and Phillips 2010). Also, the small scale aquaculture activities are in vulnerable condition due to the effects of various factors like climate change, low infrastructure facility, market demands, export facilities, and value chain, etc. Small and medium scale aquaculture producers are facing many problems due to globalization as well as increasing cost of aquaculture inputs which raise the question of future aquaculture development.

Cluster farming can be the good technological approach through application of better management practices with effective managerial process and technological interventions. The cluster farming approach can improve the farming activities with higher economic gain toward employment opportunities and livelihood generations. In India, several steps have been initiated for small aquaculture development through cluster management by area development mode viz. cage culture in reservoirs, composite fish farming, Genetically Improved Farmed Tilapia (GIFT) culture, freshwater prawn culture, Pangas culture in low land and upland hill areas and culture of white leg shrimp, brackish water fish species in coastal aquaculture as well as in inland saline water. Nonfood fish like ornamental fish culture in freshwater and coastal water in many states of India is providing a big opportunity for the producers, entrepreneurs, and trading companies through cluster management using value chain development mode. About 87% farmers are having less than 2 ha. cultivable which shows the importance of cluster aquaculture through formation of Farmer Producer Organizations (FPOs) for upscaling the production performance by adopting better management practices (NABARD 2018). The Marine Products Export Development Authority (MPEDA) has already encouraged and disseminated the cluster aquaculture system in all the maritime states and has formed about 918 aquaculture farming societies involving 19 854 farmers in India. Development of model aquaculture village for promoting culture aquaculture is also the major thrust area of development in Asian countries. The community-based aquaculture in village ponds can be utilized for sustaining aquaculture development through formation of cluster model village by utilization of water resources, management of village communities, and self-help groups (Mohapatra and Barik 2018).

8.7.2 Flow Through Aquaculture

Flow through or race-way aquaculture system is the new dimension in culture profitability point of view (Arana et al. 2018a) based on principles of water and heat conservation through recycle and reuse of water. The system enables high stocking density of fish where freshwater passes through the system and the originated wastes and metabolites are flushed away to maintain congenial environment. The most common raceways are rectangular shaped and constructed from concrete material where water flows through the series of tanks along the axis (Lawson 1995). In a typical raceway, the water flows with uniform water velocity across the tank cross section with approximate plug flow. But, friction losses at the tank-water and air-water boundary layers cause water velocities to vary across the width and depth of the raceway. The highest water velocity is at mid-depth, while the slightly reduced velocity is at the air-water interface and the lowest velocity is found along the bottom of the raceway (Fornshell 2019). Plug flow of water in the culture unit which enables better displacement and removal of waste and metabolites over throated flow design are increasingly being popular in the changed environmental context toward water conservation and meeting up of increasing market demand of high value fishes. In India, Central Institute of Freshwater Aquaculture, Bhubaneswar, studied the feasibility of culturing a number of species including carps in the flow through system for industrial aquaculture using irrigation canal water. As India is naturally endowed with varied agroclimatic regions, this system is mostly suitable for high altitudes regions where flow through arrangement can be easily achieved through gravity, basically below dams or hill streams as widely being practiced in Indonesia with red tilapia and others. However, the system has not been commercially promoted in India due to lack of package of practice and public awareness.

Although promising as on growing systems in many respects, traditional raceway systems face problems of excess feed and feces removal (Cripps and Poxton 1993). Water flow introduced at the bottom of the tank through baffles (Wagner 1993), and mechanical brushing (Tipping 1994), are well-known methods for assisting removal of solids from the rearing chamber. Water flow introduced at the bottom of the tank through baffles (Wagner 1993), and mechanical brushing (Tipping 1994), are well-known methods for assisting removal of solids from the rearing chamber. However, the major constraints involved in these systems in the plains are the excessive use of water and the high capital investment required during establishment (Singh et al. 2016). But the system can have the huge potential to enable additional income generation and livelihood generation. Recently in India, ICAR- Directorate of Coldwater Fisheries Research (DCFR), Bhimtal has developed mass seed production and culture technology of mahseer and rainbow trout in flow through hatchery system. The system is farmers friendly for breeding, egg incubation, and larval rearing with continuous water flow.

The major drawbacks of the traditional raceway are the limited localized body and the constant discharge of wastes into open water bodies (Masser and Woods 2008). The In-Pond Raceway System (IPRS) is a recently introduced strategy that allows high stocking density with reduced environmental impact (Arana et al. 2018b). The IPRS is similar to cages that can be adapted in any water body with advantage of controlled water circulation for maintaining optimal water quality and to improve feed management. In the IPRS model, originally developed at Auburn University, USA during 1990s, rectangular enclosures were suspended from floating pier where airlifts were placed at the one end of multiple raceways to pumped water into the raceways at the surface. The water is discharged from the raceways along with the bottom on the opposite sides of the airlifts through a solid settling chamber and then it is flown back to the impoundments (Caillouet 1995; Masser 1997). The solid waste is collected through quiescent-cone or tube like settlers attached on the backside of the raceway. The advantage of airlift system is not only easy movement of high volumes of water but also provide some degree of aeration when dissolved oxygen concentration is low (Boyd 1990). Subsequently, several models have been developed and patented with the idea of stocking fish in close enclosures, where water passes through it in suspended water bodies (Masser 2012).

8.7.3 Recirculatory Aquaculture System (RAS)

In the changed perspective, out of climate change and gradual shifting of paradigm from mere fish farming to becoming aquaculture entrepreneur, RAS is recently gaining momentum through with inherent risks involved. The system allows rearing of fish at high stocking densities in indoor tanks with customized and controlled culture environment formatted as per the species of culture. Development of modern RAS has been started nearly a century ego, but novel changes in designing and operation of the traditional RAS has resulted in substantial improvements in solid capture, biofiltration, and gas exchange (Espinal and Matulic 2019). The RAS can enhance fish production through utilization of limited water supply and land usage and also the system has great flexibility to locate production facility itself adjacent to markets. In 1/8th of a hectare and 1/6th of water use, the RAS was stated to be capable in producing up to 60 MT of fish per annum with suitable species of culture and in India, such production may reach up to 500 MT in the same area. In spite of these, commercial application of RAS are still lacking behind due to lack of proper scientific knowledge of farming, complexity in designing, financial involvement, high input and operational costs, risk factors, and nonavailability of trained expertise.

In recent years, National Fisheries Development Boards (NFDBs), Hyderabad has developed backyard small scale RAS design. The Ministry of Fisheries, Government of India is promoting RAS through the PMMSY (Pradhan Mantri Matsya Sampada Yojona) among the prospective farmers. ICAR–DCFR also has highlighted the initiation of large-scale outdoor RAS for rainbow trout culture at Champawat, Uttarakhand (Singh et al. 2016). Considering the huge potential and advantages, many states in India are identified for RAS application in aquaculture sector under the initiative taken by the Government of India under Blue Revolution viz. low-cost RAS, medium range RAS, and innovative activities (Sultan et al. 2017) where Pangasius and tilapia are considered as the suitable candidate species for culture in RAS. The management procedure in RAS is mainly dependent upon the quantity and quality of feed, and filtration efficiencies. Backyard RAS facility should be promoted to encourage small and marginal farmers and entrepreneurs to enhance fish production in urban and semi-urban areas in the changing consumer demand and market expansion (Figure 8.2).

8.7.4 Aquaponics

Aquaponics is an amalgamation of aquaculture and hydroponics, where aquaculture refers to the farming of aquatic organisms and hydroponics refers to cultivation of soilless terrestrial plants. In aquaponics system, aquaculture effluent undergoes microbial transformations to be used as source of nutrients for plant growth, while nutrient harvesting by the plants remediates water in situ for aquaculture (Yep and Zheng 2019). Advanced aquaponics systems have been transitioned into RAS in which the waste produced by aquatic organisms is filtered through tanks of naturally occurring microbes that mineralizes organic compounds and make them inorganically available for plant uptake. The main nutrient conversion occurs when the ammonia (NH_3) is transformed into nitrate (NO_3), via nitrifying bacteria. The aquaculture effluent filled with nutrients is then filtered into a

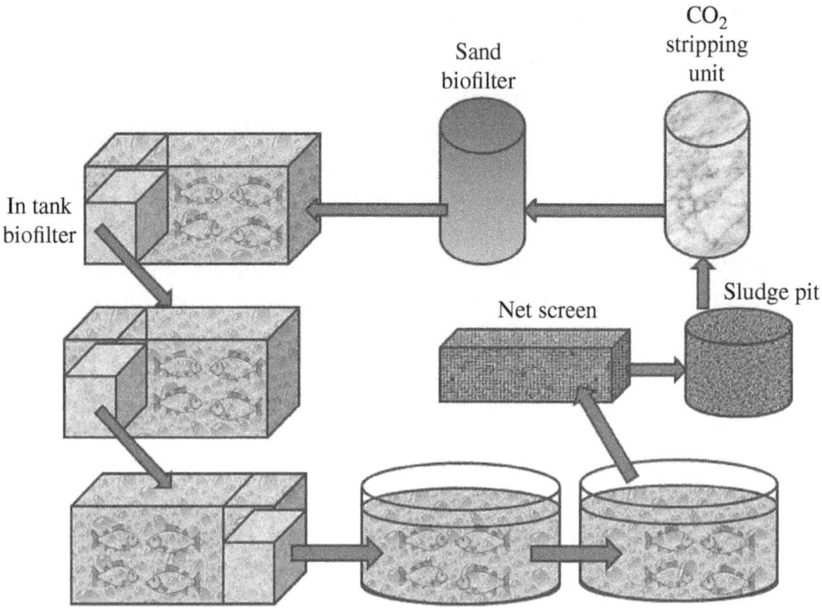

Figure 8.2 Design of small scale recirculatory aquaculture unit.

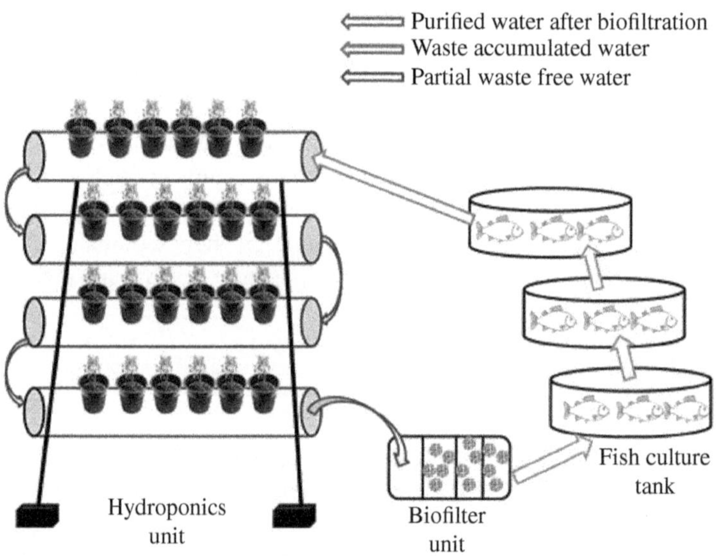

Figure 8.3 Design of small scale aquaponics unit.

hydroponic system as manure which is subsequently assimilated by the plant roots and microbes. Then water is recycled back to the aquatic rearing tanks, remediated of accumulated nutrients. This form of cultivation is considered as one of the most efficient and sustainable animal protein production systems (Figure 8.3). Compared to domesticated animal products such as beef, goat, and sheep, fish requires less feed per kilogram added growth (Tilman and Clark 2014) and in aquaponics, feed loss and fish waste can be recycled and reused.

Furthermore, in terms of water efficiency, aquaponics is probably more efficient than the stand-alone system either of conventional RAS and or hydroponics. Aquaponics system utilizes around 50% of nutrients through originally supplemented fish feed as plant fertilizer thereby supplementing substantially lower amount of fertilizer than hydroponics and in some cases, total withdrawal of fertilizer supplementation. Reduction in the use of fertilizer has a substantial impact on agriculture as nitrogen fertilizer production alone has been estimated to account for 57% of all agriculture energy demands (Mudahar and Hignett 1985) and phosphate reserves are predicted to be depleted during the next six to seven decades (Oelkers and Valsami-Jones 2008). Alongside, aquaculture system retains only 25% of the nitrogen in fish tissue leaving behind the rest 75% as excretory product into the environment (Krom et al. 1995; Hargreaves 1998).

Aquaculture operations require a high-quality filtration system to remove autochthonously generated toxic compounds like ammonia, nitrite, and suspended particles from the culture systems. If not properly managed, they may leach into the ambience and cause soil degradation and, eutrophication of waterbodies. According to Lennard and Leonard (2006), Water in the aquaponics system is replaced only to replenish water loss due to plant-mediated evapotranspiration, where evaporative loss is the principal pathway. Aquaculture, being the fastest growing food-producing sector, is predicted to contribute 54% of the estimated 200 million tonnes of fish requirement by 2030, this is a sector that can have an immense impact on the environment (FAO 2018). Closed-loop ecosystem approach has already drawn significant attention among the people as it mitigates several collateral fallouts of conventional agriculture. Therefore, aquaponics system can be optimized in controlled environments to decrease production losses due to urbanization.

8.7.5 Biofloc Technology (BFT)

BFT is a promising future technology which promotes the retention of waste and its conversion to biofloc as natural food for fish/shrimp in the aquaculture system (Panigrahi et al. 2014; Mandal and Das 2018; Dinda et al. 2020). Biofloc is composed of microorganisms such as algae (dinoflagellates and diatoms), zooplankton, bacteria, protozoan, nematodes, and particulate organic matter such as feces and uneaten feed that conglomerate together and perform symbiotic processes to maintain the water quality, maintain biosecurity, support high density of fish culture and reduce water exchange in the aquaculture system (Figure 8.4). It is a protein-rich live feed formed as a result of the conversion of unused feed and excreta into a natural feed in a simple and low-cost technique unlike the others.

In BFT, the heterotrophic microbe in the biofloc, through in situ bioremediation process, converts the organic nitrogenous waste to inorganic form and subsequent transfer to single cell protein and microbial protein which are utilized by both the microbes and filter feeding herbivorous and omnivorous fish species (Avnimelech 2009; Panigrahi et al. 2014). BFT system is associated with high fish stocking density and zero or minimal water exchange with the addition of external low-cost carbon sources to encourage the development of heterotrophic bacteria (Avnimelech 2009). By addition of cheap carbohydrate

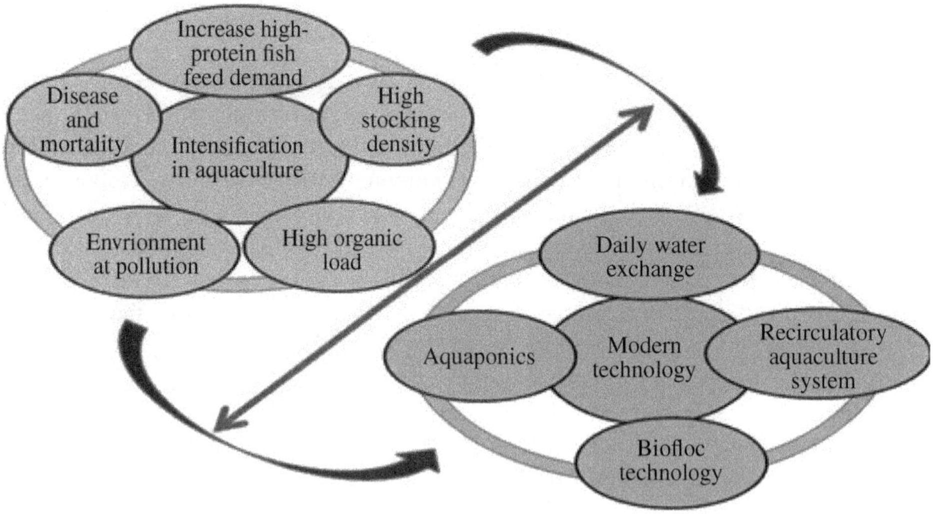

Figure 8.4 Sustainable approach in aquaculture system.

sources such as molasses or tapioca usually in a ratio around C: N 12–15 : 1 in the water column, biofloc converts the toxic nutrients into beneficial food sources. Usually in BFT, heterotrophic bacteria are more dominant and show 10 times more growth than autotrophic nitrifying bacteria (Hargreaves 2006). The main principle is to recycle nutrient by maintaining a high carbon/nitrogen (C/N) ratio in the water in order to stimulate heterotrophic bacterial growth that converts ammonia into microbial biomass (Figure 8.5). The microorganisms have two major roles as maintenance of water quality, by the uptake of nitrogen compounds generating in situ microbial protein; and increasing culture feasibility by reducing feed conversion ratio and a decrease of feed costs (Avnimelech 1999; Mandal and Das 2018).

In BFT system, species should be able to tolerate poorer water quality with high solids, as well as obtain nutrition from the biofloc through filter feeding. Tilapia is the major species cultured in BFT systems. Emerenciano et al. (2013) made an attempt to evaluate the effectiveness of BFT in larval rearing of the giant freshwater prawn. Reports are available on the production potential of biofloc-based culture systems of *Oreochromis mossambicus*, *Cyprinus carpio* var. communis, and *L. vannamei* (Ekasari et al. 2015; Dinda 2015; Mandal 2017). NFDB, India has given emphasis to adopt and discriminate BFT duly stocking fish seed using collapsible/Fibre - Reinforced Plastic (FRP) tanks with 100 nos/1000 l water (min. 1000 nos per 15 000 l tank – depending on species). The farmers who want to start biofloc system for the first time can stock food fish at 50–100 nos. fingerlings/1000 l water (depending on species) and ornamental fish at 35–50 nos. fingerlings/1000 l water (depending on species). NFDB has proposed to stock different varieties of filter feeding species like tilapia (*O. niloticus*), common carp (*C. carpio*), pangas (*Pangasianodon hypophthalmus*), white leg shrimp (*L. vannamei*), etc. with the objectives: to promote high yielding intensive fish farming in small area using BFT and to encourage farmers and unemployed youth into income earning. In India, farmers are practicing biofloc culture system with species like magur (*C.*

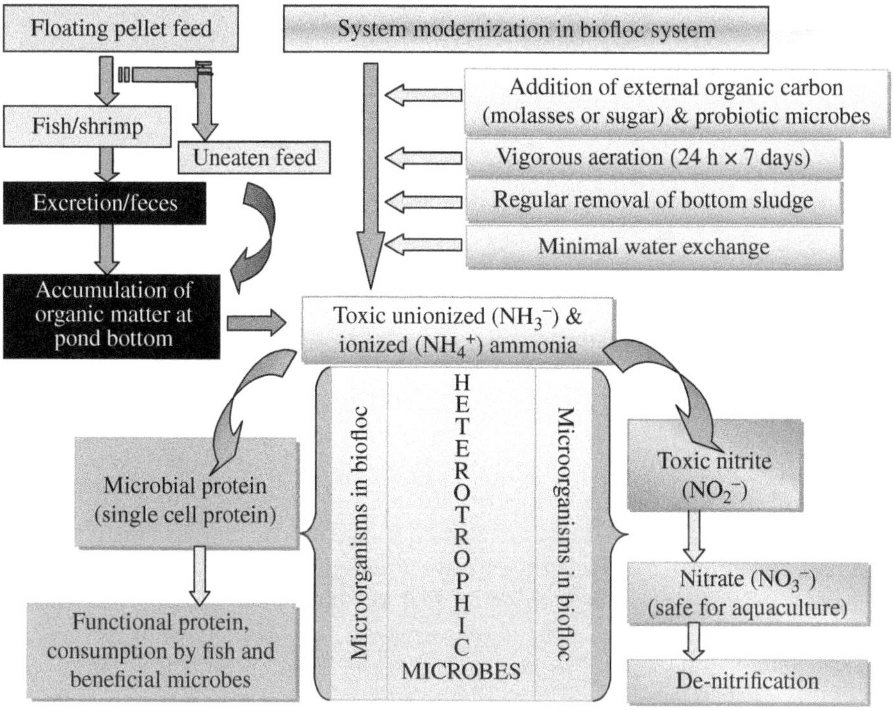

Figure 8.5 Mechanism of biofloc system.

batrachus), singhi (*H. fossilis*), Vietnam koi (*A. testudineus*), pabda (*O. pabda*) in freshwater and milk fish (*Chanos chanos*), pearl spot (*E. suratensis*), etc. in brackish water culture system.

8.7.6 Integrated Multi-Trophic Aquaculture (IMTA)

IMTA is the farming of species incorporating vertical integration of different trophic levels with complementary ecosystem functions in a way that enables one species' uneaten feed and feces, nutrients and by-products to be recaptured and transformed into fertilizer, feed, and energy for the other crops in the subsequent trophic level. This is intended to take advantage of synergistic interactions among species thereby bio-mitigation results in (Chopin 2013). Therefore, IMTA is the explicit incorporation of species from different trophic positions or nutritional levels within the same system (Chopin and Robinson 2004). This is distinctly different from the traditional polyculture system which very often denotes coculture of variable fish species from the same trophic level and from overlapping feeding niches. Therefore, organisms may collectively share the same biological and chemical processes with few synergistic benefits which could potentially lead to significant shifts in the ecosystem functioning.

The monoculture farming systems such as fed-aquaculture species and the organic/inorganic extractive aquaculture species (herbivorous fishes, bivalves, and aquatic plants) are independently farmed in different geographical locations resulting in pronounced

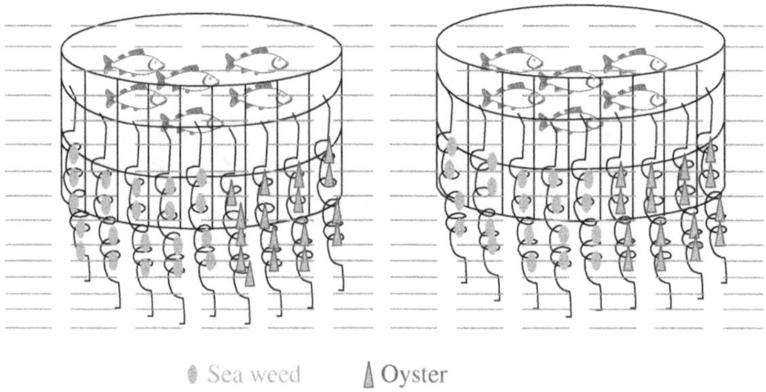

Figure 8.6 IMTA: Coculture of fish, sea weed, and oyster in cage system. (*See insert for color representation of the figure.*)

shift in the environmental processes (Figure 8.6). IMTA involves simultaneous culturing fed species (finfish/shrimp) with organic extractive aquaculture species (shellfish/herbivorous fish) and inorganic extractive species (seaweed) that utilize the inorganic wastes from aquaculture for their growth (Barrington et al. 2009). The multi-trophic subsystems are integrated in IMTA that refers to the more intensive cultivation of the different species in proximity of each other, linked by nutrient and energy transfer through water. Ideally, the biological and chemical processes in an IMTA system should balance and this can be achieved through the appropriate selection and proportions of different species providing different ecosystem functions. The cocultured species should be more than just biofilters; they should also be harvestable crops of commercial value (Chopin 2006). A working IMTA system should result in greater production for the overall system, based on mutual benefits to the cocultured species and improved ecosystem health; even if the individual production of some of the species is lower compared to what could be reached in monoculture practices over a short-term period (Neori et al. 2004).

8.8 Strategies

Aquaculture contributes significantly to the food security, employment opportunity, and livelihood generation. Several efforts have been initiated to create the suitable sustainable framework on the basis of environmental and socioeconomic development. Optimum utilization of land, water, energy, and available resources and effective adoption of the modern technological changes can be the best way toward enhancement of aquaculture production in inland freshwater and coastal aquaculture systems (Figure 8.7).

Majority of fish farmers are facing lots of challenges from quality seed and feed availability as well as lack of participatory approaches, extension communication, effective monitoring, and evaluations are also the other barriers for fish production. To increase aquaculture production and farmer's income in a changed environment, effective strategies required are:

- availability and optimal use of water
- rational use of lands and landscaping

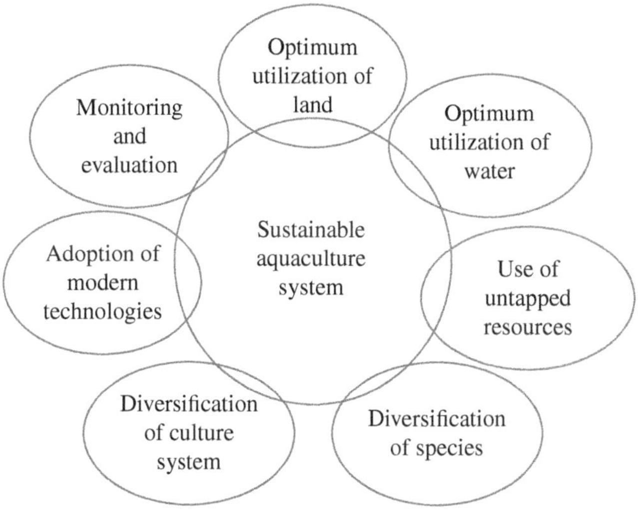

Figure 8.7 Dimensions of sustainable aquaculture system.

- use of untapped resources in aquaculture system.
- Geographic Information System (GIS)-based mapping of land and water resources
- species diversification redesigning of cropping pattern based on agroclimatic and socio-economic conditions
- effective dissemination of fish breeding and culture technologies
- setting up of fish brood bank
- development of low-cost model for rural areas
- conservation of aquaculture resources; species-specific and location specific
- development of quarantine and mobile fish health laboratory
- development of mobile water quality monitoring laboratory
- effective mitigation measures with reference to climate change and vulnerability
- regular monitoring and evaluation of culture system
- application and scientific validation of indigenous technical knowledge to cope up with changing environment
- application policies, guidelines for maintaining ecosystem-based sustainable approaches, and,
- public and consumer awareness.

8.9 Conclusion

Importance of aquaculture practices to meet the nutritional security and food demand of the ever increasing population has been well established. With the increasing demand and changing scenarios out of climate related factors, economic upsurge, globalization of market, and technological advancement, aquaculture also has been transforming from traditional subsistence activity to technologically sound production system incorporating scientific designs and package of practices. However, along with this transition through intensification, overexploitation of resources, generation of wastes, environmental degradation and at times lack of suitable technologies, and/or limitations of culture technologies have become imminent. Therefore, creative strategies should be taken to fulfill the demand and to fetch high economical returns through conserving ecological means and rational use of resources. Selection of techniques and level of intensification should be logically pragmatic and situation specific toward overall development of aquaculture as well as maintenance of ecological sustainability in the changed perspective.

References

Ahmed, N. and Diana, J.S. (2015). Threatening "white gold": impacts of climate change on shrimp farming in coastal Bangladesh. *Ocean and Coastal Management* 114: 42–52.

Allan, G.L. and Maguire, G.B. (1992a). Effects of stocking density on production of *Penaeus monodon* Fabricius in model farming ponds. *Aquaculture* 107: 49–66. https://doi.org/10.1016/0044-8486(92)90049-Q.

Allan, G.L. and Maguire, G.B. (1992b). Effects of pH and salinity on survival, growth and osmoregulation in *Penaeus monodon* Fabricius. *Aquaculture* 107: 33–47. https://doi.org/10.1016/0044-8486(92)90048-P.

Allan, G.L., Banens, B., and Fielder, S. (2001). Developing commercial inland saline aquaculture in Australia: Part 1 R&D Plan. Final report to FRDC Project No. 98/335, NSW Fisheries Final Report Series No. 31, NSW Fisheries, Cronulla, NSW.

Allan, G.L., Fielder, D.S., Fitzsimmons, K.M. et al. (2009). Inland saline aquaculture. In: *New Technologies in Aquaculture* (eds. G. Burnell and G. Allan), 1119–1147. Woodhead Publishing.

Anand, P.E.V. (2019). The fish farming industry of India. Global Aquaculture Alliance. https://www.aquaculturealliance.org/advocate/the-fish-farming-industry-of-india (accessed 15 September 2021).

Anon (2001). Dryland salinity in Australia: a summary of the National Land and Water Resources Audit – extent impacts, processes and management options, Australian Natural Resources Atlas, Department of the Environment, Water, Heritage and the Arts, Canberra. pp. 1–12. www.insidecotton.com › jspui › bitstream (accessed 15 September 2021).

Ansal, M.D. and Singh, P. (2019). Development of inland saline-water aquaculture in Punjab, India. https://www.aquaculturealliance.org/advocate/development-inland-saline-water-aquaculture-punjab-india (accessed 15 September 2021).

Ansal, M.D., Dhawan, A., and Kaur, V.I. (2010). Duckweed based bio-remediation of village ponds: an ecologically and economically viable integrated approach for rural development through aquaculture. *Livestock Research for Rural Development* 22 (7): 129.

Ansal, M.D., Dhawan, A., Singh, G., and Kaur, K. (2016). Species selection for enhancing productivity of freshwater carps in Inland Saline Water of Punjab-A Field Study. *Indian Journal of Ecology* 43 (1): 45–49.

Araki, H. and Schmid, C. (2010). Is hatchery stocking a help or harm?: evidence, limitations and future directions in ecological and genetic surveys. In: *Aquaculture, Supplement:*

Genetics in Aquaculture X 308 (eds. M. Vandeputtee, B. Chatain and G. Hulata), S2–S11. Aquaculture, Elsevier, Supplement 1 https://doi.org/10.1016/j.aquaculture.2010.05.036.

Arana, E., Chappell, J., Hanson, T., et al. (2018a). In-Pond Raceway System, IPRS, tilapia, demonstration project, Mexico. pp. 1–9. https://ussec.org › wp-content › uploads › 2019/10 (accessed 15 September 2021).

Arana, E., Chappell, J, Hanson, T. et al. (2018b). Commercial demonstration of In-Pond Raceways. pp. 1–9. https://ussec.org › wp-content › uploads › 2019/10 (accessed 15 September 2021).

Avnimelech, Y. (1999). Carbon/nitrogen ratio as a control element in aquaculture systems. *Aquaculture* 176 (3–4): 227–235. https://doi.org/10.1016/S0044-8486(99)00085-X.

Avnimelech, Y. (2009). *Biofloc Technology*, 2e, 182. Baton Rouge, LA: World Aquaculture Society.

Bailly, D. and Paquotte, P. (1996). Aquaculture and environment interactions in the perspective of renewable resource management theory. *Coastal Management* 24 (3): 251–269. https://doi.org/10.1080/08920759609362294.

Barange, M. Bahri, T. Beveridge, M.C.M. et al. (eds.) (2018). Impacts of climate change on fisheries and aquaculture: synthesis of current knowledge,adaptation and mitigation options. FAO Fisheries and Aquaculture Technical Paper No. 627. Rome, FAO. 628 pp.

Barrington, K., Chopin, T., and Robinson, S. (2009). Integrated multi-trophic aquaculture (IMTA) in marine temperate waters. Integrated mariculture: a global review. *FAO Fisheries and Aquaculture Technical Paper* 529: 7–46.

Biswas, B., Das, S.K., and Mandal, A. (2018a). Socio-economic dimensions and their impacts upon productivity of composite fish farming in North-24 Parganas district, West Bengal. *Journal of Entomology and Zoology Studies* 6 (2): 1131–1135.

Biswas, B., Das, S.K., Mondal, I., and Mandal, A. (2018b). Composite fish farming in West Bengal, India: redesigning management practices during the course of last five decades. *International Journal of Aquaculture* 8 (12): 90–97. https://doi.org/10.5376/ija.2018.08.0012.

Boyd, C.E. (1990). *Water Quality in Ponds for Aquaculture*. Auburn, Alabama: Alabama Agricultural Experiment Station, Auburn University.

Caillouet, E.W. (1995). Floating fish cultivating system and related method. US Patent 5,450,818, 19 September 1995. https://patents.google.com/patent/US5450818A/en (accessed 15 September 2021).

Chopin, T. (2006). Integrated multi-trophic aquaculture. What it is and why you should care and don't confuse it with polyculture. *Northern Aquaculture* 12 (4): 4.

Chopin, T. (2013). Integrated multi-trophic aquaculture–ancient, adaptable concept focuses on ecological integration. *Global Aquaculture Advocate* 16 (2): 16–19.

Chopin, T. and Robinson, S. (2004). Defining the appropriate regulatory and policy framework for the development of integrated multi-trophic aquaculture practices: introduction to the workshop and positioning of the issues. *Bulletin of the Aquaculture Association of Canada* 104 (3): 4–10.

Chowdhury, S.R., Nayak, A.K., Brahmanand, P.S. et al. (2018). *Delineation of Waterlogged Areas Using Spatial Techniques for Suitable Crop Management in Eastern India*. Bulletin No.79, 44. Bhubaneswar, Odisha, India: ICAR Indian Institute of Water Management.

Costa-Pierce, B.A. (ed.) (2008). *Ecological Aquaculture: The Evolution of the Blue Revolution*, 400. Wiley.

Cripps, S.J. and Poxton, M.G. (1993). A method for the quantification and optimization of hydrodynamics in culture tanks. *Aquaculture International* 1 (1): 55–71.

DAHDF. (2017). Annual report 2016–17, Department of Animal Husbandry Dairying and Fisheries. Ministry of Agriculture, Government of India. pp. 91–153

Das, M. K. and Sharma, A. P. 2010. Add fisheries and aquaculture management to our solutions for climate change and food security. *Bulletin*, 167:1–24. http://www.cifri.res.in/Bulletins/Bulletin%20No.167.pdf (accessed 15 September 2021).

De Silva, S.S. and Soto, D. (2009). Climate change and aquaculture: potential impacts, adaptation and mitigation. In: *Climate Change Implications for Fisheries and Aquaculture: Overview of Current Scientific Knowledge*. FAO Fisheries and Aquaculture Technical Paper. No. 530 (eds. K. Cochrane, C. De Young, D. Soto and T. Bahri), 151–212. Rome: FAO.

Devaraj, M. and Appukuttan, K.K. (2000). Perspective on coastal aquaculture in India. pp. 1–12. http://eprints.cmfri.org.in/4321/1/44.pdf (accessed 15 September 2021).

DFAHD (2018). *Handbook of Fisheries Statistics*, 1–190. New Delhi: Fisheries Statistics Division, Department of Fisheries, Ministry of Fisheries, Animal Husbandry & Dairying Government of India.

DFAHD (2020). *Handbook of Fisheries Statistics*, 1–96. New Delhi: Fisheries Statistics Division, Department of Fisheries. Ministry of Fisheries, Animal Husbandry & Dairying Government of India.

Dhawan, A. and Kaur, S. 2002. Pig dung as pond manure: effect on water quality, pond productivity and growth of carps in polyculture system. *NAGA: the ICLARM Quarterly*, 25(1), 11–14. http://www.worldfishcenter.org/Naga/Naga25-1/pdf/N (accessed 15 September 2021).

Dinda, R. (2015). Neem (Azadirachta indica) supplemented biofloc medium in common carp (Cyprinus carpio, L.) culture:growth response and environmental health. MFSc thesis. West Bengal University of Animal and Fishery Sciences, Kolkata, India. pp. 65.

Dinda, R., Mandal, A., and Das, S.K. (2020). Neem (*Azadirachta indica* A. Juss) supplemented biofloc medium as alternative feed in common carp (*Cyprinus carpio* var. communis Linnaeus) culture. *Journal of Applied Aquaculture* 32 (4): 361–379. https://doi.org/10.1080/10454438.2019.1645076.

Dudgeon, D., Arthington, A.H., Gessner, M.O. et al. (2006). Freshwaterbiodiversity: importance, threats, status and conservation challenges. *Biological Reviews* 81 (2): 163–182. https://doi.org/10.1017/S1464793105006950.

Dumbauld, B.R., Ruesink, J.L., and Rumrill, S.S. (2009). The ecological role of bivalve shellfish aquaculture in the estuarine environment: a review with application to oyster and clam culture in West Coast (USA) estuaries. *Aquaculture* 290 (3–4): 196–223.

Easterling, D.R., Meehl, G.A., Parmesan, C. et al. (2000). Climate extremes: observations, modeling, and impacts. *Science* 289 (5487): 2068–2074. https://doi.org/10.1126/science.289.5487.2068.

Ekasari, J., Zairin, M. Jr., Putri, D.U. et al. (2015). Biofloc-based reproductive performance of Nile tilapia *Oreochromis niloticus* L. broodstock. *Aquaculture Research* 46 (2): 509–512. https://doi.org/10.1111/are.12185.

Emerenciano, M., Gaxiola, G., and Cuzon, G. (2013). Biofloc technology (BFT): a review for aquaculture application and animal food industry. In: *Biomass Now-Cultivation and*

Utilization (ed. M.D. Matovic), 301–328. Croatia: InTech, Janeza Trdine http://dx.doi.org/10.5772/53902.

Espinal, C.A. and Matulic, D. (2019). Recirculating aquaculture technologies. In: *Aquaponics Food Production Systems* (eds. S. Goddek, A. Joyce, B. Kotzen and G. Burnell), 35–76. Cham: Springer.

FAO. (2018). The State of World Fisheries and Aquaculture 2020. Meeting the sustainable development goals. Rome. p. 194. http://www.fao.org/3/i9540en/I9540EN.pdf (accessed 15 September 2021).

FAO. (2020). The State of World Fisheries and Aquaculture 2020. Sustainability in action. Rome. p. 184. https://doi.org/10.4060/ca9229en

Ficke, A.D., Myrick, C.A., and Hansen, L.J. (2007). Potential impacts of global climate change on freshwater fisheries. *Reviews in Fish Biology and Fisheries* 17 (4): 581–613.

Field, C.B., Barros, V.R., Mach, K.J. et al. (2014). Technical summary. In: *Climate Change 2014: Impacts, Adaptation, and Vulnerability*, Part A: Global and Sectoral Aspects. Contribution of Working Group II to the Fifth Assessment Report of the Intergovernmental Panel on Climate Change. (eds. C.B. Field, V.R. Barros, D.J. Dokken, et al.), 35–94. Cambridge, United Kingdom; New York, NY, USA: Cambridge University Press. http://www.ipcc.ch/report/ar5/wg2.

Fornshell, G. (2019). Raceways-Freshwater aquaculture. https://freshwater-aquaculture.extension.org/raceways (accessed 15 September 2021).

Froehlich, H.E., Gentry, R.R., and Halpern, B.S. (2017). Conservation aquaculture: shifting the narrative and paradigm of aquaculture's role in resource management. *Biological Conservation* 215: 162–168. http://dx.doi.org/10.1016/j.biocon.2017.09.012.

Ghosh, D. and Biswas, J.K. (2015). Impact of jute retting on native fish diversity and aquatic health of roadside transitory water bodies: an assessment in eastern India. *Journal of Ecological Engineering* 16 (4): 14–21.

Goswami, B. and Samajdar, T. 2011. Knowledge of fish growers about fish culture practices, *Indian Research Journal of Extension Education*, 11 (2): 25–30 http://seea.org.in/ojs/index.php/irjee/article/download/523/518 (accessed 15 September 2021).

Goswami, C. and Zade, V.S. (2015). Analysis of international trade and economical commercial scope of ornamental fishes. *International Journal of Engineering and Applied Sciences* 2 (5): 2394–3661.

Gowen, R.J. and Rosenthal, H. (1993). The environmental consequences of intensive coastal aquaculture in developed countries: what lessons can be learnt. In: *Environment and Aquaculture in Developing Countries* (eds. R.S.V. Pullin, H. Rosenthal and J.L. MacLean), 102–112. Metro Manila: International Center for Living Aquatic Resources Management.

Greiner, J.T., McGlathery, K.J., Gunnell, J., and McKee, B.A. (2013). Seagrass restoration enhances "blue carbon" sequestration in coastal waters. *PLoS One* 8: e72469. https://doi.org/10.1371/journal.pone.0072469.

Hait, D. (2021). Socio-economic status of the small-scale fish farmers and impact of covid-19 on farming activities in Hooghly district, West Bengal. MFSc. Thesis. West Bengal University of Animal and Fishery Sciences, Kolkata, India. pp.1–79.

Hargreaves, J.A. (1998). Nitrogen biogeochemistry of aquaculture ponds. *Aquaculture* 166 (3–4): 181–212. https://doi.org/10.1016/S0044-8486(98)00298-1.

Hargreaves, J.A. (2006). Photosynthetic suspended-growth systems in aquaculture. *Aquacultural Engineering* 34 (3): 344–363. https://doi.org/10.1016/j.aquaeng.2005.08.009.

Hehre, E.J. and Meeuwig, J.J. (2016). A global analysis of the relationship between farmed seaweed production and herbivorous fish catch. *PLoS One* 11: e0148250. https://doi.org/10.1371/journal.pone.0148250.

Humphries, A.T., Ayvazian, S.G., Carey, J.C. et al. (2016). Directly measured denitrification reveals oyster aquaculture and restored oyster reefs remove nitrogen at comparable high rates. *Frontiers in Marine Science* 3 http://dx.doi.org/10.3389/fmars.2016.00074.

Hussein, M.S. (2012). Effect of feed, manure and their combination on the growth of *Cyprinus carpio* fry and fingerlings. *Egyptian Journal of Aquatic Biology and Fisheries* 16 (2): 153–168. www.ejabf.eg.net/pdf/vol-16-n-2/Issue%2014.pdf.

ICAR (2000). *Technologies from ICAR (for Industrialization)*. New Delhi: Indian Council of Agricultural Research 350 p.

ICAR (2011). *Handbook of Fisheries and Aquaculture*, 755. Directorate of Information and Publications of Agriculture, Indian Council of Agricultural Research.

Jana, B.B. (1998). Sewage-fed aquaculture: the Calcutta model. *Ecological Engineering* 11 (1–4): 73–85. https://doi.org/10.1016/S0925-8574(98)00024-X.

Jones, T., Parrish, J.K., Peterson, W.T. et al. (2018). Massive mortality of aplanktivorous seabird in response to a marine heatwave. *Geophysical Research Letters* 45 (7): 3193–3202.

Katiha, K., Sharma, A.P., Ekka, A. et al. (2014). *Economic Valuation of Inland Open Water Fisheries Resources*. Bulletin No. 187, 1–50. Barrackpore, West Bengal: Central Inland Fisheries Research Institute.

Kim, S.Y., Kurapov, A.L., and Kosro, P.M. (2015). Influence of varying upper ocean stratification on coastal near-inertial currents. *Journal of Geophysical Research, Oceans* 120 (12): 8504–8527. https://doi.org/10.1002/2015JC011153.

Krom, M.D., Ellner, S., Van Rijn, J., and Neori, A. (1995). Nitrogen and phosphorus cycling and transformations in a prototype'non-polluting'integrated mariculture system, Eilat, Israel. Marine Ecology Progress Series, pp. 25–36. https://www.jstor.org/stable/24849761

Kumlu, M. and Kir, M. (2005). Food consumption, moulting and survival of *Penaeus semisulcatus* during over-wintering. *Aquaculture Research* 36 (2): 137–143. https://doi.org/10.1111/j.1365-2109.2004.01196.x.

Lawson, T.B. (1995). *Fundamentals of Aquacultural Engineering*. New York: Chapman & Hall 355 pp.

Lennard, W.A. and Leonard, B.V. (2006). A comparison of three different hydroponic sub-systems (gravel bed, floating and nutrient film technique) in an aquaponic test system. *Aquaculture International* 14 (6): 539–550. https://doi.org/10.1007/s10499-006-9053-2.

Lin, N. (2019). Tropical cyclones and heat waves. *Nature Climate Change* 9: 579–580. https://doi.org/10.1038/s41558-019-0537-2.

Lu, C.L., Chen, S.N., and Hung, S.W. (2019). Application of novel technology in aquaculture. In: *Emerging Technologies, Environment and Research for Sustainable Aquaculture* (eds. Q. Lu and M. Serahuddin), 424. Intech Open https://doi.org/10.5772/intechopen.82887.

Majumdar, P., Pegu, C., Behera, B.K., and Das, B.K. (2018). Future aspects of integrated fish farming. *Acta Scientific Agriculture* 2 (12): 45–47.

Mandal, A. (2017). Environmental amelioration and production potential of biofloc based culture systems of Litopenaeusvannamei, *Oreochromis mossambicus* and *Cyprinus carpio* var. communis. PhD thesis, West Bengal University of Animal and Fishery Sciences, Kolkata, India. pp. 327.

Mandal, A. and Das, S.K. (2018). Comparative efficacy of neem (Azadirachta indica) and non-neem supplemented biofloc media in controlling the harmful luminescent bacteria in natural pond culture of Litopenaeusvannaemei. *Aquaculture* 492: 157–163. https://doi.org/10.1016/j.aquaculture.2018.04.006.

Masser, M. (1997). In-pond raceways. Southern Regional Aquaculture Center, Stoneville, Mississippi, USA. Publication 170. pp. 1–8. https://appliedecology.cals.ncsu.edu › 170.

Masser, M.P. (2012). In-pond raceways. In: *Aquaculture Production System* (ed. J.H. Tidwell), 387–394. https://doi.org/10.1002/9781118250105.ch15.

Masser, M.P. and Woods, P. (2008). Cage Culture Problems. Southern Regional Aquaculture Center (SRAC), United States Department of Agriculture (USDA), Cooperative State Research, Education, and Extension Service (CSREES). Publication no. 165. pp. 1–4. http://fisheries.tamu.edu › files › 2013/09 › SRAC.

Mills, K.E., Pershing, A.J., Brown, C.J. et al. (2012). Lessons from the 2012 ocean heat wave in the Northwest Atlantic. *Oceanography* 26: 60–64. https://doi.org/10.5670/oceanog.2013.27.

Mohapatra, B.C. and Barik, N.K. (2018). Development of model village cluster for aquaculture: a case in Begunia block of Khordha District, Odisha, India. *International Journal of Fisheries and Aquatic Studies* 6 (2): 534–540.

Mudahar, M.S. and Hignett, T.P. (1985). Energy efficiency in nitrogen fertilizer production. *Energy in Agriculture* 4: 159–177. https://doi.org/10.1016/0167-5826(85)90014-2.

Murugesan, V.K., Manoharan, S., and Palaniswamy, R. (2005). Pen fish culture in reservoirs: an alternative to land based nurseries. *NAGA World Fish Center Newsletter* 28 (1 & 2): 49–53.

Mylliemngap, B.K. and Ramanujam, S.N. (2011). Icthyodiversity in the coal mining and adjacent non-coal mining drainages of Jaintia Hills, India. *Asian Fisheries Society* 24: 177–185.

NABARD (2018). *Sectoral Paper on Fisheries and Aquaculture*, 1–86. Mumbai: National Bank for Agriculture and Rural Development https://www.nabard.org/auth/writereaddata/file/Fisheries%20and%20Aquaculture.pdf.

Nelson, E.J., Kareiva, P., Ruckelshaus, M. et al. (2013). Climate change's impact on key ecosystem services and the human well being they support in the US. *Frontiers in Ecology and the Environment* 11: 483–893. https://doi.org/10.1890/120312.

Neori, A., Chopin, T., Troell, M. et al. (2004). Integrated aquaculture: rationale, evolution and state of the art emphasizing seaweed biofiltration in modern mariculture. *Aquaculture* 231 (1–4): 361–391. https://doi.org/10.1016/j.aquaculture.2003.11.015.

NFDB. (2020). National Fisheries Policy. pp. 1–32. https://nfdb.gov.in/PDF/National_Fisheries_Policy_2020.pdf

Oelkers, E.H. and Valsami-Jones, E. (2008). Phosphate mineral reactivity and global sustainability. *Elements* 4 (2): 83–87. https://doi.org/10.2113/GSELEMENTS.4.2.83.

Oguntuga, O.A., Adesina, B.T., and Akinwole, A.O. (2009). The challenges of climate change in fisheries andaquaculture: possible adaptation measures in Nigeria. Capacity Development and Career Prospects in Conservation Science, p. 56.

Otchere, F.A., Veiga, M.M., Hinton, J.J. et al. (2004). Transforming open mining pits into fish farms: moving towards sustainability. *Natural Resources Forum* 28 (3): 216–223.

Panigrahi, A., Sundaram, M., Ravichandran, P., Gopal, C. 2014. Microbial based approach for shrimp culture and management. *ENVIS Newsletter Microorganisms and Environment Management*, (12):1–12. http://krishi.icar.gov.in/jspui/handle/123456789/12139 (accessed 15 September 2021).

Pentreath, R.J. (1994). The discharge of waters from active and abandoned mines. In: *Mining and its Environmental Impacts* (eds. R.E. Hester and R.M. Harrison), 121–131. UK: Royal Society of Chem.

Poff, N.L., Brinson, M.M., and Day, J.W. (2002). Aquatic Ecosytems and Global Climate Change: Potential Impacts on Inland Freshwater and Coastal Wetland Ecosystems in the United States. Pew Center on Global Climate Change. pp. 1–57.

Pradeepkiran, J.A. (2019). Aquaculture role in global food security with nutritional value: a review. *Translational Animal Science* 3 (2): 903–910.

Radhakrishnan, K., Aanand, S., Padmavathy, P., and Biswas, I. (2019). Current status of freshwater cage aquaculture in India: towards blue revolution. *Aquaculture* 23 (1): 1–9.

Raut, S.B. and Gawande, S. (2019). Environmental impact assessment of Kharland scheme. *International Journal of Innovative Research in Science, Engineering and Technology* 8 (7): 1–9. https://doi.org/10.15680/IJIRSET.2019.0807006.

Ruthrof, K.X., Breshears, D.D., Fontaine, J.B. et al. (2018). Subcontinental heat wave triggers terrestrial and marine, multi-taxa responses. *Scientific Reports* 8 (1): 13094. https://doi.org/10.1038/s41598-018-31236-5.

Sarkar, U.K., Das, B.K., Mishal, P., and Karnatak, G. (2019). *Perspectives on Climate Change and Inland Fisheries in India*, 1–428. Barrackpore, India: ICAR-CIFRI publication.

Sarkar, U.K., Mishal, P., Borah, S. et al. (2020). Status, potential, prospects, and issues of floodplain wetland fisheries in India: synthesis and review for sustainable management. *Reviews in Fisheries Science & Aquaculture*: 1–32. https://doi.org/10.1080/23308249.2020.1779650.

Seitz, R.D., Wennhage, H., Bergström, U. et al. (2014). Ecological value of coastal habitats for commercially and ecologically important species. *ICES Journal of Marine Science* 71: 648–665. http://dx.doi.org/10.1093/icesjms/fst152.

Selvaraj, C., Murugesan, V.K., and Aravindakshan, P.K. (1990). Impact of stocking of advanced fingerlings on the fish yield from Aliyar reservoir. In: *Contributions to the Fisheries of Inland Open Water Systems in India Part I* (eds. A.G. Jhingran, V.K. Unnithan and A. Ghosh), 109–114. Barrackpore, India: IFSI, CIFRI.

Seneviratne, S.I., Nicholls, N., Easterling, D. et al. (2012). Changes in climate extremes and their impacts on the natural physical environment. In: *Managing the Risks of Extreme Events and Disasters to Advance Climate Change Adaptation* A Special Report of Working Groups I and II of the Intergovernmental Panel on Climate Change (IPCC) (eds. C.B. Field, V. Barros, T.F. Stocker, et al.), 109–230. Cambridge, UK; New York, NY, USA: CambridgeUniversity Press.

Singh, A.K., Kamalam, B.S., and Kumar, P. (2016). Charting ways to invigorate rainbow trout production in India. *Journal of Fisheries Sciences* 10 (2): 25–32.

Smale, D.A., Wernberg, T., Oliver, E.C. et al. (2019). Marine heat waves threaten global biodiversity and the provision of ecosystem services. *Nature Climate Change* 9 (4): 306–312. https://doi.org/10.1038/s41558-019-0412-1.

Subasinghe, R.P. and Phillips, M.J. (2010). Small-scale aquaculture: organization, clusters and business. pp. 1–4. http://www.fao.org/3/al363e/al363e17.pdf (accessed 15 September 2021).

Sultan, S.A., Essa, K.S.A.T., Khalil, M.H. et al. (2017). Evaluation of groundwater potentiality survey in south Ataqa-northwestern part of Gulf of Suez by using resistivity data and site-selection modeling. *NRIAG Journal of Astronomy and Geophysics* 6 (1): 230–243. https://doi.org/10.1016/j.nrjag.2017.02.002.

Taylor, M.D., Chick, R.C., Lorenzen, K. et al. (2017). Fisheries enhancement and restoration in a changing world. *Fisheries Research* 186 (2): 407–412. https://doi.org/10.1016/j.fishres.2016.10.004.

Tilman, D. and Clark, M. (2014). Global diets link environmental sustainability and human health. *Nature* 515 (7528): 518–522. https://doi.org/10.1038/nature13959.

Tipping, J.M. (1994). Effects of raceway cleaning frequency on growth and freshwater survival of hatchery steelhead. *The Progressive Fish-Culturist* 56 (4): 293–295. https://doi.org/10.1577/1548-8640(1994)056<0293:EORCFO>2.3.CO;2.

Tisdell, C.A. and Leung, P.S. (1999). Overview of environmental and sustainability issues in aquaculture. *Aquaculture Economics and Management* 3 (1): 1–5.

Vass, K.K., Das, M.K., and Srivastava, P.K. (2009). Assessing the impact of climate change on inland fisheries in River Ganga and its plains in India. *Aquatic Ecosystem Health & Management* 12 (2): 138–151. https://doi.org/10.1080/14634980902908746.

Vignesh, M. (2021). Impact of COVID-19 on small scale coastal aquafarming in Cuddalore District, Tamil Nadu, India. MFSc thesis. West Bengal University of Animal and Fishery Sciences, Kolkata, India. pp. 1–93.

Vijayan, K.K. and Balasubramanian, C.P. (2019). Brackishwater aquaculture in India: a driver for blue economy. In: *Aquatic Resources and Blue Economy* (eds. C. Pradhan, R. Fernandez, A.T. Anna, et al.), 64–72. AQUABE, 28–30 November 2019.

Wagner, E.J. (1993). Evaluation of a new baffle design for solid waste removal from hatchery raceways. *The Progressive Fish-Culturist* 55 (1): 43–47. https://doi.org/10.1577/1548-8640(1993)055<0043:EOANBD>2.3.CO;2.

Westlund, L., Poulain, F., and Bage, H. (2007). *Disaster Response and Risk Management in the Fisheries Sector*, 1–56. Rome, Fisheries Technical Paper No, 479: Food and Agriculture Organization of the United Nations.

Wortley, L., Hero, J.M., and Howes, M. (2013). Evaluating ecological restoration success: a review of the literature. *Restoration Ecology* 21 (5): 537–543. https://doi.org/10.1111/rec.12028.

Yep, B. and Zheng, Y. (2019). Aquaponic trends and challenges – a review. *Journal of Cleaner Production* 228: 1586–1599. https://doi.org/10.1016/j.jclepro.2019.04.290.

Section 3

Water Management in Agricultural Systems

9

An Approach to Understand Conservation Agriculture

Anwesha Dey, Shubhi Patel, and H.P. Singh

Department of Agricultural Economics, Institute of Agricultural Sciences, Banaras Hindu University, Varanasi, Uttar Pradesh, India

CONTENTS

Sustainable Agriculture Systems and Technologies, First Edition. Edited by Pavan Kumar, A.K. Pandey, Susheel Kumar Singh, S.S. Singh, and V.K. Singh.

9.1 Introduction

The current cropping system dominated by intensive tillage, monoculture, and injudicious use of chemicals has jeopardized the future agricultural production potential. The global population is estimated to rise from 7.7 billion in 2019 to 9.7 billion in 2050 (United Nations 2019), thus, providing food to every human being on this earth and also safeguarding our environment at the same time is a real challenge to modern agriculture. Hence, to alleviate poverty and hunger worldwide, the yield has to be enhanced as land is limited. According to an estimate, land degradation has affected 2 billion people and 1.9 billion ha of global land area (Ahmad and Pandey 2018). Land degradation costs the world economy between US$18 and US$20 trillion annually (Albaladejo et al. 2021). At all levels of development, intensification of agriculture has negative impacts on natural resources such as soil, water, air, and biodiversity (Montgomery 2007; Kassam et al. 2013; Dumansky et al. 2014). The conventional tillage-based cropping system has increased crop production but has threatened the world by its effects such as land degradation, depletion of natural resources, a decline in soil organic matter, soil erosion, salinization, water scarcity, and climate change (Singh et al. 2013; Bisht et al. 2014; Tomar et al. 2014; Yadav et al. 2014; Pandey et al. 2015; Mandal et al. 2020; Rani et al. 2021). Traditional tillage-based agriculture is highly mechanized giving rise to problems such as soil erosion, surface and underground water pollution and increased water consumption by crop (Wolff and Stein 1998; Bhan and Behera 2014). Thus, to build a regenerative agro-ecosystem and achieve sustainable intensification, a paradigm shift from conventional agriculture to conservation agriculture is deemed to be one of the most suitable options.

9.2 Definition

Conservation agriculture is defined by the Food and Agricultural Organization (FAO) "as a farming system that promotes minimum soil disturbance (i.e. no tillage), maintenance of a permanent soil cover, and diversification of plant species." It enriches the biodiversity and also boosts various natural biological processes above and below the ground surface, which in return increases both water and nutrient efficiency and promotes sustainable crop production (FAO 2016; Smith et al. 2016). In other words, conservation agriculture is an approach to manage agro-ecosystems for improved and sustained productivity, increased profits, and food security while preserving and enhancing the resource base and the environment (FAO; Friedrich et al. 2012). Conservation agriculture stands on three pillars: minimum mechanical soil disturbance (i.e. no tillage), permanent soil organic cover (at least 30%) and species diversification through crop rotations (FAO 2016; Somasundaram et al. 2020). Conservation agriculture improves the soil's physical, chemical, and biological parameters, and thus enhances its quality (Basavanneppa et al. 2017; Yadav et al. 2017; Shrestha et al. 2020). Thus, conservation agriculture is the face of modern agriculture as it combines sustainable crop production with the protection of the environment.

9.3 Principles of Conservation Agriculture

Conservation agriculture is majorly focused to lessen the excessive soil mixing and sustain a layer of crop residues on the surface of the soil to reduce environmental damage. Despite high crop variability, the three major principles of conservation agriculture (as stated by FAO) apply universally to all cropping systems with locally adaptable practices. The principles are as follows:

9.3.1 Minimum Tillage and Soil Disturbance

The conventional agriculture system involves tillage as major practice for crop production, in contrary conservation agriculture involves minimum or no tillage. Conservation agriculture promotes direct sowing of the crop with the least soil disturbance after the harvest of the previous crop. It can be practiced both manually and mechanically. The major advantages of minimum tillage are such as it restores the surface soil by protecting it against water and wind erosion, minimizes costs in long term by the reduction in fuel use, less labor requirement and by saving time; it also helps to conserve soil moisture by increasing the soil infiltration rate; and it assists in increasing soil organic matter. In long term, the use of fertilizers per ha is decreased as the yield per unit of fertilizer/manure applied increases.

9.3.2 Permanent Soil Covers with Crop Residues and Live Mulches

Crop residues are the plant materials such as leaves, stalks, roots, etc., that are left behind after harvesting the crop. Mulch is a layer of organic matter spread over the soil surface that helps to conserve soil moisture, improve soil fertility, reduce weed growth, etc. Live mulches are cover crops that are intercropped or under-sown with the main crop to fulfill the purpose of a mulch and protect the topsoil layer. The major advantages of permanent soil cover are, it lowers the risk of soil erosion by wind and water, reduces weed growth by suppressing weed germination, increases nutrient efficiency by recycling nutrients, and leads to growth of soil organic matter and carbon sequestration.

9.3.3 Crop Rotation and Intercropping

Monocropping is one of the key features of traditional or conventional agriculture, which involves sowing of a single crop year after year on the same piece of land resulting in depletion of nutrients and organic matter in the soil. On contrary, conservation agriculture involves crop rotation (i.e. growing different crops alternatively on the same piece of land) and intercropping (i.e. growing two or more crops all together on the same land). The major advantages of crop diversification are such as it improves water-use efficiency because crops with different root lengths absorb soil water at different depths in the soil, enriches soil fertility, and enhances crop production as different crops take nutrients at different soil depth due to variations in their rooting length, and soil nutrients are utilized efficiently. Some additional benefits includes rise in nitrogen content in soil by inclusion of legumes in crop rotation aids to boost the successive cereal production. Diversification reduces pest and disease incidences as different crops are susceptible to varied causing agents, thus, cultivating such crops in rotation reduces the incidence of the same pest and disease.

9.4 History of Conservation Agriculture

Conventional agriculture is characterized as tillage-based agriculture. For producing food, humans had been tilling the soil since 3000 BCE. It was first started in Mesopotamia (Hillel 1998; Friedrich et al. 2012) in the valleys of the Euphrates and Nile rivers. The early tillage implements were mostly hand or animal-drawn; mechanical tillage implements were available after the industrial revolution in the nineteenth century (Friedrich et al. 2012). When in the 1930s the dustbowl caused mass destruction in areas of mid-west United States (Friedrich et al. 2012; Farooq et al. 2011), it was for the first time tillage was questioned by Edward H. Faulkner in the manuscript titled "Plowman's Folly" (Farooq et al. 2011). This incident raised concerns regarding soil protection and the term conservation tillage was introduced. In the 1940s, sowing of seed with help of seeding machinery showed the world that without tillage sowing is possible (Friedrich et al. 2012; Farooq et al. 2011). Despite conservation agriculture in the United States came into the limelight in the 1940s, it took 20 years for no-tillage to enter into the country's farming practice. In the 1970s, increased fuel prices and drought-induced soil erosion led the farmers to shift to conservation agriculture commercially (Farooq et al. 2011). At the same time in Brazil, no-tillage farming got started, and in West Africa, no-tillage and mulching techniques were in the testing stage (Greenland 1975; Lal 1976; Friedrich et al. 2012). In South Africa and South America, conservation agriculture took 20 years to reach significant adoption levels (Derpsch et al. 2010; Farooq et al. 2011). By the 1990s, conservation agriculture became so popular that it got spread exponentially across the globe and resulted in an agricultural revolution in regions of southern Brazil, Argentina, and Paraguay (Friedrich et al. 2012; Kassam et al. 2019). The rising acceptance of conservation agriculture fascinated the attention of many international organizations such as FAO, CIRAD, World Bank, IFAD, GIZ, NORAD, ACIAR, and some CGIAR centers for further development of this agriculture system (Kassam et al. 2019). With the promotion of conservation agriculture, it made its way in several parts of Asia and Africa and also in the countries like Canada, Australia, Spain, and Finland (Farooq et al. 2011). Currently, conservation agriculture is one of the most popular agricultural systems across the globe, as it is being practiced in the majority of the nations. Conservation agriculture is being suitable for almost all crops, major soil types, and climatic conditions, thus making it the most viable option for increasing production while protecting the environment.

9.5 How Conservation Agriculture Is Beneficial?

For a technology or practice to get adopted, it has to be advantageous in comparison to existing technology or practice. To make farmers shift from what they are doing to what they should do, knowledge about the benefits they can derive helps in better adoption. In this context, the advantages of conservation agriculture are highlighted below:

9.5.1 Agronomic Advantages

Conservation agriculture has numerous positive impacts on soil and land, and some major agronomic benefits derived from this conservation system are increase in soil organic matter due to use of crop residues and mulches, reduction in surface run-off and enhancement in groundwater levels due to improvement in water infiltration rate (Shrestha et al. 2020), improvement in nutrient-use efficiency and water-use efficiency (Jat et al. 2012; Saharawat et al. 2012), increment in production and yield (4–10%) (Gathala et al. 2011), weed incidence is reduced (Malik et al. 2005; Bhan and Behera 2014), and it also advances sowing date (Malik et al. 2005; Bhan and Behera 2014).

9.5.2 Economic Advantages

The sustainable practice of conservation agriculture has positive impression on farmer's pocket proving it to be an efficient and feasible option. The major reasons that contribute to help farmer earn the extra penny are the conservation technique is time-saving and requires fewer inputs, requires less labor, reduces overall cost of production (Malik et al. 2005), and promotes less use of machinery resulting in cut in fuel costs.

9.5.3 Environmental Advantages

The major advantage conservation system has over the traditional system is that it helps to protect our environment and conserves the natural resource base from excessive exploitation. Out of the many ways in which our environment benefits from this system, some advantages are it helps avoid and lower the risk of soil erosion, improves air and water quality, enhances carbon sequestration in long term leading to a reduction in the greenhouse effect and global warming, helps to replenish the loss of biodiversity, and is also resilient to climate change.

9.6 Global Scenario of Conservation Agriculture

9.6.1 World

In the twenty-first-century, conservation agriculture is globally adopted by farmers. In 2015–2016, it was practiced on 12.5% of the world's total cropland (180 m ha), which was 69% more than in 2008–2009. It is being practiced in about 125 million ha including the United States (26.5 million ha), Brazil (25.5 million ha) and Argentina (25.5 million ha) (Kumar et al. 2017). Conservation agriculture is largely adopted and practiced in South and North America, followed by Australia and New Zealand, Asia, Russia and Ukraine, Europe and Africa (Kassam et al. 2019). The footprints of conservation agriculture can be found all over the globe from Arctic Circle (e.g. Finland) in the north to 50° south (e.g. the Falkland Islands), extending over the tropics (e.g. Kenya, Tanzania, Uganda), to countries at sea-level to countries at 3000 m altitude (Friedrich et al. 2012; Kassam et al. 2019). Conservation agriculture is practiced over a range of different climatic variations such as from extreme dry regions with 250 mm rainfall (e.g. Morocco, Syria, Western Australia) to regions with 3000 mm rainfall (e.g. Chile) (Friedrich et al. 2012; Kassam et al. 2019). Farm size is not a constraint in this system as it is equally effective on half a hectare to thousand hectares of land. Soil conservation is one of the major outcomes of conservation agriculture as it can be practiced on any type of soil varying from 90% sand (e.g. North Africa, southern Mediterranean zone, coastal zones in tropical Africa, Australia) to 80% clay (e.g. Brazil's Oxisols and Alfisols) (Friedrich et al. 2012). Conservation agriculture is a boon to the environment; in southern Brazil, it has aided the restoration of destroyed savanna and forest soils – *the cerrado* – to fertile lands.

9.6.2 India

Conservation agriculture in India is trying to spread its roots for a long time, but the rise in its adoption is observed recently in the last decade. In India, the spread of this system of agriculture is confined to the irrigated regions of Indo-Gangetic plains, covering an area of 2 million ha (Singh et al. 2018). The Indo-Gangetic plains are the heartland of the Indian green revolution, which cover the states of Punjab, Haryana, Uttar Pradesh, Bihar, and West Bengal. In this fertile region, the rice–wheat cropping system prevails (Bhan and Behera 2014) giving it the names "bread basket" and "rice bowl" of India (Singh et al. 2018). The major conservation technologies majorly adopted in this region are zero-tillage, laser land leveling, bed planting, surface seeding, rotary tillage, use of leaf color chart, mechanical rice transplanter, etc. (Singh et al. 2011). The application of zero-tillage technology in wheat in North-Western India has reduced the production cost by ₹2000 to ₹3000 per ha (Malik et al. 2005; Bhan and Behera 2014). Indian agriculture holds tremendous potential for the adoption and promotion of conservation agriculture in other agro-eco regions also such as rain-fed semi-arid tropics and the arid regions of the mountain agroecosystems (Bhan and Behera 2014).

9.7 Conventional vs Conservation Agriculture

Conventional and conservation agriculture differs from each other in the following ways (Table 9.1).

9.8 Different Types of Conservation Agriculture Practices

Conservation agriculture includes many different types of practices like zero tillage/no tillage, minimum tillage, surface seeding, precision farming, bed planting, direct seeded rice.

9.8.1 Zero Tillage/No Tillage

Zero tillage or no tillage is the placement of crop seeds directly into the soil without any prior land preparation with the previous crop stubbles already present in the field. No tillage enhances the soil structure by improving the biological activity by soil microbes, nutrient recycling, increasing water-use efficiency, water infiltration rate, and soil water holding

Table 9.1 Comparison between conventional and conservation agriculture.

Particulars	Conventional	Conservation
Tillage	Excessive mechanical tillage is practiced	No-tillage or minimum tillage is practiced
Soil erosion	High rate of soil erosion by wind and water	Low rate of soil erosion by wind and water
Water infiltration	Low rate of water infiltration	High rate of water infiltration
Residue management	Burning or removal of crop residues	Crop residues are used as permanent soil cover
Weed management	Promotes weed growth by providing conditions for weed seed germination.	Prevents weed growth by suppressing weed seed germination
Cropping system	Monocropping culture is prevalent with less or no crop rotations.	Diversification of crop with crop rotations and intercropping
Labor requirement	More manual labor is required	Less labor is required
Adaptation to stress	Adaptation to stress is poor, high yield losses due to stress conditions	Adaptation to stress is comparatively higher, less yield loss due to stress conditions
Soil organic carbon (SOC)	The capacity to preserve soil organic carbon is comparatively low	High capacity to preserve soil organic carbon
Productivity	In long run, the productivity benefits will decline	In long run, the productivity benefits will increase
Sustainability	Conventional farming is a less sustainable practice.	The sustainability quotient of conservation farming is high.
Soil fertility	Reduced soil fertility	Soil fertility increases
Environmental damage	Causes huge damage to natural resources and thus to the environment	Helps to preserve the natural resources and causes low damage to the environment
Cost of production	In long run the cost of production is high	The cost of production will reduce in the long run

capacity (Hobbs et al. 2008; Bhatt 2017). In other words, zero tillage is a combined package of farm management practices involving crop rotation, residue management, weed control, pest management, harvesting (Laxmi et al. 2007). According to Landers, the principal impacts of zero tillage are like soil compaction is reduced, pulverization of soil aggregates and prevention of pan formation, binding of soil particles tightly, conservation of soil moisture, soil biological activity increases by a factor of 2–4, increment in soil organic matter resulting in positive carbon sink and better water and nutrient retention, reduction in weed incidence, and production and yield increase by multiple folds. Zero tillage advantages in several ways such as it reduces cost of cultivation and makes the food available at cheaper rates, lowers the risk of famines caused due to floods and droughts, a potential reduction in NO_2 emission, reduces air pollution by eliminating crop residue burning, and lowers farm diesel consumption.

In regions of South Asia, where farmers are burdened to produce more from the small cultivable land with a target of maximum resource utilization and minimum environmental damage, adoption of practices such as zero tillage is an efficient way for sustainable intensification. Zero tillage is commonly practiced in rice–wheat cropping system (CSISA 2018), especially in the Indo-Gangetic plains of India, which majorly involves sowing of wheat directly into the field having standing rice stubbles (Bhatt 2017) by a tractor-driven zero tillage seed-cum-fertilizer drill, sometimes with the basal fertilizers. Though this conservation technology is common in wheat, it can also be used in sowing crops such as chickpea, green gram, lentil, mustard, rice, and maize (CSISA 2018). In wheat crop, zero tillage has some major benefits such as early planting (helps to avoid terminal heat at the time of grain filling), nutrient management, less incidence of weed *Phalaris minor*, better grain quality, and less water requirement (CSISA 2018). It also helps farmers to avoid burning rice stubbles, thus leading to a reduction in air pollution. Studies suggest that the use of conventional tillage in wheat leads to a yield loss of 1–1.5% when grown in late November (Mehala 2015); zero tillage is a solution to the above problem as it reduces the sowing time of wheat immediately after paddy harvesting.

Adoption of zero tillage on large scale is limited by few factors such as farmer's ability to mechanically control the weeds by tillage is lost, inappropriate incorporation of plant residues increases the risk of carrying over plant diseases, and it is a time-consuming process, taking many years for the soil to regain its structure; thus, for a farmer to realize all the benefits of zero tillage, it may take 50 years.

9.8.2 Minimum Tillage

The conservation practice of minimum tillage is mainly aimed at reducing tillage activities to the minimum to ensure good seedbed, rapid germination, satisfactory crop stand, and favorable growing condition. This tillage can be applied in two ways: high-cost operations can be eliminated, which do not give additional benefits when compared to their cost; and operations that can be combined should be carried out together to save both cost and time.

Minimum tillage has many positive agronomic impacts such as soil structure is improved due to in situ crop residue decomposition, the layer of vegetation on the topsoil and channels formed by old roots increases the water infiltration rate, smooth root growth is facilitated due to improved soil structure, less soil compaction due to avoidance of the use of heavy tillage machinery, and reduced soil erosion by wind and water. The above effects are distinctly visible after two to three years of continuous practice of minimum tillage. Despite being an environmental-friendly option, minimum tillage practice has some limitations such as lower rate of seed germination, slow rate of decomposition adds more nitrogen, in leguminous plants such as peas and beans, the process of nodulation gets affected, the use of ordinary equipment make the sowing process difficult, and perennial weeds dominance leads to continuous use of herbicides.

9.8.3 Surface Seeding

This conservation method is majorly adopted in the rice–wheat cropping system. The seeds of Rabi cereals (usually wheat) are broadcasted on the wet soil surface either before or after the standing Kharif crop (usually rice) is harvested. The soil condition is not disturbed when seeds are broadcasted. The moisture present in the soil helps the seeds to germinate. This method is best suited to regions where the soil is moist, fine-textured, and poorly drained. These soils are normally unsuitable for tillage before planting. This extreme form of zero tillage is commonly practiced in the low-lying Indo-Gangetic regions of India, Pakistan, and Nepal.

Surface seeding method is an advantageous practice because it is an appropriate method for resource-poor farmers as no machinery is used, input use is condensed, it can be practiced on any size of field, mulching helps to prevent weeds and keep the soil surface moist, advances sowing time, reduces irrigation water used for seedbed preparation due to residual soil moisture, and lastly it doubles the cropping intensity of wheat in areas where one crop is only possible after paddy harvest because of prolonged wet conditions. Seeds used for planting can be dry or soaked (in freshwater for 6–10 hours). The soaked seeds germinate rapidly and uniformly. Dry seeds perform best when seeded in the afternoon; this is because it promotes water exposure and seed swelling in a warmer environment (Yadav 2019).

9.8.4 Precision Farming

According to the International Society of Precision Agriculture (ISPA), "Precision Agriculture is a management strategy that gathers, processes and analyzes temporal, spatial and individual data and combines it with other information to support management decisions according to estimated variability for improved resource use efficiency, productivity, quality, profitability, and sustainability of agricultural production." In other words, it is a farming approach that involves the use of information technology for the application of inputs in precise amounts to soil and crop to maintain optimum health and productivity. Precision agriculture is also known as satellite agriculture, site-specific crop management, and as-needed farming. The advent of GPS and GNSS has promoted the practice of precision farming among the farming community. Precision farming is a technology-enabled approach that aids the specific needs of individual crops and plots (Cropin n.d.). Sustainable development is augmented by proper utilization of inputs leading to less damage to the environment. "Doing the right thing at right time at the right place in the right way" is the key character of precision agriculture.

Precision farming has numerous pros that makes it the future of modern agriculture; some of the major pros are it augments agricultural productivity, reduces soil degradation, lowers excessive use of chemicals in the form of pesticides, herbicides, etc., effective use of water resources, surveying of fields can be done easily by the use of GPS as geo-mapping helps in estimating yield and analyzes soil characteristics, enhances quality and quantity of agricultural production, increases resource-use efficiency, and thus, reduction in cost of cultivation, enables scientists and government organizations disseminate the information at right time to farmers, reduces the risk of groundwater contamination and nitrate leaching by optimal use of fertilizers, and better resource management as wastage of resources is trimmed down.

In developing and underdeveloped countries, small landholdings due to fragmentation of land is a big problem for proper agricultural development. In India, more than 58% of operational landholdings are of size less than 1 ha, and only a few states such as Punjab, Haryana, Rajasthan, and Gujarat comprising 20% of agricultural lands have operational landholding size of 4 ha or more (Mungarwal and Mehta 2019). Thus, smart farming concepts such as precision farming is pill to India's ailments (fragmented land and small operational landholding size). Site-specific application of inputs helps to minimize the use of fertilizers, water, herbicides, pesticides, and farm implements, thus making precision farming an eco-friendly and cost-effective solution (Solomon 2020). In water-stress regions, use of remote-sensing technology provides data related to the quantity of soil moisture helping farmers to grow crops accordingly and also increases irrigation efficiency. Big data analytics helps to forecast and mitigate weather, water stress, pests/diseases, and nutrient deficiency in advance (Kumar 2020). Precision agriculture has some additional benefits such as it helps to regulate the emission of greenhouse gases in agricultural fields and amplifies employment opportunities for skilled labor (Kumar 2020).

The major issues that had limited the adoption of precision farming on a commercial scale are high capital costs, discouraging the farmers to adopt the technology, many precision farming techniques are still under development and may require expert advice or aid for proper implementation; full implementation of this technology requires actual and sufficient data, which in itself is a long and time-consuming process; and analysis and collection of data is a difficult task involving skilled labor. Lack of proper infrastructure, unique landholding pattern, diverse demography, distinct social and economic conditions, lack of proper information, and absence of risk-taking attitude in farmers are some other reasons that hinder the expansion of this smart farming technique in India.

9.8.5 Bed Planting

The conventional agriculture system involves the flat planting of crops under irrigated conditions. The flat planting method has numerous negative impacts such as (Tripathi and Das 2017) wastage of water and formation of soil crust due to excessive flood irrigation, increased pest and disease incidence due to excessive crop canopy humidity caused by flood irrigation and high dose of nitrogen fertilizers, intercropping cannot be practiced in flatbed planting as this cropping system does not support crops with different moisture requirement, and higher crop lodging is observed due to no passage of air. Post-green revolution, the major challenge was to increase production with limited land and water resources and maintain the environmental sustainability. Bed planting was a solution to the above problems. In the bed planting system, the land is prepared at first and then raised beds and furrows are formed manually or with help of raised bed planting machine (Mollah et al. 2009), seeding and fertilizer placement is done together in one go (Tripathi and Das 2017). Seeds are sown on top of the bed, and furrows are used for irrigation, drainage, weeding, and other intercultural operations (Tripathi and Das 2017). The ridge-furrow system of planting is useful to grow high-value crops that are sensitive to waterlogging stress. Crops such as cotton, maize, wheat, and soybean are grown on raised beds (Pandey et al. 2013). The furrows in raised bed system are comparatively wider than the conventional system; two furrows are merged and the ridge's width is twice wide as the traditional one (FAO 2016). The merging of ridges helps to grow the same number of crops with half the amount of irrigation water (FAO 2016). Raised bed planting is a boon as it opens up opportunities for proper fertilizer placement and mechanical control of weeds (Sharma et al. 2002; Singh et al. 2010), whereas a permanent bed system is an advantage in water-stress areas where the groundwater levels are continuously falling. The bed planting method saves 18% to 30–50% of irrigation water (Hossain 2001; Naresh et al. 2010; Singh et al. 2010) in comparison with the traditional flat planting method. Crop yield is improved by more than 20% (Pandey et al. 2013). Crop diversification and intercropping are facilitated by this conservation system. Some of the major intercropping patterns that are practiced are wheat with chickpea, and Indian-mustard with sugarcane, maize with potato, mint with wheat, rice with soybean, and pigeon pea with sorghum or green gram (Pandey et al. 2013). The popularity of bed planting among farmers rose from 6 to 75% in high yielding and irrigated wheat-growing regions of north-western Mexico (Sayre and Moreno Ramos 1997; Tripathi and Das 2017).

9.8.5.1 Furrow Irrigated Raised Bed

In the furrow irrigated raised bed (FIRB) system, the crops are grown on top of beds or ridges, and the horizontally flowing irrigation water through the furrows is diverted toward the beds through infiltration and percolation. Factors such as spacing between tyres of the tractor, type of soil, rainfall range, groundwater level, salinity and quality of irrigation water, and other requirements of crops grown in rotation determine the dimensions of the bed (Pandey et al. 2013). Beds of height 12–15 cm and width 37–107 cm are maintained depending on the crop (Naresh et al. 2012).

Bed planting technique is preferred because of the reasons such as it helps in simple and efficient management of irrigation water resulting in increased water use efficiency, enhances production of upland crops due to improved drainage, augments fertilizer use efficiency, 20–25% lower seed rate is required, better plant stand establishment is allowed, crop productivity is increased by 5–10%, crop lodging is reduced, weeds can be easily controlled mechanically or by hand weeding and rouging, reducing dependence on herbicides, easy application of herbicides on the beds, raised beds enable the better penetration of solar radiation, thus strengthening the plant from its base, on non-rainy days moisture stress is reduced, and the time for irrigation is saved by 25–30%.

9.8.6 Direct-Seeded Rice

The conventional puddled transplanting system involves sowing paddy seeds in nurseries, and after 20–25 days, the young seedlings are transplanted to the puddled field. Transplanting involves heavy water losses through puddling, surface evaporation, and percolation (Farooq et al. 2011). A huge amount of water is used for puddling, which results in breakage of capillary pores, destruction of soil aggregates, formation of hardpans, and reduction in soil permeability (Farooq et al. 2011; Kaur and Singh 2017). The production of 1 kg of rough rice requires 5000 l of water (Bouman 2009). The profit margins in transplanted paddy are observing a declining trend due to increased water inputs and labor costs (Pandey and Velasco 1999; Kaur and Singh 2017). Moreover, the transplanted rice is a major contributor to the emission of greenhouse gas (i.e. methane). To enhance water productivity, minimize the labor requirement, and mitigate greenhouse gas emission, an alternative crop establishment such as direct seeding is needed. In direct seeding, seeds are directly sown in the field instead of transplanting seedlings from the nursery. The conservational technique of direct seeding saves irrigation water, labor, time, and energy, decreases greenhouse gases emission, and promotes better growth of succeeding crops, etc. (Kaur and Singh 2017). Direct seeding can be done by three methods:

1) Wet direct seeding – In the wet direct seeding method, pre-germinated seeds are sown directly into wet puddled soil. The environment of the puddled seedbed can be aerobic or anaerobic. This method is usually practiced in well-irrigated areas with a good drainage system.
2) Dry direct seeding – In the dry direct seeding method, dry seeds are sown directly into dry soil. Aerobic soil is preferred for this method. The sowing of seeds is done in three ways: broadcasting, drilling, and dibbing. This method of direct seeding is most suitable for rain-fed areas and can also be practiced in irrigated areas with precise water control. In developing nations, dry direct seeding has been the prime method of sowing paddy since the 1950s (Pandey and Velasco 2005; Farooq et al. 2011).
3) Water seeding – In the water seeding method, seeds are directly sown in standing water having anaerobic soil condition. This method is practiced in irrigated areas with good land leveling (Balasubramanian and Hill 2002). This method is popular in areas with red rice problem (Balasubramanian and Hill 2002; Farooq et al. 2011).

Major reasons for adoption of direct-seeded rice over puddled transplanted rice:

1) Water conservation – Traditional rice crop consumes 50% of total irrigation water in Asia (Kaur and Singh 2017). Direct seeding conserves a substantial amount of water (Dawe 2005) and also utilize them efficiently. According to various studies, irrigation requirement in DSR in non-puddled soils of Northwestern India decreased by 35–57% (Sharma et al. 2002; Singh et al. 2002; Farooq et al. 2011), in small plot trials reduced by 20%, and in raised bed DSR reduced by 12–60% (Gupta et al. 2003).
2) Less labor requirement – DSR is comparatively a labor-saving technology. The traditional rice planting method requires 37% more labor than DSR (Kaur and Singh 2017). In DSR, labor demand is evenly spread out over the whole period, leading to proper utilization of family labor and reduce dependence on hired labor, thus reducing labor costs.
3) Higher grain yield – Higher grain yields are obtained with good management practices. Augmented yield is attributed to increased panicle number, higher thousand kernel weight, and lower sterility percentage (Dingkuhn et al. 1991; Sarkar and Das 2003; Farooq et al. 2011).
4) Lower greenhouse gas emission – The anaerobic condition, i.e. absence of oxygen caused due to prolonged standing water in traditional puddled rice, is the prime source of methane (a greenhouse gas) emission adding to the global warming crisis. Rice cultivation accounts for 10–20% of total global annual methane emissions (Reiner 2000). The use of the dry direct seeding method reduces the methane emission by 30–58%, thus helping to reduce the environmental damage (Pathak et al. 2009; Singh et al. 2009).
5) Reduced cost of cultivation – Less labor requirement leads to a reduction in labor costs. In regions with high labor wage rate, the saving rate for DSR establishments was the US$50/ha (Kumar et al. 2009). According to a study, the income from direct-seeded rice and transplanted rice was ₹34 953 per ha and ₹30 420 and returns per rupee investment were ₹3.12 and ₹2.66, respectively.

The advantages that makes direct-seeded rice more adoptable to farming community are it is labor saving, planting is done easily and fast, crop matures early by 7–10 days allowing timely sowing of the subsequent crops, efficient use of water, soil conditions are improved, and reduction in greenhouse gas emission. This technique of conservation agriculture too possess some major limitations such as direct sowing of seeds exposes them to various pests such as rats, birds, and snails, increased crop-weed competition due to collision of their maturity time, high weed infestation especially in dry field conditions, higher susceptibility to diseases, water stress during the reproductive phase may lead to panicle sterility, micronutrient deficiencies occur due to imbalanced application of nitrogenous fertilizer, and lodging of the crop.

9.9 Impact of Conservational Agriculture on Crop Production

Conservation agriculture has different impacts on different crops, evident from the Table 9.2.

9.10 Future Prospect of Conservation Agriculture in India

A developing nation like India faces a major challenge of fulfilling its growing needs for food and energy. In the journey of meeting its needs, the country faces threats such as natural resource degradation, climate change, and accelerating production costs. Thus, there is a dire need for a crop management system adapted to local needs that play a major role in boosting the socio-economic profile of the local farmers and is also ecological. The prime reasons why conservation agriculture practices should be promoted in India are as follows:

1) Cost of production – In India where the majority of farmers falls in the small and marginal farmer's category, cost of production plays a key role in determining their income levels. Adoption of resource-conserving technologies such as zero tillage or no tillage helps to reduce the cost of production enabling the farmers to earn more profits. Reduction in costs is credited to savings from diesel, labor costs, and input costs.
2) Yields – Crop yields under the conservation agriculture system are comparatively higher than the conventional system. Increased yields are attributed to factors such as prevention of soil degradation, enhanced soil fertility, augmented soil moisture regime, and crop rotation (Bhan and Behera 2014).
3) Resource efficiency – Conservation agriculture practices enhances water-use efficiency and nutrient-use efficiency. Most of the conservation practices save water and helps to replenish groundwater levels. Improved soil structure helps in the efficient absorption of nutrients leading to increased yields.
4) Environment – The principle of permanent cover with crop residue and mulches eliminates crop residue burning to help to reduce greenhouse gases emission, and it also avoids erosion of surface soil. Conservation practices such as direct seeding reduce methane emission.

Table 9.2 Impact of conservation agriculture on different crops.

Crop	Impact of conservation agriculture	Reference
Rice	Increase in grain yield by 12%	Bell et al. (2019)
	Minimum tillage and no-tillage save 16 and 19% of the total cost, respectively, in comparison to conventional tillage	Calcante and Oberti (2019)
Wheat	Increase in crop yield by 4–10% due to zero tillage	Gathala et al. (2011)
	Grain yield increased by 79%	Shrestha et al. (2020)
	Moisture loss by evaporation was reduced by 65% in land covered with stubble mulches	Klocke et al. (2009)
	Increment in production by 27%	Bashour et al. (2016)
Maize	The benefit–cost ratio is higher in plot having no tillage with residue (2.43) than in plot having no tillage without residue (2.38) in Chitwan, Nepal	Santa Bahadur and Shrestha (2014)
	A higher benefit–cost ratio was obtained from zero tillage (2.32) as compared with traditional tillage (2.20)	Pariyar et al. (2019)
	A higher benefit–cost ratio was obtained from conservation tillage with residue (2.5) as compared to conventional tillage without residue (1.7)	Karki et al. (2014)
Barley	Average yields from the practice of zero tillage and reduced tillage were more than conventional moldboard tillage in the Pampas region	Buschiazzo et al. (1998)
Sorghum	Enhanced production of sorghum as a biofuel feedstock in semi-arid regions of South Africa	Malobane et al. (2018)
Potato	Increment in productivity and net profit by 25 and 24%, respectively	Mosquera et al. (2019)
	Soil carbon content rose by 29% by the practice of conservation tillage for seven years	Quintero and Comerford (2013)
Cotton	Enhanced water conservation, reduced soil erosion, and increased soil organic matter	Blanco and Lal (2008), Patel et al. (2019)
	Higher profits from no tillage with residue retention	Naveen Kumar and Babalad (2017)
	Increased soil moisture and reduced weed incidence	Bilalis et al. (2003)
Lentil	Yield in experimental plots under conservation tillage increased by 27.7%, and the cost of cultivation got reduced by 17.5%	Bashour et al. (2016)
	Grain yield increased by 20–30%	Wahbi et al. (2014)
Soybean	20% higher average yields were obtained under no tillage with a better germination rate	Thiagalingam et al. (1991)
Beans	Straw yield and grain yield increased by 13 and 7%, respectively, in the maize-bean cropping system	Liben et al. (2018)
	Higher net profits of dry bean	Micheni et al. (2014)

9.11 Challenges and Constraints in Conservation Agriculture

The total area under conservation agriculture is still smaller than the total arable area under tillage. But in recent decades, conservation agriculture has gained some popularity and is being preferred by farmers across the globe. In many countries such as the United States, Canada, Australia, Brazil, Argentina, Paraguay, and Uruguay, conservation agriculture has become part of the mainstream agriculture (Kassam et al. 2019). In developing countries like India, China, Zimbabwe, Kazakhstan, etc., the adoption rate of conservation agriculture has increased in the past few decades, but there is a huge scope for more adoption. Conservation agriculture is truly the future of agriculture, but the spread of potential benefits of this sustainable system is still limited and surrounded by several constraints. Some of the major constraints that impede the adoption of this economic and environment-friendly system on a broad scale are that crop residue in many parts of the world, especially in developing and underdeveloped countries, is used to feed the livestock, thus competing its use for permanent cover in conservation agriculture. Farmers in rain-fed areas face the problem of crop residue scarcity due to less production of biomass. The crop residue has another alternate use, it is used as a fuel. Then, crop residue burning is a practice that farmers adopt to timely sow the subsequent crop. It is a common practice in the rice–wheat cropping system where farmers prefer to burn the rice stubbles to sow the wheat crop on time. The burning of residue causes damage to the environment. Apart from this, lack of knowledge about the potential benefits of conservation agriculture is one of the major hindrances in the broad adoption of this system. Moreover, conservation agriculture requires site-specific knowledge about the appropriate technology or practice based on local set of situation adds to the constraints. The performance of this system depends on a basic understanding of the processes and components and their interaction. Proper management strategies are needed to acknowledge conservation agriculture as a system. There is lack of appropriate seeding facility for small and marginal farmers. The major challenge is to develop, standardize, and promote farm machinery for seeding that is focused on a particular array of crop or cropping system. The machinery should also manage the harvesting operation. The research and development in conservation agriculture need to create a linkage between the core group including scientists, farmers, village level workers, extension agents, and other stakeholders related to farming for the development and promotion of new technologies and advancement of existing techniques. Highly skilled and scientifically trained manpower is required to address the problems and develop the conservation system from the perspective of local farmers.

Finally, the major constraint that hinders the adoption of this sustainable agriculture system is it gives results in long term. The impact of conservation agriculture may not be noticeable in the initial years of evaluation. To develop an enhanced soil–water and nutrient management system, a basic understanding of the physical, biological, and chemical interactions is a needed (Abrol and Sangar 2006).

9.12 Conclusion and Policy Implications

The condition of Indian agriculture, with rising issues of soil degradation, groundwater depletion, pollution of land, water, and soil, climate change calls for suitable solutions that could cope up with the rising issues in a sustainable manner. One of the solutions that repair the quality of soil and also conserve the resources is conservation agriculture. Conservation agriculture is one of the promising solutions for that with varied options like zero tillage, minimum tillage, surface seeding, precision farming, bed planting, direct-seeded rice, etc. These techniques are good for saving the input cost, reduce the number of seeds required, save water requirement, and also increase the yields, thus proving as a profitable option with lesser cost and higher returns. Despite this, challenges exist in form of a lower rate of adoptions, need for skill and machinery, and majorly extension of the technology. Conservation agriculture has the potential to cope up with the deteriorating condition of resource scarcity. With better awareness and solutions for skill development, this can serve as a strong tool for sustainable agriculture.

References

Abrol, I.P. and Sangar, S. (2006). Sustaining Indian agriculture–conservation agriculture the way forward. *Current Science* 91 (8): 1020–1025.

Ahmad, N. and Pandey, P. (2018). Assessment and monitoring of land degradation using geospatial technology in Bathinda district, Punjab, India. *Solid Earth* 9 (1): 75–90.

Albaladejo, J., Díaz-Pereira, E., and de Vente, J. (2021). Eco-holistic soil conservation to support land degradation neutrality and the sustainable development goals. *Catena* 196: 104823.

Balasubramanian, V. and Hill, J.E. (2002). Direct seeding of rice in Asia: emerging issues and strategic research needs for the 21st century. *Direct Seeding: Research Strategies and Opportunities*: 15–39.

Basavanneppa, M.A., Gaddi, A.K., Chittapur, B.M. et al. (2017). Yield maximization through resource conservation technologies under maize-chickpea cropping system in vertisols of Tunga Bhadra command project area of Karnataka. *Research on Crops* 18 (2): 225–231.

Bashour, I., Al-Ouda, A., Kassam, A. et al. (2016). An overview of conservation agriculture in the dry Mediterranean environments with a special focus on Syria and Lebanon. *AIMS Agriculture and Food* 1 (1): 67–84.

Bell, R.W., Haque, M., Jahiruddin, M. et al. (2019). Conservation agriculture for rice-based intensive cropping by smallholders in the eastern Gangetic plain. *Agriculture* 9 (1): 5.

Bhan, S. and Behera, U.K. (2014). Conservation agriculture in India–problems, prospects and policy issues. *International Soil and Water Conservation Research* 2 (4): 1–12.

Bhatt, R. (2017). Zero tillage impacts on soil environment and properties. *Journal of Environmental and Agricultural Sciences* 10: 01–19.

Bilalis, D., Sidiras, N., Economou, G., and Vakali, C. (2003). Effect of different levels of wheat straw soil surface coverage on weed flora in *Vicia faba* crops. *Journal of Agronomy and Crop Science* 189 (4): 233–241.

Bisht, P., Kumar, P., Yadav, M. et al. (2014). Spatial dynamics for relative contribution of cropping pattern analysis on environment by integrating remote sensing and GIS. *International Journal of Plant Production* 8 (1): 1–17.

Blanco, H. and Lal, R. (2008). *Principles of Soil Conservation and Management*, vol. 167169. New York: Springer.

Bouman, B. (2009). How much water does rice use. *Management* 69: 115–133.

Buschiazzo, D.E., Panigatti, J.L., and Unger, P.W. (1998). Tillage effects on soil properties and crop production in the subhumid and semiarid Argentinean Pampas. *Soil and Tillage Research* 49 (1–2): 105–116.

Calcante, A. and Oberti, R. (2019). A technical-economic comparison between conventional tillage and conservative techniques in paddy-rice production practice in Northern Italy. *Agronomy* 9 (12): 886.

Cropin (n.d.). An introduction to precision farming. https://www.cropin.com/precision-farming (accessed 09 January 2021).

CSISA (2018). Zero tillage wheat: Training of trainer's module.

Dawe, D. (2005). Increasing water productivity in rice-based systems in Asia–past trends, current problems, and future prospects. *Plant Production Science* 8 (3): 221–230.

Derpsch, R., Friedrich, T.,Kassam, A., and Li, H. (2010). Current status of adoption of no-till farming inthe world and some of its main benefits. International journal of agricultural and biological engineering, 3(1): 1–25.

Dingkuhn, M., de Vries, F.P., De Datta, S.K., and Van Laar, H.H. (1991). Concepts for a new plant type for direct seeded flooded tropical rice. *Direct Seeded Flooded Rice in the Tropics* 1: 17–18.

Dumansky, J., Reicosky, D.C., and Peiretti, R.A. (2014). Pioneers in soil conservation and conservation agriculture. Special issue. *International Soil and Water Conservation Research* 2 (1).

FAO (2016). Raised beds for improving crop water productivity and water efficiency in irrigated dryland agriculture, Egypt. http://www.fao.org/family-farming/detail/en/c/1040392/.

Farooq, M., Siddique, K.H., Rehman, H. et al. (2011). Rice direct seeding: experiences, challenges and opportunities. *Soil and Tillage Research* 111 (2): 87–98.

Friedrich, T., Derpsch, R., and Kassam, A. (2012). Overview of the global spread of conservation agriculture. *Field Actions Science Reports. The Journal of Field Actions* (Special Issue 6): 1–7.

Gathala, M.K., Ladha, J.K., Kumar, V. et al. (2011). Tillage and crop establishment affects sustainability of South Asian rice–wheat system. *Agronomy Journal* 103 (4): 961–971.

Greenland, D.J. (1975). Bringing the green revolution to the shifting cultivator. *Science* 190 (4217): 841–844.

Gupta, R.K., Naresh, R.K., Hobbs, P.R. et al. (2003). Sustainability of post-green revolution agriculture: the rice–wheat cropping systems of the Indo-Gangetic Plains and China. *Improving the Productivity and Sustainability of Rice-Wheat Systems: Issues and Impacts* 65: 1–25.

Hillel, D. (1998). *Environmental Soil Physics: Fundamentals, Applications, and Environmental Considerations*. Elsevier.

Hobbs, P.R., Sayre, K., and Gupta, R. (2008). The role of conservation agriculture in sustainable agriculture. *Philosophical Transactions of the Royal Society B: Biological Sciences* 363 (1491): 543–555.

Hossain, M.I. (2001). Performance of bed planting and nitrogen fertilizer under rice-wheat-mungbean cropping systems in Bangladesh. http://www.cimmyt.org/bangladesh (accessed 12 January 2021).

Jat, M.L., Malik, R.K., Saharawat, Y.S., et al. (2012). *Proceedings of Regional Dialogue on Conservation Agricultural in South Asia.*

Karki, T.B., Shrestha, J., and Tripathi, M.P. (2014). Preface to special issue on conservation agriculture for sustainable intensification. *World* 2 (6A).

Kassam, A., Gottlieb Basch, T.F., Shaxson, F. et al. (2013). 14 Sustainable soil management is more than what and how crops are grown. In: *Principles of Sustainable Soil Management in Agroecosystems* (eds. R. Lal and B.A. Stewart), 337–399. CRC Press.

Kassam, A., Friedrich, T., and Derpsch, R. (2019). Global spread of conservation agriculture. *International Journal of Environmental Studies* 76 (1): 29–51.

Kaur, J. and Singh, A. (2017). Direct seeded rice: prospects, problems/constraints and researchable issues in India. *Current Agriculture Research Journal* 5 (1): 13.

Klocke, N.L., Currie, R.S., and Aiken, R.M. (2009). Soil water evaporation and crop residues. *Transactions of the ASABE* 52 (1): 103–110.

Kumar, S. (2020). Precision farming in India-features, merits, demerits and challenges. https://www.iasexpress.net/precision-farming-in-india-features-merits-demerits-and-challenges (accessed 14 January 2021).

Kumar, V., Ladha, J.K., and Gathala, M.K. (2009). Direct drill-seeded rice: A need of the day. *Annual Meeting of Agronomy Society of America*, Pittsburgh (2 November 2009). pp. 1–5.

Kumar, A., Singh, R., Singh, S. et al. (2017). Impact of resource conservation technologies in Haryana. *Journal of Community Mobilization and Sustainable Development* 12 (2): 257–264.

Lal, R. (1976). No-tillage effects on soil properties under different crops in Western Nigeria. *Soil Science Society of America Journal* 40 (5): 762–768.

Laxmi, V., Erenstein, O., and Gupta, R.K. (2007). *Impact of Zero Tillage in India's Rice-Wheat Systems.* CIMMYT.

Liben, F.M., Tadesse, B., Tola, Y.T. et al. (2018). Conservation agriculture effects on crop productivity and soil properties in Ethiopia. *Agronomy Journal* 110 (2): 758–767.

Malik, R.K., Gupta, R.K., Singh, C.M. et al. (2005). Accelerating the adoption of resource conservation technologies in rice wheat system of the Indo-Gangetic Plains. *Proceedings of Project Workshop, Directorate of Extension Education, Chaudhary Charan Singh Haryana Agricultural University (CCSHAU)*, Hisar, Haryana, India. pp. 1–2.

Malobane, M.E., Nciizah, A.D., Wakindiki, I.I., and Mudau, F.N. (2018). Sustainable production of sweet sorghum for biofuel production through conservation agriculture in South Africa. *Food and Energy Security* 7 (3): e00129.

Mandal, V.P., Rehman, S., Ahmed, R. et al. (2020). Land suitability assessment for optimal cropping sequences in Katihar district of Bihar, India using GIS and AHP. *Spatial Information Research* 28 (5): 589–599.

Mehala, V. (2015). Economic analysis of resource conservation technologies in Haryana. Doctoral thesis, Chaudhary Charan Singh Haryana Agricultural University, Hissar, Haryana, India. https://krishikosh.egranth.ac.in/handle/1/2034118.

Micheni, A., Ouma, J., Kanampiu, F. et al. (2014). Maize-bean farming system under conservation agriculture: Assessing productivity and sustainability in Eastern Kenya, 2012–2014. www.ctic.org (accessed 10 February 2021).

Mollah, M.I.U., Bhuiya, M.S.U., and Kabir, M.H. (2009). Bed planting a new crop establishment method for wheat in rice-wheat cropping system. *Journal of Agriculture & Rural Development* 7 (1&2): 23–31.

Montgomery, D. (2007). *Dirt: The Erosion of Civilizations*. University of California Press, 287pp.

Mosquera, V.B., Delgado, J.A., Alwang, J.R. et al. (2019). Conservation agriculture increases yields and economic returns of potato, forage, and grain systems of the Andes. *Agronomy Journal* 111 (6): 2747–2753.

Mungarwal, A.K. and Mehta, S.K. (2019). Why farmers today need to take up precision farming. *Down To Earth*. www.downtoearth.org.in (accessed 14 March 2021).

Naresh, R.K., Singh, B., Kumar, A. et al. (2010). Diversifying the intensive cereal cropping systems of the Indo-Gangetic plains through horticulture. *Annals of Horticulture* 3 (1): 1–11.

Naresh, R.K., Singh, B., Singh, S.P. et al. (2012). Furrow irrigated raised bed (FIRB) planting technique for diversification of rice-wheat system for western IGP region. *International Journal of Life Science Biotechnology and Pharmaceuticals Research* 1 (3): 134–141.

Naveen Kumar, B.T. and Babalad, H.B. (2017). Responseof cotton (*Gossypium hirsutum* L.) to different conservation agricultural practices under rainfed situations. *International Journal of Current Microbiology and Applied Sciences* 6: 2279–2286.

Pandey, S. and Velasco, L.E. (1999). Economics of alternative rice establishment methods in Asia: a strategic analysis. *Social Sciences Division Discussion Paper, International Rice Research Institute, Los Banos, Philippines* 1 (1): 12–18.

Pandey, S. and Velasco, L. (2005). Trends in crop establishment methods in Asia and research issues. *Rice Is Life: Scientific Perspectives for the 21st Century*, Tsukuba, Japan (4–7 November 2004). pp. 178–181.

Pandey, V.P., Singh, B., and Tripathi, H.P. (2013). Planting of crops with furrow irrigated raised bed system and advantages of raised bed planting in crop production. *Krishisewa*. https://www.krishisewa.com/miscellaneous-articles/resource-management/272-firb.html (accessed 14 March 2021).

Pandey, P.C., Mandal, V.P., Katiyar, S. et al. (2015). Geospatial approach to assess the impact of nutrients on Rice equivalent yield using MODIS sensors'-based MOD13Q1-NDVI data. *IEEE Sensors Journal* 15 (11): 6108–6115.

Pariyar, K., Chaudhary, A., Sapkota, P. et al. (2019). Effects of conservation agriculture on productivity and economics of maize-wheat based cropping systems in midwestern Nepal. *SAARC Journal of Agriculture* 17 (1): 49–63.

Patel, B.D., Chaudhari, D.D., Patel, V.J. et al. (2019). Cotton-greengram system productivity as influenced by tillage and weed management practices. *Crop Research* (0970–4884) 54 (1 and 2): 20–27.

Pathak, H., Saharawat, Y.S., Gathala, M. et al. (2009). Simulating environmental impact of resource-conserving technologies in the rice-wheat system of the Indo-Gangetic Plains.

Quintero, M. and Comerford, N.B. (2013). Effects of conservation tillage on total and aggregated soil organic carbon in the Andes. *Open Journal of Soil Science* 2013: 361–373.

Rani, M., Joshi, H., Kumar, K. et al. (2021). Climate change scenario of hydro-chemical analysis and mapping spatio-temporal changes in water chemistry of water springs in Kumaun Himalaya. *Environment, Development and Sustainability 23*: 4659–4674.

Reiner, W. (2000). The role of rice plants in regulating mechanisms of methane missions. *Biology and Fertility of Soils* 31: 20–29.

Saharawat, Y.S., Ladha, J.K., Pathak, H. et al. (2012). Simulation of resource-conserving technologies on productivity, income and greenhouse gas GHG emission in rice-wheat system. *Journal of Soil Science and Environmental Management* 3 (1): 9–22.

Santa Bahadur, B.K. and Shrestha, J. (2014). Effect of conservation agriculture on growth and productivity of maize (*Zea mays* L.) in Terai region of Nepal. *World* 2 (4): 168–175.
Sarkar, R.K. and Das, S. (2003). Yield of rainfed lowland rice with medium water depth under anaerobic direct seeding and transplanting. *Tropical Science* 43 (4): 192–198.
Sayre, K.D. and Moreno Ramos, O.H. (1997). *Applications of Raised-Bed Planting Systems to Wheat*. CIMMYT.
Sharma, P.K., Bhushan, L., Ladha, J.K. et al. (2002). Crop-water relations in rice-wheat cropping under different tillage systems and water-management practices in a marginally sodic, medium-textured soil. *Water-Wise Rice Production* (8–11 April 2002). Los Baños, Philippines: International Rice Research Institute, 8. pp. 223–235.
Shrestha, J.I.B.A.N., Subedi, S.U.B.A.S.H., Timsina, K.P. et al. (2020). Conservation agriculture as an approach towards sustainable crop production: a review. *Farming Management* 5: 7–15.
Singh, A.K., Choudhury, B.U., and Bouman, B.A.M. (2002). Effects of rice establishment methods on crop performance, water use, and mineral nitrogen. *Water-Wise Rice Production* (8–11 April 2002). Los Baños, Philippines: International Rice Research Institute. pp. 237–246.
Singh, S.K., Bharadwaj, V., Thakur, T.C. et al. (2009). Influence of crop establishment methods on methane emission from rice fields. *Current Science* 97 (1): 84–89.
Singh, B., Naresh, R.K., Singh, K.V. et al. (2010). Influence of permanent raised bed planting and residue management on sustainability of vegetable based farming system in Western Indo Gangetic Plains. *Annals of Horticulture* 3 (2): 129–140.
Singh, N.P., Singh, R.P., Kumar, R. et al. (2011). Adoption of resource conservation technologies in Indo-Gangetic Plains of India: scouting for profitability and efficiency. *Agricultural Economics Research Review* 24 (1): 15–24.
Singh, K., Kumar, P., and Singh, B.K. (2013). An associative relational impact of water quality on crop yield: a comprehensive index analysis using LISS-III sensor. *IEEE Sensors Journal* 13 (12): 4912–4917.
Singh, O.P., Gautam, Y., Singh, P.K., and Singh, H.P. (2018). Enhancing inputs use efficiency through resource conservation technologies: empirical evidences from different agro-climatic zones of Uttar Pradesh. *Indian Journal of Agricultural Economics* 73 (3): 249–268.
Smith, P., House, J.I., Bustamante, M. et al. (2016). Global change pressures on soils from land use and management. *Global Change Biology* 22 (3): 1008–1028.
Solomon, R. (2020). Precision agriculture in India: New technologies are here, but wide scale adoption is far off. https://www.precisionag.com/in-field-technologies/precision-agriculture-in-india-new-technologies-are-here-but-wide-scale-adoption-is-far-off (accessed 24 March 2021).
Somasundaram, J., Sinha, N.K., Dalal, R.C. et al. (2020). No-till farming and conservation agriculture in South Asia – issues, challenges, prospects and benefits. *Critical Reviews in Plant Sciences* 39 (3): 236–279.
Thiagalingam, K., Gould, N., and Watson, P. (1991). Effect of tillage on rainfed maize and soybean yield and the nitrogen fertilizer requirements for maize. *Soil and Tillage Research* 19 (1): 47–54.
Tomar, V., Mandal, V.P., Srivastava, P. et al. (2014). Rice equivalent crop yield assessment using MODIS sensors' based MOD13A1-NDVI data. *IEEE Sensors Journal* 14 (10): 3599–3605.

Tripathi, S.C. and Das, A. (2017). Bed planting for resource conservation, diversification and sustainability of wheat based cropping system. *Journal of Wheat Research* 9 (1): 1–11.
United Nations (2019). Growing at a slower pace, world population is expected to reach 9.7 billion in 2050 and could peak at nearly 11 billion around 2100. *United Nations News*.
Wahbi, A., Miwak, H., and Singh, R. (2014). Effects of conservation agriculture on soil physical properties and yield of lentil in Northern Syria. *EGU General Assembly Conference Abstracts*, Vienna (27 April–2 May). p. 3280.
Wolff, P. and Stein, T.M. (1998). Water efficiency and conservation in agriculture: opportunities and limitations. *Agriculture and Rural Development* 2: 2–20.
Yadav, R.S. (2019). Comparative economics of surface seeding and conventional method of wheat cultivation in Mirzapur district, U.P, India. Master of Science thesis. Banaras Hindu University, Varanasi, U.P, India.
Yadav, M., Sharma, M.P., Prawasi, R. et al. (2014). Estimation of wheat/rice residue burning areas in major districts of Haryana, India, using remote sensing data. *Journal of the Indian Society of Remote Sensing* 42 (2): 343–352.
Yadav, M.R., Parihar, C.M., Kumar, R. et al. (2017). Conservation agriculture and soil quality - an overview. *International Journal of Current Microbiology and Applied Sciences* 6: 1–28.

10

Quality of Irrigation Water for Sustainable Agriculture Development in India

Bharat Lal[1], Abhishek Kumar Shukla[1], Pavan Kumar[2], and Susheel Kumar Singh[1]

[1] College of Agriculture, Rani Lakshmi Bai Central Agricultural University, Jhansi, Uttar Pradesh, India
[2] College of Horticulture and Forestry, Rani Lakshmi Bai Central Agricultural University, Jhansi, Uttar Pradesh, India

CONTENTS

Sustainable Agriculture Systems and Technologies, First Edition. Edited by Pavan Kumar, A.K. Pandey, Susheel Kumar Singh, S.S. Singh, and V.K. Singh.

10.1 Introduction

Water is the most important natural resource, which plays a pivotal role in all the living body and it is one of the fundamental needs of the globe (Kumbhar and Salkar 2014) or in other words, we can say "Water is life." The whole water resources are not consumable because only 2.5% of it is fresh water in the world, while the rest is saline in nature. Furthermore, the freshwater is present in different forms, i.e. ground ice and permafrost (69.0%), groundwater (30.1%), and surface water (1.2%) globally (USGS). In India, the average annual precipitation of water is about 4000 billion cubic meter (BCM), in which most of the precipitation (75%) occurs during the monsoon season (June to September). The per capita water availability in India is 1720.29 m^3 (CWC 2014). Water is used for various purposes ranging from drinking, irrigation, industrial, and other allied sectors (Upadhay 2013). Approximately, 70% of the fresh water is used by human beings for agricultural sectors (FAO 2011) whereas, 80% groundwater is consumed by rural population to fulfill their domestic use in the developing countries like India (Karthikeyan et al. 2010; Kumbhar and Salkar 2014).

Today, water scarcity in terms of quality and quantity is a big challenge due to the pressure of increasing human population and other indirect human activities like urbanization, industrialization, and deforestation which result in deterioration of the water quality. Water is very essential for the all living beings including human beings, organisms, and other inhabitants. Water is basically used for many purposes such as drinking, cleaning, agriculture, industry, and other commercial uses (FAO 2011). Only 2.5% water is fresh water, a majority of which is in ice form and only 0.26% water is present in the river, lakes, soil, and aquifers. Groundwater level is depleted drastically because of the exploitation of groundwater without returning it in the form of recharge and the concentration of different salts, heavy metals, cations, and anions are increasing in the groundwater that reduce the availability of quality water for drinking and irrigation purpose (Hoogeveen et al. 2015). While, most of the Indian population lives in rural areas and totally depends on the agricultural sector with limited cultivable lands, the continuous application of such type of polluted or bad quality irrigation water degrades the soil quality due to the presence of different salts, especially, in soil physical properties (Singh et al. 2013; Bisht et al. 2014; Tomar et al. 2014; Yadav et al. 2014). Determination of irrigation water quality will help in identifying the dominant cations and anions in irrigation water which degrade the quality of irrigation water and also gives an idea about the best way to manage dominating ions in the irrigation water. In this chapter, major focus is on the current scenario and irrigation potential of water resources which are used for the irrigation purpose. The criteria for classification of irrigation water quality and standards are also discussed. Better management of poor quality irrigation water to achieve sustainable approaches through implementation of various agricultural activities is also discussed.

10.2 Global Water Resources and Their Scarcity

The world water resources are estimated to be about 43 750 km^3/year which is distributed based on the climatic and physiographic conditions of the world. The freshwater is available in different forms, i.e. ground ice and permafrost (69.0%), groundwater (30.1%), and surface water is 1.2% (USGS). Freshwater resources in continental level: the United States secures the first position with 45% followed by Asia (28%), Europe (15.5%), and Africa covers only 9% of total fresh water resources. The water resources availability per habitant is largest in the United States (24 000 m^3/year), Europe (9300 m^3/year), and Africa (5000 m^3/year) while the least was in Asia, i.e. 3400 m^3/year. The availability of freshwater varies from the 10 m^3/year in Kuwait to more than 100 000 m^3/year in Canada. Falkenmark (1986) has proposed the threshold level of water which is 500 m^3/habitant water availability for scarcity level and 1000 m^3/year is the water stress level. Nineteen countries in the world fall under the freshwater water resources is less than the 500 m^3/habitant and 29 counties were having less than 1000 m^3/year water. It has been observed that 1000 m^3/year water resources availability is necessary to sustain life and meet the irrigation requirements for the agricultural production system. The renewable water resources are the amount of water resources which are either internal (surface and groundwater) or external water (river and other water), mainly originating through the hydrological cycles. The total water resources were highest in Brazil (8647 km^3/year) and lowest in Nauru (0.01 km^3/year). Meanwhile the total water resources availability per capita is the largest in Iceland and the lowest in Kuwait, i.e. 508383.8 m^3/inhabitant/year and 4.93 m^3/inhabitant/year, respectively. Top 20 richest and poorest countries of water resources are presented in Tables 10.1 and 10.2 (Aquastat).

The world population is increasing day-by-day and it will reach up to 8.4–8.6 billion by 2030 (United Nations 2015). The growing population requires more food grain production and it creates pressure on land, water, and other natural resources. The world's total water availability is 1400 million cubic km, whereas only 0.003% water is freshwater resources which is used for drinking, agriculture, and industrial purposes. Agricultural water withdrawal is almost more than 70% (95% in developing country) of the total freshwater resources for the food grain production. However, the food grain production is increased 100% in the last 30 years and its future need will increase 50% by 2050 (FAO 2021, water at a glance), while the agricultural water withdrawal rises up to 10% only for irrigation and other agricultural activities. Agricultural system consumes more water than the others like industry and domestic use (Figure 10.1).

The huge pressure of increasing world population demands more water for its utilization and creates water stress in terms of quantity and quality or natural phenomena like climatic conditions or lack of freshwater resources due to the uneven distribution of rainfall, droughts, and other natural disasters. Fifteen countries in the world have been observed as more water stress (>80%) in 1992, which is simultaneously increased in 1997 (20 countries), 2002 (22 countries), 2007 (21 countries), 2012 (23 countries), and 2017 (24 countries). The water stresses represent that we are continuously facing low water availability and need to conserve for future generations (Figure 10.2).

10.3 Water Resources in India

In India, average annual precipitation of water is about 4000 BCM in which most of the precipitation (75%) occurs during the monsoon season (June to September) and 1869 BCM volume of water is lost through runoff. Total utilizable groundwater resource in India is

Table 10.1 Top 15 countries poor in water resources in the world.

	National Rainfall Index (NRI) (mm/yr)		Total internal renewable water resources (IRWR) (km^3/yr)		Water resources: total external renewable (km^3/yr)		Total renewable water resources (km^3/yr)		Total renewable water resources per capita ($m^3/inhab/yr$)	
SN	Poor country		Poor country		Poor country		Poor country		Poor country	
1	Oman	23.44	Kuwait	0	Qatar	0.002	Nauru	0.01	Kuwait	4.930846
2	Qatar	36	Bahrain	0.004	Kuwait	0.02	Kuwait	0.02	United Arab Emirates	15.81077
3	United Arab Emirates	51.95	Nauru	0.01	Palestine	0.025	Saint Kitts and Nevis	0.024	Qatar	21.28653
4	Kuwait	67.35	Saint Kitts and Nevis	0.024	Estonia	0.096	Maldives	0.03	Maldives	60.43489
5	Turkmenistan	71.38	Maldives	0.03	Bahrain	0.112	Malta	0.0505	Saudi Arabia	72.50497
6	Egypt	106.6	Malta	0.0505	Jordan	0.255	Antigua and Barbuda	0.052	Yemen	75.44507
7	Djibouti	107.4	Antigua and Barbuda	0.052	Bulgaria	0.3	Qatar	0.058	Bahrain	77.63996
8	Saudi Arabia	128.6	Qatar	0.056	Spain	0.3	Barbados	0.08	Jordan	95.75057
9	Libya	140.6	Barbados	0.08	Angola	0.4	Saint Vincent and the Grenadines	0.1	Singapore	105.1149
10	Cabo Verde	144	Saint Vincent and the Grenadines	0.1	Algeria	0.42	Bahrain	0.116	Libya	106.3713
11	Jordan	148.8	United Arab Emirates	0.15	Tunisia	0.42	United Arab Emirates	0.15	Malta	115.3144
12	Mongolia	177.5	Dominica	0.2	Armenia	0.91	Dominica	0.2	Palestine	176.3135
13	Afghanistan	185	Grenade	0.2	Burkina Faso	1	Grenade	0.2	Israel	215.9186
14	Uzbekistan	188.5	Cabo Verde	0.3	North Macedonia	1	Cabo Verde	0.3	Barbados	279.4936
15	Mauritania	190.3	Djibouti	0.3	Haiti	1.015	Djibouti	0.3	Algeria	281.8852

Table 10.2 Top 15 countries rich in water resources in the world.

	National Rainfall Index (NRI) (mm/yr)		Total internal renewable water resources (IRWR) (km^3/yr)		Water resources: total external renewable (km^3/yr)		Total renewable water resources (km^3/yr)		Total renewable water resources per capita (m^3/inhab/yr)	
SN	**Rich country**		**Rich country**		**Rich country**		**Rich country**		**Rich country**	
1	Brunei Darussalam	3564	Brazil	5661	Brazil	2986	Brazil	8647	Iceland	508383.8
2	Papua New Guinea	3250	Russian Federation	4312	Bangladesh	1122.032	Russian Federation	4525.445	Guyana	349577.3
3	Malaysia	3194	Canada	2850	Congo	610	US America	3069	Suriname	173533.2
4	Solomon Islands	2945	United States of America	2818	Argentina	584.24	Canada	2902	Congo	162795.9
5	Indonesia	2929	China	2812.9	Viet Nam	524.7	China	2840.22	Bhutan	104618.9
6	Costa Rica	2865	Colombia	2145	Venezuela	520	Colombia	2360	Papua New Guinea	94927.3
7	Colombia	2862	Indonesia	2018.7	India	464.9	Indonesia	2018.7	Gabon	80394.3
8	Philippines	2640	Peru	1641	Congo	383	India	1910.9	Canada	79004.48
9	Bhutan	2609	India	1446	Cambodia	355.5	Peru	1879.8	Norway	74202.38
10	Ecuador	2575	Myanmar	1002.8	Paraguay	270.77	Venezuela	1325	Solomon Islands	70278.71
11	Sierra Leone	2546	Congo	900	Bolivia	270.5	Congo	1283	New Zealand	69544.37
12	Chile	2483	Chile	885	US America	251	Bangladesh	1227.032	Peru	59781.9
13	Panama	2461	Venezuela	805	Peru	238.8	Myanmar	1167.8	Belize	57838.73
14	Equatorial Guinea	2447	Papua New Guinea	801	Colombia	215	Chile	923.06	Paraguay	56468.12
15	Liberia	2413	Malaysia	580	Thailand	214.1	Viet Nam	884.12	Bolivia	51282.72

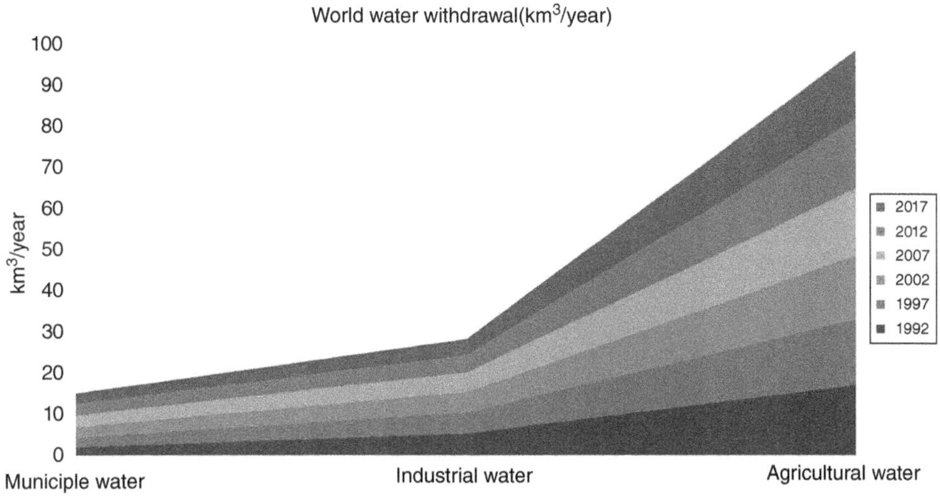

Figure 10.1 World water consumption by different sectors.

about 433 BCM. Table 10.3 presented the water resources in India (CWC 2014). The per capita water availability in India is 1720.29 m^3 (CWC 2005). Water is used for the various purposes ranging from drinking, irrigation, industrial, and other allied sectors (Upadhay 2013). Approximately 70% of the fresh water is used by human for agricultural sectors whereas, 80% groundwater is needed for rural population to fulfill their domestic use in the developing country like India (Karthikeyan et al. 2010; Kumbhar and Salkar 2014). Different kinds of water resources have been observed in India which are the main sources of water.

10.3.1 Surface Water

The total surface water in the country is estimated about 1869 BCM Suhag 2016. This amount is actually surface runoff water originated from the rainfall and snowmelt. The surface water includes river, canal, ponds and lakes, whereas a little amount of surface water is utilized and most of the surface water about 690 (BCM) is mobilized from upstream to downstream direction which ultimately mix with the ocean water because of lacking water storage facilities and appropriate water harvesting structures to conserve this moving water.

10.3.2 Rainfall Water

Rainfall is the only source of water that recharges the surface and groundwater sources. In India, basically four types of monsoon are dominant namely winter monsoon which fall in between the month of January and February and it has 37% long period average (LPA), pre-monsoon (March to May) – 89% LPA, and during monsoon periods (June to September) – 91% of LPA while the post-monsoon season (October to December) – 50% of LPA. The average annual rainfall of India is 1150 mm which is unevenly distributed all over India.

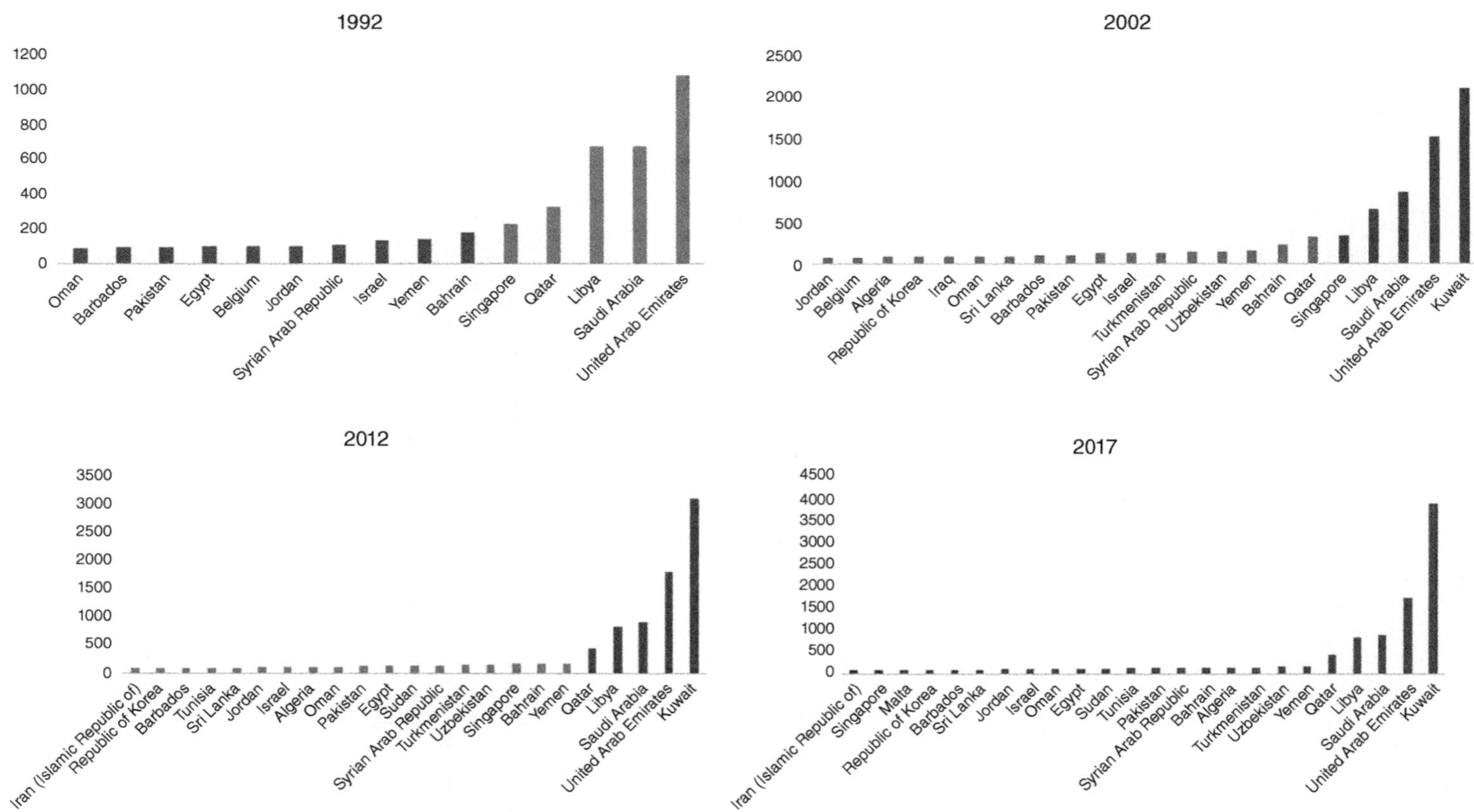

Figure 10.2 Temporal changes in heavy water stress country (>80%) in the world. *Source*: Data source is Aquastat.

Table 10.3 Indian water resources.

SN	Parameters	Coverage (billion cubic m/yr)
1	Ultimate irrigation potential	140 Mha
2	Net irrigated area	50 Mha
3	Natural runoff (surface and groundwater)	1869 BCM
4	Estimated utilizable surface water potential	690 BCM
5	Groundwater resources	433 BCM
6	Available groundwater resources for irrigation	361 BCM
7	Net utilizable groundwater resources for irrigation	325 CM

South west monsoon is the main source of water because almost 80% rainfall occurs in this monsoon from June to September in most parts of the country and is the only hope for agriculture in rainfed and dry regions of India (Agricultural Statistics at a Glance 2018).

10.3.3 Groundwater

Groundwater is the fraction of water that is stored below the ground in between the pores of rocks and soil particles. The depth of water availability from the surface is called water level, it is namely shallow and deep-water level. The heavy rains are the only source for recharging the groundwater and increase the water level and reversely, extraction of groundwater results in depletion of the water level. The annual water availability in the country, i.e. natural runoff water in rivers is 1869 BCM/year in which usable water is 1123 BCM/year. Out of 1869 BCM/year whole water is not usable due to loss of water from uneven topography and moved through rivers. Surface water is 690 BCM/year and groundwater are only 433 BCM/year (CGWB 2019), whereas, the availability of groundwater is 398 BCM. Rainfall contribution is 68% for groundwater recharge and remaining is 32% losses through canal seepage or stored in tank, pond, and other water conservation structures.

The agricultural sectors utilized 89% of the groundwater resources for the crop production, irrigation and meet other agricultural demands followed by 9% in domestic purpose for drinking, bathing, swimming, clothing, and toilet, etc. 9% while only 2% groundwater is utilized by the industrial sectors (Annual report 2014). Different water sources were used for the irrigation purpose in agricultural sectors but major exploitation of groundwater from tube well are increasing in the present scenario (Figure 10.3).

10.4 Status of Groundwater Quality of India

Groundwater level is depleted periodically due to heavy exploitation of groundwater without giving back, therefore invisible chemical substances like physicochemical (pH and EC), cations (Ca^{2+}, Mg^{2+}, Na^{+}, K^{+}, Fe^{2+}, total hardness, and other cations), and anions (HCO_3^-, CO_3^{2-}, Cl^-, SO_4^{2-}, NO_3^-, F^-, B, and PO_4^{3-}) were increasing in groundwater and polluting its

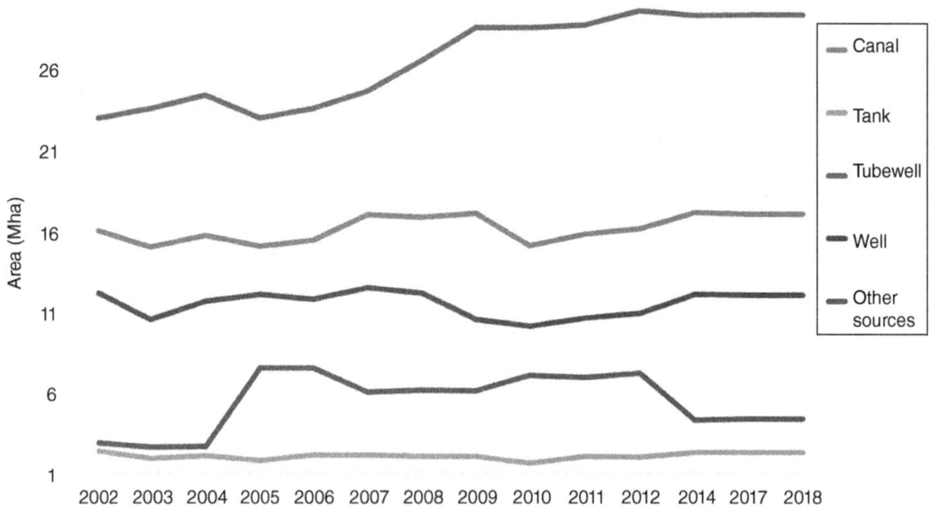

Figure 10.3 Utilization of different water resources in agriculture sectors in India. *Source*: Agriculture and Horticulture statistics 2014 and 2018.

quality and making it unfit for consumption and irrigation purposes. In this section, various research articles of different state and union territories in India have been studied and the obtained mean concentrations are presented in Table 10.4. There are 13 parameters used to explain the groundwater of India namely pH, EC, total hardness, calcium, magnesium, sodium, potassium, chloride, bicarbonate, sulfate, nitrate, fluoride, and total dissolved solids. The pH of water shows the acidity or alkaline nature which ranges from 5.6 to 8.8. Electrical conductivity of groundwater represents the presence of soluble salts in water, the EC of groundwater varied from 97.13 μS/cm to as high as 4017.65 μS/cm. The hardness of the Indian groundwater is observed to be a maximum of 572.94 mg/l. The total hardness develops due to the calcium and magnesium ions which varied between 10.53–300 and 2.09–326.41 mg/l, respectively. Maximum sodium and potassium ions reported are 339.22 and 163.8 mg/l, respectively. The maximum value of different anions was also observed, i.e. chloride (743.4 mg/l), bicarbonate (631.83 mg/l), sulfate (263.46 mg/l), nitrate (71.31 mg/l), and fluoride content (4.4 mg/l). Piper diagram used to identify the Indian groundwater type (Figure 10.4) using the collected research articles from different states and union territories.

10.5 Impact of Poor-Quality Irrigation Water

The irrigation water quality-related problems are mainly a concern about the quantity and type of salts present in the irrigation water. These salts mainly originated from the weathering of the rocks and dissolution of primary and secondary minerals (calcite, gypsum, hematite, goethite, etc.); the dissolved salt is supplied through irrigation water and remains on the soil surface after the evaporation and may deteriorate the soil's physical quality and ultimately reduce the crop yield (Pandey et al. 2015; Rani et al. 2019; Mandal et al. 2020). The following problems should be considered in the irrigation water quality.

Table 10.4 Average physicochemical characterization of groundwater from different states of India.

SN	States	pH	EC	TH	TDS	Ca	Mg	Na	K	CO_3^{2-}	HCO_3^-	Cl	SO_4^{2-}	NO_3^-	F^-	References
1	Andhra Pradesh	7.9	878.58	233.69	571.08	48.89	27.03	84.35	20.07	—	347.57	77.06	35.53	—	3.04	Adimalla and Qian (2019)
2	Assam	7.6	425	185	272	48	16	10	4.7	—	—	9	7.1	2.5	0.17	Jain et al. (2018)
3	Bihar	7.57	327		288	34.9	22.9	35.9	11.7	21.5	311	30.8	204	39.4	—	Maity et al. (2020)
4	Chhattisgarh	7.94	598.6	225.95	—	58.12	21.43	25.47	5.05	0	168.85	55	—	15.6	0.43	Jhariya et al. (2017)
5	Goa	6.7	307	105	151	28.2	8.5	25.7	5.5	—	—	255	23.33	62.5	—	Haritash et al. (2017)
6	Gujarat	7.07	378.2	431.69	307.51	108.13	326.41	—	—	—	—	365.89	5.53	0.23	4.02	Rupal et al. (2012)
7	Haryana	7.89	2011.33		1074.54	74.31	35.19	339.22	19.24	14.85	288.78	371.39	227.1	32.76	2.06	Sheikh et al. (2017)
8	Jharkhand	7	855	272	434	184	106	98.4	8.9	—	—	8	152	16.9	1.63	Tirkey et al. (2017)
9	Karnataka	6.64	163.64	63.3	101.45	14.09	6.85	3.17	1.78	0	45.07	22.6	18.93	1.4	0.19	Ravikumar and Somashekar (2017)
10	Kerala	6.13	447	206	289	44.66	21.37	24.37	5.14	0	101.14	342.22	11.45	1.62	0.15	Manjusree et al. (2017)
11	Madhya Pradesh	7.3	817.37	450.38	572.15	140.21	48.31	119.89	40.64	—	631.83	140.57	59.8	71.31	1.09	Srivastava (2017)
12	Maharashtra	8.3	2973.4	463.1	1932.7	78	65.2	104.2	5.9	20.8	186.4	255.3	109.8	40.1	—	Mukate et al. (2019)
13	Manipur	6.77	477	106	234.3	45.99	11.73	189.02	50.81	—	17.02	59.23	58.13	40.47	0.68	Wazir et al. (2020)
14	Meghalaya	5.9	214	530	137	300	43.74	299	163.8	—	—	743.4	47.5	31	0.09	Jain et al. (2021)
15	Mizoram	7.3	278		179	25.49	13.8	31	4	—	165	165.39	19.4	—	—	Thambidurai et al. (2014)
16	Odisha	7.9	4017.65	572.94	2490.94	123.41	208.78	312.18	28.79	—	—	516.11	157.76	16.44	—	Agrekar et al. (2012)
17	Punjab	7.2	2016.36	363	1108.42	153.12	51.26	276.54	15.32	—	—	187.16	263.46	5.07	2.31	Kaur et al. (2016)
18	Rajasthan	8.05	1630	383	656.28	55.6	58.56	96.6	10.14	—	378.2	162.59	56.16	25.08	4.4	Chaudhary and Kumar (2018)
19	Tamil Nadu	7.3	1802.14	505.45	1260.97	111.25	53.86	157.17	20.05	—	—	231.95	219.53	12.23	0.44	Duraisamy et al. (2019)

(*Continued*)

Table 10.4 (Continued)

SN	States	pH	EC	TH	TDS	Ca	Mg	Na	K	CO_3^{2-}	HCO_3^-	Cl	SO_4^{2-}	NO_3^-	F^-	References
20	Telangana	8.8	1469	365	928	94	32	90	3	3	278	160	65	62	1.43	Laxman et al. (2021)
21	Tripura	6.3	97.13	—		10.53	4.71	3.52	1.21	—	26.73	20.85	8.54	0.57	0.26	Paul et al. (2019)
22	Uttar Pradesh	6.6	560	—	360	74.47	2.09	45.29	6.78	—	0.28	45.35	7.98	9.35	—	Singh and Singh (2018)
23	Uttarakhand	5.6	935	—	565	117	74	45	35	—	375	187.5	60	—	—	Dudeja et al. (2011)
24	West Bengal	7.1	705	250.67	343.14	49.09	31.21	41.57	4.01	24.75	175.21	60.26	19.22	—	—	Nag and Das (2017)
25	Chandigarh	7.15	623	360	481.7	—	—	37.28	9.67	—	—	43.98	229	0.29	—	Ravindra et al. (2019)
26	Jammu and Kashmir	7.35	504	216	323.66	64.32	13.46	16.66	1.87	0	201.65	24.6	37.6	12.64	0.11	Jasrotia et al. (2019)
27	Ladakh	7.5	441.51	135	291.22	32.15	14.04	7.54	1.6	—	—	3.51	10.69	0.16	—	Dolma et al. (2015)
28	Lakshadweep	7.8	1507.2	229.62	979.68	44.2	29.05	66.7	27.28	30.24	173.75	136.19	11.58	10.05	—	Antony et al. (2020)
29	Puducherry	7.4	911.1	—	1423.6	76.21	49.6	298.54	40.58	—	320.08	544.67	53.64	44.98	0.31	Gopinath et al. (2016)

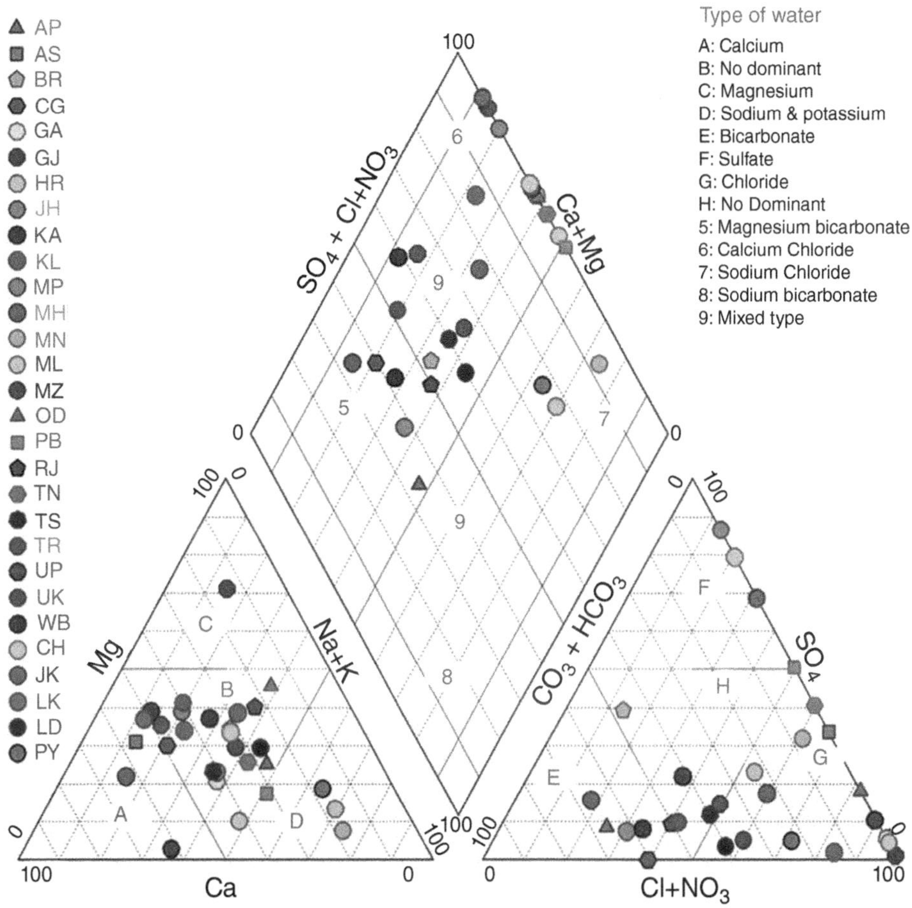

Figure 10.4 Ground water hydro chemistry of different states of India. (*See insert for color representation of the figure.*)

10.5.1 Salinity

Continuous application of salt enriched irrigation water may accumulate the salt content in the crop root zone area, saline water is mainly found in those area in arid and semiarid conditions where lack of sufficient rainfall to leach out the soluble salt present in the soil, shallow water table (<2 m from the surface) is also responsible for salinity, the basic nature of parent material like basalt and use of some base forming fertilizers (sodium nitrate, calcium nitrate) can cause the soil salinity. In saline soil or salt affected soil, plants are unable to uptake sufficient amount of water due to osmotic effect. Salinity conditions create water stress which causes slow plant growth and produces some symptoms similar to drought, such as wilting, bluish green color, thicker, and waxy leaf, etc.

10.5.2 Infiltration Rate

Infiltration is the entry of water into the soil and movement toward the lower horizon of soil and supply of water for the crop; it depends on soil texture, structure, pore size, and the presence of different salt contents in soil, degree of compaction, and irrigation water quality. Irrigation water may be categorized into two types on the basis of presence of salt content, i.e. saline water and sodic water, saline water mainly containsg Ca^{2+} and Mg^{2+} ions which promote the soil aggregation and infiltration instead of sodic water which is maximum saturated with Na^{+} ions, it act as a dispersing agent of the soil particle therefore pore space has been clogged by the smaller particle like silt and clay which inhibit the entry of water into the soil and decreases the infiltration rate of soil.

10.5.3 Specific Ion Toxicity

The toxicity of specific ions is specially a concern about sodium, chlorine, and boron, which are taken up by plant and deposited in high concentration into the plant body and cause damage to the plant tissue, thereby diminishing the crop yield. The toxic ions can be absorbed by a plant through water uptake and is translocated from root to the upper plant portion and accumulates in the place where more transpiration occurs, mainly in the leaf portion (leaf tips and leaf margins). Accumulation of toxic concentrates usually produces damaging symptoms of leaf burn or intervened chlorosis. Direct leaf burning is also noticed in sprinkler irrigation system whereas, salt is directly sprayed on the leaf surface and may cause severe damage. The degree of crop damage depends on the type of toxic ion concentration of toxic ion, exposure duration, concentration of toxic ion, crop sensitivity, and climatic conditions.

10.5.4 Miscellaneous Ion Toxicity

Other ions are also present in the irrigation water which affects indirectly and damages the crop yield in term of deficiency of other elements like Fe, Mn, Zn, and Cu, which are mostly available in acid soil condition. The application of NO_3^- enriched irrigation water will cause the more vegetative growth of crops and therefore crop lodging are responsible to reduce crop yield. Bicarbonate content in irrigation water was also damage crop and soil and ultimately reduces their yield.

10.6 Irrigation Water Quality Parameters

The quality of irrigation water is an important concept for the sustainable agriculture especially when the irrigated water is having a greater number of cations and anions (Mohamed et al. 2020) which are harmful and deteriorate the soil quality on continuous application of salt enriched irrigation water (BIS 1987). There are some basic criteria which are useful to categorize the quality of irrigation water are

10.6.1 Salinity Hazards

The salinity of irrigation water is mainly because of the dominance of invisible soluble salt like Cl^-, SO_4^{2-}, CO_3^{2-}, HCO_3^- of Ca^{2+}, Mg^{2+}, Na^+, and K^+. The application of high salinity water continuously into the soil directly affects the nutrient imbalance and causes deficiency

Table 10.5 Suitability of irrigation water on the basis of electrical conductivity.

SN	Irrigation water category Wilcox 1955	Water type	Suitability of irrigation water
1	Low-salinity water (C1) <250	Excellent	Suitable for all types of crops and soils
2	Medium salinity water (C2) 250–750	Good	Can be used if a moderate amount of leaching occurs normal salt-tolerant plant can be grown without much salinity control
3	High salinity water (C3) 750–2250	Doubtful	Unsuitable for soil with restricted drainage
4	Very high salinity water (C4) >2250	Unsuitable	Unsuitable for irrigation
	PS (Doneen 1962) <3	Suitable	—
	3–5	Good to injurious	—
	>5	Injurious to unsatisfactory	—

Source: Wilcox (1955) and Doneen (1962).

of Fe, Mn, Zn, and Cu etc., and the toxicity of Ca^{2+}, Mg^{2+}, Cl^-, SO_4^{2-}, and Na^+ and indirectly effects like osmosis, salt accumulation on soil near crop root zone so plant water moves to the soil and plant becomes physiologically dry and ultimately reduces the crop production. This soluble ion can be measured with the help of conductivity meter and is represented as milli siemens per meter (mS/m) or desi siemens per meter (dS/m) or micro siemens per centimeter (μS/cm) or parts per million (ppm). We can categorize the tested irrigation water by using Table 10.5 on the basis of electrical conductivity. The potential salinity (PS) is also calculated and used to determine the salinity level in irrigation water quality.

$$PS = Cl^- + \left(\frac{SO_4^{2-}}{2}\right)$$

In some, salty water is having >2250 μS/cm electrical conductivity, therefore Shahid and Mahmoudi (2014) modified the USSL staff classification for higher salinity in water.

10.6.2 Sodium Hazard

It can be understood by its name; basically it represents the dominancy of Na^+ in irrigation water. On continuous application of high-Na^+ content irrigation water, it accumulates in the soil layers where crops can be grown. Sodium ion is responsible for the deflocculation in soil particle which breaks the aggregation of soil particle and keeps it on dispersed condition because it has high hydration energy. The soil particle adheres by water molecule which reduces the attraction or cohesion of soil particle (sand, silt, and clay) to form aggregation. Meanwhile, applied irrigation water with sodium rich creates a waterlogged condition in soil and low permeability of water into the soil horizons. The physical quality of soil is disturbed due to sodium and counter effect of nutrient imbalance for plant. The sodium ion in irrigation water can be measured by use of flame-photometers and calculated by using the following equations of sodium adsorption ratio (SAR) and sodium percentage, Kelly ratio and permeability index can be calculated and we can also classify the irrigation water quality as shown in Table 10.6.

$$\mathrm{SAR} = \frac{\mathrm{Na}^{+}}{\sqrt{\mathrm{Ca}^{2+} + \mathrm{Mg}^{2+} / 2}}$$

$$\mathrm{Na}(\%) = \frac{\mathrm{Na}^{+}}{\left(\mathrm{Ca}^{2+} + \mathrm{Mg}^{2+} + \mathrm{Na}^{+}\right)} \times 100$$

Table 10.6 Suitability of irrigation water based on sodium ion activity.

Parameters	Water type	Quality	Suitability for irrigation
SAR (Richards 1954)	Low sodium water (S1) <10	Excellent	Suitable for all crops and soil except sodium sensitive crops
	Medium sodium water (S2) 10–18	Good	Suitable for coarse texture/organic soil with good permeability
	High sodium water (S3) 18–26	Doubtful	Harmful for almost all types of soil, required good drainage, high leaching and gypsum addition
	Very high sodium water (S4) >26	Unsuitable	Unsuitable for irrigation
% Na (Wilcox 1955)	<20	Excellent	—
	20–40	Good	—
	40–60	Permissible	—
	60–80	Doubtful	—
	>80	Unsuitable	—
% Na (Eaton 1950)	<60	Safe	—
	>60	Unsafe	—
KR (Kelly 1963)	<1	Suitable	—
	>1	Unsuitable	—
PI (Doneen 1964)	>75	Excellent	—
	75–25	Good	—
	<25	Unsuitable	—

$$KR = \frac{Na^+}{\left(Ca^{2+} + Mg^{2+}\right)}$$

$$PI(\%) = \frac{\left(Na^+ + \sqrt{HCO_3^-}\right)}{\left(Ca^{2+} + Mg^{2+} + Na^+\right)} \times 100$$

10.6.3 Carbonate and Bicarbonate

The CO_3^{2-} and HCO_3^- are the important ions to characterize the quality of irrigation water. Irrigation water with high CO_3^{2-} and HCO_3^- ions precipitate the salt like Ca^{2+} as $CaCO_3$ and Mg^{2+} as a $MgCO_3$, whereas in the drier region evaporation was maximum it stimulates and vaporizes the water molecule to the atmosphere so far there is a chance to increase the sodium ion concentration that can be measured as residual sodium carbonate (RSC) and it can be calculated and interpret the irrigation water quality as per equation and Table 10.7.

$$Ca^{2+} + CO_3^{2-} = CaCO_3(\text{precipitate})$$

$$RSC\left(\frac{me}{1}\right) = \left(HCO_3^- + CO_3^{2-}\right) - \left(Ca^{2+} + Mg^{2+}\right)$$

10.6.4 Specific Ions Toxicity

Irrigation water also contains other ions like boron, chloride magnesium, nitrate, and fluoride which deteriorate the irrigation water quality.

10.6.4.1 Magnesium Hazard

It is an important qualitative criteria for judging the quality of irrigation water, and it is a simple relation between Mg^{2+} to Ca^{2+} and Mg^{2+}; application of heavy magnesium

Table 10.7 Suitability of irrigation water as per RSC and MHR value.

Parameters	Category	Water types
RSC (Eaton 1950)	<1.25	Good
	1.25–2.25	Doubtful
	>2.25	Unsuitable
MHR (Raghunath 1987)	<50	Suitable
	>50	Unsuitable

irrigation water into the soil causes an imbalance in the Ca : Mg ratio and when it reached <50, deteriorates the soil physical properties (Table 10.7).

$$\text{MHR} = \frac{\text{Mg}^{2+}}{\left(\text{Ca}^{2+} + \text{Mg}^{2+}\right)} \times 100$$

10.6.4.2 Boron Hazard

Boron is required in small quantity by plant that is why it is considered as a micronutrient; it acts for the flower and pollen grain formation in plant. Boron is toxic and it directly influences the plant's growth and development when it reaches high concentration in irrigation water but all irrigation water does not have sufficient boron and it has been rarely observed to cause toxicity (Table 10.8).

10.6.4.3 Chlorine Hazard

Chlorine ions do not affect the physical quality of soil and are not adsorbed on the exchange site of soil but some plants are very sensitive to chloride ion concentration, particularly when high chloride irrigation water is applied to such crops which reduces their yield, such as lemon, blackberry, grape orange, and onion. Chloride ion concentration can be calculated by using the formula (Table 10.9).

$$\text{Cl}^-\left(\text{meql}^{-1}\right) = \frac{\text{Cl}^-}{\text{CO}_3^{2-} + \text{HCO}_3^- + \text{SO}_4^{2-} + \text{Cl}^- + \text{NO}_3^-}$$

10.7 Irrigation Water Quality of Indian Groundwater

For testing the irrigation water quality, the collected papers of Indian groundwater were used to calculate the salinity hazards and Wilcox diagram. The salinity is the major problem of groundwater and falls under the C1–C4 category as presented in Figure 10.5 and the Wilcox diagram is the ratio between electrical conductivity and sodium ions which four

Table 10.8 Boron concentration on crops.

Boron concentration (ppm)	Suitability for crop
<0.5	All crops can be grown
0.5–1.0	Most crops can be grown except some sensitive crops
1.0–2.0	Semi-tolerant crops only survive
2.0–4.0	Only tolerant crops can be grown

Source: Follett and Soltanpour (2002) and Bauder et al. (2011).

Table 10.9 Chloride content in irrigation water and their suitability for crops production.

Cl^- content		
$Meql^{-1}$	ppm	Crop suitability
<2	<70	Safe for all crops
2–4	70–140	Slight to moderately injurious for sensitive crop
4–10	141–350	Slight to substantial injury on moderately tolerant plant
>10	>350	Cause severe problem and not suitable for irrigation

Source: Ludwick et al. (1990) and Bauder et al. (2011).

groundwater samples fall under the doubtful to unsuitable for irrigation purpose due to high salinity and sodium ions are depicted in Figure 10.6. Both the figures drawn on collected research article on groundwater of different states of India (Table 10.4).

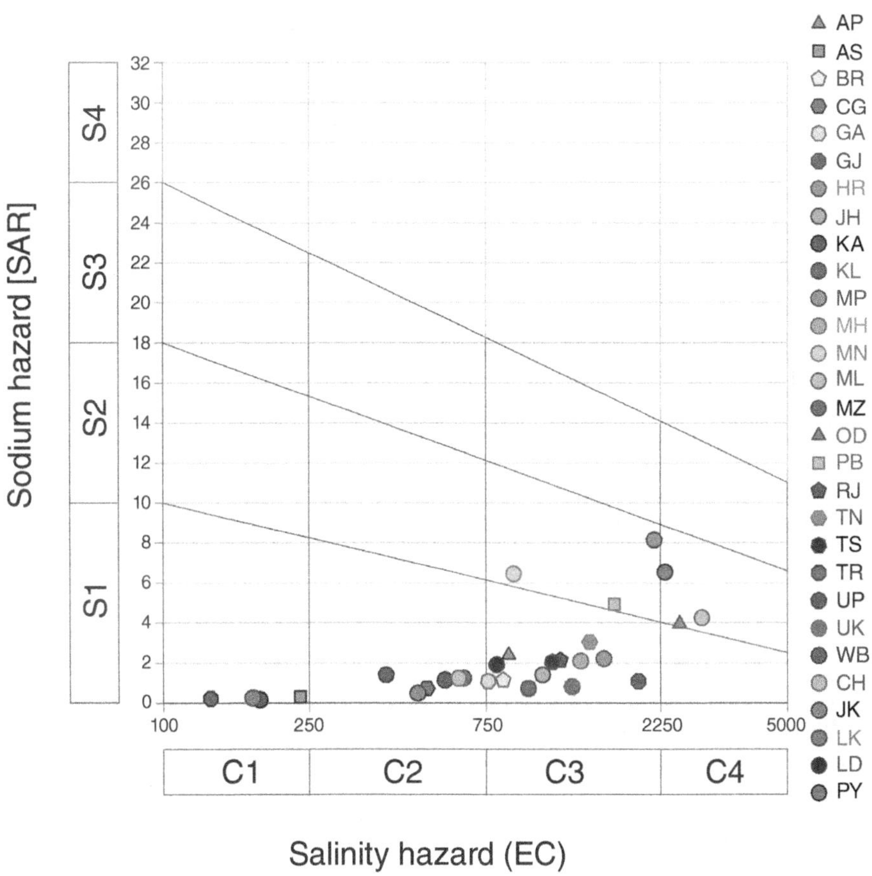

Figure 10.5 Salinity and sodium hazards of the Indian groundwater. (*See insert for color representation of the figure.*)

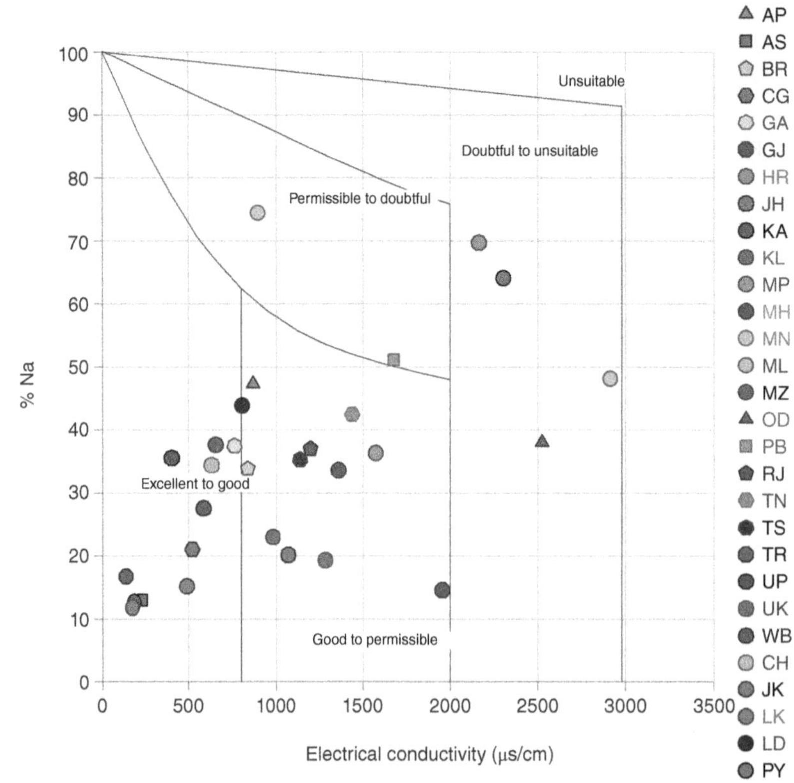

Figure 10.6 Wilcox diagram of the Indian groundwater.

10.8 Sustainable Irrigation Water Management Options in Agriculture

Several types of problem have been observed in the application of poor-quality irrigation water. In this topic, we focus on some management aspects to reuse the bad quality of water or other soil and cropping management to reduce the harmful effects on soil and plant growth. The following methods can achieve the sustainable goals of agriculture.

10.8.1 Leaching Requirement

Leaching requirements are the fraction of irrigation water for removing the accumulated salt in the root zone to control the salinity level. The leaching requirement can be calculated by using equation

$$\mathrm{LR} = \frac{D_{\mathrm{dw}} \times 100}{D_{\mathrm{iw}}} = \frac{\mathrm{EC}_{\mathrm{iw}} \times 100}{\mathrm{EC}_{\mathrm{dw}}}$$

where LR = is the leaching requirement (%), D_{dw} is the depth of drainage water (inches), =Depth of irrigation water (inches)

EC_{iw} is the electrical conductivity of the irrigation water (dS/m), and EC_{dw} is the electrical conductivity of the drainage water (dS/m).

10.8.2 Proper Drainage

In shallow water, table water is able to move toward the soil surface through capillary action; finally, salt accumulated in the surface soil when water evaporated. Drainage water helps to move the dissolved salts from the root zone, a slight disc or tillage in ponded water is more effective to remove the salt. The drainage can depend on the topography, slope, soil texture, soil structure, presence of salt, and quality of applied water for drainage.

10.8.3 Growing of Salt Tolerance Crops

All crops are not suitable in the salinity condition but only few selected plants can give profitable yield in this condition because of their ability to take water in osmotic condition for their physiological needs and metabolic activity (Table 10.10).

10.8.4 Blending of Poor-Quality Irrigation Water

The blending of irrigation water is maximizing the irrigation water volume by mixing of poor-quality irrigation water with good quality water and salinity hazards of irrigation water can be reduced by dilution and reduce the severity of damage.

Table 10.10 List of tolerant crops in irrigation water salinity (ECw) and soil salinity (ECe) condition.

	100%		75%		50%		0%	
Crops	ECe	ECw	ECe	ECw	ECe	ECw	ECe	ECw
Barley	8	5.3	13	8.7	18	12	28	19
Cotton	7.7	5.1	13	8.4	17	12	27	18
Sugarbeet	7	4.7	11	7.5	15	10	24	16
Sorghum	6.8	4.5	8.4	5.6	9.9	6.7	13	8.7
Wheat	6	4	9.5	6.3	13	8.7	20	13
Soybean	5	3.3	6.3	4.2	7.5	5	10	6.7
Rice	3	2	5.1	3.4	7.2	4.8	11	7.6
Maize	1.7	1.1	3.8	2.5	5.9	3.9	10	6.7
Tomato	2.5	1.7	5	3.4	7.6	5	13	8.4
Broccoli	2.8	1.9	5.5	3.7	8.2	5.5	14	9.1
Cucumber	2.5	1.7	4.4	2.9	6.3	4.2	10	6.8
Beans	1	0.7	2.3	1.5	3.6	2.4	6.3	4.2
Date palm	4	2.7	11	7.3	18	12	32	21
Orange	1.7	1.1	3.3	2.2	4.8	3.2	8	5.4
Strawberry	1	0.7	1.8	1.2	2.5	1.7	4	2.7

10.8.5 Other Cultural Practices

Land leveling is the smoothening of land as per slope aspects, the undulating land promotes the waterlogging condition due to lack of proper drainage and more chances of salt accumulation. The irrigation should be applied in appropriate frequent manner as compared to long gap between two irrigation and irrigation should be done when minimum evaporation losses occur at evening or cool time. Ridge and furrow system of seed sowing is good because it reduces the salt accumulation surrounding the seed materials. Fertilizer contact with seed material should be avoided as respect to the broadcast of fertilizers. Deep tillage helps to break the subsurface compaction which increases the water permeability, soil aeration, and provides the favorable condition for root growth. Scrapping is the removal of accumulated salt from the soil surface which decreases the level of salinity and yield can be maximized. The flood method of irrigation has been replaced by sprinkler in sandy soil condition and drip or trickle irrigation system for clay soil, meanwhile less water is required and frequent irrigation can be provided to avoid the damage of salinity (Juan et al. 2019). Application of organic matter either from the previous crop residue or extern source can be added into the soil which not only reduces the salinity but also improves the soil physical properties like water holding capacity, infiltration rate, soil structure, aggregation, etc. When sodic water is applied to the field, soil became sodium enriched in that condition certain chemical ameliorates have been used to improve the soil quality such as gypsum, sulfur, iron pyrite iron sulfate and lime sulfur can be used (Konstantinos and Maria 2015).

10.9 Government and Public Awareness to Sustainable Water Use in Agriculture

Central government has focused on micro-irrigation-based irrigation system to improve the efficiency of applied water (CWC 2014). In this respect, the government has taken certain initiatives in 1992; in 2006, it implemented centrally sponsored scheme on micro-irrigation; in 2010, works for national mission on micro-irrigation and national mission for sustainable agriculture have been implemented in 2014; Pradhan Mantri Krishi Sinchayee Yojana (PMKSY) was launched in 2015 by adopting the slogan of "per drop more crop" (Jain et al. 2019 and Table 10.11).

10.9.1 Government Initiatives on Micro-Irrigation

The micro-irrigation system was implemented in India with having 4 779 604 ha area for drip and 5 474 803 ha for the sprinkler irrigation system which saves the maximum irrigation water are presented in Figure 10.7 (Agricultural Statistics at a Glance 2018).

For public awareness, we searched a number of published research papers on two different topics, i.e. irrigation water quality and quantity of irrigation water for sustainable agriculture from 1990 to 2021 up to 15 April 2021 in the library of science direct at globally. Figure 10.8 shows that the number of publications has increased from 1990 to 2020.

Table 10.11 Institution for ground water management.

Organizations	Major role
Central Water Commission	Initiation and coordination of different scheme of water conservation in state of country level and monitoring water quality
Central Ground Water Board	Groundwater monitoring, developing new technologies and disseminate policies for the groundwater management
Central Groundwater Authority	It has been constructed under the environment protection act 1986 to monitor and control the groundwater resources and penal activities
Central Pollution Control Board	Implementation of water (prevention and pollution control Act 1974 to restore the water quality)
Central Soil Salinity Research Institute (CSSRI)	It works on salinity water management and how to restore the irrigation water quality

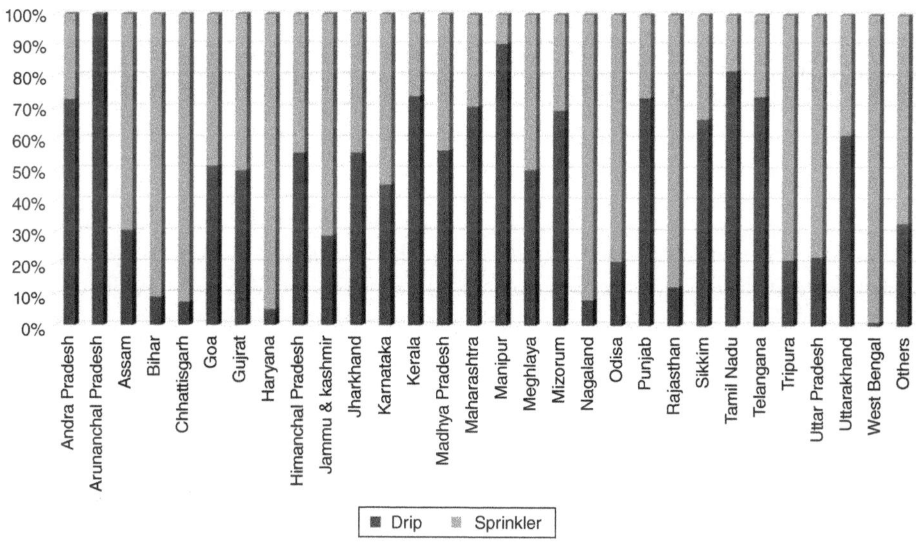

Figure 10.7 Area of drip and sprinkler irrigation system in different states of India.

10.10 Conclusion

In the current situation, we are facing the water crisis in terms of scarcity of water with respect to quality and quantity. Agriculture is the prime sector utilizing surface or groundwater resources for irrigation purposes to produce the food, fiber, and fuel. Continuous application of poor-quality irrigation water will deteriorate the soil quality as well as reduce the food grain production which is not sufficient to supply the food demands of current growing population and it will create huge pressure on land. In this chapter, we focused the certain criteria to evaluate the irrigation water quality and their parameters helps to minimize the hazard of irrigation water and improve the soil quality while maintaining sustainability approach without compromising the future generation's needs.

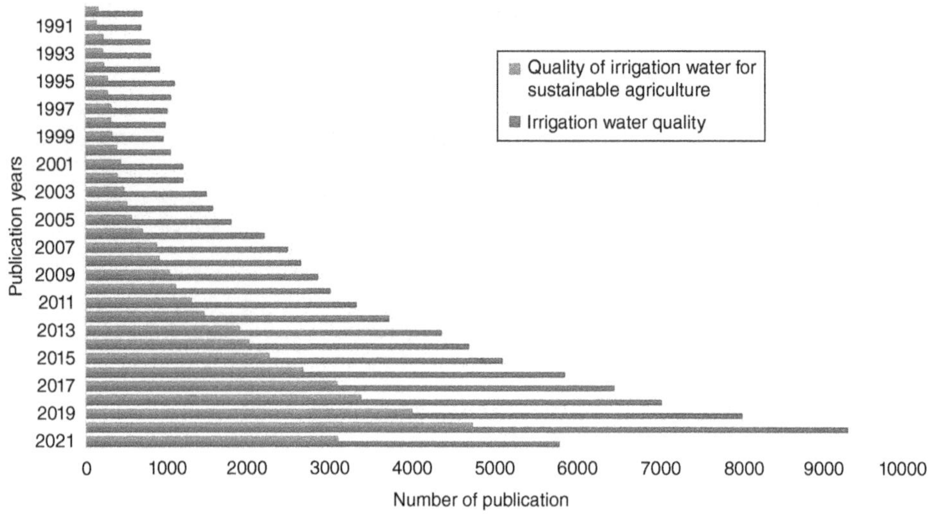

Figure 10.8 Number of publications on irrigation water quality and quality of irrigation water for sustainable agriculture from 1990 to 2021 at global level. *Source*: Modified from Agricultural Statistics at a Glance (2018).

References

Adimalla, N. and Qian, H. (2019). Groundwater quality evaluation using water quality index (WQI) fordrinking purposes and human health risk (HHR) assessment in an agricultural region of Nanganu South India. *Ecotoxicology and Environmental Safety* 176: 153–161.

Agrekar, S.H., Singh, N.B., Rau, T.Y. et al. (2012). Study of seasonal variation in ground water quality using multivariate statistical technique. *International Journal of Environmental Engineering and Management* 3 (2): 61–68.

Agricultural Statistics at a Glance (2018). Government of India Ministry of Agriculture & Farmers Welfare Department of Agriculture, Cooperation & Farmers Welfare Directorate of Economics and Statistics.

Annual Report (2014). River development and Ganga rejuvenation, Ministry of Water Resources. http://wrmin.nic.in/writereaddata/AR_2013-14.pdf (accessed 20 April 2021).

Antony, S., Dev, V.V., Kaliraj, S. et al. (2020). Seasonal variability of groundwater quality in coastal aquifers of Kavaratti Island, Lakshadweep Archipelago, India. *Groundwater for Sustainable Development* 11: 100377.

Bauder, T.A., Waskom, R.M., Sutherland, P.L., and Davis, J.G. (2011). *Irrigation Water Quality Criteria*, Crop series/irrigation, no. 0.506, 4. Colorado State University Extension Publication.

Bisht, P., Kumar, P., Yadav, M. et al. (2014). Spatial dynamics for relative contribution of cropping pattern analysis on environment by integrating remote sensing and GIS. *International Journal of Plant Production* 8 (1): 1–17.

Bureau of Indian Standards (1987). *Indian Standard Guidelines for the Quality of Irrigation Water*. New Delhi: Bureau of Indian Standards.

Central Ground Water Board (2019). *Dynamic Ground Water Resources of India*. Government of India Ministry of Jal Shakti Department of Water Resources, RD and GR.
Chaudhary, V.S. and Kumar, S. (2018). Assessment of groundwater quality for drinking and irrigation purposes in arid areas of Rajasthan, India. *Applied Water Science* 8: 218.
CWC (2005). *General Guidelines for Water Audit and Water Conservation*. New Delhi: Central Water Commission Evaluation of water utilisation Directorate.
CWC (2014). *Guidelines for Improving Water Use Efficiency in Irrigation, Domestic and Industrial Sectors*. Ministry of Water Resources Central Water Commission.
Dolma, K., Rishi, M.S., and Lata, R. (2015). Evaluation of groundwater quality and its suitability for drinking purposes – a case of Lehtown, Ladakh (J&K), India. *International Journal of Scientific and Engineering Research* 6: 1–5.
Doneen, L.D. (1962). The influence of crop and soil on percolating water. *Proceedings of the Biennial Conference on Groundwater Recharge*. pp. 156–163.
Doneen, L.D. (1964). *Notes on Water Quality in Agriculture*. Water Science and Engineering, University of California, Davis.
Dudeja, D., Bartarya, S.K., and Biyani, A.K. (2011). Hydrochemical and water quality assessment of groundwater in Doon Valley of outer Himalaya, Uttarakhand, India. *Environmental Monitoring and Assessment* 181: 183–204.
Duraisamy, S., Govindhaswamy, V., Duraisamy, K. et al. (2019). Hydrogeochemical characterization and evaluation of groundwater quality in Kangayam taluk, Tirupur district, Tamil Nadu, Indiausing GIS techniques. *Environmental Geochemistry and Health* 41: 851–873.
Eaton, E.M. (1950). Significance of carbonates in irrigation waters. *Soil Science* 69: 123–133.
Falkenmark, M. (1986). Fresh water: time for a modified approach. *Ambio* 15: 192–200.
FAO (2011). *The State of the World's Land and Water Resources for Food and Agriculture: Managing Systems at Risk*. London: Food and Agriculture Organization of the United Nations, Rome and Earthscan.
FAO (2021). Water at a glance: The relationship between water, agriculture, food security and poverty. Rome-15. http://www.fao.org/nr/water/docs/waterataglance.pdf (accessed 24 April 2021).
Follett, R.H. and Soltanpour, P.N. (2002). *Irrigation Water Quality Criteria*, Publication no. 0.506. Colorado State University.
Gopinath, S., Srinivasamoorthy, K., Vasanthavigar, M. et al. (2016). Hydrochemical characteristics and salinity of ground water in parts of Nagapattinam district of Tamil Nadu and the Union Territory of Puducherry, India. *Carbonates and Evaporites* https://doi.org/10.1007/s13146-016-0300-y.
Haritash, A.K., Mathur, K., Singh, P., and Singh, S.K. (2017). Hydrochemical characterization and suitability assessment of groundwater in Baga–Calangute stretch of Goa, India. *Environmental Earth Sciences* 76: 341.
Hoogeveen, J., Faurès, J.M., Peiser, L. et al. (2015). Globwat – a global water balance model to assess water use in irrigated agriculture. *Hydrology and Earth System Sciences* 19: 3829–3844.
Jain, C.K., Sharma, S.K., and Singh, S. (2018). Physico-chemical characteristics and hydrogeological mechanisms in groundwater with special reference to arsenic contamination in Barpetadistrict, Assam (India). *Environmental Monitoring and Assessment* 190: 417.

Jain, R., Kishore, P., and Singh, D.K. (2019). Irrigation in India: status, challenges and options. *Journal of Soil and Water Conservation* 18: 4.

Jain, C.K., Sharma, S.K., and Singh, S. (2021). Assessment of groundwater quality and determination of hydro-chemical evolution of groundwater in Shillong, Meghalaya (India). *Applied Sciences* 3: 33.

Jasrotia, A.S., Taloora, A.K., Andotrac, U., and Kumar, R. (2019). Monitoring and assessment of groundwater quality and its suitability for domestic and agricultural use in the Cenozoic rocks of Jammu Himalaya, India: a geospatial technology-based approach. *Groundwater for Sustainable Development* 8: 554–566.

Jhariya, D.C., Kumar, T., Dewangan, R. et al. (2017). Assessment of groundwater quality index for drinking purpose in the Durgdistrict, Chhattisgarh using geographical information system (GIS) and multi-criteria decision analysis (MCDA) techniques. *Journal Geological Society of India* 89: 453–459.

Juan, F., Munoz, V., Jose, A. et al. (2019). Sustainable irrigation in agriculture: an analysis of global research. *Water* 11: 1758.

Karthikeyan, K., Nanthakumar, K., Velmurugan, P. et al. (2010). Prevalence of certain inorganic constituents in groundwater samples of Erode district, Tamilnadu, India, with special emphasis on fluoride, fluorosis and its remedial measures. *Environmental Monitoring and Assessment* 160: 141–155.

Kaur, T., Bhardwaj, R., and Arora, S. (2016). Assessment of groundwater quality for drinking and irrigation purposes using hydrochemical studies in Malwa region, southwestern part of Punjab, India. *Applied Water Science* https://doi.org/10.1007/s13201-016-0476-2.

Kelly, W.P. (1963). Use of saline irrigation water. *Soil Science* 95: 355–391. https://doi.org/10.1097/00010694-196306000-00003.

Konstantinos, C. and Maria, B. (2015). Sustainable water management in agriculture under climate change. *Agriculture and Agricultural Science Procedia* 4: 88–98.

Kumbhar, V.S. and Salkar, V.D. (2014). Use of PAC as a substitute for alum in Nalgonda technique. *International Journal of Emerging Technology and Advanced Engineering* 4 (10): 154–161.

Laxman, K.D., Dhakate, R., Guguloth, S., and Srinivas, B. (2021). Hydrochemical appraisal of groundwater quality for drinking and agricultural utility in a granitic terrain of Maheshwaram area of Ranga reddy district, Telangana state, India. *Hydro Research* 4: 11–23.

Ludwick, A.E., Campbell, K.B., Johnson, R.D. et al. (1990). Water and plant growth. In: *Western Fertilizer Handbook – Horticulture Edition*, 15–43. Illinois: Interstate Publishers Inc.

Maity, S., Biswas, R., and Sarkar, A. (2020). Comparative valuation of groundwater quality parameters in Bhojpur, Bihar for arsenic risk assessment. *Chemosphere* 259: 127398.

Mandal, V.P., Rehman, S., Ahmed, R. et al. (2020). Land suitability assessment for optimal cropping sequences in Katihar district of Bihar, India using GIS and AHP. *Spatial Information Research* 28 (5): 589–599.

Manjusree, T.M., Sabu, J., Marco, P., and Thomas, J. (2017). Integrated approach for identifying the factors controlling groundwater quality of a tropical coastal zone in Kerala, India. *Environment and Earth Science* 76: 486.

Mohamed, K.A.F., Sameh, K.A.E., Ali, A.A. et al. (2020). Multivariate analysis for assessing irrigation water quality: a case study of the Bahr Mouise canal, Eastern Nile delta. *Water* 12: 2537.

Mukate, S.V., Panaskae, B.D., Wagh, M.V., and Baker, J.S. (2019). Comprehensive hydrochemistry and geothermal water quality of Konkan, Maharashtra, India for sustainable industrial development. *Groundwater for Sustainable Development* 12: 100518.

Nag, S.K. and Das, S. (2017). Assessment of groundwater quality from Bankura I and II blocks, Bankura District, West Bengal, India. *Applied Water Science* 7: 2787–2802.

Pandey, P.C., Mandal, V.P., Katiyar, S. et al. (2015). Geospatial approach to assess the impact of nutrients on Rice equivalent yield using MODIS sensors'-based MOD13Q1-NDVI data. *IEEE Sensors Journal* 15 (11): 6108–6115.

Paul, R., Brindha, K., Gowrisankar, G. et al. (2019). Identification of hydrogeochemical processes controlling groundwater quality in Tripura, Northeast India using evaluation indices, GIS, and multivariate statistical methods. *Environmental Earth Sciences* 78: 470.

Raghunath, H.M. (1987). *Groundwater*, 563. New Delhi: Wiley Eastern Ltd.

Rani, M., Joshi, H., Kumar, K. et al. (2019). Climate change scenario of hydro-chemical analysis and mapping spatio-temporal changes in water chemistry of water springs in Kumaun Himalaya. *Environment, Development and Sustainability* 18 (4): 1–16.

Ravikumar, P. and Somashekar, R.K. (2017). Principal component analysis and hydrochemical facies characterization to evaluate groundwater quality in Varahi river basin, Karnataka state, India. *Applied Water Science* 7: 745–755.

Ravindra, K., Thind, P.S., Mor, S. et al. (2019). Evaluation of groundwater contamination in Chandigarh: source identification and health risk assessment. *Environmental Pollution* 255: 113062.

Richards, L.A. (1954). *Diagnosis and Improvement of Saline and Alkali Soils*, Agricultural Handbook 60, 98–99. New Delhi: USDA and IBH Pub. Co. Ltd.

Rupal, M., Tanushree, B., and Sukalyan, C. (2012). Quality characterization of groundwater using water quality index in Surat city, Gujarat, India. *International Research Journal of Environment Science* 1 (4): 14–23.

Shahid, S.A. and Mahmoudi, H. (2014). National strategy to improve plant and animal production in the United Arab Emirates. Soil and water resources annexes.

Sheikh, A.M., Azad, C., Mukherjee, S., and Rina, K. (2017). An assessment of groundwater salinization in Haryana statein India using hydrochemical tools in association with GIS. *Environment and Earth Science* 76: 465.

Singh, A.L. and Singh, V.K. (2018). Assessment of groundwater quality of Ballia district, Uttar Pradesh, India, with reference to arsenic contamination using multivariate statistical analysis. *Applied Water Science* 8: 95.

Singh, K., Kumar, P., and Singh, B.K. (2013). An associative relational impact of water quality on crop yield: a comprehensive index analysis using LISS-III sensor. *IEEE Sensors Journal* 13 (12): 4912–4917.

Srivastava, S.K. (2017). Assessment of groundwater quality for the suitability of irrigation and itsimpacts on crop yields in the Guna district, India. *Agricultural Water Management* 216: 224–241.

Suhag, R. (2016). *Overview of Ground Water in India*. PRS Central Ground Water Board.

Thambidurai, P., Chandrasekharam, D., and Chandrashekhar, A.K. (2014). Hydrogeochemistry and groundwater quality in Champhai, Mizoram, North Eastern India. *International Journal of Earth Ccience and Engineering* 7: 2.

Tirkey, P., Bhattacharya, T., Chakraborty, S., and Barica, S. (2017). Assessment of groundwater quality and associated health risks: a case study of Ranchi city, Jharkhand, India. *Groundwater for Sustainable Development* https://doi.org/10.1016/j.gsd.2017.05.002.

Tomar, V., Mandal, V.P., Srivastava, P. et al. (2014). Rice equivalent crop yield assessment using MODIS sensors' based MOD13A1-NDVI data. *IEEE Sensors Journal* 14 (10): 3599–3605.

United Nations (2015). *World Population Prospects: The 2015 Revision, Key Findings and Advance Tables*, ESA/P/WP.241. Department of Economic and Social Affairs, Population Division https://esa.un.org/unpd/wpp/Publications/Files/Key_Findings_WPP_2015.pdf.

Upadhay, M. (2013). High fluoride incidence in groundwater and its potential health effects in parts of Sarguja area Chhattisgarh, India. *Recent Research in Science and Technology* 5 (5): 47–50.

Wazir, A., Singh, K.S., Yumnam, G. et al. (2020). Hydrogeochemical assessment of groundwater quality for few habitations of Chandel district Manipur (India). *Applied Water Science* 10 (123): 2–13.

Wilcox, L.V. (1955). *Classification and Use of Irrigation Waters*, Circular 969. Washington, DC: USDA.

Yadav, M., Sharma, M.P., Prawasi, R. et al. (2014). Estimation of wheat/rice residue burning areas in major districts of Haryana, India, using remote sensing data. *Journal of the Indian Society of Remote Sensing* 42 (2): 343–352.

11

Agricultural Water Footprint and Precision Management

V.K. Singh[1], G.A. Rajanna[2], V. Paramesha[3], and Pravin Kumar Upadhyay[4]

[1] *ICAR-Central Research Institute for Dryland Agriculture, Hyderabad, Telangana, India*
[2] *ICAR-Directorate of Groundnut Research, Anantapur, Andhra Pradesh, India*
[3] *Natural Resource Management Division, ICAR-Central Coastal Agricultural Research Institute, Old Goa, Goa, India*
[4] *Division of Agronomy, ICAR-Indian Agricultural Research Institute, New Delhi, Delhi, India*

CONTENTS

11.1 Introduction

Water shortage is one of the biggest challenges fronting the world's 2.8 billion inhabitants in the near future, and almost 1.2 billion people lack access to safe drinking water (Chartres and Varma 2010). However, water scarcity has emerged as the major constraint to sustainable crop production, especially in arid and semi-arid agro-ecologies around the world (Simsek and Comlekcioglu 2011). India's per capita water supply has dropped from 5177 m^3 in 1951 to 1441 m^3 in 2015, with a further decline to 1174 m^3 by 2051 (GOI 2018). The main challenge of this century is to provide a balanced diet to the world's growing population in an equitable and sustainable manner. Hunger and starvation are now synonymous with malnutrition, food shortages, and other health issues. We would need to feed 9.3 billion

Sustainable Agriculture Systems and Technologies, First Edition. Edited by Pavan Kumar, A.K. Pandey, Susheel Kumar Singh, S.S. Singh, and V.K. Singh.

people by 2050, which will only be possible through changes in agricultural production, such as closing the yield gap and sustainable intensification, as well as a shift in consumption patterns (Beddington et al. 2012). Overexploitation of water supplies as a result of rapid population growth, rising food demand, industrialization, and urbanization resulted in increased demand for water and causing freshwater shortage. According to a recent estimate, only 1.2% of the world's fresh water is available for all human activities (Mishra and Dubey 2015). Freshwater is needed for not only human activities but also agricultural production and industrial processes. According to published data from previous studies in Asia and elsewhere, global freshwater demand will increase over the next few decades, but supply will decrease (Sachan and Haq 2014). Agriculture uses 70% of all water consumed globally, followed by industry (20%) and domestic use (10%). Consumptive water usage for grain and fodder production is projected to rise by 0.7% each year from its current amount of 6400 Gm^3/year in 2000 to 9060 Gm^3/year by 2050, in order to feed the world's 9.2 billion people (Rosegrant et al. 2009). However, climate change combined with decreased water availability from surface and groundwater supplies are causing greater concern in India for sustainable water resource management. Water availability for the agricultural sector in India is limited due to rising water demand from all sectors. Therefore, indiscriminate and inefficient utilization of available water resources would create severe water scarcity for agricultural activities. In research communities, new paradigms and approaches to water usage, such as virtual water content (VWC) and WF, have emerged to encourage effective and safe water use, as well as water supply policy and management (Aldaya and Llamas 2008; Zeng et al. 2012).

Hoekstra (2003) coined the term "water footprint" to describe a measure of freshwater usage that quantifies sector-specific water use and contamination during development or consumption. The WF of a country is identified as the amount of water needed for the development and provision of goods and services by the country's residents (Hoekstra 2003). As a result, it refers to a producer's or consumer's direct and indirect water use. The WF has three elements, according to Hoekstra et al. (2011): (i) the consumption/loss of water from surface and groundwater supplies during the development process is referred to as blue WF, (ii) the consumption/loss of rainwater for crop production is referred to as green WF.·In an assumed catchment area, "consumption" refers to the depletion of water from land and surface sources. It mainly includes evapotranspiration, water usage by crops and goods, and water flows back to another catchment area or the sea, and (iii) gray WF, which refers to water contamination and the amount of freshwater needed to dilute a load of contaminants without affecting the water's characteristics. As a result, every crop's WF is expressed in terms of economic value (either in rupee/$/£/¥/€). Allan (2003) coined the word "virtual water" to describe the water used in the processing of food, fiber, and nonfood commodities based on water use indicators. Thus, the term virtual water (VW) has the potential to resolve water uncertainty and political pressure, especially in areas where there is a scarcity of water (Allan 2003). VW, according to Falkenmark (1995), is the difference between green and blue water. As a result, the WF study is enhanced by variations in blue-green waters due to various opportunity costs and environmental impacts. As a result, both the WF and VW definitions commonly refer to the use of water supplies to produce products. According to Garrido et al. (2010) and Harris et al. (2017), acute water scarcity reduces crop and water productivity, and consumers become more technologically superior in the efficient use of blue water. However, this structural adaptation occurs only when blue water is the main source of consumption and water is scarce. In the sense of integrated water resource management, the WF may play a complementary role. When it comes to the environmental impact of anthropogenic activities, WF evidence raises concerns among the general public, administrators, and stakeholders.

To meet the challenges of feeding, the growing population from declining water resources with the threat of climate change will need knowledge and suitable water management strategies. Thus, irrigation practices established for sustainable agricultural development goals require careful planning and management by reducing water use, especially for irrigated agricultural systems. Adoption of precision water management techniques is the need of the hour. At the same time, growing more crops per unit area with less water application or using water via evapotranspiration could result in more crop per drop of water (Kijne et al. 2003; Playan and Matoes 2006; Molden 2007; Zwart and Bastiaanssen 2004). Surface irrigation methods where the irrigation efficiency is 30–40% as compared with the attainable level of 60–70% due to higher non-beneficial evapotranspiration (Rajanna et al. 2017, 2018; Rajanna and Dhindwal 2019). In the surface irrigation method, about 2000 mm of water is applied in puddle transplanted rice (PTR) with a very low irrigation efficiency of 30–35% (Mandal et al. 2019). Kar et al. (2014) opined that WFs of agricultural crops may be used to compare water use. To address the current annual water deficit and potential escalating water demand, it is vital to work on sustainable crop production and increase water use efficiency (WUE) (Jain and Kumar 2007; Minhas et al. 2010). Therefore, well-designed efficient water management systems like drip irrigation, sprinkler irrigation, sub-surface irrigation, and adopting sensor-based irrigation scheduling are vital for sustaining production with adequate quality, more specifically for field crops and horticultural crops. A mini-sprinkler irrigation system, according to Singh et al. (2020), is a viable choice for direct-seeded rice (DSR) cultivation in India's groundwater-scarce Indo-Gangetic plains (IGP). Criteria for the application of irrigation should be based on when, where, and how much to apply. In this regard, knowing WF helps to achieve sustainability of freshwater resources through lowering the WFs could be possible. Among the precision management techniques, drip and sprinkler irrigation are proved fruitful in unveiling high water productivity by saving enough irrigation water (35–75%) in crops. Malhotra and Das (2016) reported 25–60% and 10–60% irrigation water saving by drip irrigation in orchard and vegetable crops, respectively, with high water productivity and crop yield than conventional irrigation. Therefore, lowering the WFs means producing more crop yield (above-ground biomass) using less amount of water. Assessing and analyzing the WFs in agriculture is essential to address the issues like climate change and overexploitation of freshwater resources. The present report enlists the per-capita water demand in India, WFs of agricultural and horticultural crops, and precision techniques for managing water resources.

11.2 Water Footprints of India and World

India is having 2.45% of the land, 4% of water resources, and 16% population, and India is the largest consumer of freshwater (FAO 2016). Freshwater is needed to sustain this increasing population. However, the amount of freshwater available is diminishing day by day. As

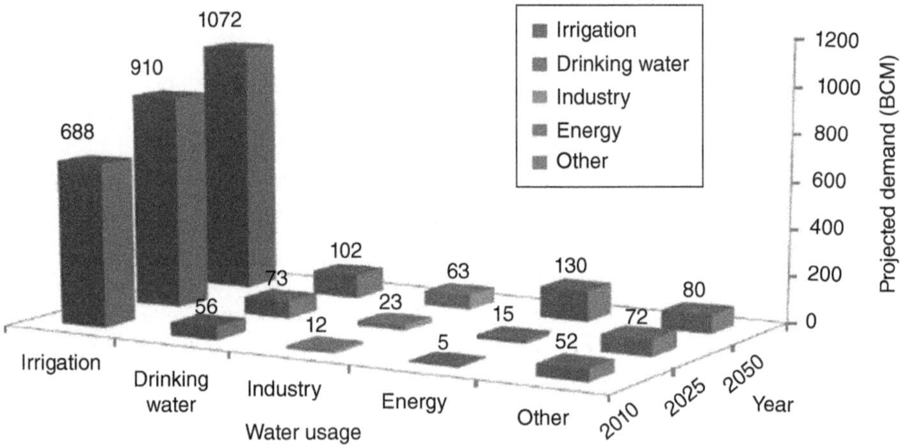

Figure 11.1 Water demand in India by 2050.

food demand rises, so does water consumption in order to increase crop production, putting more pressure on sustainable water supplies. In India, 91% of freshwater is withdrawn for agricultural purposes. However, the total exploitable water resources in India are ~1123 billion cubic meters (BCM) water, Of which, surface water is~690 BCM and groundwater is ~433 BCM. Irrigation uses approximately 85% of water (688 BCM) and is expected to rise to 1072 BCM by 2050 (Figure 11.1). Groundwater is the primary source of irrigation water; annual groundwater recharge is nearly 433 BCM, including 212.5 BCM used for irrigation and 18.1 BCM for domestic and industrial use (CGWB 2011). The per capita average annual freshwater availability has been decreasing since 1951, falling from 5177 m^3 in 1951 to 1441 m^3 in 2015, and is expected to fall to 1174 m^3 by 2050 due to rising population, urbanization, and rapid development in the region. As a result, efficient water use and wastewater recycling are important for successful water resource management (CWC 2012). About 13% of total water is used in Indian agriculture and is termed as the total VW supply; 35% is used for rice, 17% is used for raw sugar, and 14% is used for edible oils. The states with the highest net export of VW were Punjab, Haryana, Madhya Pradesh, and Uttar Pradesh. Bihar, Jharkhand, and Kerala had the largest total VW inputs. Punjab, Uttar Pradesh, Rajasthan, Tamil Nadu, and Haryana have the most water shortage for consumption. Water reserves will be exhausted in these states in the near future, posing a threat to food security.

The total WF of the United States is higher than that of any other nation, and it is higher than the global average (Figure 11.2). India, Japan, South Africa, Bangladesh, China, and Ethiopia have lower WFs than the global average. However, the United Kingdom (UK) and Japan depend on outside sources for 60–70% of their WF (Figure 11.2). Likewise, India's WFs are lower than the global average. Because of abundant fresh water resources, water footprints and reliance on water footprints from outside contributors are 1–3% lower in India as compared to other countries. Total water withdrawal for agricultural purposes in India was estimated to be 688 km^3/a in 2010 (FAO 2016) and is steadily increasing over the years. Water shortage is affecting most parts of the African continent, India, and South America. Likely, southern India is also facing physical water scarcity. Thus, WF of India or any other nation depends on (i) consumptive characteristics like consumptive volume

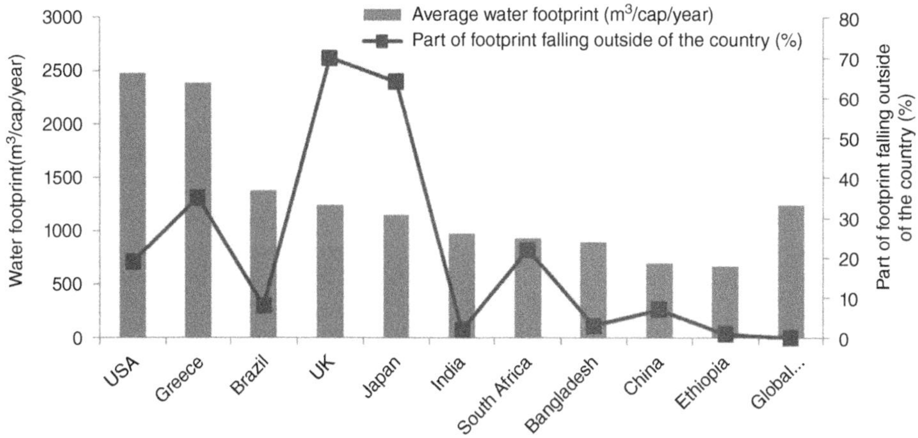

Figure 11.2 Global average water footprint (m^3/capita/year).

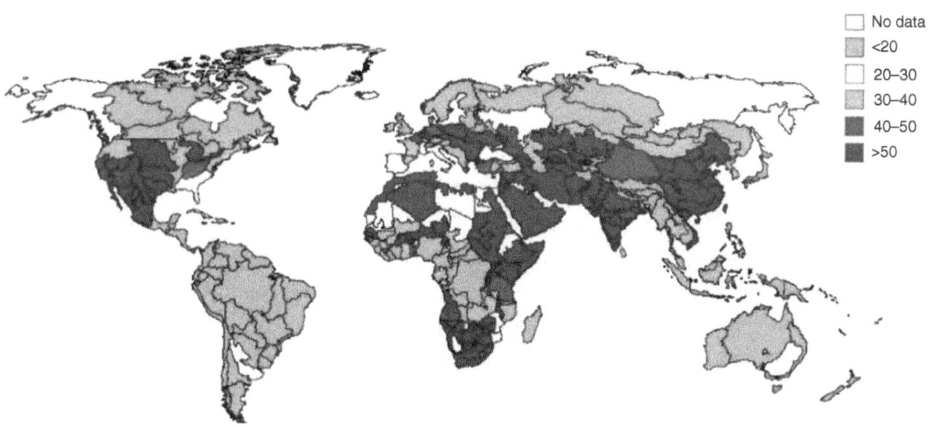

Figure 11.3 Water scarcity faced by the world by 2050. *Source*: Modified from Hoekstra and Chapagain (2007). © John Wiley & Sons. (*See insert for color representation of the figure.*).

(gross national income) and consumptive pattern, and (ii) production circumstances like climate and agricultural practices. Figure 11.3 shows that countries with green backgrounds have a WF that is equal to or greater than the global average. Countries in red have a higher WF than the global average, and by 2050, water scarcity will be apparent. By 2050, half of the world's population will be facing water shortage, with India being one of many countries affected.

11.3 Analysis of Water Footprint in Agriculture

Assessing and analyzing the WFs in agriculture is essential to address the issues like climate change and overexploitation of freshwater resources. This also assists in deciding how much water is required to manufacture a commodity, enhancing environmental integrity and rising long-term efficiency, as well as recognizing and resolving global water shortage challenges. WF exchange research is paving the way for new approaches to global water policy and sustainable water resource management. It is important to determine the evapotranspiration rate, which measures the amount of water lost by crops through evapotranspiration, to understand the water requirements of each crop. However, variables such as crop type, environment, and the growth process are carefully studied to determine the crop's water requirements. As a result, most studies use a modeling approach to estimate WFs (Chukalla et al. 2017). WFs can be measured at the field level by evaluating the soil and crop types to enable optimal water management. Based on differences in water-absorbent properties, inaccurate results may be obtained based on soil and crop type assessment. As a consequence, the amount of water needed varies depending on the crop and soil type. Gray WFs, on the other hand, are evaluated using a field data set that includes yield maps, soil texture, and crop operations. Leaching and associated gray WFs are measured using application rates and various environmental variables. Rice, wheat, sorghum, millet, and maize were measured and seasonal state-level blue and green WFs were extracted using Cool Farm Tool Water (CFTW), a new online water assessment tool (Kayatz et al. 2019). Using a spatially and temporally explicit data set of crop production and irrigation area, the CFTW model was used to estimate cereal water use in India (Kayatz et al. 2019). The agricultural water assessment tool CFTW can be used to produce data on crop production, location, and irrigation to quantify total water usage in cereal production (Hillier et al. 2011; Kayatz et al. 2019). The variability in WFs can be analyzed for all states and seasons to better understand the drivers of total water use. Similarly, Chapagain et al. (2006) calculated total water consumption (m^3/year) for crops in each country by multiplying the national cultivated area (ha/year) by the water depth (mm/year). Since the initial soil moisture before land preparation is believed to be negligible, residual soil moisture after the harvest is not included in the calculation of WFs.

11.4 Water Footprints of Agricultural and Horticultural Crops

Among agricultural crops, rice, and sugarcane are considered water-guzzling crops in India, as they need a greater amount of water to grow. The type of staple food consumed in an area, as well as the amount of oil and sugar consumed, have a significant impact on its WF. The mean VWC in the crops consumed primarily in that state determines the magnitude of WF in that state (Kampman 2007). Cereals were responsible for 50% of India's dietary WF, implying a possible opportunity to reduce water usage in Indian agriculture (Kayatz et al. 2019). Concurrently, groundwater depletion increased by 23% between 2000 and 2010 (Dalin et al. 2017) and is a major concern for long-term production and self-sufficiency (Barik et al. 2017). Rice, cotton, sugarcane, and maize are water-intensive crops with higher crop water demands. Higher crop yield and a lower rate of evapotranspiration are the main reasons for the decrease of WF in cereals. The production season has also shifted from *Kharif* to *Rabi*, which will help to reduce total WFs (33.4–45.0% compared to Kharif) for all cereal crops except rice, as well as ground and surface water supplies. It is assumed that the water is distributed evenly over all irrigated fields (63% of which are used for rice, wheat, maize, sorghum, and millet production), and that approximately 433 km^3/a of water is removed for cereal crop production (Government of India 2018). As a result, 105 km^3/a of gross blue water use is measured, meaning that 75.7% of the water removed is lost during conveyance, drainage, or deep percolation. According to Chakrabarti et al. (2011), the total WF of rice production was determined to be 1071.1 m^3/t for transplanted rice and 953.8 m^3/t for DSR. Application of irrigation water is less in DSR by around 25% and thus having less total WF in DSR.

As a result of the change in demand, cereal consumption trends have changed, with more rice and wheat being consumed and less coarse cereals such as millet, maize, and sorghum being consumed. However, the Indian population is severely deficient in micronutrients, and increasing consumption of nutrient-dense coarse cereals has been proposed as a beneficial public health nutrition intervention (DeFries et al. 2015; Rao et al. 2018). Changes in crop yields and cropping practices are developed to evaluate important factors for increased cereal production in India. As cereal production has increased over the last few decades, increased agricultural land area, fertilizer, and water use have all had an effect on the local environment (Barik et al. 2017). According to Kayatz et al. (2019), the average annual total water usage for cereal production decreased from 393.2 to 367.1 km^3 (6.6%) between 2005 and 2014. Wheat and rice production used the most water (80.6% of total water use), with Uttar Pradesh, Punjab, and Rajasthan (all in North India) accounting for 20.0, 8.4, and 8.4% of total Indian water use, respectively.

The national WF of rice production and consumption is estimated using data from foreign trade and domestic production. Rice cultivation has a 784 km^3/year global WF, with an average of 1325 m^3/t, with 48% green, 44% blue, and 8% gray water (Chapagain and Hoekstra 2011). As a result, rice production in south Asian countries has a large WF. To reduce the WF, several technologies have been developed, including the rice intensification method, aerobic rice/DSR, precision nutrient control, and micro-irrigation. DSR will help to better use rainwater by reducing groundwater and surface water use (Chakrabarti et al. 2011). According to Kayatz et al. (2019), wheat and maize have similar annual total WFs and are only slightly lower than rice. Sorghum and millet, with yields of 2894 and 2884 l/kg, respectively, had the lowest yields and the highest annual WFs across all seasons.

Seasonal variations in WFs were also apparent. Rice has the highest WF in *Rabi*, followed by sorghum. Rice is the only cereal with a higher WF in *Rabi* than in *Kharif*, at 10.1%. All other cereals, on the other hand, had lower WFs in *Rabi* than *Kharif*, ranging from 45.0% (maize) to 33.4% (rice and sorghum). The footprints of cereal blue water were generally low during the Kharif season (1.5% of total WF, except for rice). However, during the *Rabi* season, cereal blue WFs ranged from 12.3 to 78.3% of total WF for sorghum and rice, respectively (Figure 11.4). Sorghum and millet had the highest total WFs, but the smallest blue WFs. The gross annual WFs for all cereals in India decreased from 2005 to 2014, according to Kayatz et al. (2019), due to higher yields and lower evapotranspiration. Millet had the greatest reduction in total WFs (44%) from 4184 to 2324 l/kg, resulting in the greatest increase in yield from 0.7 to 1.3 t/ha (Figure 11.4).

Despite the fact that the majority of water diverted to agriculture in India is used to grow staple food crops and only about 10% of agricultural water is used for horticultural crops, there is a huge potential to increase overall water productivity on a basin level by improving

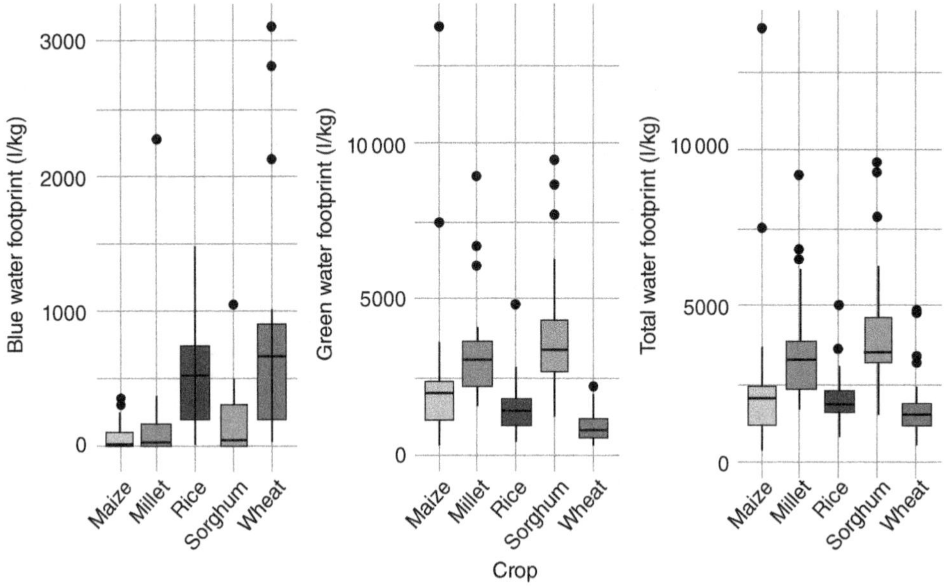

Figure 11.4 Between 2005 and 2014, average state-level blue, green, and total water footprints. *Source*: Kayatz et al. (2019). © John Wiley & Sons.

horticultural crop water productivity. Water productivity improvements in horticultural crops have the potential to minimize water usage in terms of providing unit energy, protein, and other nutrients, as well as economic returns. The majority of horticultural crops are perennial, with deep and extensive root systems capable of extracting water from deeper layers and a thick canopy to capture as much natural resources as possible, resulting in higher green water productivity. It is thus possible to increase both blue water and green water horticultural crop production through effective interventions (Malhotra and Das 2016). When it comes to fruits, tomato, and onion crops have a higher WUE than other vegetables (Table 11.1).

Among horticulture crops, fruit crops have seen a twofold increase in area and production over the last decade. With a 6.38 million hectare area, India is the world's second-largest fruit producer (74.9 million tonnes), accounting for 11% of global fruit output. India is the world's leading producer of mango, banana, papaya, pomegranate, sapota, and aonla. However, the water requirement of major fruit crops like mango, guava, grapes, and papaya was much lower than rice and sugarcane. Similarly, WUE of vegetable crops are much higher than agricultural crops (Table 11.1).

11.5 Precision Management of Water Resources

Water management by well-designed systems is essential for horticultural crops to maintain production and quality. If water is applied when it is not required, crop loss occurs, and if a water shortage occurs during the active growth process or fruit development stages, significant production and quality losses occur. As a result, it is critical to monitor the water

Table 11.1 Number of irrigations, water requirement, and WUE of vegetable crops in India.

Crop	No. of irrigations	Water requirement (mm)	WUE (kg/ha)
Brinjal (June–October)	7–8	486	43.2
Chili (July–November)	3–5	640	19.4
Onion (July–November)	3–5	500	77.2
Okra (July–October)	8–9	345	50.1
Cauliflower (October–January)	20	541	31.3
Carrot (Winter)	14	441	59.1
Tomato (October–February)	12	336	165.6
	Daily	336	209.3
Capsicum	16	480	25.6
	Daily	330	50.6
Garden Pea (October–February)	08	240	41.2
Water melon (December–April)	Daily	506	53.6
	14	469	68.4

Source: Adopted from Malhotra and Das (2016).

in order to determine when, where, and how to achieve optimal efficiency and productivity. As a result, scheduling should be dependent on plant water balance in relation to soil and the environment. To avoid damages, water must be added to the root zone region. Since several losses occur during transportation, an effective delivery method that improves performance must be used. Water was applied in a way that allowed for maximum output. When, where, and how water should be applied has been determined. As compared to conventional irrigation, drip irrigation has shown to be successful in demonstrating high water efficiency by saving irrigation water by 25–60% in various orchard crops and vegetables while increasing yield by 10–60%. It is a relatively new irrigation system that is gaining traction in areas where there is a scarcity of water and salt problems.

11.5.1 Adoption of Sensor-Based Micro-Irrigation System

Although the area under micro-irrigations systems (drip irrigation and sprinkler irrigation) is increasing significantly, its adoption is confined to limited areas in India because of the high initial cost. Further increasing water productivity by improving the yield level is possible through sensor-based irrigation. The working principles of sensors are that the automated system releases the water as and when the crop needs to be based on the soil moisture content depletion. Automatic irrigation control was first implemented using tensiometers, soil moisture probes, and granular matrix sensors (GMS). Phene and Howell (1984) used a soil matric potential sensor to track subsurface drip irrigation in a tomato crop. van-Iersel et al. (2009) used soil moisture probes to open and close solenoid valves based on the amount of water in the substrate. Precision sensor-controlled irrigation based on volumetric moisture content measurements is an effective tool for increasing overall WUE and improving the quality of soilless-grown lettuce by acting on the substrate moisture level

(Montesano et al. 2016). Automation adoption is minimal, however, because tension-based irrigation necessitates frequent maintenance and refilling. According to van-Iersel et al. (2013), capillarity fracturing causes tensiometers to become dislodged, resulting in inaccurate readings. Volumetric water-content-based soil moisture probes have become a viable method of automation due to low-cost sensors (Jones 2007). By comparing soil moisture content to matric potential, precision irrigation thresholds can be determined using these sensor-based probes. By matching water applications to moisture availability and plant uptake rates, sensor-based irrigation can reduce irrigation water consumption while preventing the negative effects of under- or overwatering (Lichtenberg et al. 2013). Soil moisture sensor systems save money on irrigation water, power, labor, and fertilizer, among other things. On the other hand, the economic benefits of investing in sensor-based irrigation systems are determined by the initial expense, as well as the relative magnitudes of the benefits.

11.5.2 Adoption of Micro-Irrigation System

Sustainable water management is possible by improving agricultural irrigation systems efficiency. In India, the major factor ascribed to lower water productivity is inefficiency of irrigation systems. The overall efficiency of canal system is about 30–40% while that of ground-water-based systems is about 60% (Malhotra and Das 2016). The contribution of the losses during application of water is about 25–40%. Sprinkler and drip irrigation systems in rice use 40% less water and yield 18% more grain than conventional flooded basin irrigation, making them a practical solution to the aforementioned concerns (Kahlown et al. 2007; Sharda et al. 2017). The use of blue water was minimized by drip irrigation with plastic mulch. In Boro rice, Karim et al. (2014) discovered a 34% reduction in irrigation water use, as well as a 7.6% increase in grain yield and a 31% increase in net benefit, by using sprinkler irrigation rather than flood irrigation. Similarly, mini-sprinkler irrigation system with high water application efficiency in DSR can save substantial amount of irrigation water (Singh et al. 2020). Flat sowing of mustard with drip irrigation in 60/30 and 30/60 cm paired row land configurations resulted in the highest seed yield (141–161%), oil productivity (139–150%), and water-use efficiency of mustard seed (141–162%), according to Rathore et al. (2020). Recently, Bhaskar and Joshi (2011) developed "tissue irrigation" a more precise irrigation system for horticultural crops. It means supplying irrigation water directly into the xylem vessels of the crop. Thus, it avoids evaporation losses and achieves 80% higher WUE than normal drip irrigation with 30% WUE.

11.5.3 Modified Crop Establishment Techniques

Crop WUE and irrigation water productivity (IWP) have been found to increase when crop bed planting and zero tillage (ZT) techniques are used. Similarly, compared to conventional tillage, using the raised bed technique to grow wheat and cluster bean crops increased WUE and IWP by 26–28 and 25–35%, respectively (Rajanna et al. 2017, 2019). Similarly, as compared to CT, overall consumptive water usage by wheat was 9–12% lower under raised beds and 5–6% lower under ZT. In comparison to the conventional system, bed-planting of wheat could save an average of 22–54% irrigation water (Rajanna and Dhindwal 2019). Wheat yield in raised beds is determined by how well the plants absorb solar radiation that falls between the outer rows. Using a conservation agriculture-based ZT and raised bed system in cluster bean, substantial irrigation water savings have been found (Rajanna et al. 2016).

11.5.3.1 Direct-Seeded Rice

PTR needs a large amount of irrigation water and caused a drop in the ground water table in rice-growing areas of India's IGP. When compared to transplanted rice, sprinkler irrigated-DSR saved 52.8% on irrigation water, resulting in 1.6–2.6 times higher grain water productivity. On the other hand, growing DSR increased the green WF by 53% while reducing the blue WF by 27% (Gerbens-Leenes et al. 2013). According to Jothimani and Thiagrajan (2005), alternate wetting and drying rice improved WUE by using less water than traditional flooding.

11.5.3.2 Scheduling Irrigation

Appropriate scheduling of irrigation results in an economy of water use and an increase in yield per mm of water applied. A differential irrigation scheduling based on the phonological stages of crop growth not only reduces the water requirement but also avoids water stress (both deficit and excess) during critical stages and thereby enhancing yield. Irrigation scheduling at critical stages benefited wheat crops by providing adequate moisture at the right time, which resulted in improved growth, yield parameters, and grain yield. The lower yield in only one irrigation at CRI only was due to lack of soil moisture, as it received only one irrigation at crown-root initiation and no irrigation thereafter at all other critical stages, where every moisture stress causes maximum yield reduction (Patil et al. 2014) by reducing the growth and yield components. Due to a higher grain yield combined with a smaller amount of irrigation water applied, irrigation applied at crown-root initiation stage in wheat with IW:CPE = 0.75 provided the highest WUE of 12.9 and 14.0 kg/ha-mm (Rajanna et al. 2018).

11.5.3.3 Use of Polymers

Precise technologies aiming at reducing consumptive use (CU) of available water without compromising productivity need to be invented and introduced in crop production. The use of specialty polymers known as super-absorbents or hydrogels has been shown to be very effective in improving water retention in the soil around the root zone by reducing percolation and evaporative losses, ensuring a stronger and longer supply of moisture to the crop (Narjary et al. 2012; Dar et al. 2017). As a result, hydrogel application in these soils may be a viable water management solution. Various studies have found that hydrogels increase soil moisture content and water productivity by 22.5% in Indian mustard (Rathore et al. 2019) and 97.1% in maize (Islam et al. 2011) as compared to no-hydrogel applied plots under deficit-irrigated conditions. Hydrogels have also been recorded to boost the quality of agricultural produce by increasing the size and color of fruits and flowers (Dar et al. 2017). Pusa hydrogel also assisted plants in surviving sustained moisture stress, delaying the onset of permanent wilting, and as a result, lowering crop irrigation and fertigation requirements (Patil et al. 2014).

11.5.3.4 Methods of Minimizing Evaporation from Soil

Mulching, alternative wetting and drying, and other water-saving techniques are available. Crop residue as a soil cover is another valuable strategic approach for reducing water loss from the soil surface and increasing WUE in semi-arid areas. Ali et al. (2018) discovered that wheat residue at 5 t/ha improved soil water supply, yield, and WUE as compared to no residue. Changing the crops grown by farmers may also help to increase water sustainability (Davis et al. 2017). By switching from rice and wheat to maize, sorghum, and millet, India's blue water needs for cereal production could be drastically reduced. Millet and sorghum cultivation are being promoted by the Indian government, and these cereals may be included in the public distribution system (Balani 2013).

11.6 Conclusion

Climate change and people's consumption habits have an effect on a country's WF. Global per capita water supply is declining due to population pressure. Rice has a higher green WF in India. The concept of WF has recently been added to the management of integrated water resources as a new and extended dimension. Drip irrigation, mulching, alternate wetting, and drying are some of the water-saving technologies available. India's agricultural system has increased cereal production while using less water due to changes in genotypes and higher crop productivity. This strategy has limited utility in resolving India's water crisis while sustaining crop production due to the increased irrigation area. Solutions for reducing pressure on freshwater resources, alleviating excessive groundwater use, and ensuring cereal production for food security are required depending on the growing season. Future research in India on cereal WFs should focus on scenarios that optimize cereal productivity and water use (both total and blue ware) in a cropping season. It may be possible to accurately educate local stakeholders by overlaying information on sustainable water supply with details on water usage trends and footprints. Person and household-level variables that influence output behaviors must also be considered when assessing policy responses. Increasing agricultural and horticultural production, which is less dependent on blue water, produces high yields and can be grown during India's growing seasons, and can help to conserve water sources while reducing water use. Millets, pulses, and oilseed crops may also help to reduce dependence on freshwater, but massive investments and research into improving yields, such as through high-yielding varieties, is needed to sustain current production levels. It is, however, crucial to establish and incorporate precision irrigation practices in order to reduce the blue WF.

References

Aldaya, M.M. and Llamas, M.R. (2008). Water Footprint Analysis for the Guadiana River Basin. Value of Water Research Report Series No. 35, UNESCO-IHE, Delft, Netherlands.

Ali, S., Jan, A., Manzoor, S.A. et al. (2018). Soil amendments strategies to improve water-use efficiency and productivity of maize under different irrigation conditions. *Agricultural Water Management* 210: 88–95.

Allan, J.A. (2003). Virtual water – the water, food, and trade nexus. Useful concept or misleading metaphor? *Water International* 28: 106–113.

Balani, S. (2013). *Functioning of the Public Distribution System*. New Delhi: PRS Legislative Research.

Barik, B., Ghosh, S., Sahana, A.S. et al. (2017). Water–food–energy nexus with changing agricultural scenarios in India during recent decades. *Hydrology and Earth System Sciences* 21 (3041–3060): 2017.

Beddington, J., Asaduzzaman, M., Clark, M. et al. (2012). *Achieving Food Security in the Face of Climate Change: Final Report from the Commission on Sustainable Agriculture and Climate Change, CGIAR Research Program on Climate Change*. Copenhagen, Denmark: Agriculture and Food Security (CCAFS).

Bhaskar, S and Joshi, M. (2011). Tissue irrigation- a new dimension in micro-irrigation. *8th International Micro Irrigation Congress and 21st International Congress on Irrigation and Drainage of ICID* (16–22 October 2011), Tehran, Iran, pp. 166–173.

CGWB (2011). Ground water year book– India 2011-12. Central Ground Water Board, Ministry of Water Resources, Government of India.

Chakrabarti, B., Mina, U., Pramanik, P., and Sharma, D.K. (2011). *Water Footprint of Transplanted and Direct Seeded Rice. Environmental Sustainability-Concepts, Principles, Evidences & Innovations*. New Delhi: Indian Agricultural Research Institute.

Chapagain, A.K. and Hoekstra, A.Y. (2011). The blue, green and grey water footprint of rice from production and consumption perspectives. *Ecological Economics* 70: 749–758.

Chapagain, A.K., Hoekstra, A.Y., and HHG, S. (2006). Water saving through international trade of agricultural products. *Hydrology and Earth Science System* 10: 455–468.

Chartres, C. and Varma, S. (2010). *Out of Water: From Abundance to Scarcity and how to Solve the world's Water Problems*. FT Press.

Chukalla, A.D., Krol, M.S., and Hoekstra, A.Y. (2017). Marginal cost curves for water footprint reduction in irrigated agriculture: guiding a cost-effective reduction of crop water consumption to a permit or benchmark level. *Hydrology and Earth System Sciences* 21: 3507–3524.

CWC (2012). *Water and Related Statistics. Water Planning and Project Wing*. India: Central Water Commission.

Dalin, C., Wada, Y., Kastner, T., and Puma, M.J. (2017). Groundwater depletion embedded in international food trade. *Nature* 543: 700–704. https://doi.org/10.1038/nature21403.

Dar, S.B., Mishra, D., Zahida, R., and Afshana, B.B. (2017). Hydrogel: to enhance crop productivity per unit available water under moisture stress agriculture. *Bulletin of Environment Pharmacology and Life Sciences* 6 (10): 129–135.

Davis, K.F., Rulli, M.C., Seveso, A., and D'Odorico, P. (2017). Increased food production and reduced water use through optimized crop distribution. *Nature Geoscience* 1 https://doi.org/10.1038/s41561-017-0004-5.

DeFries, R., Fanzo, J., Remans, R. et al. (2015). Metrics for land scarce agriculture. *Science* 349 (6245): 238–240.

Falkenmark, M. (1995). Coping with water scarcity under rapid population growth. *Conference of SADC Ministers* (23–24 November), Pretoria.

FAO (2016). AQUASTATWebsite. Food and Agriculture Organization of the United Nations (FAO). http://www.fao.org/nr/water/aquastat/countries_regions/IND/ (accessed 16 March 2021).

Garrido, A., Llamas, M.R., Varela-Ortega, C. et al. (2010). *Water Footprint and Virtual Water Trade in Spain*. New York: Springer.

Gerbens-Leenes, P.W., Mekonnen, M.M., and Hoekstra, A.Y. (2013). The water footprint ofpoultry, pork and beef: a comparative study in different countries and produc-tion systems. *Water Resources and Industry* 1–2: 25–36.

GOI (2018). EnviStats India 2018. Government of India. www.mospi.gov.in

Government of India (2018). Open Government Data (OGD) Platform India. https://data.gov.in/.

Harris, F., Green, R.F., Joy, E.J.M. et al. (2017). The water use of Indian diets and socio-demographic factors related to dietary blue water footprint. *Science of the Total Environment* 587–588: 128–136. https://doi.org/10.1016/j.scitotenv.2017.02.085.

Hillier, J., Walter, C., Malin, D. et al. (2011). A farm focused calculator for emissions from crop and livestock production. *Environmental Modelling and Software* 26: 1070–1078. https://doi.org/10.1016/j.envsoft.2011.03.014.

Hoekstra, A.Y. (ed.) (2003). Virtual water trade: proceedings of the International Expert Meeting on Virtual Water Trade. Value of Water Research Report Series No.12, UNESCO-IHE, Delft.

Hoekstra, A. and Chapagain, A. (2007). Water footprints of nations: water use by people as a function of their consumption pattern. *Water Resources Management* 21 (1): 35–48. https://doi.org/10.1007/978-1-4020-5591-1_3.

Hoekstra, A.Y., Chapagain, A.K., Aldaya, M.M., and Mekonnen, M.M. (2011). *The Water Footprint Assessment Manual. Setting the Global Standard*. London: Earthscan.

van-Iersel, M., Seymour, R.M., Chappell, M. et al. (2009). Soil moisture sensor-based irrigation reduces water use and nutrient leaching in a commercial nursery. *SNA Research Conference* 54: 17–21.

van-Iersel, M., Chappell, M., and Lea-Cox, J.D. (2013). Sensors for improved irrigation management in greenhouse and nursery production. *HortTechnology* 23: 735–746.

Islam, M.R., Hu, Y., Mao, S. et al. (2011). Effectiveness of a water-saving super-absorbent polymer in soil water conservation for corn (*Zea mays* L.) based on eco-physiological parameters. *Journal of the Science of Food and Agriculture* 91: 1998–2005. https://doi.org/10.1002/jsfa.4408.

Jain, A.K. and Kumar, R. (2007). *Water Management Issues–Punjab, North-West India.(In) Indo-US Workshop on Innovative E-technologies for Distance Education and Extension/ Outreach for Efficient Water Management*. Hyderabad: ICRISAT.

Jones, H.G. (2007). Monitoring plant and soil water status: Es¬tablished and novel methods revisited and their relevance to studies of drought tolerance. *Journal of Experimental Botany* 55: 2427–2436.

Jothimani, S. and Thiagrajan, T.M. (2005). Water and nitrogen use efficiency of rice under system of rice intensification. Water Policy Research Highlight 24. IWMI-TATA, Colombo, Srilanka.

Kahlown, M.A., Raoof, A., Zubair, M., and Kemper, W.D. (2007). Water use efficiency and economic feasibility of growing rice and wheat with sprinkler irrigation in the Indus Basin of Pakistan. *Agricultural Water Management* 87: 292–298.

Kampman, D.A. (2007). The water footprint of India. A study on water use in relation to the consumption of agricultural goods in the Indian states, pp. 1–152.

Kar, G., Singh, R., Kumar, A., and Sikka, A.K. (2014). Farm level water footprints of crop production: concept and accounting. Bulletin No. 67. Directorate of Water Management, Indian Council of Agricultural Research, Chandrasekharpur, Bhubaneswar, India, 56p.

Karim, M.R., Alam, M.M., Ladha, J.K. et al. (2014). Effect of different irrigation and tillage methods on yield and resource use efficiency of boro rice (*Oryza sativa*). *Bangladesh Journal of Agricultural Research* 39 (1): 151–163.

Kayatz, B., Harris, F., Hillier, J. et al. (2019). "More crop per drop": exploring India's cereal water use since 2005. *Science of the Total Environment* 673 (2019): 207–217.

Kijne, J.W., Barker, R., and Molden, D. (2003). *Water Productivity in Agriculture: Limits and Opportunities for Development*. Wallingford, UK: IWMI and CABI Publisher.

Lichtenberg, E., Majsztrik, J., and Saavoss, M. (2013). Profitability of sensor-based irrigation in greenhouse and nursery crops. *HortTechnology* 23 (6): 770–774.

Malhotra, S.K. and Das, M. (2016). Augmenting water use efficiency in horticulture. *Souvenir, World Aqua Congress* (28-30 December 2016), India Habitat Centre, New Delhi, pp. 45–60.

Mandal, K.G., Thakur, A.K., and Ambast, S.K. (2019). Current rice farming, water resources and micro-irrigation. *Current Science* 116 (4): 568–576.

Minhas, P.S., Jalota, S.K., Arora, V.K. et al. (2010). Managing water resources for ensuing sustainable agriculture: situational analysis and options for Punjab. *Research Bulletin* 2: 40.

Mishra, R.K. and Dubey, S.C. (2015). Fresh water availability and it's global challenge. *International Journal of Engineering Science Invention Research & Development* 2 (4): 2015.

Molden, D.J. (2007). *Water for Food, Water for Life, Comprehensive Assessment of Water in Agriculture*. Earthscan London: International Water Management Institute.

Montesano, F.F., van Iersel, M.W., and Parente, A. (2016). Timer versus moisture sensor-based irrigation control of soilless lettuce: effects on yield, quality and water use efficiency. *Horticultural Science (Prague)* 43 (2): 67–75.

Narjary, B., Aggarwal, P., Singh, A. et al. (2012). Water availability in different soils in relation to hydrogel application. *Geoderma* 187–188: 94–101. https://doi.org/10.1016/j.geoderma.2012.03.002.

Patil, M.D., Dhindwal, A.S., and Rajanna, G.A. (2014). Integrated moisture stress management in wheat (*Triticum aestivum*). *Indian Journal of Agronomy* 59 (4): 629–633.

Phene, C.l. and Howell, T.A. (1984). Soil sensor control of high-frequency irrigation systems. *Transactions of ASAE* 27: 392–396.

Playan, E. and Matoes, L. (2006). Modernization and optimization of irrigation systems to increase water productivity. *Agricultural Water Management* 80: 100–116.

Rajanna, G.A. and Dhindwal, A.S. (2019). Water dynamics, productivity and heat use efficiency responses in wheat (*Triticum aestivum*) to land configuration techniques and irrigation schedules. *Indian Journal of Agricultural Sciences* 89 (6): 912–919.

Rajanna, G.A., Dhindwal, A.S., and Sriharsha, V.P. (2016). Performance of clusterbean (*Cyamopsis tetragonoloba*) under variable irrigation schedules and crop establishment techniques. *Indian Journal of Agronomy* 61 (2): 223–230.

Rajanna, G.A., Dhindwal, A.S., and Nanwal, R.K. (2017). Effect of irrigation schedules on plant – water relations, root, grain yield and water productivity of wheat [*Triticumaestivum*(L.)] under various crop establishment techniques. *Cereal Research Communications* 45 (1): 166–177.

Rajanna, G.A., Dhindwal, A.S., Narender, P.M.D., and Shivakumar, L. (2018). Alleviating moisture stress under irrigation scheduling and crop establishment techniques on productivity and profitability of wheat (Triticumaestivum) under semi-arid conditions of western India. *Indian Journal of Agricultural Sciences* 88 (3): 372–378.

Rajanna, G.A., Dhindwal, A.S., Rawal, S., and Pooniya, V. (2019). Energetics, water and crop productivity of wheat (*Triticumaestivum*)–cluster bean (*Cyamopsistetragonoloba*) sequence under land configuration and irrigation regime in semi-arid agro-ecosystem. *Indian Journal of Agronomy* 64 (4): 450–457.

Rao, N.D., Min, J., DeFries, R. et al. (2018). Healthy, affordable and climate-friendly diets in India. *Global Environmental Change* 49: 154–165.

Rathore, S.S., Shekhawat, K., Dass, A. et al. (2019). Deficit-irrigation scheduling, super absorbent polymers-hydrogel enhance seed yield, water productivity and economics of Indian mustard under semi-arid ecologies. *Irrigation and Drainage* 68 (3): 531–541. https://doi.org/10.1002/ird.2322.

Rathore, S.S., Shekhawat, K., and Rajanna, G.A. (2020). Land configurations in surface drip irrigation for enhancing productivity, profitability and water-use efficiency of Indianmustard (*Brassica juncea*) under semi-arid conditions. *Indian Journal of Agricultural Sciences* 90 (8): 1538–1543.

Rosegrant, M.W., Ringler, C., and Zhu, T. (2009). Water for agriculture: maintaining foodsecurity under growing scarcity. *Annual Review of Environment and Resources* 34 (1): 205–222.

Sachan, B. and Haq, N.L. (2014). Importance of water for sustainable agriculture production and socio-economic development of India. *International Journal of Research in Agriculture and Food Sciences* 2 (6) http://www.ijsk.org/ijrafs.html (accessed 21 March 2021 and 8 September 2021).

Sharda, R., Mahajan, G., Siag, M. et al. (2017). Performance of drip-irrigated dry-seeded rice (*Oryza sativa* L.) in South Asia. *Paddy and Environment* 15: 93–100.

Simsek, M. and Comlekcioglu, N. (2011). Effects of different irrigation regimes and nitrogen levels on yield and quality of melon (*Cucumis melo* L.). *African Journal of Biotechnology* 10 (49): 10009–10018.

Singh, R., Singh, A., Kumar, S. et al. (2020). Feasibility of mini-sprinkler irrigation system in direct seeded rice (*Oryza sativa*) in Indo-Gangetic plains of India. *Indian Journal of Agricultural Sciences* 90 (10): 1946–1951.

Zeng, Z., Liu, J., Koeneman, P.H. et al. (2012). Assessing water footprint at river basin level: a case study for theHeihe River Basin in northwest China. *Hydrology and Earth System Sciences* 16: 2771–2781.

Zwart, S.J. and Bastiaanssen, W.G.M. (2004). Review of measured crop water productivity values for irrigated wheat, rice, cotton and maize. *Agricultural Water Management* 69: 115–133.

12

Drip Fertigation for Enhancing Crop Yield, Nutrient Uptake, Nutrient, and Water Use Efficiency

V. Paramesha[1], G.A. Rajanna[2], Parveen Kumar[1], M.S. Sannagoudar[3], and H.M. Halli[3]

[1] *Natural Resource Management Division, ICAR-Central Coastal Agricultural Research Institute, Old Goa, Goa, India*
[2] *ICAR-Directorate of Groundnut Research, Regional Station, Anantapur, Andhra Pradesh, India*
[3] *ICAR-Indian Grassland and Fodder Research Institute, Jhansi, Uttar Pradesh, India*

CONTENTS

12.1 Introduction

The traditional method of unregulated surface irrigation often results in leaching of mobile nutrients, which pollutes surface water sources, ground water, and degrades soil health. Drip irrigation, on the other hand, minimizes such losses, resulting in higher yields (25–30%) and resource productivity over traditional irrigation. The primary factor regulating efficient use of nutrients when implemented through a subsurface drip irrigation system is maintaining nutrients at optimal levels within the crop rhizosphere (Shukla et al. 2009). The higher installation cost of a drip irrigation system, on the other hand, is a significant deterrent to its widespread adoption. However, reducing this issue by adjusting lateral spacing and switching to a paired-row planting pattern, determining the best time and amount of water/fertilizer application, and increasing planting densities within the furrows (Goel et al. 2005) would significantly

Sustainable Agriculture Systems and Technologies, First Edition. Edited by Pavan Kumar, A.K. Pandey, Susheel Kumar Singh, S.S. Singh, and V.K. Singh.

improve water use efficiency (WUE). It is well established that maintaining optimal soil moisture conditions during the crop growing season is critical to achieving higher yields (Rajanna et al. 2016, 2017). Due to uncontrolled use, rapid commercialization of agriculture, and urbanization, water is a prime and scarce resource that is also overexploited. Furthermore, the country's and the world's ground water resources have been overexploited to the tune of 80–85% of their capacity.

Agriculture uses 81% of India's water (WRI 2007), so allowing more efficient use of water in agriculture should be a top priority. The significant loss of water and nutrients (up to 50%) in traditional irrigation and fertilizer application methods resulted in lower crop production and inefficient resource usage. Despite these drawbacks, the flood irrigation system is commonly used in Indian agriculture, with significant water loss due to evaporation and other distribution losses such as percolation, leaching, and runoff (Rosegrant 1997). Water input per unit irrigated area will have to be reduced due to water shortages and environmental concerns (INCID 2006). Traditional flood irrigation systems cause excessive wetting of the soil profile, which has an impact on root aeration and plant development (Rajanna et al. 2018, 2019). Furthermore, the distribution of plant nutrients in moist soil, volume is determined by their mobility, sorption, and precipitation responses with the soil particles (Bresler 1977). However, under surface application methods, this distribution is non-uniform (Rajanna and Dhindwal 2019). In such circumstances, farm-level advanced technologies, such as drip fertigation, will be needed to increase production while improving water and nutrient use performance.

In this context, fertigation is a modern method of applying nutrients by irrigation water that has proven to be very successful in increasing yield and reducing resource consumption. Drip fertigation has been scientifically proven to increase crop yield per unit volume of water and nutrients used, making it a viable option for more efficient water and nutrient use (Patel and Rajput 2009). Since water and nutrients are critical inputs, they are distributed precisely in the effective crop root zone as required during the crop development stages. Subsurface drip fertigation, on the other hand, is an effective method of applying water and nutrients below the surface soil through drip irrigation. Aside from saving water and nutrients, the fertigation method has a huge potential for increasing crop yield. By burying the dripper lines or placing them at or below the plant root region, subsurface or surface drip irrigation system can be used permanently or temporarily. As a result of the irrigation water serving as a carrier for the nutrients, the combined application of water and nutrients is optimal for proper crop development. Water-soluble fertilizers are delivered directly to the feeding root zone by applying small amounts of fertilizer frequently through the irrigation method. When properly treated, it opens up new possibilities for growing crops in nutrient solution-like conditions. Hence, farmers must implement fertigation on a large scale in order for agriculture to be profitable and economically viable. In commercial and high-value crops, fertigation through surface and subsurface drip irrigation has already become commonplace. Similarly, depending on resource availability and knowledge level requirements, it is becoming increasingly common in field crops. However, the research on these methods that have been conducted elsewhere are briefly listed under the headings below.

12.1.1 Drip Fertigation Has a Number of Advantages

- **More effective water use:** Soil evaporation, surface runoff, and deep percolation are decreased or removed entirely. Drier soils with less soil crusting can help with infiltration and preservation of seasonal precipitation. A smaller fraction of the soil volume may be wetted in widely spaced crops, eliminating excessive irrigation water losses even further.

- **Lower water quality risks:** Deep percolation reduces or eliminates runoff into streams, as well as nutrient and chemical leaching.
- **Increased opportunities for degraded water use:** Smaller, more frequent irrigation applications can help to maintain a more stable and lower soil matric capacity, which can help to minimize salinity risks.
- **Improved water application uniformity:** Improved in-field uniformities will lead to better water, nutrient, and salt management.
- **Improved plant growth, yield, and quality:** Many crops react favorably.
- **Improved plant health:** Since crop canopies are drier and less humid, there is less disease and fungal strain. Some forms of soil fumigation may also be done with the method.
- **Better fertilizer and pesticide management:** More precise and timely fertilizer and pesticide application through the system will result in greater effectiveness and, in certain instances, a reduction in their usage.
- **Better weed control:** In drier areas, weed germination and growth are often reduced.
- **Better double cropping opportunities:** Crop timing can be improved because the device does not need to be removed during harvest and reinstalled before planting the second crop.
- **Improved farming operations and management:** During irrigation activities, many field operations will take place. Soil compaction is minimized as a result of field operations, and irrigation-induced soil crusting is significantly reduced.
- **Automation:** The system's closed-loop pressurized function, which reduces application variability as well as soil water and nutrient redistribution variability, makes it an excellent candidate for automation and advanced irrigation control technologies.
- **Lower energy costs:** Operating stresses are typically lower than those used in sprinkler irrigation.
- **System integrity issues:** Compared to mechanical-move sprinkler irrigation systems, there are less mechanized parts in a system. Since the majority of the parts are made of plastic, they are less likely to corrode in an irrigation system.
- **Design versatility:** As opposed to center pivot sprinkler irrigation systems, there is more flexibility in balancing field shape and field size. The device can be sized to suit the available water supply quickly and affordably. Drip lines may be installed in widely spaced crops for optimal water and nutrient uptake.
- **System longevity:** When correctly planned and operated, the installations can have a long economic life. Long system life allows investment costs to be amortized for several years, enabling lower-value commodity crops to be grown economically.

12.2 Effect of Drip Fertigation on Crop Productivity

Table 12.1 illustrates the yield advantage under drip fertigation over surface drip and surface method of irrigation. The improved output under drip fertigation was attributed

Table 12.1 Effect of drip irrigation and fertigation on yield (t/ha) of different crops.

Crops	SDI + fertigation	Surface drip	Surface irrigation	References
Maize (Grain yield)	7.53	—	5.94	Honnappa et al. (2018)
	6.57	5.60	5.06	Fanish et al. (2011)
	7.87		6.88	Shruthi et al. (2018)
Wheat	3.91	3.61	3.28	Abdullah and Pawar (2013)
Aerobic rice (Grain yield)	5.45	4.56	3.46	Gururaj (2016)
Groundnut (pod yield)	2.42	2.01	1514	Sanju et al. (2014)
	2.24)	1.71	1.00	Rajwade and Dwivedi (2018)
Mustard (pod yield)	1.97	1.73	1.41	Oma (2019)
Castor (seed yield)	2.11	2.04	1.58	Pushpanathan et al. (2018)
Sugarcane	193.94	175.14	—	Mahesh et al. (2016)
Cotton	1.56	—	0.65	Hadole et al. (2012)
Tomato	79.27	71.92	59.50	Hebbar et al. (2004)
Capsicum	0.87	0.72	0.64	Patil and Das (2015)
Chili	12.09	11.48	—	Chanthai and Wonprasaid (2016)

to the crop root zone's ability to retain a favorable soil water status, allowing the plants to more efficiently use moisture and nutrients from the limited wetted area (Hebbar et al. 2004). Furthermore, the availability of nutrients and soil moisture at the active root growth area was due to the higher crop yield in subsurface drip fertigation over drip and furrow irrigation. Drip-fertigated maize with 50% P and K from water-soluble fertilizers yielded 7.3 t/ha, a significant improvement in grain yield. Drip fertigation increased yield by 35% as compared to drip irrigation with fertilizer application to the soil (Fanish et al. 2011). According to Tasal and Pawar (2015), 100% drip fertigation with foliar sprays was found to be more efficient than conventional irrigation and fertilization in raising wheat yield (25.6%) thus saving 44.5% water and taking 0.8 ha more area under irrigation. Furthermore, when drip fertigation was used during the grain production stage, Jabran et al. (2011) observed a higher number of spikelets per spike in wheat. Drip fertigation with 150% of the recommended NPK and biofertigation resulted in higher yield attributes such as the maximum number of sympodial branches (18.1), fruiting points (68.5), number of bolls (29.5), and boll weight (4.8 g/boll) as well as a 43.0% increase in cotton seed yield over conventional surface irrigation with soil fertilizer application (Jayakumara et al. 2014).

12.3 Effect of Drip Fertigation on Water Use Efficiency (WUE)

Due to the higher yield advantage obtained with limited irrigation water usage under this form of irrigation, drip fertigation has the ability to increase crop WUE. The comparison of

Table 12.2 Effect of drip irrigation and fertigation on WUE (kg ha/cm) of different crops.

Crop	SDI + fertigation	Surface drip	Surface irrigation	References
Maize	206	175	94	Fanish et al. (2011)
	326		118	Shruthi et al. (2018)
Winter wheat	176	163	82	Abdullah and Pawar (2013)
Aerobic rice	64.6	54.1	26.9	Gururaj (2016)
	91.01	—	31.45	Rekha et al. (2015)
Groundnut	59.89	49.76	25.36	Sanju et al. (2014)
Mustard	77.7	68.4	55.8	Oma (2019)
Cotton	31.9	31.3	27.6	Singh et al. (2012)
Chili	343	331	—	Chanthai and Wonprasaid (2016)
	321	289	—	Singh et al. (2017)

WUE of different crops under various irrigation methods is shown in Table 12.2. While surface drip irrigation increases crop yield, the combined effect of nutrients and water under drip fertigation is critical for improving yield and water use performance. Earlier researchers recorded higher WUE in maize (Fanish et al. 2011), winter wheat (Abdullah and Pawar 2013), aerobic rice (Gururaj 2016), groundnut (Sanju et al. 2014), and cotton (Singh et al. 2012). Drip fertigation had a higher WUE due to simultaneous availability of water and nutrients to crops (Chanthai and Wonprasaid 2016), decreased evaporation (Oma 2019), weed problems (Singh et al. 2017), and other losses. Drip fertigation produces about 2.7 times more gross dry matter than furrow irrigation and fertilizer application on soil with the same volume of fertilizer but different methods of application, according to Sagheb and Hobbi (2002). Phene et al. (1989) discovered that high-frequency subsurface and surface drip irrigation (several irrigations per day) resulted in higher WUE than low-frequency surface drip irrigation (one irrigation at two to three days). Phene et al. (1989); Sagheb and Hobbi (2002); Tanaskovic et al. (2007) recorded higher WUE as a result of drip fertigation over furrow irrigation and fertilizer broadcasting. Controlling transpiration during crucial crop growth cycles and maintaining adequate crop nutrition, will increase WUE (Rajanna et al. 2017, 2018, 2019). These concepts have been integrated into the drip irrigation system, which is much more beneficial than the surface irrigation method.

12.4 Effect of Drip Fertigation on Nutrient Uptake

The soil solution process, which is primarily determined using soil moisture availability, determines the concentration and availability of various nutrients in the soil for plant uptake. The higher availability of soil moisture provided by drip fertigation's continuous water supply resulted in higher nutrient availability in the soil, which improved nutrient uptake by the crop. Increased biomass production due to continuous availability of water and nutrients to the crop resulted in increased nutrient uptake under drip fertigation. The increased uptake may also be attributable to the split application of N and K under drip

Table 12.3 Effect of drip irrigation and fertigation on nutrient uptake (kg/ha) in different crops.

Crop	SDI + fertigation	Surface drip	Surface irrigation	References
Maize	N (225.6), P (64.01), and K (218.3)	—	N (195.1), P (53.53), and K (187.6)	Honnappa et al. (2018)
	N (198.41), P (79.83), and K (157.92)	—	N (158.33), P (55.17), and K (123.0)	Shruthi et al. (2018)
Aerobic rice	N (126.9), P (34.87), and K (80.87)	N (116.1), P (29.29), and K (71.34)	N (104.1), P (25.08), and K (64.60)	Gururaj (2016)
	N (141.20), P (19.47), and K (97.17)	—	N (83.40), P (13.0), and K (61.20)	Rekha et al. (2015)
Groundnut	N (98.3), P (22.1), and (41.5)	N (78.1), P (17.3), and K (31.6)	N (60.3), P (14.9), and K (22.4)	Sanju et al. (2014)
Mustard	N (77.35)	N (58.63)	N (42.31)	Oma (2019)
Tomato	N (165.7), P (16.5), and K (113.5)	N (14.1), P (13.3), and K (94.3)	N (109.3), P (9.5), and K (69.1)	Hebbar et al. (2004)
	N (44.28)	—	40.28	Singandhupe et al. (2003)
Capsicum	N (212.6), P (199.2), and K (186.9)	N (199.9), P (34.76), and K (158.0)	N (194.1), P (32.44), and K (152.9)	Patil and Das (2015)
Chili	N (92.04), P (19.21), and K (136.2)	N (84.63), P (17.11), and K (125.4)	—	Singh et al. (2017)

fertigation, which resulted in minimal nutrient loss and made nutrients available to the crop continuously. Table 12.3 shows that soil drip fertigation has a higher nutrient absorption than surface irrigation and surface drip fertigation. As compared to soil application, fertigation produces higher yields due to increased nutrient absorption, fertilizer consumption quality, and the percentage of nutrient extracted from fertilizer (Rathore et al. 2020). With the maximum N, P, and K uptake of 110.9, 28.2, and 110.6 kg/ha at 120 DAS, drip fertigation with 150% of the recommended dose of NPK and biofertigation is significantly superior (Jayakumara et al. 2014). In comparison to low frequency (two weeks intervals) drip irrigation, they found that high frequency drip irrigation prevented cyclic water stress and root deterioration, which could contribute to higher nutrient uptake.

12.5 Effect of Drip Fertigation on Nutrient Use Efficiency

The growing area under micro-irrigated crops provides a great opportunity to test new methods for providing complete and balanced plant nutrient systems, which could improve plant health and yields. Drip fertigation is a low-cost method of fertilizer delivery that, when used correctly, can reduce overall fertilizer application rates while also reducing the negative environmental impact on agricultural productivity. Drip fertigation allows for accurate fertilizer distribution and consistency. It is a cost-effective and agronomically sound method of delivering soluble plant nutrients to the active root zone (Rathore

Table 12.4 Effect of drip irrigation and fertigation on nutrient use efficiency (kg yield/kg nutrient) in different crops.

Crop	SDI + fertigation	Surface drip	Surface irrigation	References
Maize	N (50), P (100), and K (188)	–	N (40), P (79), and K (149)	Honnappa et al. (2018)
	N (34.51), P (69.02), and K (129.42)		N (27.61), P (53.35), and K (103.56)	Shruthi et al. (2018)
Aerobic rice	N (54.5), P (109), and K (109)	N (45.7), P (91.3), and K (54.1)	N (34.7), P (69.3), and K (69.3)	Gururaj (2016)
	N (65.03), P (130.05), and K (130.05)		N (33.75), P (67.49), and K (67.49)	Rekha et al. (2015)
Tomato	226.5	205.5	170.0	Hebbar et al. (2004)
Chili	N (3.36), P (120), and K (13.4)	N (3.18), P (114), and K (12.7)	—	Chanthai and Wonprasaid (2016)

et al. 2020). Table 12.4 shows that drip fertigation has a higher nutrient usage efficiency than surface irrigation in corn, aerobic rice, tomato, and chili. The data showed an increase in the nutrient use efficiency (NUE) of applied nutrients in maize by 10, 21, and 40 kg yield/kg nutrient, resulting in nutrient savings and lower production costs. Drip fertigation increases NUE, which reduces not only fertilizer consumption but also the environmental impact associated with higher fertilizer use. Hebbar et al. (2004) found that drip fertigation increased NUE (226 kg/kg nutrient) in tomato, reducing nitrogen usage in high-value vegetable crops that need more nutrients. Applying DAP at a lower rate (33 kg P/ha) produced nearly the same wheat grain yield as applying it at a higher rate of 44 kg P/ha. As compared to fertilizers applied in a single application, drip fertigation systems facilitate multiple small dose fertilizer injections at different intervals, reducing the risk of leaching and improving nutrient usage quality, yield, and nutrient uptake.

12.6 Effect of Drip Fertigation on Soil Nutrient Dynamics

The mobility of nutrients has been demonstrated using a drip fertigation process. Understanding the effects of such transitions on nutrient mobility is crucial for figuring out how soil fertility interactions function. Under drip fertigation, the nitrogen content in the soil increased as the distance from the emitter increased up to a point, then decreased. When using a drip fertigation method, the nitrogen content in the soil profile does not accumulate at the periphery of the wetting front and is not leached from the root zone (Rathore et al. 2020). Because of its higher mobility in the soil, N is well distributed in the top layer from 0 to 30 cm. The availability of P was restricted to the 0–15 cm soil layer under drip fertigation due to reduced P mobility. The amount of available phosphorus decreased as distance and soil depth increased. Just below the dripper, the phosphorus supply peaked. When drip fertigation was used, the soil K content was significantly higher in the surface soil than in the subsoil; this may be because the majority of the applied K was retained in

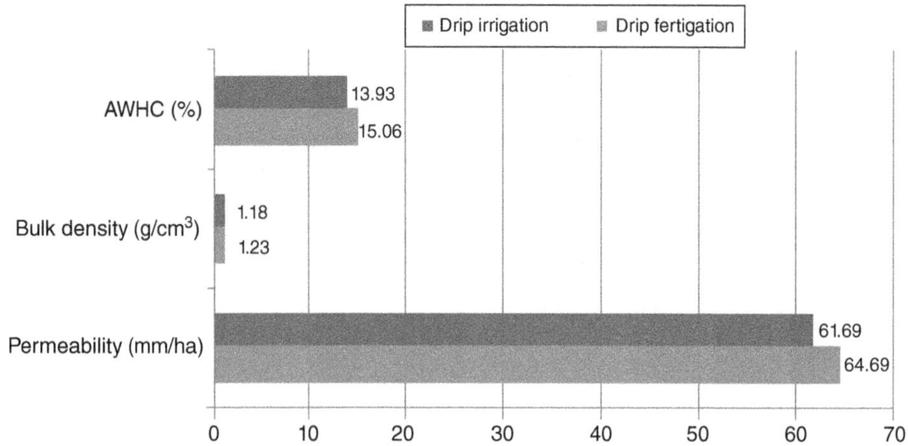

Figure 12.1 Effect of drip irrigation and fertigation on soil physical properties in chili. *Source*: Modified from Chanthai and Wonprasaid (2016). © John Wiley & Sons.

the surface soil and downward movement was slow. So, nutrient availability would increase with drip fertigation because the method delivers nutrients directly to the crop root region, and improved water availability contributes to better root growth and nutrient uptake. When comparing drip fertigation to drip irrigation, Chanthai and Wonprasaid (2016) found that drip fertigation had a higher soil permeability (64.69 mm/ha) and a higher available water holding capacity (15.06%) (Figure 12.1). In comparison to surface irrigation, drip fertigation results in less leaching loss and greater nutrient flow in the soil, which can explain the higher residual available N, P, and K in the soil. By providing water at shorter intervals and providing N and K through drip irrigation, a favorable situation was created, resulting in more nutrients being present in the soil.

12.7 Constraints in Adoption of Drip Irrigation

While drip irrigation has many advantages, it also has a number of limitations, such as a smaller wetting pattern. Monitoring and analyzing drip fertigation irrigation events necessitates a high level of expertise in improved technology, soil resources, and crop water requirements. On soils with vertical cracking, reduced upward water movement can be especially problematic. Above the drip line, salinity levels can rise, posing a risk to emerging seedlings, or small transplants. Drip line placement can restrict primary and secondary tillage operations. Irrigation and fertilization become more important issues with smaller crop root areas, both in terms of timing and number. Even when the root zone is well watered, smaller crop root zones can be inadequate to escape diurnal crop water stresses. Since drip fertigation systems are spatially set, accommodating crops with different row spacing – be more difficult. In some soils and climates, some crops cannot grow properly. Peanuts may not be able to peg into dry soil properly. A larger wetting pattern can be beneficial to tree crops. When compared to other irrigation systems, it has a high initial investment expense. The device has no resale value or just a marginal salvage value in certain situations. Water filtration, as in all micro-irrigation systems, is essential to ensuring proper system operation and system longevity. Maintenance and repairs must be performed on a regular basis. Rodent-caused leaks can be more difficult to find and fix, particularly in deeper systems. Root intrusion in the drip lines must be controlled, and system operational and design procedures must provide precautions to restrict or avoid further intrusion. Visual indicators of system operation and uniformity of system implementation are scarce. This is especially true in some areas where growers have little experience with these systems.

12.8 Conclusion

Drip fertigation has translated some of the most applicable principles into recent field practice advances, such as matching the water needs of crops and cropping systems with adequate water sources, fertilization, tillage, and mulching. Drip fertigation, according to the study, not only increases productivity but also ensures higher efficiency of the two most essential inputs in crop production, namely water and nutrients, by increasing the WUE and NUE. The high crop yield achieved with drip fertigation emphasizes the value of an adequate supply of nutrients and water in order to resolve soil and climatic constraints that vary by soil type. Drip fertigation can provide farmers with advantages such as increased yields, resource productivity, and profitability. Fertigation technology provides many advantages as a result of a shortage of water and cultivated fields, as well as higher fertilizer prices. By minimizing the use of fertilizers and gasoline, this technology has the ability to reduce pollution. The following questions should be addressed in future drip fertigation studies: (i) Fertilizers that can be used for drip fertigation, (ii) scheduling fertilization in drip-irrigated crops, (iii) drip fertigation irrigation and fertilizer dose optimization, (iv) emitter clogging assessment, and (v) economic viability of drip fertigation in various agro-climatic regions.

References

Abdullah, T. and Pawar, D.D. (2013). Production and economics of wheat (*Triticum aestivum* L.) under drip fertigation. *International Journal of Science and Research* 4 (5): 1907–1909.

Bresler, E. (1977). Trickle-drip irrigation: principles and application to soil-water management. *Advances in Agronomy* 29: 343–393.

Chanthai, S. and Wonprasaid, S. (2016). Effects of fertigation and water application frequency on yield, water and fertilizer use efficiency of chili (*Capsicum annuum* L.). *International Journal of Research in Chemical, Metallurgical and Civil Engg.* 3 (2): 2349–1450.

Fanish, S.A., Muthukrishnan, P., and Jeyaraman, S. (2011). Effect of drip fertigation and intercrops on yield and water use efficiency of maize (*Zea mays*) under maize based intercropping system. *Journal of Crop and Weed* 7 (2): 41–45.

Goel, A.C., Vijaykumar and Dhindsa, J.P.S. (2005). Feasibility of drip irrigation in sugarcane at Haryana. *Proc. Nation. Sem. Relevance of Micro irrigation in Sugarcane*, VSI Pune, Maharashtra, India, pp. 9–10.

Gururaj, K. (2016). Optimization of irrigation and fertilizer levels in sugarcane under drip fertigation. PhD thesis. University of Agricultural Sciences Bengaluru.

Hadole, S.S., Bhagat, G.J., Nagone, A.H., and Thakur, V.R. (2012). Nutrient management through drip system of irrigation in cotton. *PKV Research Journal* 36 (2): 36–42.

Hebbar, S.S., Ramachandrappa, B.K., Nanjappa, H.V., and Prabhakar, M. (2004). Studies on NPK drip fertigation in field grown tomato (*Lycopersicon esculentum* Mill.). *Europian Journal of Agronomy* 21: 117–127.

Honnappa, H.M., Denesh, G.R., Patil, L., and Hamsa, N. (2018). Studies on drip fertigation under different planting geometry in maize (*Zea mays* L.). *Journal of Pharmacognosy and Phytochemistry* 7 (6): 295–299.

INCID (2006). *Drip Irrigation-Prospects and Coverage in India*. New Delhi: Indian National Committee on Irrigation and Drainage, Ministry of Water Resources.

Jabran, K., Cheema, Z.A., Farooq, M., and Khan, M.B. (2011). Fertigation and foliar application of fertilizers alone and in combination with canola extracts enhances yield in wheat. *Crop & Environment* 2 (1): 42–45.

Jayakumara, M., Surendranb, U., and Manickasundaram, P. (2014). Drip fertigation effects on yield, nutrient uptake and soil fertility of Bt cotton in semi arid tropics. *International Journal of Plant Production* 8 (3): 375–390.

Mahesh, R., Raja, A.N., and Archana, H.A. (2016). Performance of surface and subsurface drip fertigation on yield and water use efficiency of sugarcane. *2nd World Irrigation Forum* (6–8 November 2016), Chiang Mai, Thailand.

Oma, S.B. (2019). Effect of drip irrigation level and micronutrient application method on productivity of Indian mustard (Brassica juncea L.). M.Sc. (Agri.) thesis. SKN Agricultural University Jobner.

Patel, N. and Rajput, T.B.S. (2009). Effect of subsurface drip irrigation on onion yield. *Irrigation Science* 27: 97–108. https://doi.org/10.1007/s00271-008-0125-0.

Patil, K.V.O. and Das, J.C. (2015). Effect of drip irrigation and fertilizer management on capsicum (*Capsicum annum* L). *IOSR Journal of Agriculture and Veterinary Science* 8 (1): 10–13.

Phene, C.J., Davis K.R., McCormick R. L. et al. (1989). Water-fertility management for subsurface drip irrigated tomato and pepper production in the Tropics. *International Symposium on Integrated Management Practices*, Tainan, Taiwan, pp. 323–338.

Pushpanathan, K.R., Pandian, B.J., Mariappan, G., and Jeyakumar, P. (2018). Effect of drip fertigation on biochemical parameter and castor equivalent yield of castor + onion inter cropping system. *Journal of Pharmacognosy and Phytochemistry* SP1: 68–72.

Rajanna, G.A. and Dhindwal, A.S. (2019). Water dynamics, productivity and heat use efficiency responses in wheat (*Triticum aestivum*) to land configuration techniques and irrigation schedules. *Indian Journal of Agricultural Sciences* 89 (6): 912–919.

Rajanna, G.A., Dhindwal, A.S., and Sriharsha, V.P. (2016). Performance of clusterbean (*Cyamopsis tetragonoloba*) under variable irrigation schedules and crop establishment techniques. *Indian Journal of Agronomy* 61 (2): 223–230.

Rajanna, G.A., Dhindwal, A.S., and Nanwal, R.K. (2017). Effect of irrigation schedules on plant – water relations, root, grain yield and water productivity of wheat [*Triticum aestivum* (L.)] under various crop establishment techniques. *Cereal Research Communications* 45 (1): 166–177.

Rajanna, G.A., Dhindwal, A.S., Narender, P.M.D., and Shivakumar, L. (2018). Alleviating moisture stress under irrigation scheduling and crop establishment techniques on productivity and profitability of wheat (*Triticum aestivum*) under semi-arid conditions of western India. *Indian Journal of Agricultural Sciences* 88 (3): 372–378.

Rajanna, G.A., Dhindwal, A.S., Rawal, S., and Pooniya, V. (2019). Energetics, water and crop productivity of wheat (*Triticum aestivum*)–cluster bean (*Cyamopsis tetragonoloba*) sequence under land configuration and irrigation regime in semi-arid agro-ecosystem. *Indian Journal of Agronomy* 64 (4): 450–457.

Rajwade, O.P. and Dwivedi, S.K. (2018). Effect of drip fertigation on growth and yield of soybean. *International Journal of Chemical Studies* 6 (5): 1738–1741.

Rathore, S.S., Shekhawat, K., and Rajanna, G.A. (2020). Land configurations in surface drip irrigation for enhancing productivity, profitability and water-use efficiency of Indian mustard (*Brassica juncea*) under semi-arid conditions. *Indian Journal of Agricultural Sciences* 90 (8): 1538–1543.

Rekha, B., Jayadeva, H.M., Kombali, G. et al. (2015). Effect of source and rate of fertilizer application through drip irrigation on yield and nutrient budgeting in aerobic rice. *Green Farming* 6 (6): 1350–1352.

Rosegrant, W.M. (1997). Water resources in the twenty first century: challenges and implication of action. *Food and Agriculture and the Environment Conference*, Washington, DC, USA.

Sagheb, N. and Hobbi, M.S. (2002). Field evaluation of urea fertilizer and water use efciency by tomato under trickle fertigation and furrow irrigation in Iran. Fertirigation for improved water use efciency and crop yield, IAEA-TECDOC-1266. Water balance and fertigation for crop improvement in West Asia. Results of a technical co-operation project organized by the Joint FAO/IAEA/Division of Nuclear Techniques in Food and Agriculture, pp. 3–13.

Sanju, H.R., Mudalagiriyappa, M., Ramachandrappa, B.K. et al. (2014). Studies on drip fertigation of NPK fertilizers in groundnut (*Arachis hypogaea* L.). *Ecology Environment and Conservation* 20 (2): 223–227.

Shruthi, M.K., Sheshadri, T., Yogananda, S.B., and Prakash, S.S. (2018). Yield and nutrient uptake of hybrid maize as influenced by different fertigation intervals, duration and fertilizer levels in Southern Dry Zone of Karnataka. *International Journal of Current Microbiology and Applied Sciences* 7: 3787–3796.

Shukla, S.K., Yadav, R.L., Singh, P.N., and Singh, I. (2009). Potassium nutrition for improving stubble bud sprouting dry matter partitioning nutrient uptake and winterinitiated sugarcane ratoon yield. *Europian Journal of Agronomy* 30: 27–33.

Singandhupe, R.B., Rao, G.G.S.N., Patil, N.G., and Brahmanand, P.S. (2003). Fertigation studies and irrigation scheduling in drip irrigation system in tomato crop (*Lycopersicon esculentum* L.). *Europian Journal of Agronomy* 19: 327–340.

Singh, J., Babar, S., Deshmukh, R. et al. (2012). Drip fertigation effect on yield, water use efficiency and economics of bt cotton in semi-arid region. *PKV Research Journal* 36 (2): 521–528.

Singh, A., Sharma, S.K., Chopra, R. et al. (2017). Impact of drip fertigation on yield and nutrients uptake by chilli (*Capsicum annuum* L.). *International Journal of Current Microbiology and Applied Sciences* 6 (8): 1409–1416.

Tanaskovic, V., Cukaliev, O., and Iljovski, I. (2007). Effect of irrigation method and regime of irrigation and fertilization on tomato yield. Anniversary Yearbook of the Faculty of Agricultural Sciences and Food, 60 years Faculty of Agricultural Sciences and Food, vol. 53. Skopje, Macedonia, pp. 137–149.

Tasal, A. and Pawar, D.D. (2015). Production and economics of wheat (*Triticum aestivum* L.) under drip fertigation. *International Journal of Science and Research* 4 (5): 1907–1909. https://www.ijsr.net/search_index_results_paperid.php?id=SUB154520.

WRI (2007). Annual Report of World Resources Institute-2006-07. publication/ wri-annual-report-2006.

Section 4

Precision Agriculture

13

Sustainable Agriculture Systems and Technologies

Amit K. Singh, Avijit Ghosh, Manjanagouda S. Sannagoudar, R.V. Kumar, Sunil Kumar, Prashant Deo Singh, and Safik Ahamad

GSM Division, ICAR-IGFRI, Jhansi, Uttar Pradesh, India

CONTENTS

13.1 Introduction

Sustainable development has been defined as a set of principles that guide us to effective utilization of natural resources without undermining their integrity and stability for future generations. Sustainable development has been major agenda in global platform like UNO, World Bank, and climate change conferences since 1987 Brundtland report. Sustainable agriculture refers to farming methods that are pleasant to the environment and allow for the crops or livestock husbandry without causing harm to human or natural systems. It entails avoiding negative consequences for land, water, wildlife, nearby or downstream

Sustainable Agriculture Systems and Technologies, First Edition. Edited by Pavan Kumar, A.K. Pandey, Susheel Kumar Singh, S.S. Singh, and V.K. Singh.

infrastructure, as well as for human population. Sustainable agriculture is a multigenerational philosophy that passes on a conserved or enhanced natural resource, biotic, and economic base rather than a depleted or polluted one. Alternate land use systems, conservation agriculture, information and communication technologies, mixed farming, multiple cropping, and crop rotation are all elements of sustainable agriculture.

In India, land degradation is not adequately discussed. In recent years, one of the main causes of declining soil quality has been land degradation (Bhattacharyya et al. 2015). According to the National Bureau of Soil Survey and Land Use Planning (NBSS and LUP 2004), of India's total geographical area (328.7 Mha), 146.8 million hectares was affected by degradation. The existing unplanned and exhaustive agricultural practices have been prolonged at the expense of our dwindling natural resources and future generation's needs seem compromised. In view of mounting human population, the farming suffers from the twin problems of meager allocation of land and water resources, which does not commensurate with the huge livestock population (Kumar et al. 2019). Resource crunch, overexploitation, population growth create undue pressure on limited natural resources. There is need for multiple options or combination of options to address the problem in each natural resource whether land, process based farming or involvement of newer technology in the agriculture. In subsequent sections of this chapter, various sustainable agricultural technologies like alternate land use system, precision farming, low-carbon landscape, conservation agriculture, organic agriculture, integrated nutrient management (INM), regenerative agriculture, and drones have been discussed.

13.2 Alternate Land Use System

Alternate land use system that enhances the resilience of human society and reduces the vulnerability of human settlements are indispensable for livelihood security of farming communities (Kumar et al. 2019). Land use systems that can reduce soil and water losses, restore ecological balance, and improve both land use productivity and economic revenues are desired (Bhattacharyya et al. 2015). Alternate land-use-based systems like silvipasture, agro forestry, and hortipastoral models not only diversify the options available to farmers but they also minimizes the impact of climate change on the livelihood opportunity.

Silvipasture system is a sustainable system of land use, which integrates forestry, horticulture, and livestock management. In the era of diminishing per capita land availability, silvipasture system with the integration of perennial woody trees with grasses/legumes/pastures provides the sustainable technology for increasing system productivity. It also helps in competent utilization of natural resources like sunlight, moisture, and plant nutrients than standalone system either by agriculture alone or forestry exclusively. Perennial woody trees are integrated with crop and pasture with the aim to provide food, fodder, and fuel wood besides conserving the agro ecosystem. Agri-silvi-pasture system and this system should preferentially include animal component to enhance profitability. Silvipasture system is adopted in the areas of poor soil fertility and low water availability. Benefits of silvipasture system not only include production from degraded lands but also conserving and soil and moisture. Integrating the livestock resources enhances the system productivity and overall economic return per unit land. Synergistic behavioral pattern of silvipasture system enhances the soil fertility along with increase in soil organic carbon (SOC). The tree-based system serves as a sustainable land use technology for livestock production by effectively using wastelands for fodder production. Silvipasture system enhances the carbon fractions and improves the biological health and hence can be considered as one of the eco-restoration strategies of the degraded lands (Ghosh et al. 2021a).

Hortipasture is a form of agroforestry system that combines fruit trees and pasture (grasses and/or legumes) on the same piece of land. This system acts as one of the best and economic alternative system for class V and VI (Singh 1996; Sharma 2005; Kumar and Choubey 2008). It can provide protective food (fruit) for humans and fodder for animals, bridging the large gap between supply and demand for fruit and fodder while also halting soil erosion (Khan and Kumar 2010). By continuous deposition of plant biomass and turnover leaf litter, this method is successful at improving and conserving soil quality. Because of the long roots of tree components that go deep into the soil, this provides a constant supply of organic material to the soil (Albrecht and Kandji 2003; Barreto et al. 2010). Biological properties can also be optimized in the soil under agroforestry system (Van Barkum and Eardly 2002; Udawatta et al. 2008; Yadav et al. 2010). The ameliorating effect of tree and organic matter inputs, and differences in litter quality and quantity and root exudates have been stated by several authors to increase soil microbial biomass and diversity in perennial systems. (Myers et al. 2001; Mungai et al. 2005; Sorensen and Sessitsch 2007). Several benefits of agroforestry practices to soil microorganisms have been identified, but little knowledge about hortipasture systems in rainfed situations is available. Because of its fallen leaves in March and April, aonla (*Emblica officinalis* Gaertn), guava (*Psidium guajava* L.), tamarind (*Tamarindus indica* L.), and bael (*Aegle marmelos* Corr.) are ideal for rainfed areas. Litter fall decomposed with the help of micro flora and build up soil health (Kumar et al. 2009). Long-term in situ moisture conservation measures in hortipastoral system improves carbon sequestration, soil structural ability to reduce soil loss (Ghosh et al. 2021b), and biological health of the degraded land (Ghosh et al. 2019). Hortipasture system with soil moisture conservation measures enhances the fruit and pasture yield showing the trend of sustainability even drought years, hence providing better benefit:cost ratio in semi-arid India (Kumar et al. 2019).

13.3 Modern Sustainable Technology

13.3.1 Precision Farming

Precision agriculture refers to the distinctive application of agricultural inputs based on soil, weather, and crop requirements in order to increase agricultural production, efficiency, and profitability. It is a modern agriculture activity that involves the use of technology in agriculture to improve productivity and profitability (Tomar et al. 2014; Pandey et al. 2015; Mandal et al. 2020; Rani et al. 2020), such as remote sensing, GPS, and Geographical Information System (GIS). It allows farmers to make more informed decision-making of crop inputs such as pesticides, fertilizers, tillage, and irrigation water. More efficient input use would result in increased crop yield and quality without polluting the environment, resulting in sustainable agriculture and production. In short, precision agriculture is all about farming in right way, at right place, and at right time. Remote sensing has been considered integral part of precision agriculture.

Remote sensing has been defined as science of retrieving information from objects not located in vicinity or located at a distance. These platforms can be ground based, airborne, or satellite based. Reflected and emitted radiation from a targeted area is studied for information extraction from behavior of electromagnetic spectrum. Application of remote sensing is suitable in the field of crop insurance, soil moisture retrieval, crop area estimation, yield estimation, crop canopy management (Singh et al. 2013; Bisht et al. 2014; Yadav et al. 2014).

To minimize the losses of impact of climate change and vagaries of monsoon on farming community, remote sensing can be used for ensuring the crop health assessment, and crop condition estimation for providing insurance benefits. Crop insurance is now being considered as important tool for ensuring livelihood opportunities of farmers. Soil moisture estimation through satellite-based sensors and their integration with local field conditions are helpful in providing agro-advisories to farmers. Knowledge of available soil moisture prepares the farmers for agriculture-related operations like sowing of seeds, application of fertilizers and pesticides in the standing crop, irrigation scheduling, etc. Normalized difference vegetation index, which is linked with the status of vegetation in the field (Tucker 1986; Jensen 2000; Hao et al. 2011), gives an idea about the standing crop, area sown under various category of crops in the field. The principle of remote sensing has been characterizing the spectral signature associated with the object, so each crop will have associated spectral signature. Multispectral remote sensing has been used for identification of color pattern by vegetation (Dakshinamurti et al. 1971). For achieving this objective supervised and unsupervised classification algorithms have been employed by the researchers (Panda and Hoogenboom 2009). Remote sensing has been employed to assess the area under particular crop for forecasting the yield. Area under crop can be identified with interlinking the ratio with crop condition. Leaf area index (LAI) has been important biophysical parameter for this purpose. FASAL (Forecasting Agricultural output using Space, Agro meteorological and Land based observations) has been launched to forecast the output of the crop by government of India. Reliable statistics regarding production of agricultural crops holds key for successful marketing of products (Deb et al. 2017). Remote sensing has been used for detection of biotic stress like pest, disease occurrence, and abiotic stress like drought, flood, salinity, etc. Spectral indices, multivariate modeling has been used by researchers for characterization of plant abiotic stress (Das et al. 2018). Machine learning techniques like partial least square, multiple adoptive regression spline, random forest has been used for analyzing plant biophysical parameters (Asner and Martin 2008; Darvishzadeh et al. 2008; Ramoelo et al. 2012; Li et al. 2014; Das et al. 2018).

13.3.2 Drones

Drones are remote-controlled aircraft with no man on board. Drones have been successfully used in fields like disaster management operations, military, etc. Precision farming combines the sensor data with raw data for mapping spatial variability of the field. Drones may perform soil health scans, crop health monitoring, irrigation schedule preparation, fertilizer application, yield estimation, and weather analysis. Multispectral and hyper spectral image analysis based on NDVI differentiates soil from grass, detects plant under stress and crop growth stages. Drone imaging and mapping can take care of herd of animals. The underlying principle for drone imaging is temperature comparison between animals.

Another area of application of drone technology in agricultural crop protection may be crop spraying. This technology with combination of variable rate technology will reduce the overall amount of pesticides and losses in the operations. Drone technology can be used for assessing soil condition with 3D mapping. This will help in modeling soil flow and managing soil quality. Drone-enabled NDVI index values analysis products aid in determining the precise harvesting timing.

In the field of agricultural drones, there is a considerable scope for expansion. With the advancement in technology, crop imaging will need to adopt as well. With the data that drones capture from the crops, the farmers are able to analyze their crops and make well-informed decisions on how to proceed given the accurate crop information. In this sector, software for analyzing and correcting crop production has the prospective to expand. Farmers would use a drone to fly over their crops, identifying an issue in a particular region and taking the appropriate steps to fix it. This allows the farmer to concentrate on the overall task of production rather than surveying their crops. Both the purchase and running costs of modern drones make them too expensive for small farms in developing nations.

13.4 Input and Process-Based Sustainable Technologies

13.4.1 Low Carbon Landscape

The two most important pillars for long-term agricultural growth are environmental sustainability and crop productivity. The theme of "low-carbon ecosystem" is suggested to fix them and to achieve a world with a low-carbon footprint. A low-carbon ecosystem is made up of materials and services that help to mitigate greenhouse gas emissions and reduce total carbon emissions.

Low-carbon landscape designs (forests, grasslands, and agricultural areas, for example) are important for preserving the balance between anthropogenic activities and natural ecosystems. As a result, better management of terrestrial ecosystems is needed to better cope with climate change. However, the idea of identifying and applying low-carbon ecosystems is still in its early stages, and the development of a low-carbon terrestrial ecosystem is still a work in progress. There are currently only a few studies that measure carbon emissions and absorption/accumulation in the terrestrial ecosystem. In addition, research into low-carbon landscape components and their social effects are very less. In order to address climate change, we plan to propose a new conceptual description of low-carbon landscape design. Theoretically, "low-carbon ecosystems" should include elements that minimize greenhouse gas emissions, mitigate climate change, and make a low-carbon farm a reality.

Based on the green house gas (GHG) emission minimization strategy, these components are primarily categorized as carbon loss reduction and carbon sequestration mechanism. Practices such as livestock and manure management (regionally suitable for ages, rotational grazing, high-quality feed, improved manure storage facilities, optimized manure application to soil), and soil conservation (crop rotations, reduced fallow, avoid open grazing), alternate land-use systems for degraded land (manure/compost application) and water management (need-based irrigation, rainwater harvesting) may help to minimize carbon emissions from terrestrial agro-ecosystems and encourage carbon sequestration. Within 30–50 years, an estimated 1–2 Gt C per year could be sequestered globally by implementing recommended management practices (Lal 2003). Carbon sequestration capacity is greatest in polluted soils, where C has been reduced to a large extent. On the other hand, if soils with high C content are already close to local C saturation, they may not be able to sequester large quantities of carbon. Preventing C losses in such soils should be a top priority. According to Sommer and Bossio (2014), the rate of carbon sequestration decreases over time, peaking after 20–40 years. SOC distribution is rather heterogeneous in India and is influenced by a variety of factors including soil type, land use, land use transition, climate, landscape, and soil management practices. As a result, soils in different agro-ecological regions have different carbon source and sink capacity, necessitating different management practices to ensure carbon flows that support soil and ecosystem services.

13.4.2 Conservation Agriculture

"Conservation agriculture", has been defined as a system having minimal soil disturbance along with permanent crop cover and crop rotation (FAO 2004). According to this definition, conservation agriculture aims at conserving and efficient utilization of the natural resources through integrated management. Agricultural operations like tillage affect the soil health by altering the soil structure. Heavy tillage breaks the macro aggregates into the micro aggregates (Angers et al. 1992), thereby pore size distribution and water holding capacity also get affected. This ultimately leads to poor infiltration and increased runoff (Ferreras et al. 2000). Conservation agriculture practices improve the soil moisture content (Saharawat et al. 2012), air filled porosity, aggregation, water use efficiency (McVay et al. 2006), water holding capacity and reduces soil bulk density (Kumar et al. 2018). Conservation agriculture improves the soil chemical properties like SOC, available N and DTPA extractable Zn. Greater residue addition as a part of conservation agriculture enhances the plant stored nitrogen (Thomas et al. 2007; Pheap et al. 2019), calcium (Chan et al. 1992), potassium (Bravo et al. 2007; Zhao et al. 2017), phosphorus (Bravo et al. 2007; Zhao et al. 2017), manganese (Rhoton 2000) and zinc (Rhoton 2000) magnesium (Chan et al. 1992) (Figure 13.1).

13.4.3 Organic Agriculture

According to Lockie (2006), organic foods are those that are produced without the use of growth hormones, pesticides, or artificial fertilizers. Natural food items are supposed to be free of pesticides, artificial fertilizers, and residue-free healthy products by the general public (Jolly et al. 1989; Huang 1996). Using organic farming practices to cultivate land provides significant environmental, social, and health benefits, including nutrient recycling and fixation, soil formation, improved erosion and flood control, and carbon sequestration. Under low-input agriculture, selected quality parameters of cereals, pluses, and oilseeds content were higher than in traditional agriculture. Organic agriculture works on four underlying principles: principle of health, ecology, fairness, and care. Principle of health

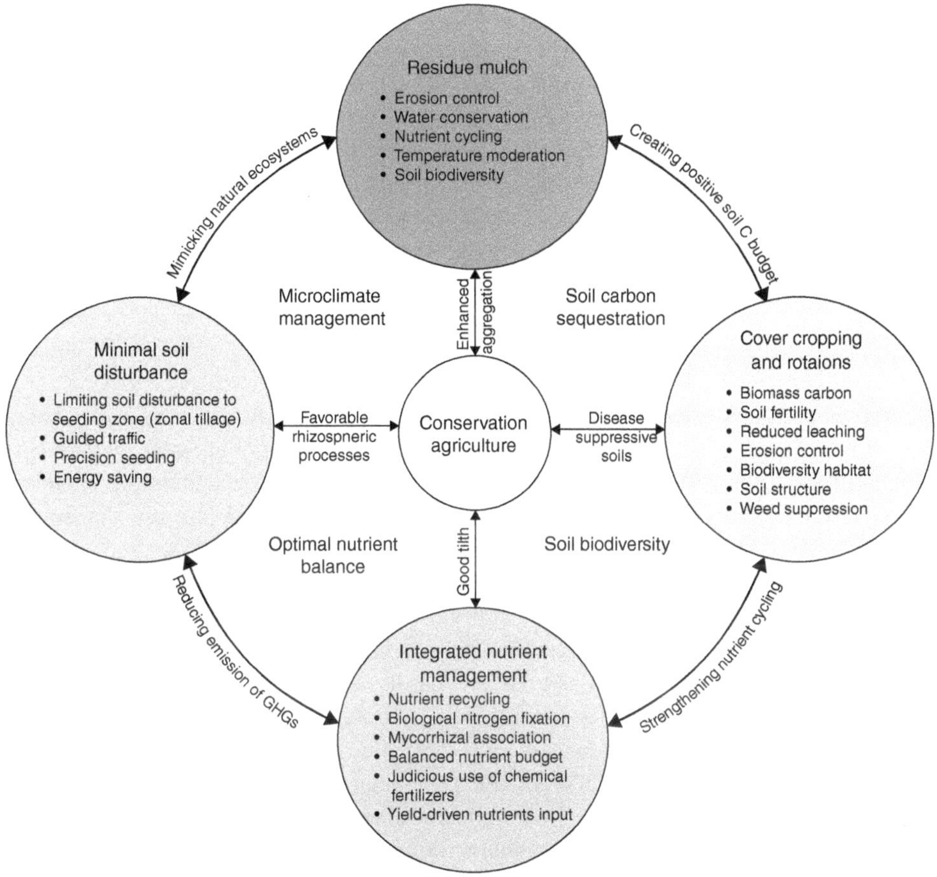

Figure 13.1 Four conceptual pillars of conservation agriculture practices.

underlies the integrity of whole ecosystem, which includes not only absence of disease but also maintenance of physical and biological health maintenance. Principle of ecology means that organic cultivation, pastoral, and wild harvesting systems can all work in harmony with natural cycles and ecological balances. These cycles are universal, but they operate differently depending on the location. Local environments, ecology, history, and scale must all be considered when implementing organic management. In order to preserve and enhance environmental quality, inputs should be minimized by reuse, recycling, and effective material and energy management. Fairness, respect, impartiality, and stewardship of the shared world characterize the fairness between people and in their interactions with other living beings. Principle of care underlies the fact that organic agriculture is a living, complex system that responds to both internal and external pressures. Organic agriculture practitioners may improve production and productivity, but this does not come at the expense of health and well-being. As a result, emerging technology must be evaluated, and current approaches must be re-examined.

Organic agriculture has risen nearly 29-fold in the last 17 years, from 42 000 ha under certified organic farming in 2003–2004. As of 31 March 2018, the total area registered under the National Programme for Organic Production (NPOP) for organic certification was 3.56 million hectares (2017–2018). This involves 1.78 million hectares (50%) of cultivable land and another 1.78 million hectares (50%) for wild harvest collection. India produced about 1.70 million MT of certified organic products in 2017–2018, which includes a wide range of food items such as oil seeds, sugar cane, cereals & millets, cotton, pulses, medicinal plants, tea, fruits, spices, dry fruits, vegetables, and coffee, among others. The organic agriculture is a holistic approach for enhancing ecosystem health including biodiversity and soil biological activity. Organic farming in long run is a way for improving nutritional quality and sustainable agricultural production.

13.4.4 Integrated Nutrient Management

When it comes to increasing agricultural production, the global food demands of a rising human population, as well as the need for an environmentally friendly strategy for a sustainable soil-plant-microbe-environmental framework, demand a lot of attention. Chemical fertilization is one way to increase crop production, but due to injudicious use of chemical input in agricultural systems, which has detonated the soil, food, climate, and human health, chemical fertilization is also rising their prices in the twenty-first century. Organic manures, on the other hand, cannot meet any of the crop's nutrient requirements. INM was created as a result of the above information being taken into account. The role of INM in resolving these problems is discussed here, as it has been suggested as a promising strategy for addressing these issues. INM has a wide range of applications for improving plant performance and resource utilization while also allowing for environmental and resource quality conservation. Advanced INM activities resulted in lower chemical fertilizer inputs and, as a result, lower human and environmental costs (such as intensity of land use, N use, reactive N losses, and GHG emissions) with no negative impact on crop yields. NM raises crop yields by 8–150% as compared with traditional methods, improves grain quality and soil health and sustainability, and increases water and nutrient usage efficiency and economic returns to farmers, according to a systematic literature review.

INM could be a creative and environmentally friendly method for sustainable agriculture around the world, according to strong and compelling evidence. Primarily, INM refers to integrating old and new nutrient management approaches into an ecologically sound and economically optimal farming system that takes advantage of all available organic, inorganic, and biological components-substrate sources.

The following are some of INM's main principles:

(a) Maximizing the use of soil nutrients to improve crop productivity and resource efficiency; as previously mentioned, the ultimate goal of INM is to maximize the use of soil nutrients to improve crop productivity and resource efficiency. (b) Spatially and temporally balancing soil nutrient supply with crop demand. The volume and timing of nutrient applications must be in accordance with the crop's nutrient requirements, according to INM, in order to achieve optimal yields and maximize nutrient use. N fertilizers applied in small amounts and on a regular basis during times of high crop demand have the ability to minimize N losses while rising crop yield and quality. (c) Reducing nitrogen losses but increasing crop yield. Major components of INM are (i) soil fertility restoring crops such as green manures, legumes, and other crops are integrated; (ii) recycling of crop residues; (iii) use of organic manures like FYM, phospho-compost vermicom post, poultry manure, biogas, slurry, biological composts, press mud cakes, compost; (iv) application of biological agent; (v) competent genotypes; (vi) stable use of fertilizer nutrients as per the requirement and target yields (Figure 13.2).

13.4.5 Regenerative Agriculture

Regenerative agriculture (RA) is a relatively new term that does not have a rigid definition rather it is a farming concept that is built upon the idea that "no-size-fits-all" and is a system-specific, holistic approach that is required to achieve sustainable development goals. The major emphasis is given to improving nutrient cycling through soil by improving soil organic matter (SOM) status and creating a positive carbon (C) budget using soil as a C sink in the terrestrial C pool assuring enhanced soil functions and minimal exploitation of natural resources. RA is primarily based on improving soil functionality, soil quality, or soil health and use RA as a tool to impart climate and economic resiliency into the agroecosystems.

As a result, RA is built on the premise of

1) Instead of indiscriminate chemical fertilizer inputs, soil fertility can be managed by increasing SOM content, biological N fixation, and nutrient recycling.
2) Rather than plowing, increasing biota behavior and species diversity (e.g. earthworms and microorganisms), and prolific plant roots improves soil structure.
3) Improving green water supply by conserving precipitation, reducing runoff and evaporation losses, moderating soil temperature, and promoting deep root systems.

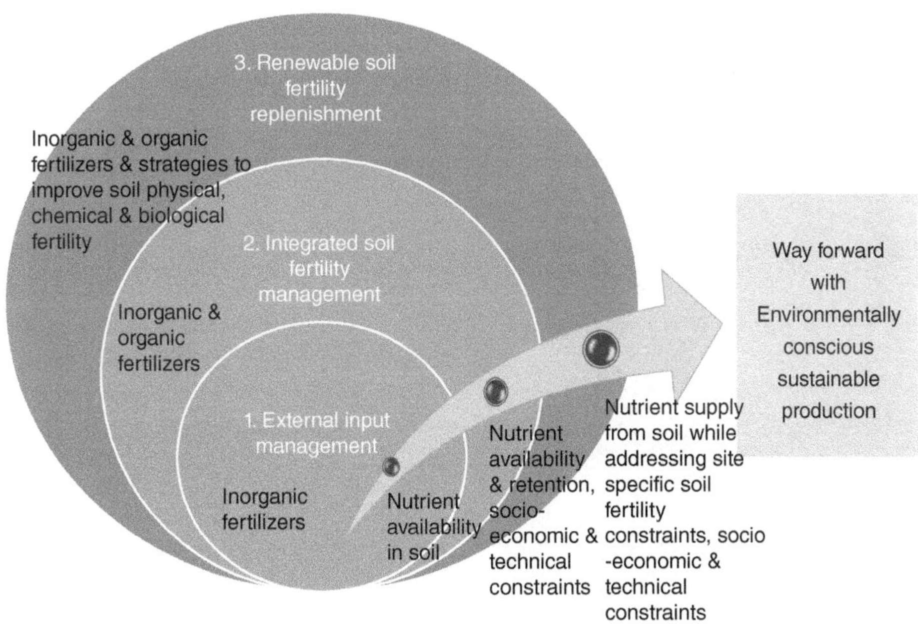

Figure 13.2 Concept, component, and pathways of sustainable production of INM.

4) Rather than using curative land shaping and engineering systems to control water and wind erosion, maintain a continuous groundcover, cover cropping, and conservation agriculture as preventative steps.
5) Using biofertilizers (e.g. compost, manure, mycorrhiza) rather than indiscriminate chemical disposal to manage soil acidification and elemental imbalance.
6) Improving water penetration rates by reducing crusting, compaction, hardening, and desiccation by retaining residual mulch, cover cropping, and bioturbation of the rhizosphere to create bio-pores (Figure 13.3).

13.5 Conclusion

Sustainable technology-led agriculture is need of the hour to enhance and maintain the ecosystem not only for present generation but also for future generation. Natural resources are facing altered kind of crisis due to overexploitation, hence strategy-based solutions are needed from scientific community. Technologies are in different level of development and

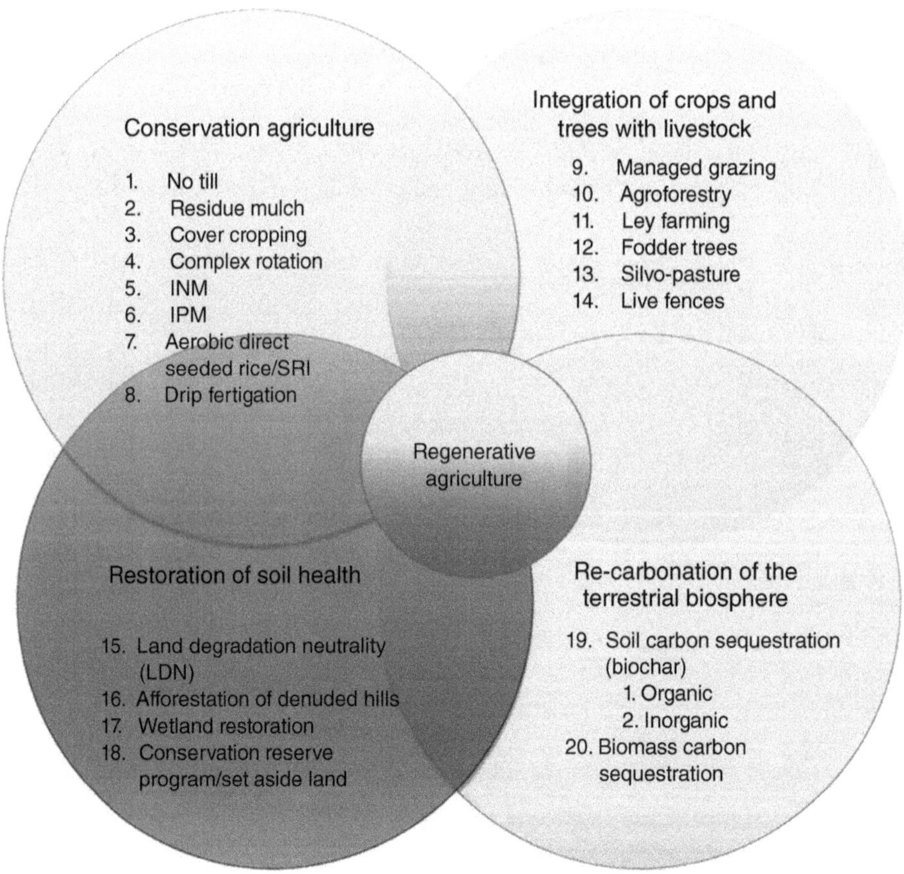

Figure 13.3 Basic views of regenerative agriculture.

evolving in due of course of time. These technologies need to be in farmers' field with appropriate adoption in their cost pattern and taking into consideration the local needs of farming community. Alternate land-use system underlies the vigilant decision-making for land resources while conservation agriculture, regenerative agriculture, and organic farming talks about holistic changes in processes for production. Precision farming along with remote sensing have better spatial coverage, which is helpful in crop area and yield assessment over the large area. Evolving over the time and lessons learnt from previous mistakes are two prerequisites for sustainable solutions, so we can assume for a better future ahead for next generations by adopting these options.

References

Albrecht, A. and Kandji, S.T. (2003). Carbon sequestration in tropical agroforestry systems. *Agriculture, Ecosystems and Environment 99* (1–3): 15–27.

Angers, D.A., Pesant, A., and Vigeux, J. (1992). Early cropping-induced changes in soil aggregation, organic matter, and microbial biomass. *Soil Science Society of America Journal*: 115–119.

Asner, G.P. and Martin, R.E. (2008). Spectral and chemical analysis of tropical forests: scaling from leaf to canopy levels. *Remote Sensing of Environment 112* (10): 3958–3970.

Barreto, P.A.R., Gama-Redrignes, F.F., Gama-Redrignes, A.C. et al. (2010). Distribution of oxidizable organic C fraction in soil under cacao agroforestry systems in Southern Bahia Brazil. *Agroforestry Systems 81* (3): 213–220.

Bhattacharyya, R., Ghosh, B., Mishra, P. et al. (2015). Soil degradation in India: challenges and potential solutions. *Sustainability 7* (4): 3528–3570.

Bisht, P., Kumar, P., Yadav, M. et al. (2014). Spatial dynamics for relative contribution of cropping pattern analysis on environment by integrating remote sensing and GIS. *International Journal of Plant Production* 8 (1): 1–17.

Bravo, C.A., Giraldez, J.V., Ordonez, R. et al. (2007). Long-term influence of conservation tillage on chemical properties of surface horizon and legume crops yield in a vertisol of southern Spain. *Soil Science 172* (2): 141–148.

Brundtland, G.H. (1987). *Our Common Future: Report of the World Commission on Environment and Development.* Geneva, UN-Dokument A/42/427: UN.

Chan, K.Y., Roberts, W.P., and Heenan, D.P. (1992). Organic carbon and associated properties of a red earth after 10 years rotation under different stubble and tillage practices. *Australian Journal of Soil Research 30* (1): 71–83.

Darvishzadeh, R., Skidmore, A., Schlerf, M. et al. (2008). LAI and chlorophyll estimation for a heterogeneous grassland using hyperspectral measurements. *ISPRS Journal of Photogrammetry and Remote Sensing* 63 (4): 409–426.

Das, B., Sahoo, R.N., Pargal, S. et al. (2017). Comparison of different uni- and multi-variate techniques for monitoring leaf water status as an indicator of water-deficit stress in wheat through spectroscopy. *Biosystems Engineering* 160: 69–83.

Das, B., Mahajan, G.R., and Singh, R. (2018). Hyperspectral remote sensing: use in detecting abiotic stresses in agriculture. In: *Advances in Crop Environment Interaction*, 317–335. Singapore: Springer.

Deb, D., Singh, A.K., and Kumar, S. (2017). Remote sensing techniques in horticulture. In: *Hi-Tech Horticulture Volume 7: Advance Techniques* (ed. S. Tyagi), 39–48. New Delhi: New India Publishing Agency.

Ferreras, L.A., Costa, J.L., Garcia, F.O., and Pecorari, C. (2000). Effect of no-tillage on some soil physical properties of a structural degraded Petrocalcic Paleudoll of the southern "Pampa" of Argentina. *Soil and Tillage Research 54* (1–2): 31–39.

Ghosh, A., Kumar, S., Manna, M.C. et al. (2019). Long-term in situ moisture conservation in horti-pasture system improves biological health of degraded land. *Journal of Environmental Management* 248: 109339.

Ghosh, A., Kumar, R.V., Manna, M.C. et al. (2021a). Eco-restoration of degraded lands through trees and grasses improves soil carbon sequestration and biological activity in tropical climates. *Ecological Engineering* 162: 106176.

Ghosh, A., Singh, A.K., Kumar, S. et al. (2021b). Do moisture conservation practices influence stability of soil organic carbon and structure? *CATENA* 199: 105127.

Huang, C.L. (1996). Consumer preferences and attitudes towards organically grown produce. *European Review of Agricultural Economics 23* (3): 331–342.

Jat, H.S., Datta, A., Sharma, P.C. et al. (2018). Assessing soil properties and nutrient availability under conservation agriculture practices in a reclaimed sodic soil in cereal-based systems of North-West India. *Archives of Agronomy and Soil Science 64* (4): 531–545.

Jolly, D.A., Schutz, G.H., Diaz-Knauf, K.V., and Johal, J. (1989). Organic foods: consumer attitudes and use. *Food Technology (Chicago) 43* (11): 60–65.

Khan, T.A. and Kumar, S. (2010). *Physical Hortipastoral Models, Their Management and Evaluation*. Jhansi: IGFRI.

Kumar, S. and Choubey, B.K. (2008). Performance of Aonla (*Emblica officinalis*) –based hortipastoral system in semi-arid region under rainfed situation. *Indian Journal of Agricultural Science 78* (9): 748–751.

Kumar, S., Kumar, S., and Chaubey, B.K. (2009). Aonla-based hortipastoral system for soil nutrient build up and profitability. *Annals of Arid Zone 48* (2): 153–157.

Kumar, V., Kumar, M., Singh, S.K., Jat, R.K., 2018. Impact of conservation agriculture on soil physical properties in rice-wheat system of eastern indo-gangetic plains.

Kumar, S., Singh, A.K., Singh, R. et al. (2019). Degraded land restoration ecological way through horti-pasture systems and soil moisture conservation to sustain productive economic viability. *Land Degradation & Development 30* (12): 1516–1529.

Lal, R. (2003). Global potential of soil carbon sequestration to mitigate the greenhouse effect. *Critical Reviews in Plant Sciences 22* (2): 151–184.

Li, F., Mistele, B., Hu, Y. et al. (2014). Reflectance estimation of canopy nitrogen content in winter wheat using optimised hyperspectral spectral indices and partial least squares regression. *European Journal of Agronomy* 52: 198–209.

Lockie, S. (2006). Capturing the sustainability agenda: Organic foods and media discourses on food scares, environment, genetic engineering, and health. *Agriculture and Human Values 23* (3): 313–323.

Mandal, V.P., Rehman, S., Ahmed, R. et al. (2020). Land suitability assessment for optimal cropping sequences in Katihar district of Bihar, India using GIS and AHP. *Spatial Information Research* 28 (5): 589–599.

McVay, K.A., Budde, J.A., Fabrizzi, K. et al. (2006). Management effects on soil physical properties in long-term tillage studies in Kansas. *Soil Science Society of America Journal 70* (2): 434–438.

Mungai, N.W., Motavalli, P.P., Kremer, R.J., and Nelson, K.A. (2005). Spatial of soil enzyme activities and microbial functional diversity in temperate alley cropping systems. *Biology and Fertility of Soils 42* (2): 129–136.

Myers, R.T., Zak, D.R., White, D.C., and Peacock, A. (2001). Landscape level patterns of microbial communities composition and substrate use in upland forest ecosystem. *Soil Science Society of America Journal 65* (2): 359–367.

Pandey, P.C., Mandal, V.P., Katiyar, S. et al. (2015). Geospatial approach to assess the impact of nutrients on rice equivalent yield using MODIS sensors'-based MOD13Q1-NDVI data. *IEEE Sensors Journal* 15 (11): 6108–6115.

Pheap, S., Lefevre, C., Thoumazeau, A. et al. (2019). Multi-functional assessment of soil health under conservation agriculture in Cambodia. *Soil and Tillage Research* 194: 104349.

Ramoelo, A., Skidmore, A.K., Cho, M.A. et al. (2012). Regional estimation of savanna grass nitrogen using the red-edge band of the spaceborne rapid eye sensor. *International Journal of Applied Earth Observation and Geoinformation* 19: 151–162.

Rani, M., Joshi, H., Kumar, K. et al. (2020). Climate change scenario of hydro-chemical analysis and mapping spatio-temporal changes in water chemistry of water springs in Kumaun Himalaya. *Environment, Development and Sustainability* 23: 4659–4674.

Rhoton, F.E. (2000). Influence of time on soil response to no-till practices. *Soil Science Society of America Journal* 64: 700–709.

Saharawat, Y.S., Ladha, J.K., Pathak, H. et al. (2012). Simulation of resource-conserving technologies on productivity, income and greenhouse gas emission in rice-wheat system. *Journal of Soil Science and Environmental Management* 3 (1): 9–22.

Sharma, S.K. (2005). Hortipastoral based land use systems for enhancing productivity of degraded lands under rainfed and partially irrigated conditions. *Uganda Journal of Agricultural Sciences* 9: 320–325.

Singh, K., Kumar, P., and Singh, B.K. (2013). An associative relational impact of water quality on crop yield: a comprehensive index analysis using LISS-III sensor. *IEEE Sensors Journal* 13 (12): 4912–4917.

Sommer, R. and Bossio, D. (2014). Dynamics and climate change mitigation potential of soil organic carbon sequestration. *Journal of Environmental Management* 144: 83–87.

Sorensen, J. and Sessitsch, A. (2007). Plant-associated bacteria lifestyle and molecular interactions. In: *Modern Soil Microbiology*, 2e (eds. J.D. van Elsas, J.K. Jansson and J.T. Trevors), 211–236. CRC Press.

Thomas, G.A., Dalal, R.C., and Standley, J. (2007). No-till effects on organic matter, pH, cation exchange capacity and nutrient distribution in a Luvisol in the semi-arid subtropics. *Soil and Tillage Research* 94: 295–304.

Tomar, V., Mandal, V.P., Srivastava, P. et al. (2014). Rice equivalent crop yield assessment using MODIS sensors' based MOD13A1-NDVI data. *IEEE Sensors Journal* 14 (10): 3599–3605.

Udawatta, R.P., Kremer, R.J., Adamson, B.W., and Anderson, S.H. (2008). Variations in soil aggregate stability and enzyme activities in a temperate agroforestry practice. *Applied Soil Ecology* 39: 153–160.

Van Barkum, P. and Eardly, B.D. (2002). The aquatic budding bacterium Blastobacter denitrificans is a nitrogen-fixing symbiont of *Aeschynomene indica*. In: *Nitrogen Fixation: Global Perspective* (eds. T.M. Finan, M.R. O'Brian, D.B. Layzell, et al.), –520. New York: CAB International.

Yadav, R.S., Yadav, B.L., Chippa, B.R. et al. (2010). Soil biological properties under different tree based traditional agro forestry systems in semi-arid regions of Rajasthan, India. *Agroforesty System* 81 (3): 195–202.

Yadav, M., Sharma, M.P., Prawasi, R. et al. (2014). Estimation of wheat/rice residue burning areas in major districts of Haryana, India, using remote sensing data. *Journal of the Indian Society of Remote Sensing* 42 (2): 343–352.

Zhao, X., Liu, S.-L., Pu, C. et al. (2017). Crop yields under no-till farming in China: a meta-analysis. *European Journal of Agronomy* 84: 67–75.

14

Geoinformatics, Artificial Intelligence, Sensor Technology, Big Data

Emerging Modern Tools for Sustainable Agriculture

Abhishek Singh[1], *Riya Mehrotra*[2], *Vishnu D. Rajput*[3], *Pavel Dmitriev*[4], *Anil Kumar Singh*[5], *Pradeep Kumar*[6], *Ram Sewak Tomar*[7], *Omkar Singh*[8], *and Awani Kumar Singh*[9]

[1] *Department of Agricultural Biotechnology, Sardar Vallabhbhai Patel University of Agriculture and Technology, Meerut, Uttar Pradesh, India*
[2] *Central Institute of Medicinal and Aromatic Crops, Lucknow, Uttar Pradesh, India*
[3] *Academy of Biology and Biotechnology, Southern Federal University, Rostov-on-Don, Russia*
[4] *The Botanical Garden, Southern Federal University, Rostov-on-Don, Russia*
[5] *University of Allahabad Senate House Campus, Allahabad, Uttar Pradesh, India*
[6] *Department of Forestry, North Eastern Regional Institute of Science Technology, Nirjuli, Arunachal Pradesh, India*
[7] *College of Horticulture and Forestry, Rani Lakshmi Bai Central Agricultural University, Jhansi, Uttar Pradesh, India*
[8] *Department of Soil Science & Agril. Chemistry, College of Agriculture, Sardar Vallabhbhai Patel University of Agriculture and Technology, Meerut, Uttar Pradesh, India*
[9] *Centre for Protected Cultivation and Technology, Indian Agricultural Research Institute, New Delhi, Delhi, India*

CONTENTS

Sustainable Agriculture Systems and Technologies, First Edition. Edited by Pavan Kumar, A.K. Pandey, Susheel Kumar Singh, S.S. Singh, and V.K. Singh.

14.1 Introduction

Agriculture has traditionally been considered as an intuitive practice in which knowledge is passed from one generation to another. However, nowadays the changing climate and lack of viable farming – has emerged as a major challenge. The United Nations has estimated that the global population will reach 9.8 billion by 2050, an increase of 2.2 billion from now (Singh et al. 2021). This means that due to increasing number of people, we need more crop production but unfortunately, rapid urbanization and climate change have generated pressure on agriculture. In the United States alone, urbanization and climate change have degraded the total area of farms to 899 million acres in 2018 from 913 million acres in 2014 (Singh et al. 2020). In the current technology-based era, the concept of big data has emerged, which is the bulk collection of various data including soil architecture, weather forecast, climate change, fertilizer recommendation, disease management, and crop mapping (Manyica et al. 2019). These data are extracted by means of several resources like Internet of Things (IoT) system softwares and web portals for useful information (Nidhi 2020). This information is also generated by Big data that is implemented by robots and some forms of artificial intelligence (AI) (Zambon et al. 2019). Traditionally, seasonal manpower is required to harvest crops and keep farms productive. However, a transition is noticed in society and people have moved from agricultural culture to cities to get good urban facilities which caused workforce shortage. One solution to facilitate this shortage of workers is agricultural robots integrating AI features. According to a Forbes study (Walch 2020), farm robots supplement the human labor workforce and can harvest crops at a higher capacity and faster pace than human-based agronomy (Walch 2020). Although there are numerous cases in which robots are not as compatible as humans, but agriculture industries are currently developing robotic systems to work in the field and help producers with tedious tasks (Rubio and Rovira-Más 2020). According to Reddy et al. (2016), the introduction of robots in agriculture drastically increased the productivity in several countries and reduced the farm operating costs. Also, robotic applications in agriculture are growing exponentially (Shamshiri et al. 2018). Robotic applications offers promising solutions for smart farming, like handling labor shortage and a long-time declining profitability; however, like most innovations, there exist limitations at the current early stages. These technologies are still too expensive for most farmers, especially those having small farms (Lamborelle and Fernández Álvarez 2019) because scale economics make small individual farms less profitable (Sonka 2014). Nevertheless, the cost of technology will decreases with time, and agricultural robots will surely be implemented in coming future as an alternative to bring high production (Rubio and Rovira-Más 2020). The world agricultural production and crop yields data declined in 2015. Robotic innovations are giving boost to the global agriculture and crop production market, as according to the Verified Market Intelligence report, agricultural robots will be capable of completing field tasks with greater efficiency as compared to the farmers (Verified Market Intelligence 2018). Agricultural tech startups have raised over 800 million dollars in the last five years (CBINSIGHTS 2019). Startups involving robotics and machine learning (ML) to work out agriculture procedures gained momentum in 2014, in line with rising interest in AI (Varadharajan 2019). In fact, the venture capital funding in AI has increased by 450% in the last five years (Murugesan et al. 2019). This kind of new agriculture pretends to do more with less, because nourishing people while increasing production sustainably and taking care of the environment will be crucial in the coming years. Advanced sensing technologies in agriculture help to meet the challenges in crop production; they provide detailed information on soil, crop status, and environmental conditions to allow precise application of phytosanitary products, resulting in improved water use efficiency, reduced usd of herbicides and pesticides and increased crop yield and quality (Zhang 2019).

14.2 Agriculture: Problems Worldwide and in India

Agriculture, an engine of economic growth for many nations, provides the most basic needs of humankind: food and fiber. Technological changes during the past century, such as the Green Revolution, have transformed the face of agriculture. The improved crop varieties, synthetic fertilizers, pesticides, and irrigation during the 1960s–1980s, known as the Green Revolution or third agricultural revolution, enhanced crop productivity and food security, especially in developing nations. Consequently, despite the doubling population and tripling food demand since the 1960s, global agriculture has been able to meet the demands with only a 30% expansion in the cultivated area. The demand for food and agricultural products is projected to further increase by more than 70% by 2050. Given the limited availability of arable land, a significant part of this increased demand will be met through agricultural intensification, i.e. increased use of fertilizers, pesticides, water, and other inputs. However, intensified use of agricultural inputs also causes environmental degradation, including groundwater depletion, reduced surface flows, and eutrophication. Excessive or inefficient use of natural resources (e.g. soil and water), fertilizers, and pesticides for agricultural production cause economic losses as well as increased water and nutrient losses from agriculture that lead to environmental degradation. For an economically and environmentally sustainable production system, there is a need to develop techniques that can increase crop production through increased efficiency of inputs use and reduced environmental losses (Sishodia et al. 2020).

14.3 GIS-Remote Sensing and Big Data in Smart Agriculture

Presently, agriculture can be considered to be going through a fourth revolution facilitated mainly by advances in information and communication technologies. Emerging technologies, such as remote sensing, global positioning systems (GPS), geographic information systems (GIS), IoT, Big Data analysis, AI development of unmanned aerial vehicles (UAV), and “small-sized spectral sensors” are promising tools being utilized to optimize agricultural operations and aimed to enhance production and reduce inputs and yield losses. Several IoTs technology systems utilizing cloud computing, wireless sensors networks, and big data analysis have been developed for smart farming operations such as automated wireless-controlled irrigation systems and intelligent disease and pest monitoring and forecasting systems (Zhao et al. 2010). GIS has been defined as “an organized collection of computer hardware, software, geographical data, and personnel designed to efficiently capture, store, update, analyze, manipulate, and display all forms of geographically referenced information” (Simoonga et al. 2009). GIS has been playing an important role in agriculture development around the globe. It helps farmers elevate crop production, reduce raw material cost, and manage land resources very efficiently through the integration of GIS technology. Agriculture mapping plays a crucial role in management of soil types and irrigation methodologies for any farm area. This helps farmers, especially in India control and manage agricultural resources. GIS allows clear understanding and interpretation of visual data through understanding of relationships and patterns for maps, globes, charts, and more. This information can be easily used by farmers through geographically referenced information or GIS, a computer-based tool that examines spatial relationships, patterns, and trends (Singh et al. 2013; Bisht et al. 2014; Tomar et al. 2014; Yadav et al. 2014; Pandey et al. 2015; Mandal et al. 2020; Rani et al. 2020). By connecting geography with data, GIS better understands data using a geographic context. Farmers use it for soil mapping, precision farming, and crop productivity (Figure 14.1).

14.4 Big Data and Agriculture

Big data and smart agriculture both are relatively a new concept of agriculture. Precision agriculture concept is extended through smart agriculture; based on big data information analysis, farmers take the decision and manage the situation as per information

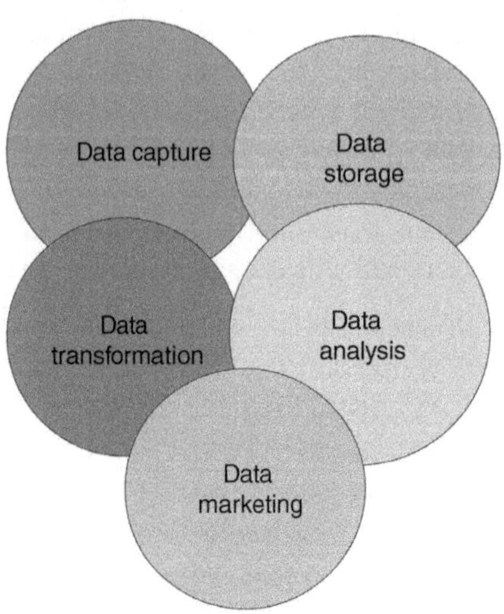

Figure 14.1 The picture depicts the steps involved; from primary data acquisition to data marketing and utilization.

provided by big data. The big data contains a key feature that is real-time assistance like suddenly changed operational conditions or other circumstances for example: weather or disease alert, through this real-time assistance crop management system(s) like weather alert system, crop sensor, pest spraying UAV carry out agile actions (Esmeijer et al. 2015). Big data assisting also have intelligent assistance which helps in the implementation, maintenance, and use of agriculture technology as well. The main role and application of big data in smart farming is to ensure in minimum cost gaining higher profit as well as sustainability. The use of AI, sensors, and smart machines in agriculture has brought agriculture to the top of the digital revolution in the modern era. Data in agriculture are a collection of various type of soil mapping containing information related to their physical and chemical properties, weather, past management practices, etc. Because of all this information, in any adverse weather or disease eruption situation(s), the farmers are warned in advance, due to which the farmers suffer less (Nidhi 2020). Big data for smart agriculture contain a broad level collection of good agriculture practices tools, these are following:

14.4.1 Digital Tools for Soil, Crop, Weeds Mapping

In digital soil and crop mapping, digital maps for soil types and their physio-chemical properties like bulk density, porosity, cationic exchange capacity (CEC), soil organic carbon (SOC), etc. are built. This technology helps in managing crops more productively using digital imagery or video images. In developed countries and some developing countries, farmers deal with commercial farming and some even practice agriculture on acres of land, it is almost complicated to get instant updates and alerts from such big planted lands without using technology. Digital soil and crop mapping strategy, tools, and techniques like GPS-based applications, help to generate accurate digital maps for various soil properties and hence are of great use to farmers community.

14.4.2 Weather Prediction Tools

Crop growth and development, yield and total agriculture production depend on weather. Besides India, in many other countries like Bangladesh, Pakistan, Japan, Korea, and China agriculture system depends on the weather conditions. In other words, agriculture system is weather-based in all countries. Weather aberrations can cause physical damage to crops and soil erosion. Immediate weather changes cause severe damage to crops. The quality of crops from agricultural land to market, their transportation and storage, all these economical procedures depend on the weather. Bad weather hampers all these aspects and cause economical losses. Agricultural weather forecasts give information about the low-pressure areas, cyclones, tornadoes, and depressions, wind speed and direction, relative humidity, max, min, and dew point temperatures, amount and type of coverage of sky by clouds, rainfall and snow, events like fog, frost, hail, thunderstorms, and wind squalls. Tools like thermometers, barometers, rain gauges, wind vanes, earth observation satellites, UAV (Drones), weather balloons, and automatic weather stations monitor agriculture weather forecasting elements.

14.4.3 Recommendation of Fertilizers for Agriculture Practices

Knowing the exact fertilizer rate for crop fields is a science and this science requires multiple factors analysis at nano level. These parameters are: crop nutrient uptake rate, soil chemical, physical and biological properties, weather and climatic factors, water composition and irrigation techniques, land type, fertilizer characteristics, interaction of fertilizers between crops, crop–crop interaction, etc. Miscalculated and excess use of fertilizers in agriculture fields accumulate toxic compounds in the soil and give rise to various pollutants. Big data tools enable researchers to advise the farmers with the right quantity of fertilizers for different parts of the field.

14.4.4 Pest Management and Disease Detection

In modern agriculture, with the help of big data tools, advance algorithms are developed which are used to identify the pattern and behavior of various microorganisms and pests, forecasting the invasion of pests and the spread of microscopic diseases can be monitored. Agricultural pests can quickly cut into a farmer's profit. Also, misusing and using higher amount of pesticides can have adverse effects on mankind, plants and other living things. UAV and crop sensors assist in pest control, mid-season crop health monitoring and reduced the use of pesticides.

14.4.5 Digital Tools for Analysis of Climate Change

Climate change due to global warming is a looming concern that affected the agriculture sector. A project of big data provided IoT sensors to Taiwanese farmers for rice production so they can collect information that is necessary about their crop. The collective information of IoT sensors will help farmers to optimize their production cycles even in adverse climatic conditions. The traditional farming is not able to analyze these climate changes and hence farmers practicing traditional techniques face huge economical losses, being the solution of all these concern; big data can revolutionize the future of farming.

14.4.6 Digital Automated Irrigation System

All the countries in the world are currently in a situation where they are required to use water in a very efficient manner. According to the recent studies, water availability is getting scarce day-by-day worldwide and more than one-third of the world population would face total water shortage by the year 2025. In agriculture too, the major problem which farmers face is the water scarcity, in order to improve the usage of water, the irrigation systems suggested are – drip irrigation which is implemented as automated irrigation system for small scale farms, another irrigation system is – automated irrigation system that utilizes weather predictions.

14.5 GIS-Remote Sensing in Agriculture

Agriculture plays a dominant role in the economies of both developed and developing countries. Satellite and airborne images and images from the UAV are used as mapping tools to classify crops, examine their health, viability, and to monitor farming practices. Agricultural applications of remote sensing include crop type classification, crop condition assessment, crop yield estimation, soil characteristic mapping, mapping of soil management practices, weed mapping, and compliance monitoring (farming practices).

Remote sensing is defined according to certain functions. It involves acquiring the information of an object/property by a device not in contact with the object under study. This involves the utilization of device at a distance for gathering information pertinent to the environment, such as measurements of force fields, electromagnetic radiation, or acoustic energy for aircraft, space crafts, or ships. Sun is the source of electromagnetic radiation that illuminate objects. The radiant energy of sun strikes the ground and is absorbed by objects on the earth's surface. The energy that is not scattered or absorbed is reflected back to the antennae; remote sensors; where data in the form of wavelengths is converted into useful information. The technique employs devices such as multispectral and hyperspectral cameras, lasers, radio frequency receivers, radar systems, sonars, seismographs, gravimeters, magnetometers, and scintillation counters (Figure 14.2). Some examples of remote sensing applications are given in the areas that have importance for the geographers. Remote sensing systems, using information and communication technologies, usually generate a large volume of spectral data due to high spatial/spectral/radiometric/temporal resolutions needed for application in smart agriculture. Emerging data processing techniques such as big sata analysis have been utilized to draw useful information

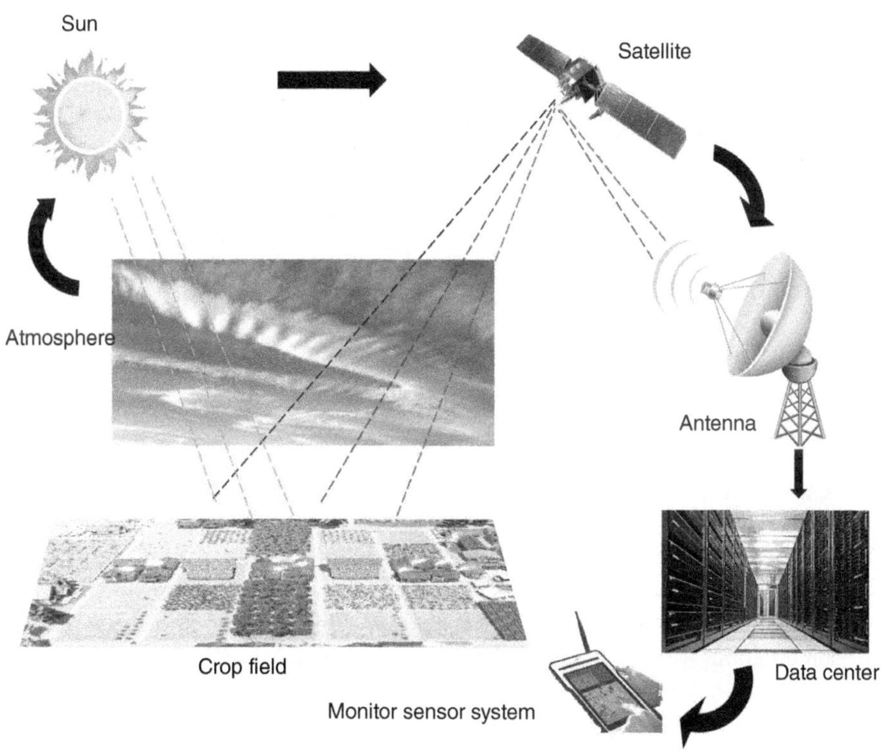

Figure 14.2 Process of data acquisition and its conversion into useful information.

from the large volume of data. Also, cloud computing systems have been used to store, process, and distribute/utilize such a large amount of data for applications in smart agriculture. All these advanced data acquisition and processing techniques have been applied globally, to aid the decision-making process for field crops, horticulture, pasture, and livestock. In the past, several studies have provided reviews of remote sensing techniques and applications in agriculture. While some studies focus on specific applications, such as soil properties estimation, evapotranspiration (ET) estimation, and disease and pest management, others include more than one area of applications (Sishodia et al. 2020). Remote sensing technology has been developing with new, high-performance sensors with higher spatial, spectral, and temporal resolutions. Agricultural remote sensing is a highly specialized field to generate images and spectral data in huge volume and extreme complexity to drive decisions for agricultural development. In the agricultural area, remote sensing is conducted for monitoring soil properties and crop stress for decision support in fertilization, irrigation, and pest management for crop production. Typical agricultural remote sensing systems include visible-NIR (visible-near infrared) (0.4–1.5 mm) sensors for plant vegetation studies, SWIR (short-wavelength infrared) (1.5–3 mm) sensors for plant moisture studies, TI (thermal infrared) (3–15 mm) sensors for crop field surface or crop canopy temperature studies, and microwave sensors for soil moisture studies (Mulla 2013). LiDAR (Light Detection and Ranging) and SAR (Synthetic Aperture Radar) have been enabled to measure vegetation structure over agricultural lands (Zhang and Kovacs 2012; Mulla 2013). Remote sensing is the cornerstone of modern precision agriculture to realize site specific crop field management to account for within-field variability of soil, plant stress, and effect of treatments. Remote sensing typically works on the bands of visible (0.4–0.7 mm), infrared (0.7–15 mm), and microwave (0.75–100 cm) in the electromagnetic spectrum. All the factors with geospatial distribution and data acquisition frequency result in remote sensing big data with huge volume and high complexity (Huang et al. 2018). Big data represents the information assets characterized by high volume, velocity, and variety to require specific technology and analytical methods for its transformation into value (De Mauro and Greco 2016). Currently, 13 operational satellites are in Sun synchronous orbit–RESOURCESAT-1, 2, 2A CARTOSAT-1, 2, 2A, 2B, OCEANSAT-2, RISAT-1 and 2, Megha-Tropiques, SARAL and SCATSAT-1, and four in Geostationary orbit- INSAT-3D, INSAT 3A & Kalpana, INSAT-3DR. Both big data and smart farming are relatively new concepts, so it is expected that knowledge about their applications and their implications for research and development is not widely spread.

14.6 Techniques and Tools Used in Big Data Analysis

Big data is highly dimensional and heterogeneous in nature, contains complex information, sophisticated tools, and techniques are needed to extract information from it (Mucherino et al. 2009; Vitolo et al. 2015). Big data is analyzed using various algorithms, single or combination of two different techniques methods, for example, ML image processing, remote sensing, cloud platforms, GIS, vegetation indices (V.I), and NDVI (normalized difference vegetation index) (Kamilaris et al. 2017).

14.6.1 Machine Learning (ML)

ML is a prevalent technology nowadays that can be used in agriculture for more sustainability. The use of ML in agriculture make it more profitable because of its micro-label management. ML also contains artificial ML which is very fast-growing areas in applied agricultural science. Artificial techniques are being used in the agricultural sector to increase the accuracy and to find the solutions to the problems.

14.6.1.1 Livestock Management

The livestock is divided into two parts, animal livestock and livestock production. Animal welfare deals with animal health, welfare, disease, and wellbeings. The main application of ML is monitoring for early detection of animal diseases and behavior. It also monitors the economically profit loss and production of animal-related goods.

1) **Livestock Production**
   In human civilization, domesticated livestock has played a crucial role and it became an integral part of society, human culture, and also an important part of local to the global economy. Domestic livestock has contributed to the rise of human societies and civilizations by increasing the amount of food and nutrition available to people in four ways: by providing sources of meat, milk, and fertilizer, and by pulling plows. Throughout history, livestock has also provided wool, leather, other raw materials, and transport. With the help of ML analysis, the accurate data prediction of rumen fermentation, suggested the diets, fatty acid percent in milk and milk production of animals (Craninx et al. 2008). ML model support vector machines (SVM) are useful for early detection and warning of problems in the commercial production of eggs, this helps the poultry industry to analyze data related to hen and egg production (Mohammadi et al. 2015). Machine-learning algorithms base method convolutional neural networks (CNNs) effectively applied in digital images for face recognition of the animal example pig and this method overcomes the problem of distressing activity for tagging of radio-frequency identification (RFID) which have a lot of limitations like low range and time-consuming (Hansen et al. 2018).
2) **Animal Welfare**
   Animal welfare is defined as "the relationship of humans with animals and the duty they have, to assure that the animals under their care are treated humanely and responsibly." Animals collar sensors with magnetometers and three-axis accelerometers data are collected and analyzed using ML modeling preceded the events such as the estrus and the recognition of dietary changes of animals (Dutta et al. 2015). ML acts on chewing patterns data with a combination of behavior data of calves like dietary supplements, such as hay and ryegrass; these data were collected by optical FBG sensors which were analyzed using ML techniques (Pegorini et al. 2015).

14.6.1.2 Water Management

In the agricultural system, water is a fundamental aspect of farming; this show the importance of water in farming, so it requires micro-management and plays a significant role in hydrological, climatological, and agronomical balance. ML is basically used in the estimation of daily, weekly, or monthly ET (Mohammadi et al. 2015). The process of evaporation is very complex for every plant, but after understanding this process, the irrigation system can be developed to provide water according to the requirement of each plant in crop field. This is also important in present scenario (twenty-first century), as according to the recent studies, due to the continuous use of groundwater more than one-third of the world population would face total water shortage by the year 2025. Temperature sensors provide data which is analyzed using ELM model of ML to estimate accurately, every week evapotranspiration in arid regions of India, this big data analyzing model helps in crop water management (Patil and Deka 2016).

14.6.1.3 Soil Management

Soil is the best media for growth and development of plants. The healthy growth of plants requires a healthy soil and ML application helps in predication identification and estimation of soil properties like soil drying, condition, temperature, and moisture content (MC). The earth's biodiversity, climate, and geographical distribution of lands makes the soil of each area a heterogeneous natural resource, with complex processes and mechanisms that are difficult to understand (Johann et al. 2016). Through ML with the combination of big data-related to soil, a method is developed for the provision of remote agricultural management decisions, which evaluated soil drying for agricultural planning. This smart method accurately evaluates the soil drying, with ET and precipitation data, in a region located in Urbana, IL of the United States (Coopersmith et al. 2014). Big data is also a master collection of soil condition data like composition, nutrient availability, mineral quantity, and type of soil, etc. Various modeling aspect of ML with combination soil data predicated soil organic carbon (OC), MC, and total nitrogen (TN), for this analysis used a visible–near infrared (VIS–NIR) spectrophotometer to collect soil spectra from 140 unprocessed and wet samples of the top layer of Luvisol soil types these unprocessed and wet samples were collected from an arable field in Premslin, Germany in August 2013, after the harvest of wheat crops (Morellos et al. 2016). Soil moisture also estimated using ML ANN (Artificial Neural Networks) model and data are obtained from force sensors (Johann et al. 2016).

14.6.2 Cloud Platforms

Cloud computing is an information technology (IT) paradigm through which users can access shared pools of configurable system resources over the internet. Cloud platforms together with big data sources (crop, weather and climate, soil, growth progress, and pattern) collected data need to be accumulated at a common platform that should be easily accessible for preprocessing, visualization, and analysis (Kamilaris et al. 2017). Cloud computing of agriculture is used to manage, to analyze the environmental factors, analyze the soil moisture, temperature, and manage the water supply functions (Murakami et al. 2013). Cloud system allows farmers to views farm or farm field information with ground sensors, devices connected, etc. Apart from this, system allows farmers to control the farm hardware remotely such as to switch on/off bulb, to switch on/off motors, etc. with the help of microcontroller (Balbudhe et al. 2015). Cloud system has following role in case of big data:

1) Cloud system stores all the agriculture-related information provided by big data sources in a centralized cloud, which will be available to all the users like a farmer, agricultural companies, etc. at anytime, anywhere.
2) Cloud system management of all big data related to land, location, area, soil and land characteristics through centralized decision support systems.

14.6.3 Vegetation Indices

It was identified that the spectral characteristics of plants have a general pattern, expressed in a decrease in reflectivity in the ranges of 0.45–0.47 and 0.68–0.69 μm, which corresponds to the absorption bands of chlorophyll a and b, while the reflection maxima correspond to green parts of the VNR spectrum and are determined by the nature of the ongoing physiological processes. UAV combined with a thermal camera and satellite sensors provide data for measuring wavelengths of light absorbed and reflected by green plants. In the biological system, every plants leaves contain a pigment system which strongly absorbs wavelengths of visible (red) light and on other hand they also strongly reflect wavelengths of near-infrared light, which is invisible to human eye. As a plant canopy changes from early spring growth to late-season maturity and senescence, these reflectance properties also change (Figure 14.3). VIs obtained from remote sensing-based canopies are effective algorithms for quantitative and qualitative evaluations of vigor, vegetation cover and growth dynamics, among other applications (Xue and Su 2017). VIs are of various types like NDVI, GDVI (Green Normalized Difference Vegetation Index) and SAVI (Soil Adjusted Vegetation Index) but NDVI is widely used to obtain and analyze the big data in the agriculture. In biological system, to determine the density of green plant on lands, it is necessary to observe the distinct colors (wavelengths) of visible and near-infrared sunlight reflected by the plants (Skakun et al. 2018).

When sunlight passes the prism, we can see different colors of light with different wavelengths called VIBGYOR (Violet–Indigo–Blue–Green–Yellow–Orange–Red) and when sunlight strikes objects, certain wavelengths of this spectrum are absorbed and other wavelengths are reflected. The pigment in plant leaves, chlorophyll, strongly absorbs visible light (from 0.4 to 0.7 μm) for use in photosynthesis (Figure 14.4). The cell structure of the leaves, on the other hand, strongly reflects near-infrared light (from 0.7 to 1.1 μm) (Daroya and Ramos 2017). The more leaves a plant has, the more these wavelengths of light are affected, respectively. NDVI values range from +1.0 to −1.0. Areas of barren rock, sand, or snow usually show very low NDVI values (e.g., 0.1 or less) (Mahajan and Raj 2016). Sparse vegetation such as shrubs and grasslands or senescing crops may result in moderate NDVI values (approximately 0.2–0.5). High-NDVI values (approximately 0.6–0.9) correspond to dense vegetation such as that found in temperate and tropical forests or crops at their peak growth stage (Shafi et al. 2019). NDVI is a key tool to obtain and analyze the big data related to crop health. Sensors like satellite, UAV

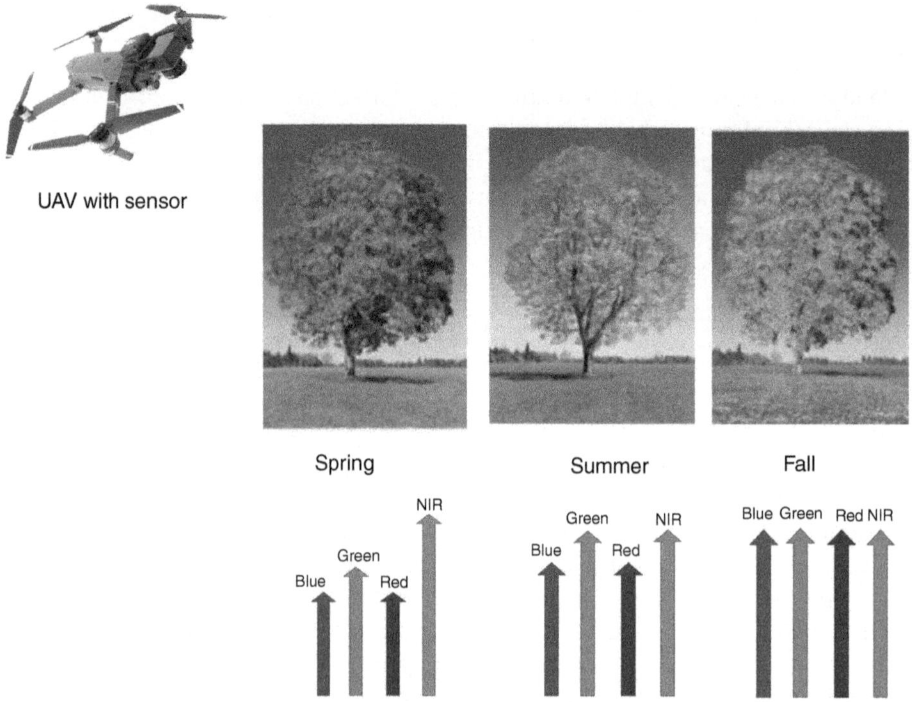

Figure 14.3 Green and healthy vegetation reflects a large portion of the near-infrared light in comparison to sparse vegetation that reflects more visible light. (*See insert for color representation of the figure.*)

collected data, are analyzed by applying a mathematical formula (Shafi et al. 2019) to quantify the density of plant growth on the Earth.

$$\text{NDVI} = \frac{(\text{NIR} - \text{VIS})}{(\text{NIR} + \text{VIS})}$$

$$\text{Normalized Difference Vegetation Index}(\text{NDVI}) = \frac{\text{Near} - \text{infrared radiation} - \text{Visible radiation}}{\text{Near} - \text{infrared radiation} + \text{Visible radiation}}$$

14.7 Role of Big Data in Agriculture Production Ecosystem: For Smart Farming

After considering all the things in the section of big data source and big data analysis sauce tool, it has been found that with the help of big data common farming will be converted into smart farming, as a result, farmers can earn good profits at low-cost wages (Wheeler and von Braun 2013). For production of the crop in a sustainable manner, it is required to utilize agricultural resources in precise way and within time for maximum resource

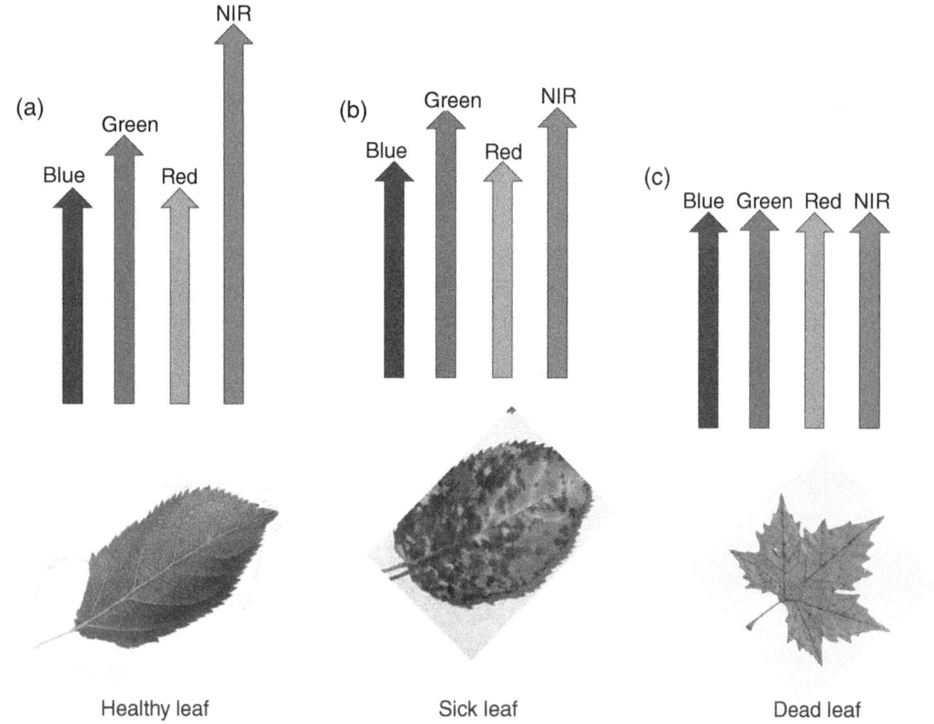

Figure 14.4 Reflection of light by healthy, sick, and dead leaves. Healthy leaf reflects more near infrared light. (a) Healthy leaf. (b) Sick leaf. (c) Dead leaf. (*See insert for color representation of the figure.*)

utilization (Fountas et al. 2015). Big data plays an important role in different aspects for smart farming like GPS for mapping, navigation, and IoT connected to remote sensors and monitoring system base autonomous driverless tractor (Reeve et al. 2011; Conesa-Muñoz et al. 2015). Smart autonomous machines and robotics base seedbed preparation to reseeding (Blackmore et al. 2005; Griepentrog et al. 2005), helicopters base smart planting from air to field is cost-effective for larger size of lands managed using GPS and IoT system (Pedersen et al. 2008; Scott 2010). Digital management of planting and sowing through automation and robotics machine monitor using IoT and cloud system (Buning 2010; Henten et al. 2013). Smart management of crop health through ML and AI, both integrated and applied to make it easy for farmers in detection of pest, weed management, and crop health through image processing like NDVI, along with all the information, UAV, RPA, UGV technology is also widely adopted in many smart farms for spraying of herbicides, pesticides, fertilizer, and weather broadcasting (Veroustraete 2015; Alimuzzaman 2016). Crop yield analysis with the help of crop mapping satellite, UAV, NDVI is a collection of data related to crop production which is stored in the cloud platform that estimates the yield of a specific location. (McBratney and Whelan 1999; Luck and Fulton 2015). Smart method of harvesting from field robotics-based efficient harvesting have an aim to analyze the maturity of crops and harvest it without damaging the grains. The robotic system has a sensor which analyses the ripening and maturing of fruit, vegetable, and grain crops before harvesting. After fruit and crop ripening and maturity check, crops are harvested using robotic systems. (Blackmore et al. 2005; Yamamoto et al. 2010; Hayashi et al. 2011).

14.8 Future Prospects

Recent satellite systems launched such as Ball Corporation's Quickbird have a four-band multispectral pushbroom sensor with a resolution of 0.8 m panchromatic and 4.5 m multispectral. With increase in applications of remote sensing future satellites will have better spatial and spectral resolutions. Launching more satellites will also improve temporal resolution. The delivery time of remote sensing data to the customer will improve. With advancement and development in remote sensing techniques, real-time satellite remote sensing systems can evolve. Research focusing cause of soil and crop variability is an area to explore instead of just measuring the variability. Technology transfer from research centers to commercial agribusiness industries with greater emphasis can be a boon to society and agriculture both. Presently, the available thermal cameras are mainly designed for nonagricultural uses, such as building and machinery inspection, they can also be used by farmers for improving irrigation scheduling under water limitations. A wide-spread use of thermography by farmers may provide incentive for manufacturers to develop modified cameras that are more appropriate for agricultural applications like routine irrigation management. Decision support systems can be worked upon and such systems can become the main link to convert the spatial data collected into detailed management recommendations at the farmer level. Decision support systems have the potential that will add the most value to remote sensing data for the farmer. By encouraging agricultural IT companies entering the marketplace, the cost of remote sensing data and other systems associated with precision agriculture are likely to come down in line with the benefits received.

14.9 Conclusion

Remote sensing techniques have additional applications in agriculture other than military applications. The agriculture industry is seen as one major market taking advantage and has potential to explore more from remote sensing applications and data providers. Remote sensing sensors collect data on energy reflected from the surface of plants and soil. Farm operators depend upon professional engineers and precision farming consultants to process the raw image data into useable information for making management decisions. There is an abundance of remote sensing technology available to measure variability in plants and soils. Also, there is a shortage of information about the causes of plant condition variability and the management solutions, needed to manage variability for improvement in crop production. The lack of knowledge to answer these variability questions is restricting the development of precision farming management decision support systems. Big data and smart farming have applications and data transformation technique as tools for crop yield prediction with high accuracies. Water stress indices can be estimated using radar data as well as optical data. Integration of various indices help in generation of information, for example: integrating satellite-based vegetation index (NDVI), land surface temperature

(LST), and rainfall (TRMM) data at an interval of every 16 days can provide effective operational means to monitor agricultural conditions over large areas. Preseason rainfall can be monitored and is significantly important to track the development of drought conditions. Time gap between rainfall and moisture condition can be removed prior to integration of the measurements. Findings reveal that the thermal images can be used by farmers to obtain detailed information about the spatial variability, in physical characteristics of the soils and potential impact of water stress. In addition, thermal images may offer useful information on the presence and distribution of abiotic, biotic, and stress factors.

Acknowledgments

The authors are thankful to the Ministry of Science and Higher Education of the Russian Federation for the financial support under project number 0852-2020-0029. The authors would also like to thank Dr. Rakesh Bhambri, Department of Geography, South Asia Institute, Heidelberg University, Germany for his valuable suggestion and correction.

References

Abhishek, S., Vishnu, R., Riya, M. et al. (2021). *Sustainable Soil Fertility Management*. Hauppauge, NY: *Nova Science Publishers*, Inc ISBN: 978-1-53619-055-7.

Alimuzzaman, M.D. (2016). *Agricultural Drone*. Department Of Computer Science and Technology University of Bedfordshire.

Balbudhe, K.S., Bulbule, A., Dhanve, N. et al. (2015). Cloud based cultivation management system. *ACSIJ Advances in Computer Science* 4 (3): 15.

Barrett, M.A., Humblet, O., Hiatt, R.A., and Adler, N.E. (2013). Big data and disease prevention: from quantified self to quantified communities. *Big Data 1* (3): 168–175. https://doi.org/10.1089/big.2013.0027.

Bechar, A. and Vigneault, C. (2016). Agricultural robots for field operations: concepts and components. *Biosystems Engineering* 149: 94–111. https://doi.org/10.1016/j.biosystemseng.2016.06.014.

Bechar, A. and Vigneault, C. (2017). Agricultural robots for field operations. Part 2: operations and systems. *Biosystems Engineering* 153: 110–128. https://doi.org/10.1016/j.biosystemseng.2016.11.004.

Becker-Reshef, I., Justice, C., Sullivan, M. et al. (2010). Monitoring global croplands with coarse resolution earth observations: the Global Agriculture Monitoring (GLAM) project. *Remote Sensing* 2 (6): 1589–1609. https://doi.org/10.3390/rs2061589.

Bisht, P., Kumar, P., Yadav, M. et al. (2014). Spatial dynamics for relative contribution of cropping pattern analysis on environment by integrating remote sensing and GIS. *International Journal of Plant Production* 8 (1): 1–17.

Blackmore S, Stout BA, Wang M, Runov B (2005), Robotic agriculture. https://www.semanticscholar.org/paper/Robotic-agriculture-the-future-of-agricultural-Blackmore-Stout/18d1d1c18fa3e58efc85c6d81496030bf54d749d

Buning, E.A. (2010). Electric drives in agricultural machinery-approach from the tractor side. *Journal of Agricultural Engineering* 47 (3): 30–35.

CBINSIGHTS (2019). Ag Tech deal activity more than triples. https://www.cbinsights.com/research/agriculture-farm-tech-startup-funding-trends (accessed 18 February 2019).

Chedad, A., Moshou, D., Aerts, J. et al. (2001). AP—animal production technology: recognition system for pig cough based on probabilistic neural networks. *Journal of Agricultural Engineering Research* 79 (4): 449–457. https://doi.org/10.1006/jaer.2001.0719.

Conesa-Muñoz, J., Gonzalez-de-Soto, M., Gonzalez-de-Santos, P., and Ribeiro, A. (2015). Distributed multi-level supervision to effectively monitor the operations of a fleet of autonomous vehicles in agricultural tasks. *Sensors* 15: 5402–5428.

Coopersmith, E.J., Minsker, B.S., Wenzel, C.E., and Gilmore, B.J. (2014). Machine learning assessments of soil drying for agricultural planning. *Computers and Electronics in Agriculture* 104: 93–104.

Corke, P.I., Roberts, J.M., and Winstanley, G.J. (1998). Robotics for the mining industry. *Autonomous Robotic Systems*: 163–181. https://doi.org/10.1007/bfb0030805.

Craninx, M., Fievez, V., Vlaeminck, B., and De Baets, B. (2008). Artificial neural network models of the rumen fermentation pattern in dairy cattle. *Computers and Electronics in Agriculture* 60: 226–238.

Daroya, R. and Ramos, M. (2017). NDVI image extraction of an agricultural land using an autonomous quadcopter with a filter-modified camera. *Proceedings of the 2017 7th IEEE International Conference on Control System, Computing and Engineering (ICCSCE)*. pp. 110–114.

De Mauro and Greco, G. (2016). A formal definition of big data based on its essential features. *Library Review* 65 (3): 122–135.

Dutta, R., Smith, D., Rawnsley, R. et al. (2015). Dynamic cattle behavioural classification using supervised ensemble classifiers. *Computers and Electronics in Agriculture* 111: 18–28.

Esmeijer, J., Bakker, T., Ooms, M., and Kotterink, B. (2015). Data-driven innovation in agriculture, case study for the OECD KBC2-programme. *TNO report TNO 2015 R10154*.

Griepentrog, H.W., Nrremark, M., Nielsen, H., and Blackmore, B.S. (2005). Seed mapping of sugar beet. *Precision Agriculture* 6: 157–165.

Gutiérrez, P. and García-Palomares, J.C. (2008). Distance-measure impacts on the calculation of transport service areas using GIS. *Environment and Planning B: Planning and Design 35* (3): 480–503. https://doi.org/10.1068/b33043.

Hansen, M.F., Smith, M.L., Smith, L.N. et al. (2018). Towards on-farm pig face recognition using convolutional neural networks. *Computers in Industry* 98: 145–152.

Hayashi, S., Yamamoto, S., Shigematsu, K. et al. (2011). Performance of movable-type harvesting robot for strawberries. *ActaHortic* 893: 317–324.

Johann, A.L., de Araújo, A.G., Delalibera, H.C., and Hirakawa, A.R. (2016). Soil moisture modeling based on stochastic behavior of forces on a no-till chisel opener. *Computers and Electronics in Agriculture* 121: 420–428.

Kamilaris, A., Kartakoullis, A., and Prenafeta Boldú, F.X. (2017). A review on the practice of big data analysis in agriculture. *Computers and Electronics in Agriculture 143*: 23–37. https://doi.org/10.1016/j.compag.2017.09.037.

Kempenaar, A., Westerink, J., Van Lierop, M. et al. (2016). "Design makes you understand"— mapping the contributions of designing to regional planning and development. *Landscape and Urban Planning 149*: 20–30. https://doi.org/10.1016/j.landurbplan.2016.01.002.

Khanal, S., Fulton, J., and Shearer, S. (2017). An overview of current and potential applications of thermal remote sensing in precision agriculture. *Computers and Electronics in Agriculture* 139: 22–32. https://doi.org/10.1016/j.compag.2017.05.001.

Lamborelle, A. and Fernández Álvarez, L. (2019). Farming 4.0: The future of agriculture. http://www.euractiv.com/section/agriculture/food/infographic/farming-4-0-the-future-of-agriculture (accessed 21 November 2019).

Lucas, M.T. and Chhajed, D. (2004). Applications of location analysis in agriculture: a survey. *Journal of the Operational Research Society* 55 (6): 561–578. https://doi.org/10.1057/palgrave.jors.2601731.

Luck, J.D. and Fulton, J.P. (2015). *Improving Yield Map Quality by Reducing Errors Through Yield Data File Post-Processing*, 9. Institute of Agriculture and Natural Resources.

Mahajan, U. and Raj, B. (2016). Drones for Normalized Difference Vegetation Index (NDVI), to estimate crop health for precision agriculture: a cheaper alternative for spatial satellite sensors. *Proceedings of the International Conference on Innovative Research in Agriculture, Food Science, Forestry, Horticulture, Aquaculture, Animal Sciences, Biodiversity, Ecological Sciences and Climate Change (AFHABEC-2016).*

Mandal, V.P., Rehman, S., Ahmed, R. et al. (2020). Land suitability assessment for optimal cropping sequences in Katihar district of Bihar, India using GIS and AHP. *Spatial Information Research* 28 (5): 589–599.

Manyica, J., Chui, M., Brown, B. et al. (2019). A big data: The next frontier for innovation, competition, and productivity | McKinsey. https://www.mckinsey.com/business-functions/mckinsey-digital/our-insights/big-data-the-next-frontier-for-innovation (accessed 21 November 2019).

McBratney, A.B. and Whelan, B.M. (1999). *The Null Hypothesis of Precision Agriculture.* Sheffield, UK: Sheffield Academic Press.

Mohammadi, K., Shamshirband, S., Motamedi, S. et al. (2015). Extreme learning machine based prediction of daily dew point temperature. *Computers and Electronics in Agriculture* 117: 214–225.

Morellos, A., Pantazi, X.E., Moshou, D. et al. (2016). Machine learning based prediction of soil total nitrogen, organic carbon and moisture content by using VIS-NIR spectroscopy. *Biosystems Engineering* 152: 104–116.

Mucherino, A., Papajorgji, P., and Pardalos, P.M. (2009). A survey of data mining techniques applied to agriculture. *Operational Research - An International Journal* 9: 121–140.

Mulla, D.J. (2013). Twenty-five years of remote sensing in precision agriculture: key advances and remaining knowledge gaps. *Biosystems Engineering* 114: 358–371.

Murakami, Y., Utomo, S.K.T., Hosono, K. et al. (2013). iFarm: Development of cloud-based system of cultivation management for precision agriculture. *IEEE 2nd Global Conference on Consumer Electronics (GCCE).*

Murugesan, R., Sudarsanam, S.K., Malathi, G. et al. (2019). Artificial Intelligence and Agriculture 5.0. *International Journal of Engineering Inventions (IJRTE)* 8: 8.

Nidhi (2020). Big data for smart agriculture. *Smart Village Technology, Modeling and Optimization in Science and Technologies* 17: 181–189.

Pandey, P.C., Mandal, V.P., Katiyar, S. et al. (2015). Geospatial approach to assess the impact of nutrients on Rice equivalent yield using MODIS sensors'-based MOD13Q1-NDVI data. *IEEE Sensors Journal* 15 (11): 6108–6115.

Patil, A.P. and Deka, P.C. (2016). An extreme learning machine approach for modeling evapotranspiration using extrinsic inputs. *Computers and Electronics in Agriculture* 121: 385–392.

Pedersen, S.M., Fountas, S., and Blackmore, S. (2008). Agricultural robots-applications and economic perspectives, service robot applications. *IntechOpen* 21: 369–382.

Pegorini, V., Karam, L.Z., Pitta, C.S.R. et al. (2015). in vivo pattern classification of ingestive behavior in ruminants using FBG sensors and machine learning. *Sensors* 15: 28456–28471.

Rani, M., Joshi, H., Kumar, K. et al. (2020). Climate change scenario of hydro-chemical analysis and mapping spatio-temporal changes in water chemistry of water springs in Kumaun Himalaya. *Environment, Development and Sustainability* 23 (3): 1–16.

Reddy, N., Reddy, A., and Kumar, J. (2016). A critical review on agricultural robots. *International Journal of Engineering Inventions (IJMET)* 7: 6.

Reeve, D.R., Eizad, Z., and Ramm, A.F. (2011). Method for decomposing task e.g. crop spraying task,to be performed on agricultural field bytractor-puller sprayer vehicle assembly, involves decomposing top-order layer based on rules to form bottom-order layer.

Rubio, V.S. and Rovira-Más, F. (2020). From smart farming towards agriculture 5.0: a review on crop data management. *Agronomy* 10: 2–21.

Scott, J. (2010). *Aerial Seeding of Cover Crops*. Des Moines: U.S. Department of Agriculture, Natural Resources Conservation Service (NRCS), Iowa State Office.

Shafi, U., Mumtaz, R., García-Nieto, J. et al. (2019). Precision agriculture techniques and practices: from considerations to applications. *Sensors* 19: 2–25.

Shamshiri, R.R., Weltzien, C., Hameed, I.A. et al. (2018). Research and development in agricultural robotics: a perspective of digital farming. *International Journal of Agricultural and Biological Engineering* 11: 1–14.

Simoonga, C., Utzinger, J., Brooker, S. et al. (2009). Remote sensing, geographical information system and spatial analysis for schistosomisis epidemiology and ecology in Africa. *Parasitology* 136 (13): 1683–1693.

Singh, K., Kumar, P., and Singh, B.K. (2013). An associative relational impact of water quality on crop yield: a comprehensive index analysis using LISS-III sensor. *IEEE Sensors Journal* 13 (12): 4912–4917.

Singh, A., Vishnu, R.D., Alok, S.K. et al. (2020). Transformation techniques and their role in crop improvements: a global scenario of GM crops. In: *Policy Issues in Genetically Modified Crops* (eds. P. Singh, A. Borthakur, A.A. Singh, et al.). Elsvier ISBN: 978012820780242.

Sishodia, R.P., Ray, R.L., and Singh, S.K. (2020). Applications of remote sensing in precision agriculture. *Remote Sensing* 12: 3136.

Skakun, S., Justice, C.O., Vermote, E., and Roger, J.C. (2018). Transitioning from MODIS to VIIRS: an analysis of inter-consistency of NDVI data sets for agricultural monitoring. *International Journal of Remote Sensing* 39: 971–992.

Sonka, S. (2014). Big data and the Ag sector: more than lots of numbers. *International Food and Agribusiness Management Review* 17: 1–20.

Tomar, V., Mandal, V.P., Srivastava, P. et al. (2014). Rice equivalent crop yield assessment using MODIS sensors' based MOD13A1-NDVI data. *IEEE Sensors Journal* 14 (10): 3599–3605.

Van Henten, E.J., Bac, C.W., Hemming, J., and Edan, Y. (2013). Robotics in protected cultivation. *IFAC Proceedings Volumes*. https://doi.org/10.3182/20130828-2-SF-3019.00070.

Varadharajan, D. (2019). AI, robotics, and the future of precision agriculture. https://www.cbinsights.com/research/ai-robotics-agriculture-tech-startups-future (accessed 21 November 2019).

Verified Market Intelligence (2018). *Global Agriculture Robots. Market Size, Status and Forecast to 2025*, 1–79. Boonton, NJ: Verified Market Intelligence Inc.

Veroustraete, F. (2015). The rise of the drones in agriculture. *EC Agriculture* 2: 325–327.

Vitolo, C., Elkhatib, Y., Reusser, D. et al. (2015). Web technologies for environmental big data. *Environmental Modelling & Software* 63: 185–198.

Walch, K. (2020). How AI is transforming agriculture. https://www.forbes.com/sites/cognitiveworld/2019/07/05/how-ai-is-transforming-agriculture (accessed 1 January 2020).

Wheeler, T. and von Braun, J. (2013). Climate change impacts on global food security. *Science* 341 (80): 508–513.

Xue, J. and Su, B. (2017). Significant remote sensing vegetation indices: a review of developments and applications. *Journal of Sensors* 2017: 1353691.

Yadav, M., Sharma, M.P., Prawasi, R. et al. (2014). Estimation of wheat/rice residue burning areas in major districts of Haryana, India, using remote sensing data. *Journal of the Indian Society of Remote Sensing* 42 (2): 343–352.

Yamamoto, S., Hayashi, S., Saito, S. et al. (2010). Development of robotic strawberry harvester to approach target fruit from hanging bench side. *IFAC Proceedings* 43: 95–100.

Zhang, Y. (2019). The role of precision agriculture. *Resource* 19: 9.

Zhang, C. and Kovacs, J.M. (2012). The application of small unmanned aerial systems for precision agriculture. *Precision Agriculture* 13: 693–712.

Zhao, J.C., Zhang, J.F., Feng, Y., and Guo, J.X. (2010). The study and application of the IOT technology in agriculture. *3rd International Conference on Computer Science and Information Technology*. pp. 462–465.

15

Investigation of the Relationship Between NDVI Index, Soil Moisture, and Precipitation Data Using Satellite Images

Shilan Felegari[1], Alireza Sharifi[2], Kamran Moravej[1], Ahmad Golchin[1], and Aqil Tariq[3]

[1] *Department of Soil Science, Faculty of Agriculture, University of Zanjan, Zanjan, Iran*
[2] *Department of Surveying Engineering, Faculty of Civil Engineering, Shahid Rajaee Teacher Training University, Tehran, Iran*
[3] *State Key Laboratory of Information Engineering in Surveying Mapping and Remote Sensing (LIESMARS), Wuhan University, Wuhan, China*

CONTENTS

15.1 Introduction

Some phenomena and effects of the land surface, such as vegetation, have changed over time due to natural or human factors that affect the condition and function of the ecosystem, so the need for detection, forecasting. And monitoring such changes in an ecosystem is of great importance (Mohajane et al. 2017). Remote sensing can estimate various parameters of earth sciences at the level of extensive coverage and an acceptable cost. One of these variables is soil moisture, which many remote sensing experts consider the combination of reflective and thermal data to be useful for obtaining information on vegetation and monitoring soil moisture (Alencar et al. 2020). Soil moisture is one of the basic parameters of the environment that directly affect the life of plants, animals, and microorganisms and

Sustainable Agriculture Systems and Technologies, First Edition. Edited by Pavan Kumar, A.K. Pandey, Susheel Kumar Singh, S.S. Singh, and V.K. Singh.

plays a major role in energy exchanges between air and soil, therefore its spatial distribution and temporal changes. are other very important components in climate and ecology models on a global, regional, and local scale (Bajoccoa et al. 2018). Soil moisture is the amount of water that exists in the space between soil particles. What is known as soil moisture is the amount of water in the soil to a depth of 10 cm, which is called soil surface moisture. Remote sensing measurements have been able to correlate with soil surface layer moisture, in this way we can understand the regional and local changes in soil surface layer moisture (Alencar et al. 2020).

One of the most important indices used to show vegetation status is Normalized Vegetation Difference Index (NDVI). The NDVI index is widely used in the study of spatio-temporal changes in vegetation. The numerical value of this index varies between −1 and +1, negative NDVI values mean the absence of vegetation on the ground. In the study of this index, two near and infrared bands are used. Differences in the spectral reflectance of red and near-infrared reflective bands, which are a function of vegetation status, constitute this indicator (Pandey et al. 2015; Bayle et al. 2019; Mandal et al. 2020; Rani et al. 2020). Due to the close relationship between vegetation status and available soil moisture, this index has received more attention for soil moisture (Blasi and Biondi 2017). The direct effects of climatic conditions on biomass and phenological patterns of vegetation have been estimated using NDVI and expressed in many ecosystems (Singh et al. 2013; Bisht et al. 2014; Tomar et al. 2014; Yadav et al. 2014). Many factors such as plant structure, interactions with canopy, plant height, species composition, plant health and vigor, leaf characteristics and plant stress, topography, and height are effective on this index (Huete 2004).

Huang et al. (2019) investigated the relationship between vegetation cover and soil percentage without vegetation cover in a watershed with vegetation indices and other remote sensing analyzes. In this study, the NDVI index was introduced as the best index in terms of correlation with vegetation. Knowing the greenness of plants at each stage of growth has important results in terms of how to provide that plant and the optimal management of the study area. Since this index is directly correlated with vegetation production, so many positive applications of this index for ecological purposes have been expressed (Chen et al. 2015).

The role of soil moisture in the root zone is widely recognized as a key parameter in meteorology, hydrology, and agriculture. Adequate knowledge of soil moisture is necessary to understand and predict the interactions between climate, and land surface processes (Koster et al. 2000). The term soil moisture generally refers to temporary storage of precipitation at depths above 1–2 m from the soil profile, although only a small percentage of precipitation is stored after evapotranspiration, surface runoff, and deep penetration into the soil. But storing the same amount of soil moisture is critical to maintaining agriculture, rangeland, and forest cover (Narasimhan et al. 2005). Moisture study in the root development zone for knowledge of water balance in arid areas and watersheds for water management planning seems necessary (Eklundh and Jonsson 2017).

Soil moisture is one of the basic soil parameters that is widely used in soil and water studies and water resources management (Behyar 2014). Soil moisture significantly affects climate, plant production, growth, soil hydrology, and ecology (carbon and nitrogen dynamics and gas emission tracking) (Robock et al. 2000). However, despite the undeniable importance of soil moisture in environmental studies due to the difficulty in continuous measurements in terms of spatial and temporal and costly and time-consuming field measurements, so far this feature is widely used in models. Biologically, exploitation is not necessary (Eea 2017). Soil moisture has spatial and temporal changes due to heterogeneity in soil properties, land cover, topography, and uneven distribution of precipitation and evapotranspiration (Narasimhan et al. 2005).

The NDVI profile acts conservatively in response to water stress. In other words, after water stress, the vegetation remains green for a while but changes rapidly with water stress against surface temperature (Goetz 1997). In recent years, climate change has had a major impact on biological systems. These changes and their effects have a profound effect on the fauna and flora of vital ecosystems today, so the effort to identify and optimally manage ecosystems is one of the most important priorities in environmental issues and their management (Gorelick et al. 2017). Vegetation indices are widely used due to their ease of calculation. The main purpose of spectral vegetation indices is to increase the information contained in spectral reflection data and to minimize the geometric effects of soil, atmosphere, and sun. NDVI is the most used in this field. The purpose of this study is to investigate the relationship between changes in NDVI index in this region with humidity data (SMAP satellite images) and precipitation (TRMM sensor images) in Gilan province over three years (from 2016 to 2018).

15.2 Methodology

15.2.1 Study Area

Gilan province with geographical coordinates of 36° 34′ to 38° 27′ north latitude and between 48° 34′ and 50° 36′ east longitude and with an area of 13 952 m^2, constitutes 0.8% of the area of Iran. The average annual rainfall in Gilan province, excluding a small area in the south, is 1402 mm, which is much higher than the national average (255) mm. From the west to the east of the province and also from the north to the south, the amount of rainfall decreases. Seasonal distribution of rainfall in the province shows that autumn receives the highest annual rainfall with a total of 42% and spring receives the lowest with a total annual rainfall of 15%. The relative humidity of the province is between 40 and 100% and its average temperature is 17.5 °C. The southernmost point of Gilan province differs from its northernmost part by almost two degrees of latitude. In meteorology, this value has little effect on temperature differences. But the temperature conditions in this small range show many differences. The main reason for this temperature difference is the role of Alborz and Talesh heights with different altitude levels.

Gilan province in terms of topography has two parts almost equal and different from each other. Its flat part with the name of Gilan plain and also with the names of Talesh and Astara plains covers the northwest, east, and center of the province and generally includes the coasts and alluvial areas of the rivers leading to the sea. The rugged and mountainous part of Gilan also includes a series of mountains that stretch from the Astara River valley in the north of the province to the mountains in the east of the province. These roughnesses are the result of orogenic movements in the late third geological period. Proper rainfall has long led to the emergence of good and extensive vegetation throughout Gilan province so that one of the natural features of this province throughout history has been the abundance of diverse forests and woodlands. However, the growth of urbanization and the indiscriminate cutting of forest trees in recent decades have led to the destruction of some forest areas in the province; however, about one-third of Gilan province is still covered with forest, which is estimated at 511 306 ha, and also has an important share in agricultural production in the country and the production of rice and nearly 100% of tea and more than half of the country's silk cocoons are produced in this province.

To conduct this research, precipitation diagrams and surface soil moisture under the title of three-year time series changes for Gilan province were obtained using coding in Google Earth Engine system as diagrams, and to estimate the NDVI index (normalized vegetation difference) was selected as a pilot part of the forest areas of Gilan province. The Google Earth Engine system is able to calculate the NDVI in areas with a maximum of 10 million pixels (Figure 15.1).

15.2.2 Precipitation, Soil Moisture, and NDVI Data

Soil moisture is divided into two parts: surface moisture (moisture up to a depth of 10 cm) and root moisture (moisture up to a depth of 2 m). Given the background of the penetration of electromagnetic waves by satellite sensors into the soil, what is measured as soil moisture in remote sensing is soil moisture to a depth of 10 cm (Entekhabi et al. 2014).

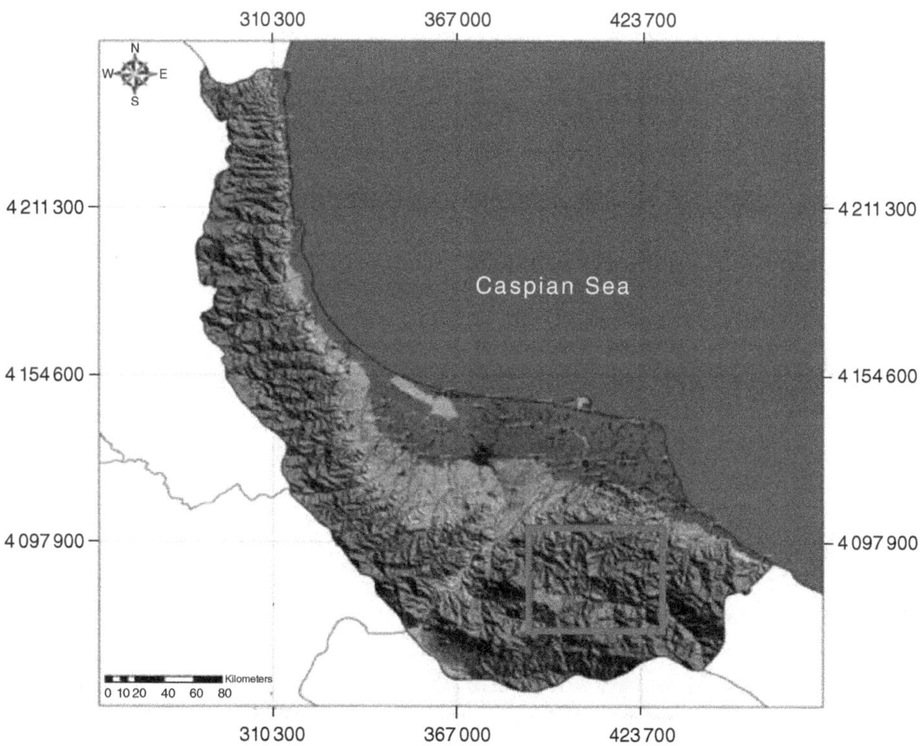

Figure 15.1 Location of study area. (*See insert for color representation of the figure.*)

The NDVI, or Normalized Vegetation Difference Index, is an indicator of the severity and weakness of vegetation status that is calculated from the red band (Rρ) and infrared (NIRρ) data using the following equation:

$$\text{NDVI} = \frac{\rho\text{NIR} - \rho R}{\rho\text{NIR} + \rho R}$$

In different sensors, the bands related to red and infrared wavelengths are different (Huang et al. 2019).

Given that this study aims to investigate the relationship between soil moisture and NDVI, the relevant data from SMAP satellite for soil moisture, TRMM satellite monthly data to assess the monthly rainfall of the region, and Landsat sensor data for NDVI index was used. The three-year data used in this study was obtained using programming in the Google Earth Engine system.

15.2.3 Introducing the Google Earth Engine System

Google Earth Engine is an application system for processing satellite images and a powerful remote sensing tool for extracting useful information from satellite images. Using this system, all areas, coverings, and land uses can be viewed and analyzed in two dimensions and three dimensions with high spatial details. Using this system, various spectral processes can be performed on different surface phenomena with different satellite data. The system is capable of carrying out projects from a local to a global scale with a spatial resolution of 10 to several kilometers (https://earthengine.google.com).

15.2.4 Dataset

15.2.4.1 Landsat-8 Sensor

Landsat 8 is used in the range of visible light, near-infrared, short-wave infrared, and thermal infrared. The Landsat 8 captures about 400 images/day, which is a significant increase from the 250 images/day on the Landsat-7 satellite. Operational Imaging Tool (OLI) and TIRS sensors have improved the signal-to-noise ratio (SNR) in radiometric performance, and as a result, this 12-bit quantization of data through these additional bits allows a better description of ground coverage. Give. Landsat OLI on Landsat 8 improves the sensors used in previous Landsat. Compared to the Whiskbroom sensors used in previous Landsat satellites, the OLI tool uses the Pushbroom sensor. With more than 7000 detectors per spectral band, the Pushbroom design makes it more sensitive, has fewer moving parts, and improves ground surface information (www.nasa.gov).

15.2.4.2 SMAP Sensor

The satellite, which consists of two active and inactive radar and radiometer sensors, was launched by NASA on 31 January 2015, and orbits the Earth in a solar orbit at an altitude of 680 km above the Earth's surface. Provides a complete ground cover of two to three days. The purpose of using these two sensors was to use a combination of radar and radiometer data to calculate the surface soil moisture. Radar and radiometer data of this satellite have a spatial resolution of 3 and 36 km, respectively. Radar data, despite their higher spatial resolution due to their high sensitivity to surface roughness and backscattering, are practically not suitable for calculating soil moisture. In contrast, radiometer data do not have the problem of surface roughness but have a low spatial resolution. Using these two sensors simultaneously, NASA provided the basis for the use of integrated algorithms to calculate soil moisture at a depth of 5 cm. The satellite crosses the equator every day at 6 a.m. and 6 p.m. and has a bandwidth of about 1000 km. It is also able to monitor changes in soil moisture up to two inches above the ground. This satellite has a spatial resolution of about 6 mi (9 km) and a time of three days (Karthikeyan et al. 2017).

15.2.4.3 TRMM Sensor

The TRMM satellite was launched on 28 November 1997, in collaboration with the US and Japanese Space Agency. Precipitation sensors on the TRMM satellite include precipitation radar, TRMM microwave imager, and visible and infrared scanners. The satellite's products are classified into three levels, called climate products. TRMM climate products are currently available on a three-hour time scale (TRMM 3B42-RT), daily (TRMM 3B42V7), and monthly (TRMM 3B43) with a spatial resolution of 0.25 × 0.25 degrees for the tropics and offerings. Middle-earth (50°S to 50°N) are available from 1998 to 2015 (https://pmm.nasa.gov/trmm). In this study, the monthly rainfall crop (TRMM 3B43) was obtained from January 2016 to December 2018. Also, the visual interpretation method was used, to be able to establish the relationship between soil moisture and vegetation with the help of the obtained change diagram.

15.3 Results and Discussion

Figure 15.2 shows the amount of monthly rainfall in three consecutive years from 2016 to 2018. Chart data show the amount of rainfall per month in millimeters per hour. The highest amount of rainfall in Gilan province according to TRMM sensor data in 2016 in September (equivalent to 10 September to 8 October) with a value of 0.137 mm/h and in

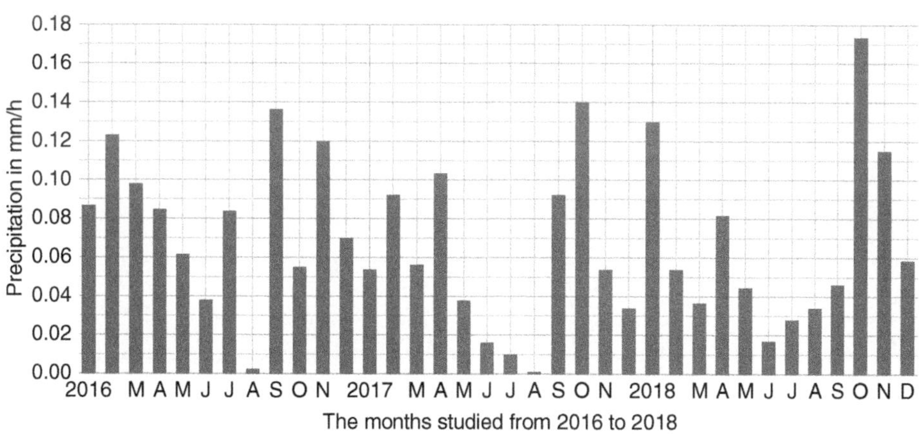

Figure 15.2 Rainfall diagram.

2017 and 2018 in October (equivalent from 9 October to 9 November) were recorded with the values of 0.141 and 0.174 mm/h, respectively.

The highest rainfall for 2016 was recorded in winter with an average of 0.1 mm/h and the lowest rainfall in spring with an average of 0.06 mm/h. The highest rainfall for 2017 was reported in autumn with an average of 0.08 mm/h and the lowest rainfall in summer with an average of 0.034 mm/h and the highest rainfall for the year was reported. 2018 was measured in autumn with an average of 0.1 mm/h and the lowest rainfall in summer with an average of 0.03 mm/h. The results of the study of rainfall in the selected three years in this study showed the highest rainfall in 2016 with an average of 0.08 mm/h and the lowest rainfall in 2017 with an average of 0.06 mm/h. The estimation of rainfall in the Caspian region can be due to the complexity of topography and the special geographical location of this region of Iran. Although the accuracy of TRMM satellite data is not high for some areas it can be considered moderate, but the spatial distribution of their accuracy is very similar to station data (Martínez-Fernández et al. 2016).

Figure 15.3 shows the soil surface moisture in three consecutive years from 2016 to 2018. The highest amount of soil surface moisture in Gilan province according to SMAP sensor data in 2016 in December (equivalent to 10 December to 10 January) with 145 mm, in 2017 in February (equivalent to 12 February to 9 March) With a value of 153.3 mm and in 2018 in January (equivalent to 11 January to 11 February) was recorded with a value of 148.6 mm. The lowest amount of soil surface moisture in Gilan province according to SMAP sensor data in 2016 in August (equivalent to 10 August to 9 September) with a value of 74 mm and in 2017 and 2018 in September (equivalent to 10 September to 8 October) was registered with the value of 72 mm.

In the visual interpretation of the two diagrams above, it can be seen that the trend of changes in precipitation and soil surface moisture is consistent. The peak of soil surface moisture figure coincides with the peak of precipitation figure in each month of the years studied in the present study. In 2016, with an average amount of 0.08 mm/h and the lowest amount of rainfall in 2017 with an average amount of 0.06 mm/h, the highest amount of surface moisture in 2016 with an average amount of 111.47 mm and the lowest amount of soil surface moisture was recorded with an average value of 108.65 mm for 2017. Soil

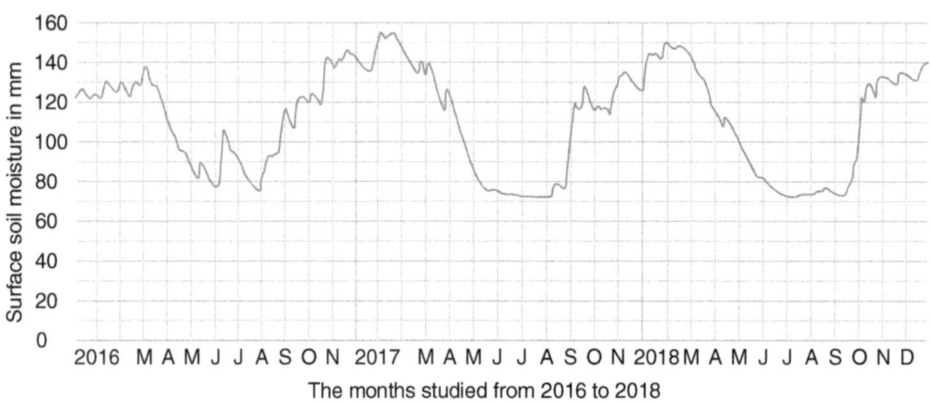

Figure 15.3 Soil surface moisture diagram.

moisture is a variable that varies in time and place, which has increased the sensitivity of monitoring these changes. Therefore, it is necessary to know the status and scope of these changes in proportion to the rainfall in the region (Colliander et al. 2017).

Figure 15.4 shows the NDVI changes over three consecutive years from 2016 to 2018. The highest amount of NDVI in Gilan province according to Landsat sensor data in 2016 in June (equivalent to June 11 to July 9) and August (equivalent to August 10 to September 9) with a value of 0.751 and in 2017 and 2018 in May (equivalent to May 11 to June 10) was recorded with values of 0.78 and 0.79, respectively. The lowest value of NDVI in Gilan province according to Landsat sensor data in 2016 in May (equivalent to 11 May to 10 June) with a value of 0.005, in 2017 in November (equivalent to 1 Novemberto29 November) With a value of 0.027 and in 2018 in February (equivalent to 12 February to 9 March) was recorded with a value of 0.029. The highest NDVI value was recorded in 2017 with an average value of 0.441 and the lowest NDVI value in 2016 with an average value of 0.392.

Comparison of Figures 15.2 and 15.3 shows that the relationship between NDVI and soil surface moisture is reversed, when the moisture index is decreasing, the NDVI diagram is ascending, and then when it is ascending, the NDVI diagram is descending. The reason for this is the delayed response of vegetation to changes in soil moisture. Plant profiles such as NDVI have a delayed response to soil moisture, and most of the time soil moisture data and other meteorological characteristics do not correlate strongly with this index in a short period. This time waste depends on the type of land cover and the method of water supply required by plants in the region (Wang et al. 2001). Comparison of average annual NDVI data with annual data of soil surface moisture and rainfall also confirms this, the average annual NDVI had the highest value in 2017, while in 2017 the lowest level of surface moisture soil and rainfall were reported but in the previous year (2016) the amount of soil surface moisture and rainfall was significant. It can be concluded that if we have a rainy year, a positive or negative effect on the NDVI index with a delay stage (in this study based on one year) is applied to this index.

These results are consistent with the results of Khazaei et al. (2017) who studied the estimation of soil surface moisture using vegetation indices using MODIS sensor images. They stated that the trend of changes in NDVI index with moisture in the sampling period

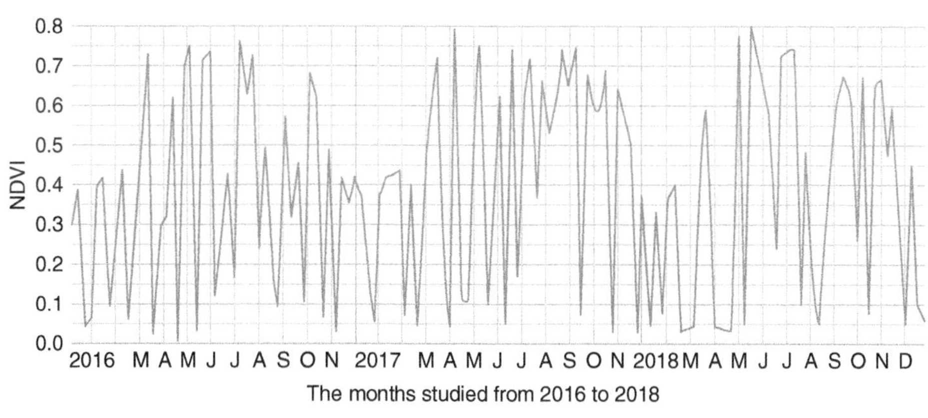

Figure 15.4 NDVI diagram.

from October 1992 to June 1993 was such that at the beginning of the sampling period when vegetation was weak, soil moisture was stored and NDVI with a delay period relative to the maximum humidity has reached its maximum value. In the final period, due to the harvest and the decrease in rainfall, the vegetation has decreased and as a result, the NDVI index has decreased.

Huete (2004) reviewed studies examining the relationship between soil and vegetation patterns in the southern United States using the AVHR Sensor NDVI Index, although there was a strong correlation between NDVI and soil-like characteristics. Percentage of alkalinity, acidity, water holding capacity in the soil, and bulk density were not reported by climatic factors after soil grouping. A better interpretation of the relationship between NDVI, rainfall, and soil moisture is obtained by subtracting the NDVI index, which is related to soil brightness or luminosity. Thus, NDVI values are related to the status and amount of NDVI vegetation.

Mccoll et al. (2017) in a study studied the relationship between soil moisture obtained from field measurements and delayed and simultaneous NDVI from MODIS sensor in the growing season of rangeland plants by linear regression and the maximum correlation of $R = 0.56$ was observed with a delay of 15 days. Al Bitar et al. (2017) also examined the direct relationship between plant index and soil moisture, and both confirmed the existence of a delay in the effect of moisture content on NDVI plant index (Beaudoing and Rodell 2020). Kerr (2016) in the study of vegetation changes under the influence of rainfall changes concluded that the trend of vegetation changes in dry years with an average rainfall below 200 mm and lack of soil moisture supply decreased by about 50%. Latham et al. (2019) stated that vegetation and soil surface temperature has a complex relationship with soil moisture. Lorenz et al. (2017) showed that the combination of NDVI and surface temperature can be used to estimate soil moisture with acceptable accuracy.

In a study by Gruber et al. (2019), study on the relationship between NDVI and global rainfall and temperature. After reviewing monthly data, they concluded that in areas where the wet season appears suddenly, vegetation grows immediately after the first rainfall occurs. In cold regions, plant growth is controlled by temperature and in warm regions, this feature is controlled by both precipitation and temperature. In other words, warm regions play a lesser role in the seasonal pattern of vegetation because in these regions there is not much need for increased temperature or rainfall.

Researchers have shown that NDVI also has a delayed response to soil moisture due to the plant's delayed response to rainfall due to soil moisture storage. The NDVI delay time to rainfall in forests and agricultural lands was two to three months. The greater the dependence on rainfall, the shorter the latency (Wang et al. 2001; Adegoke and Carleton 2002). In a study by Wang (2005) in the semi-arid region of New Mexico during the growing season (May–August) at depths of 10–20 cm soil profile with increasing latency (more than 30 and 50 days, respectively), the correlation coefficient between NDVI and soil moisture decreased, and at greater soil depths, the opposite occurred. Satellite sensor data is available continuously and extensively, providing extensive information on environmental phenomena throughout space compared to point locations such as meteorological stations (Kogan 2000). This feature makes satellite imagery efficient for moisture monitoring. The Normalized Differential Cover Index (NDVI) (since the early 1980s) has been widely used to monitor and assess vegetation cover and soil moisture availability (Townshend 1994; Tucker 1996).

15.4 Conclusion

Today, in many parts of the world, satellite data with high spatial and temporal resolution are used to know the rainfall conditions of areas without statistics, especially for mountainous and desert areas. Ensuring the accuracy of network and telemetry data versus stationary data is one of the most important conditions for the use of network and satellite data in various climatic and hydrological researches. Soil moisture is one of the issues that plants are directly affected by and combining and interpreting field results with satellite images provides users with valuable results. It is suggested that to better understand the relationship between humidity and satellite profiles, it is better to conduct such surveys with different sensors with higher spatial and temporal accuracy, as well as for different land uses. The results of this study showed that plant indices such as NDVI have a delayed response to soil moisture and in most cases, soil moisture data and other meteorological characteristics have a strong correlation with these indices in a short period, also, the trend of changes in precipitation and soil surface moisture was consistent.

References

Adegoke, J.O. and Carleton, A.M. (2002). Relations between soil moisture and satellite vegetation indices in the U.S. corn belt. *Journal of Hydrometeorology* 3: 395–405.

Al Bitar, A., Mialon, A., Kerr, Y.H. et al. (2017). The global SMOS level 3 daily soil moisture and brightness temperature maps. *Earth System Science Data* 9 (1): 293–315.

Alencar, A., Shimbo, J., Lenti, F. et al. (2020). Mapping three decades of changes in the Brazilian savanna native vegetation using landsat data processed in the Google Earth engine platform. *Remote Sensing* 12: 924–945.

Bajoccoa, S., Ferrarab, C., Aliverninib, A. et al. (2018). Remotely-sensed phenology of Italian forests: going beyond the species. *International Journal of Applied Earth* 74: 314.

Bayle, A., Carlson, B.Z., Thierion, V. et al. (2019). Improved mapping of mountain shrublands using the sentinel-2 red-edge band. *Remote Sensing* 11: 207–220.

Beaudoing, H. and Rodell, M.. (2020). NASA/GSFC/HSL, GLDAS Noah Land Surface Model L4 3 Hourly 0.25 × 0.25 Degree V2.1. Goddard Earth Sciences Data and Information Services Center (GES DISC), Maryland, USA. https://doi.org/10.5067 E7TYRXPJKWOQ. Accessed: [14 June 2019].

Behyar, M.B. (2014). Evaluation of soil moisture in Isfahan province by E-AMS method. *Journal of Geographical Research* 29 (1): 1–8.

Bisht, P., Kumar, P., Yadav, M. et al. (2014). Spatial dynamics for relative contribution of cropping pattern analysis on environment by integrating remote sensing and GIS. *International Journal of Plant Production* 8 (1): 1–17.

Blasi, C. and Biondi, E. (2017). La flora in Italia. Ministerodell'Ambiente e dellaTutela delTerritorio e del Mare. Brivio, P., Lechi, G., Zilioli, E., 2006. Principi e metodi di Telerilevamento, Citt' aStudied.

Chen, Y., Liu, T., Tian, X. et al. (2015). Effects of plastic film combined with straw mulch on grain yield and water use efficiency of winter wheat in Loess Plateau. *Filed Crops Research* 175: 53–58.

Colliander, A., Jackson, T.J., Bindlish, R. et al. (2017). Validation of SMAP surface soil moisture products with core validation sites. *Remote Sensing of Environment* 191: 215–231.

Eea (2017). Copernicus land monitoring service - high resolution layer forest. Product specifications document 1–38.

Eklundh, L. and Jonsson, P. (2017). *Timesat with Seasonal Trend Decomposition and Parallel Processing*. Software Manual. Lund University.

Entekhabi, D., Yueh, S., O'Neill, P.E., Kellogg, K.H., Allen, A., Bindlish, R., Brown, M., Chan, S., Colliander, A., Crow, W.T., Das, N., 2014. SMAP Handbook–Soil Moisture Active Passive: Mapping Soil Moisture and Freeze/Thaw from Space.

Goetz, S.J. (1997). Multi-sensor analysis of NDVI, surface temperature and biophysical variables at a mixed grassland site. *International Journal of Remote Sensing* 18 (1): 71–94.

Gorelick, N., Hancher, M., Dixon, M. et al. (2017). Google earth engine: planetary-scale geospatial analysis for everyone. *Remote Sensing of Environment* 202: 18–27.

Gruber, A., Scanlon, T., Schalie, R.V.D. et al. (2019). Evolution of the ESA CCI soil moisture climate data records and their underlying merging methodology. *Earth System Science Data* 11 (2): 717–739.

Huang, X., Liu, J., Zhu, W. et al. (2019). The optimal threshold and vegetation index time series for retrieving crop phenology based on a modified dynamic threshold method. *Remote Sensing* 11: 275–287.

Huete, A. (2004). *Remote Sensing for Natural Resources Management and Environmental Monitoring: Manual of Remote Sensing*, 3e, vol. 4. University of Arizona.

Karthikeyan, L., Pan, M., Wanders, N. et al. (2017). Four decades of microwave satellite soil moisture observations: part 2. Product validation and inter-satellite comparisons. *Advances in Water Resources* 109: 236–252.

Kerr, Y. (2016). Assessment of the SMAP passive soil moisture product. *IEEE Transactions on Geoscience and Remote Sensing* 54 (8): 4994–5007.

Khazaei, S., Raeini, M., Valizadeh, A., and Ghorbani, K. (2017). Estimation of soil surface moisture using vegetation indices and land surface temperature using MODIS sensor images (case study: Gonbad Kavous). *Iranian Journal of Irrigation and Drainage* 2 (11): 162–151.

Kogan, F.N. (2000). Contribution of remote sensing to drought early warning. In: *Early Warning Systems for Drought Preparedness and Drought Management* (eds. D.A. Wilhite and D.A. Wood), 75–87. Geneva: Word Meteorological Organization.

Koster, R.D., Suarez, M.J., and Heiser, M. (2000). Variance and predictability of precipitation at seasonal to interannual timescales. *Journal of Hydrometeorology* 1: 26–46.

Latham, B., O'Neill, P., and Yueh, S. (2019). Comparison of high-resolution airborne soil moisture retrievals to SMAP soil moisture during the SMAP validation experiment 2016 (SMAPVEX16). *Remote Sensing of Environment* 227: 137–150.

Lorenz, D.J., Otkin, J.A., Svoboda, M. et al. (2017). Predicting US drought monitor states using precipitation, soil moisture and evapotranspiration anomalies. Part-I development of a non-discrete USDM index. *Journal of Hydrometeorology* 18: 1943–1962.

Mandal, V.P., Rehman, S., Ahmed, R. et al. (2020). Land suitability assessment for optimal cropping sequences in Katihar district of Bihar, India using GIS and AHP. *Spatial Information Research* 28 (5): 589–599.

Martínez-Fernández, J., González-Zamora, A., Sanchez, N. et al. (2016). Satellite soil moisture for agricultural drought monitoring: assessment of the SMOS derived soil water deficit index. *Remote Sensing of Environment* 177: 277–286.

Mccoll, K.A., Alemohammad, S.H., Akbar, R. et al. (2017). The global distribution and dynamics of surface soil moisture. *Nature Geoscience* 10: 100–104.

Mohajane, M., Essahlaoui, A., Oudija, F. et al. (2017). Mapping forest species in the central middle atlas of Morocco (Azrou Forest) through remote sensing techniques. *Geographical Information* 6: 275–289.

Narasimhan, B., Srinivasan, R., Arnold, J.G., and Di Luzio, M. (2005). Estimation of long-term soil moisture using a distributed parameter hydrologic model and verification using remotely sensed data. *American Society of Agricultural Engineers* 48 (3): 1101–1113.

Pandey, P.C., Mandal, V.P., Katiyar, S. et al. (2015). Geospatial approach to assess the impact of nutrients on rice equivalent yield using MODIS sensors'-based MOD13Q1-NDVI data. *IEEE Sensors Journal* 15 (11): 6108–6115.

Rani, M., Joshi, H., Kumar, K. et al. (2020). Climate change scenario of hydro-chemical analysis and mapping spatio-temporal changes in water chemistry of water springs in Kumaun Himalaya. *Environment, Development and Sustainability* 23: 4659–4674.

Robock, A., Vinnikov, K.Y., Srinivasan, G. et al. (2000). The global soil moisture data bank. *Bulletin of the American Meteorological Society* 81: 1281–1299.

Singh, K., Kumar, P., and Singh, B.K. (2013). An associative relational impact of water quality on crop yield: a comprehensive index analysis using LISS-III sensor. *IEEE Sensors Journal* 13 (12): 4912–4917.

Tomar, V., Mandal, V.P., Srivastava, P. et al. (2014). Rice equivalent crop yield assessment using MODIS sensors' based MOD13A1-NDVI data. *IEEE Sensors Journal* 14 (10): 3599–3605.

Townshend, J.R.G. (1994). Global data sets for land application from the advanced very high resolution radiometer. *International Journal of Remote Sensing* 15 (17): 3319–3332.

Tucker, C.J. (1996). History of the use of AVHRR data for land applications. In: *Advances in Use of NOAA AVHRR Data for Land Applications* (eds. G. D'Souza, A.S. Selward and J.-P. Malingreau), 1–19. Dordrecht: Kluwer Academic Publishers.

Wang, J., Price, K.P., and Rich, P.M. (2001). Spatial patterns of NDVI in response to precipitation and temperature in the central Great Plains. *International Journal of Remote Sensing* 22: 3827–3844.

Wang, J., Sammis, T. W., Meier, C. A., Simmons, L. J., Miller, D. R., Samani, Z. 2005. A Modified SEBAL Model For Spatially Estimating Pecan Consumptive Water Use for Las Cruces, New Mexico.

Yadav, M., Sharma, M.P., Prawasi, R. et al. (2014). Estimation of wheat/rice residue burning areas in major districts of Haryana, India, using remote sensing data. *Journal of the Indian Society of Remote Sensing* 42 (2): 343–352.

16

Artificial Machine Learning–Based Classification of Land Cover and Crop Types Using Sentinel-2A Imagery

Ram Kumar Singh[1], *Pavan Kumar*[2], *Manoj Kumar*[3], *Keshav Tyagi*[4], *and Harshi Jain*[4]

[1] *Department of Natural Resources, TERI School of Advanced Studies, New Delhi, India*
[2] *Department of Forest Biology and Tree Improvement, College of Horticulture and Forestry, Rani Lakshmi Bai Central Agricultural University, Jhansi, Uttar Pradesh, India*
[3] *GIS Centre, Forest Research Institute (FRI), Dehradun, Uttarakhand, India*
[4] *Forest Research Institute Deemed to be University (FRIDU), Dehradun, Uttarakhand, India*

CONTENTS

16.1 Introduction

The remote sensing data availability pertains to the use of images for various application including identification, mapping, classification, and time-based changes (Joshi et al. 2006; Singh et al. 2020b). This decade will also be known for the availability of big voluminous time-series data of remote sensing. This opens a wide range of unprecedented challenges to accurately mapping (Singh et al. 2020a), automate processing (Sun et al. 2013) and modeling (Faivre et al. 2009; Kumar et al. 2019b; Singh et al. 2020c) of the images for solving various complex problem, i.e. crop field identification (Murmu and Biswas 2015), crop area mapping (Ndikumana et al. 2018), crop yield estimation (Launay and Guérif 2003), agriculture, forestry, and other land use (AFOLU) (Tubiello et al. 2014) mapping. Mapping crop

Sustainable Agriculture Systems and Technologies, First Edition. Edited by Pavan Kumar, A.K. Pandey, Susheel Kumar Singh, S.S. Singh, and V.K. Singh.

type and land cover are some of the working essential data to deal with crop monitoring, environmental and food security (Kussul et al. 2017). The crop mapping at various of its stages is difficult using a single data set, and time-series images are required for the identification and segregation of crop types (Singh et al. 2020b). For example, it is not able to identify two broad classes of the forest, crop after post-monsoon Rabi season and barren land.

The machine learning methods are superior in their efficacy and even due to various controlling of conditional parameters to perform classification, is also known as a soft classifier. The other hard classifiers are typical supervised and unsupervised classification methods. The supervised hard classification methods use the training signature of the classes to classify the output; however, controlling of parameters are very limited. The supervised (hard) classification few methods are the spectral distance (Sah et al. 2012), Bayesian (Mellert et al. 2015), maximum likelihood (Cameron and Windmeijer 1997), and others. The unsupervised hard classifiers are typically a clustering method, it clusters the image based on image available pixel-based statistical value. It has limited scope to control the clustering parameters. The hard few unsupervised clustering methods are K-mean (Iounousse et al. 2015) and Iterative Self Organizing (ISO) (Metzger et al. 2013). The machine learning methods need a large number of training data and an independent layer to classify the dependent classes output (Cutler et al. 2007). The independent layers are non-correlated with each other and are used as input for classification. The image data layers together used for machine learning classification, so it should be verified for multi-collinearity issue (Hirosawa et al. 1996). The occurrence of high collinearity between the each other of data layers is represented as redundant data set, i.e. time-series Normalized Difference Vegetation Index (NDVI) product data of crop and forest region. It is important to overcome to resolve an issue related to multi-collinearity, data redundancy, image data anomaly, minimize cloud effects, noise and data dimensionality. The principal component (PC) analyses reduce the data to fewer component layers and maintain the variability between the layers (Singh et al. 2021) and improve the signal-to-noise ratio (SNR) (Bruce et al. 2006; Singh et al. 2016), but other data filtration and averaging methods do not resolve and make the images data independent. The agriculture crop productivity (Singh et al. 2019, 2020) also increases with various local and administrative measures (Kumar et al. 2021) with the support of environmental factors (Singh et al. 2021).

16.1.1 Random Forest Classifier

The Random forest (RF) classifier was proposed by Leo Breiman, (2000) created a set of the decision tree, which predicts and enhances randomly decided data subspaces. Despite growing interest in machine learning applications, testing is limited to few fields. It is a statistical and decision-making property between the classes explained by Breiman (2004). The accuracy and classification gains can be achieved by arrangements of the decision tree; however, decision tree depth has grown following random data parameters. The outcomes predictions are obtained by aggregating over the decision tree arrangements. As the prime data parameters are arranged on structured tree predictors, and each of these trees is constructed using a term of randomness, the whole procedures are known as RFs.

The machine learning–based supervised classification methods for land cover mapping (Yousefi et al. 2015). The Classification and Regression Trees (CART) is used for broad few classes and the assessed overall accuracy is 80% (Wu et al. 2008). The CART methods are used for classification between few classes and its decision tree of small depth. It is good for the classification into few classes and for faster prediction. Yousefi et al. (2015) used a support vector machine (SVM) for crop species mapping and found an overall accuracy of above80%. They also tried with neural network-based classification and achieved efficacy above 85%. The RF classification method used by Belgiu and Csillik (2018), multiple classes are used to train the model with various independent data for efficient use. The RF testing results show a proven classified image with an overall accuracy of c. 85%. The RF is built of multiple CART-like decision trees for the classification of multiple classes. It arranges the independent data into a decision tree for the prediction of the classes. The RF machine learning prediction is based on training sample learning, and different from the average-based forecast (Kumar et al. 2020; Singh et al. 2020; Shumway et al. 2019) and trend line prediction (Singh et al. 2020). The RF method shows outperformance result for crop-based identification using remote sensing data. The method RF procedures are robust, consistent, manages to adapt sparsity in data, rate of convergence is good, depends upon training strong features, and results from outcomes are less deviated with any noisy data (Kussul et al. 2017).

The research has taken up for the classification of the land cover and type of croplands using time-series Sentinel-2A open data. The accuracy of the land cover classification is randomly tested using given training samples.

16.2 Methodology

The study is taken up for a region falling in between longitude 77.20–77.48°E and latitude 30.33–30.71°N of Saharanpur district of Uttar Pradesh, India, for land cover and crop type identification. The Sentinel-2A optical satellite image of spatial resolution 10–60 m is varying for all different 13 bands of data from the European Copernicus program and images are downloaded using earth explorer website (https://earthexplorer.com) open access images. The time-series (October 2019, November 2019, and February 2020 year) images for post-monsoon Rabi crop time images are used for machine learning–based predictive analysis. The Rabi post-monsoon crops such as wheat, gram, and other crops are primarily grown in this region.

16.2.1 Data Preprocessing

The Sentinel-2A time-series image for the post-monsoon Rabi crop season images for October 2019, November 2019, and February 2020 shown in false-color composite band combination (Figure 16.1 (first row)). The time-series images color band near-infrared (NIR) (8) and red (5) are used to create Soil Adjusted Vegetation Index (SAVI) product (Singh et al. 2020a). SAVI is a data product like NDVI, and it minimizes soil brightness influences of NIR and red color wavelength, it leads to influences of spectral vegetation indices, i.e. NDVI and non-corrected vegetation indices. The SAVI transformation involves a shifting of the origin of reflectance NIR and red color band wavelength space by the first order of soli vegetation interactions (Huete 1988; Thenkabail 2015). The

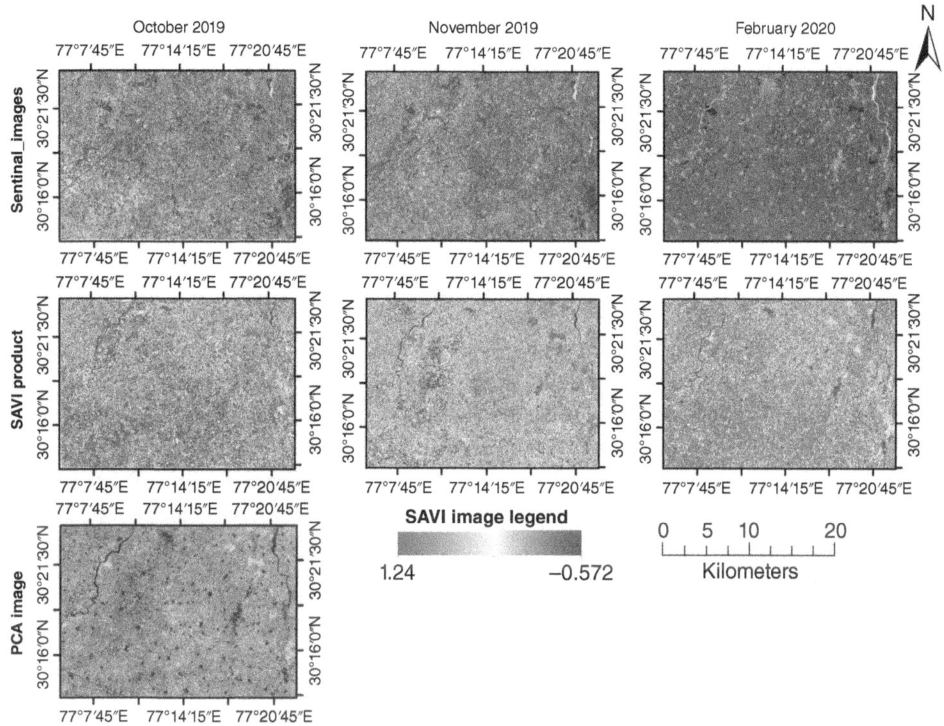

Figure 16.1 Data preprocessing downloaded and derived images: the first row for input Sentinel-2A time-series images of October 2019, November 2019, and February 2020for post-monsoon Rabi crop season, second-row-derived Soil Adjusted Vegetation Index (SAVI) product of time-series used images, and third-row Principal Component Images (PCA) of all SAVI images to create a single-independent image.

first-order differentiated NIR-red color flux for mapping of the vegetated region (Eq. (16.1)). Its advantage over NDVI is it is used to map and identify the small crop and various stages of crops, growth and changes. It was tested by Huete (1988) for cotton plantation growth mapping; SAVI methods provide far better results in comparison to the NDVI method. Its value also ranges from −2 to 2. The SAVI method describes soil-vegetation system values from remote sensing images.

$$\text{SAVI} = \frac{(\text{nir} - \text{red})}{(\text{nir} + \text{red} + L)} \cdot (1 + L) \tag{16.1}$$

The L is denoted by soil-adjusted correction parameter, and usually, its value is taken as 0.5 to accommodate most of the land cover types and vegetation mapping (Huete 1988). The available SAVI product data are used to derive its component using principal component analysis (PCA). The PCA is carried out to identify the significant components used for analysis. The PC layers have higher variability between the layers, and the PC layers that have non-correlation and higher variability between the layers are known as independent layers (Figure 16.1 (last third row)). The PC layers are used for RF-based machine learning using training samples. The vector layer for all the features, i.e. wheat field, gram field, trees, settlement, water, and other land cover signatures, are provided. The software ERDAS Imagine 2020 version is used for machine learning purpose training and testing of the model, and then for classified map random sample accuracy testing and reporting.

16.2.2 Random Forest Model

The preprocessed PC layers are used for RF machine learning training and the class samples are using vector shape file, which is a representation of class on training image. It build the statistics to learn about the different classes. The decision tree is built using PC values and the statistical value of the signature class. The RF machine learning model creates a training intellect model (Figure 16.2b). The testing is performed on PC layer image using the trained model, the artificial intelligence machine learning classifier classifies the classes on the given PC layer image (Figure 16.2c). The model performs on the same type of data for future classification of images, so testing of RF model was carried out on PC layers image and then testing also. Further, the output accuracy is tested using random sampling point all over the study region image.

16.3 Accuracy Assessment

The classified map consistency is checked using Copernicus global land cover data of spatial resolution 100 m for the year 2019. It was used as a reference to know the area covered with the same land cover class. The Copernicus land cover data is downloaded from the website (https://lcviewer.vito.be), and its accuracy was tested by Buchhorn et al. (2020), a considerable very good accuracy. The accuracy assessment of the RF machine learning method-predicted land cover is tested by using stratified random sampling point and

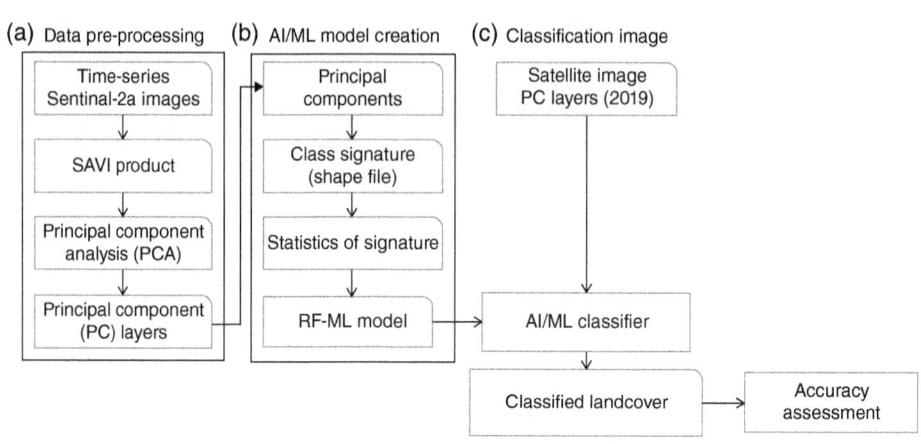

Figure 16.2 Flow diagram for land cover and crop mapping (a) preprocessing of times series images, (b) random forest (RF) training model intellect, (c) classification and accuracy assessment of land cover and crop map.

matched with available same year Google Earth Pro high spatial resolution photos. The random stratified point number 210 is generated on RF machine learning–based land cover and verified concerning Google Earth Pro photos. The accuracy is observed of the created map with reference of stratified random Google Earth Pro points to access overall and K-hat accuracy with ERDAS Imagine software error contingency matrix.

16.4 Results and Discussion

16.4.1 Data Processing

The time-series SAVI data product is transformed into three PC layers (Table 16.1). The starting three components (PC-1, PC-2, and PC-3) of the layers shows the full 100% of variability, and its representation is like a multicolor image. So, all PC layers are retained as independent variables to develop the RF machine learning intellect model.

16.4.2 Random Forest Machine Learning

The RF machine learning intellect model is developed using the class signature feature vector and PC layer. The model is used further to classify the same kind of image data. The classified land cover and crop-identified image are shown (Figure 16.3).

16.4.3 Assessment of Land Cover Accuracy and Its Consistency

The software ERDAS Imagine creates a contingency error matrix using stratified random 210 points for seven classes (Agro-forestry, Wheat, Gram, Other crops, Water, Settlement, and Sandy region) on Google Earth Pro image. The overall accuracy is 83% and K-hat 0.78. The accuracy of vegetation (Agro-forestry, Wheat, Gram, and Other crops) and water classes are higher in comparison to settlement, and sandy region. The full validation of RF machine learning classified image with Copernicus land cover image of 2019. The Copernicus image land cover has 23 classes, which are finally constituted into 5 classes: forestry, crop, settlement, water and other classes for comparison to RF machine learning classified reconstituted to 5 classes. The full validation accuracy for Agro-forestry, Crop, Water classes are in the range of 93–98%, however settlement and other areas in the range

Table 16.1 Eigenvalues and principal component analysis (PCA).

	2019	
PCs	Eigen Value	(%)
PC1	39 325.6	98.1
PC2	1 150.4	1.1
PC3	781.7	0.9

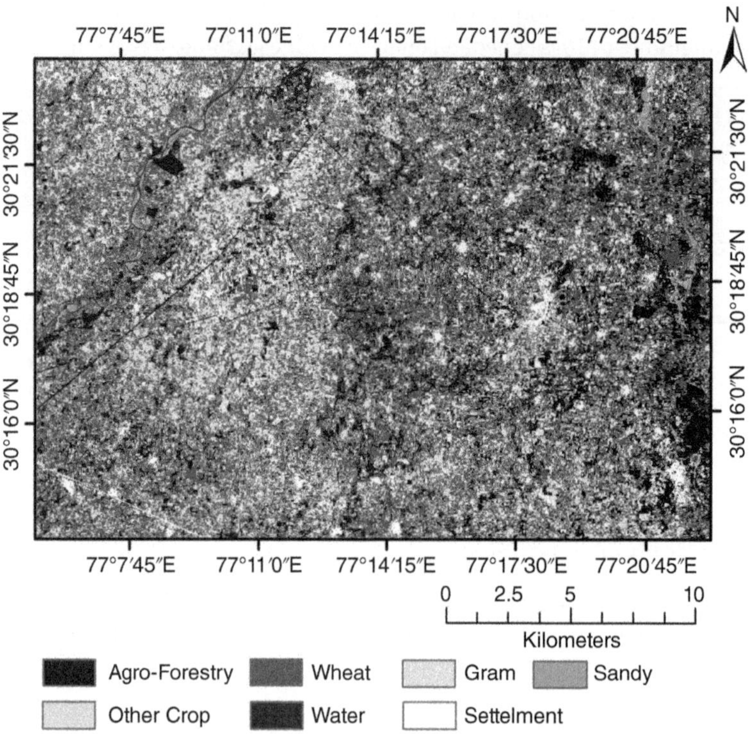

Figure 16.3 Random forest machine learning-based mapped cropland cover of the year 2019. (*See insert for color representation of the figure.*)

of 86–89%. The RF results proven to accurate and consistent use for crop and land cover mapping assessments (Belgiu and Csillik 2018; Pérez-Vega et al. 2012).

The novel approach RF machine learning method creates a complex decision tree based on given image data and class samples. The RF method creates decision on various classes of different categories. It reads various types of data to perform its model training and classification testing on the same image data. Its decision tree consists of multiple CART decision tree to become a complex classifier. The CART tree classifier used for a few classes of decision-based classification (Wu et al. 2008). Many researchers prefer machine learning classifiers, its model reused for predication-based classification on other region and upcoming data images (Iounousse et al. 2015; Sun et al. 2013). The other hard classifier whole processes of classification process need to do again with other data images, and time-taking processes on large data are time-consuming comparatively with machine learning methods. The artificial neural network (ANN)-based deep learning methods are used for object mapping, identification, predication, and tracing of objects, i.e. Semantic segmentation, Res-Net, U-Net, Seg-Net, and others. The machine learning method performs accurately for mapping of land covers and deep learning methods perform for identification. Machine learning is based on various classification for land cover (Wardlow et al. 2007), change mapping (Wardlow et al. 2007), forest vulnerability (Kumar et al. 2019a), crop mapping (Kalra and Kumar 2018), and water change mapping (Singh et al. 2020), and in other fields for accurate and consistent mapping.

16.5 Conclusion

The machine learning–based classification and mapping explain the problems of updated land cover data for the planning and management of land resources. The cropland cover-based production supports the socio-economic activities of the agrarian country. The crop production supports gross domestic product (GDP) (World Bank 2011) of the India country more than 20%, although of many industrialization activities. The updated annual mapped land cover and crop-based field identification helps to maintain food availability issues. The remote sensing time-series images of finer spatial resolution provides an opportunity for more precise and accurate mapping of cropland cover using the RF machine learning method. Machine learning works on various other data including images; however, data preprocessing becomes critical to use along with various methods. In general, the research study helps to bridge the gap of land cover data availability for spatial cropland cover for regional planning and monitoring, since it makes toward self-sustainable ("Aatma-Nirbhar") process and country.

Acknowledgments

The authors acknowledge the European Copernicus Sentinel data agency to provide Sentinel-2A images, the Copernicus land cover image and the google earth pro for photo for verification. We are very thankful to ERDAS software team for machine learning-based classification.

References

Belgiu, M. and Csillik, O. (2018). Sentinel-2 cropland mapping using pixel-based and object-based time-weighted dynamic time warping analysis. *Remote Sensing of Environment* 204: 509–523. https://doi.org/10.1016/j.rse.2017.10.005.

Breiman, L. (2000). Some Infinity Theory for Predictor Ensembles. *Technical Report-577*, Statistics Department, University of California at Berkeley.

Breiman, L. (2004). Consistency for a Simple Model of Random Forests. *Technical Report-670*, Statistics Department, University of California at Berkeley.

Bruce, L.M., Mathur, A., and Byrd, J.D. (2006). Denoising and wavelet-based feature extraction of MODIS multi-temporal vegetation signatures. *GIScience & Remote Sensing* 43 (1): 67–77. https://doi.org/10.2747/1548-1603.43.1.67.

Buchhorn, M., Smets, B., Bertels, L. et al. (2020). Copernicus global land service: land cover 100 m. *Globe* https://doi.org/10.5281/zenodo.3939050.

Cameron, A.C. and Windmeijer, A.G.F. (1997). An R-squared measure of goodness of fit for some common nonlinear regression models. *Journal of Econometrics* 77: 329–342. https://doi.org/10.1016/S0304-4076(96)01818-0.

Cutler, D.R., Edwards, T.C., Beard, K.H. et al. (2007). Random forests for classification in ecology. *Ecology* 88 (11): 2783–2792. https://doi.org/10.1890/07-0539.1.

Faivre, R., Leenhardt, D., Voltz, M. et al. (2009). Spatialising crop models. *Sustainable Agriculture* https://doi.org/10.1007/978-90-481-2666-8_42.

Hirosawa, Y., Marsh, S.E., and Kliman, D.H. (1996). Application of standardized principal component analysis of land-cover characterization using multitemporal AVHRR data. *Remote Sensing of Environment* 58 (3): 267–281. https://doi.org/10.1016/S0034-4257(96)00068-5.

Huete, A.R. (1988). A soil-adjusted vegetation index (SAVI). *Remote Sensing of Environment* 25 (3): 295–309.

Iounousse, J., Er-Raki, S., El Motassadeq, A., and Chehouani, H. (2015). Using an unsupervised approach of probabilistic neural network (PNN) for land use classification from multitemporal satellite images. *Applied Soft Computing Journal* 30: 1–13. https://doi.org/10.1016/j.asoc.2015.01.037.

Joshi, P.K.K., Roy, P.S., Singh, S. et al. (2006). Vegetation cover mapping in India using multi-temporal IRS wide field sensor (WiFS) data. *Remote Sensing of Environment* 103 (2): 190–202.

Kalra, N. and Kumar, M. (2018). Simulating the impact of climate change and its variability on agriculture. In: *Climate Change and Agriculture in India: Impact and Adaptation* (ed. S. Sheraz Mahdi), 21–28. Springer International Publishing https://doi.org/10.1007/978-3-319-90086-5_3.

Kumar, M., Singh, H., Pandey, R. et al. (2019a). Assessing vulnerability of forest ecosystem in the Indian Western Himalayan region using trends of net primary productivity. *Biodiversity and Conservation* 28 (9): 2163–2182.

Kumar, M., Padalia, H., Nandy, S. et al. (2019b). Does spatial heterogeneity of landscape explain the process of plant invasion? A case study of *Hyptis suaveolens* from Indian Western Himalaya. *Environmental Monitoring and Assessment* 191 (3 Supplement): 794. https://doi.org/10.1007/s10661-019-7682-y.

Kumar, P., Kalita, H., Patairiya, S. et al. (2020). Forecasting the dynamics of COVID-19 pandemic in top 15 countries in april 2020 through ARIMA model with machine learning approach. *MedRxiv* https://doi.org/10.1101/2020.03.30.20046227.

Kumar, P., Singh, S.S., Pandey, A.K. et al. (2021). Multi-level impacts of the COVID-19 lockdown on agricultural systems in India: the case of Uttar Pradesh. *Agricultural Systems* 187: 103027.

Kussul, N., Lavreniuk, M., Skakun, S., and Shelestov, A. (2017). Deep learning classification of land cover and crop types using remote sensing data. *IEEE Geoscience and Remote Sensing Letters* 14 (5): 778–782. https://doi.org/10.1109/LGRS.2017.2681128.

Launay, M. and Guérif, M. (2003). Ability for a model to predict crop production variability at the regional scale: an evaluation for sugar beet. *Agronomie* https://doi.org/10.1051/agro:2002078.

Mellert, K.H., Deffner, V., Küchenhoff, H., and Kölling, C. (2015). Modeling sensitivity to climate change and estimating the uncertainty of its impact: a probabilistic concept for risk assessment in forestry. *Ecological Modelling* 316: 211–216. https://doi.org/10.1016/j.ecolmodel.2015.08.014.

Metzger, M.J., Bunce, R.G.H., Jongman, R.H.G. et al. (2013). A high-resolution bioclimate map of the world: a unifying framework for global biodiversity research and monitoring. *Global Ecology and Biogeography* 22 (5): 630–638. https://doi.org/10.1111/geb.12022.

Murmu, S. and Biswas, S. (2015). Application of fuzzy logic and neural network in crop classification: a review. *Aquatic Procedia* 4: 1203–1210. https://doi.org/10.1016/j.aqpro.2015.02.153.

Ndikumana, E., Minh, D.H.T., Baghdadi, N. et al. (2018). Deep recurrent neural network for agricultural classification using multitemporal SAR Sentinel-1 for Camargue, France. *Remote Sensing* https://doi.org/10.3390/rs10081217.

Pérez-Vega, A., Mas, J.F., and Ligmann-Zielinska, A. (2012). Comparing two approaches to land use/cover change modeling and their implications for the assessment of biodiversity loss in a deciduous tropical forest. *Environmental Modelling and Software* 29 (1): 11–23. https://doi.org/10.1016/j.envsoft.2011.09.011.

Sah, S., van Aardt, J.a.N., McKeown, D.M., and Messinger, D.W. (2012). A multi-temporal analysis approach for land cover mapping in support of nuclear incident response. *Proceedings of SPIE Vol. 8390, Algorithms and Technologies for Multispectral, Hyperspectral Anb Ultraspectral Imagery XVIII*, 8390 (May), pp. 1–7. doi:https://doi.org/10.1117/12.919276.

Shumway, R.H., Stoffer, D.S., Shumway, R.H., and Stoffer, D.S. (2019). ARIMA models. In: *Time Series: A Data Analysis Approach Using R*, 89–212. New York: Springer https://doi.org/10.1201/9780429273285-5.

Singh, R.K., Govil, H., and Shweta (2016). Comparison of signal-to-noise ratio and its features variation: SPIE 9880. In: *Multispectral, Hyperspectral, and Ultraspectral Remote Sensing Technology, Techniques and Applications*. https://doi.org/doi:10.1117/12.2223773.

Singh, R.K., Sinha, V.S.P., Kumar, M. et al. (2019). Assessment of agriculture net primary productivity vulnerability based on JULES simulation model under climate change scenarios for the year 2050 SAARC nations. *AGU Fall Meeting*, San Francisco (December 2019).

Singh, R.K., Sinha, V.S.P., Joshi, P.K., and Kumar, M. (2020a). A multinomial logistic model-based land use and land cover classification for the South Asian Association for Regional Cooperation nations using moderate resolution imaging spectroradiometer product. *Environment, Development and Sustainability* https://doi.org/10.1007/s10668-020-00864-1.

Singh, R.K., Sinha, V.S.P., Joshi, P.K., and Kumar, M. (2020b). Mapping of agriculture productivity variability for the SAARC nations in response to climate change scenario for the year 2050. *GIScience & Remote Sensing*: 249–262. https://doi.org/10.1007/978-3-030-55092-9_14.

Singh, R.K., Sinha, V.S.P., Joshi, P.K., and Kumar, M. (2020c). Modelling agriculture, forestry and other land use (AFOLU) in response to climate change scenarios for the SAARC nations. *Environmental Monitoring and Assessment* 192 (4): 236. https://doi.org/10.1007/s10661-020-8144-2.

Singh, R.K., Drews, M., De La Sen, M. et al. (2020d). Short-term statistical forecasts of COVID-19 infections in India. *IEEE Access* 8: 186932–186938. https://doi.org/10.1109/access.2020.3029614.

Singh, R.K., Rani, M., Bhagavathula, A.S. et al. (2020e). The prediction of COVID-19 pandemic for top-15 affected countries using advance ARIMA model. *JMIR Public Health and Surveillance* 6: 1–10. https://doi.org/10.2196/19115.

Singh, R.K., Kumar, P., Mukherjee, S., and Suman, S. (2020f). Application of geospatial technology in agricultural water management. In: *Agricultural Water Management* (ed. P.K. Srivastava), 31–45. Elsevier Inc https://doi.org/doi:10.1016/B978-0-12-812362-1.00003-5.

Singh, R.K., Drews, M., De la Sen, M. et al. (2021). Highlighting the compound risk of COVID-19 and environmental pollutants using geospatial technology. *Scientific Reports* 11 (1): 1–12. https://doi.org/10.1038/s41598-021-87877-6.

Sun, J.B., Yang, J.Y., Zhang, C. et al. (2013). Automatic remotely sensed image classification in a grid environment based on the maximum likelihood method. *Mathematical and Computer Modelling* 58 (3–4): 573–581. https://doi.org/10.1016/j.mcm.2011.10.063.

Thenkabail, P.S. (2015). Remote sensing of land resources: monitoring, modeling, and mapping advances over the last 50 years and a vision for the future. In: *Land Resources Monitoring, Modeling, and Mapping with Remote Sensing* (ed. P.S. Thenkabail), 791–828. CRC Press https://doi.org/10.1201/b19322.

Tubiello, F.N., Salvatore, M., Cóndor Golec, R.D. et al. (2014). Agriculture, forestry and other land use emissions by sources and removals by sinks. *ESS Working Paper No.* 2, 2, 4–89. https://doi.org/10.13140/2.1.4143.4245.

Wardlow, B.D., Egbert, S.L., and Kastens, J.H. (2007). Analysis of time-series MODIS 250 m vegetation index data for crop classification in the U.S. Central Great Plains. *Remote Sensing of Environment* 108 (3): 290–310. https://doi.org/10.1016/j.rse.2006.11.021.

World Bank. (2011). http://databank.worldbank.org/data/reports.aspx?source=population-estimates-and-projections

Wu, X., Kumar, V., Ross, Q.J. et al. (2008). Top 10 algorithms in data mining. *Knowledge and Information Systems* 14 (1) https://doi.org/10.1007/s10115-007-0114-2.

Yousefi, S., Khatami, R., Mountrakis, G. et al. (2015). Accuracy assessment of land cover/land use classifiers in dry and humid areas of Iran. *Environmental Monitoring and Assessment* https://doi.org/10.1007/s10661-015-4847-1.

17

Geoinformatics and Nanotechnological Approaches for Coping Up Abiotic and Biotic Stress in Crop Plants

Abhishek Singh[1], Vishnu D. Rajput[2], Sapna Rawat[3], Ragini Sharma[4], Anil Kumar Singh[5], Pradeep Kumar[6], Awani Kumar Singh[7], Tatiana Minkina[2], Rudra Pratap Singh[8], and Shashank Singh[1]

[1] *Department of Agricultural Biotechnology, Sardar Vallabhbhai Patel University of Agriculture and Technology, Meerut, Uttar Pradesh, India*
[2] *Academy of Biology and Biotechnology, Southern Federal University, Rostov-on-Don, Russia*
[3] *Department of Botany, University of Delhi, New Delhi, Delhi, India*
[4] *Department of Zoology, Punjab Agricultural University, Ludhiana, Punjab, India*
[5] *University of Allahabad Senate House Campus, Allahabad, Uttar Pradesh, India*
[6] *Department of Forestry, North Eastern Regional Institute of Science Technology, Nirjuli, Arunachal Pradesh, India*
[7] *Centre for Protected Cultivation and Technology, Indian Agricultural Research Institute, New Delhi, Delhi, India*
[8] *College of Agriculture, Acharya Narendra Dev University of Agriculture and Technology, Ayodhya, Uttar Pradesh, India*

CONTENTS

Sustainable Agriculture Systems and Technologies, First Edition. Edited by Pavan Kumar, A.K. Pandey, Susheel Kumar Singh, S.S. Singh, and V.K. Singh.

17.1 Introduction

The significance of sustainable agriculture is the production of animal and plant products to accomplish the requirement of the present population and to protect the resources for future generations (Singh et al. 2021). Sustainable agriculture is a contemporary agricultural practice that supports protecting environmental goods for public health without changing their nutritive value. At the beginning of the twentieth century, various new crop varieties were developed through plant breeding and other advanced technologies. But in the twenty-first-century agricultural field is facing various environmental issues similar to climate change, global warming, pollution, insect, and paste are some major factors come under abiotic and biotic stresses that reduced grain production and even improved varieties of crops. Less production of grain can lead to various issues such as food security, and for this, enhancement of grain production is necessary, which follows three key steps such as detection, determination, and diagnosis. These three major steps require a precise analysis of agricultural production elements or raw material, for example, soil architecture and the requirement of fertilizer, herbicide, fungicides, insecticides, pesticides for every single crop. This precise analysis of these agricultural production elements is united with modern tools such as geoinformatics and nanoparticles. The agricultural output, poverty reduction, and food security are boosted up in the long run by these agricultural applications. Deviation from the optimal conditions limits the growth of plants and can cause biotic, as well as abiotic stresses. Plants may respond to these stresses in various ways like chlorosis, stunted growth, wilting, reduction of leaf area, and many more. It is difficult to visually calculate the levels and speed of these reactions of plants to stress conditions. Thus, newly emerged technologies like geoinformatics and nanotechnology, with the help of electromagnetic radiations reflected from plant canopies and physiochemical attributes of nanoparticles, respectively, can assess and help in minimizing the negative effects of abiotic, as well as biotic stresses. Geoinformatics approaches aid in determining the kind of soil and its structure (Panda et al. 2010). Nanoparticles, on the other hand, can be used as fertilizers, herbicides, fungicides, insecticides, pesticides, and sensors to help determine and diagnose crop needs and health in real time (Figure 17.1). These recent tools can support the utilization of these raw materials for crop production that can directly reduce the production cost of farmers and increase their income.

17.2 "3-T" Concept for Crop Management

Climate change is dramatically every year continuously affecting both crop patterns and their production. Due to this climate change, droughts, and flood salinity, diseases have affected the crops (FAO 2019). This climate change directly affects crop production and protection, so that a country whose population is completely dependent on agriculture may also face food shortage or starvation (Zhao et al. 2017). And this problem may become even more severe in the future. To solve this problem, agriculture needs such technology that can effectively solve climate-change-related problems in a short time. The 3-T concept seems effective for agriculture in the context of resolving this problem (Figure 17.2). The first T stands for trace, tracing, or real-time monitoring of yield, soil, water, and weather

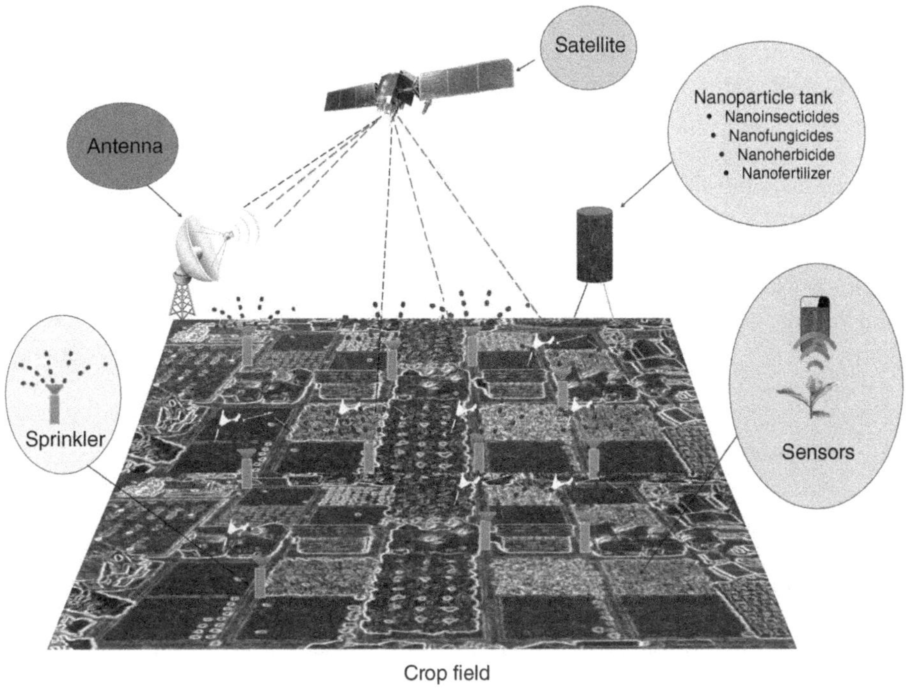

Figure 17.1 Integrated farming systems that have geoinformatics, nanoparticles, like emerging tools for coping up abiotic for sustainable agriculture.

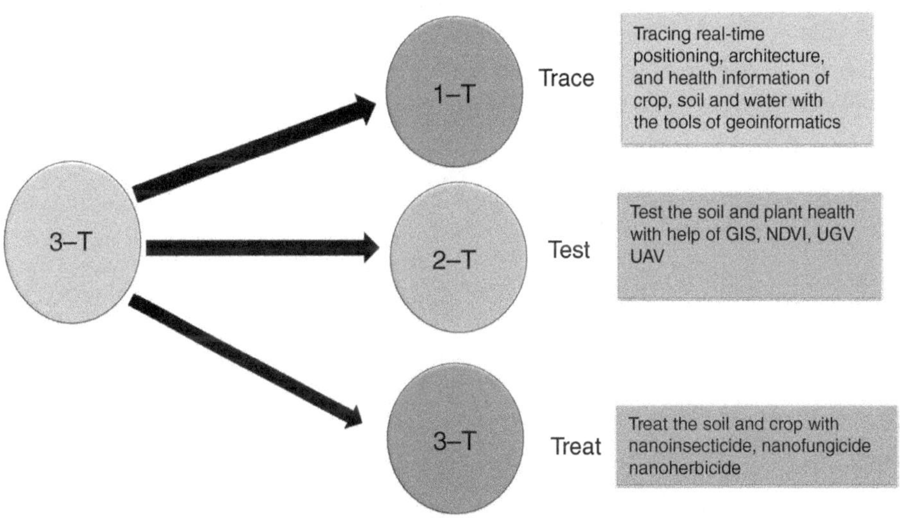

Figure 17.2 Diagrammatic representation of 3-T concept.

using geoinformatics tools such as earth observation satellites, UAV (Unmanned Aerial Vehicles), automatic weather stations, location-based mobile applications using data sets from base stations/ground sensors, and UGV (Unmanned Ground Vehicles), among others. These geoinformatics tools gathered information on digital soil and crop mapping, weather forecasting, fertilizer recommendations, disease detection, and pest control. These data help farmers to manage their fields at the micro-level and avoid the climate change advert effect on their crop. Second T means Test, after getting geoinformatics data. It is important to validate and confirm with ground data. We primarily examine each crop plant and soil at a micro level to calculate the number of fertilizers, insecticides, and herbicides that should be applied to them. After the test results come, it is necessary to take quick action on it for pre-diagnosis or treatment. Conventional fertilizers, pesticides, herbicides, insecticides, and nano fertilizer pesticides, herbicides, insecticides are used for this action, which are more effective in a short time, and it increases the health and nutrition of both the soil and the crop all these aspects come under third T that means Treat (Beneduzi et al. 2012).

17.3 Geoinformatics

Geoinformatics is the combination of two words, geo means "earth" and informatics means "the study of information processing." Thus, geoinformatics is a combination of the study of "Earth sciences" and "Informatics." Geoinformatics largely deals with the collection, analysis, storage, retrieval, representation, and dissemination of information about the earth with the assistance of information technology (Panda et al. 2010). Geographical information systems (GIS), Global Positioning System (GPS), and Remote sensing (RS) are various tools and techniques that are used in agriculture in geoinformatics. All the tools can be used together to improve soil fertility, crop productivity, reduce water waste, and increase the amount of resources employed in agriculture for crop production and protection (Munier et al. 2018). Disease identification, crop mapping, weather prediction, soil mapping, fertilizer suggestion, and pest control are all part of the geoinformatics-based agricultural production and protection (Figure 17.3).

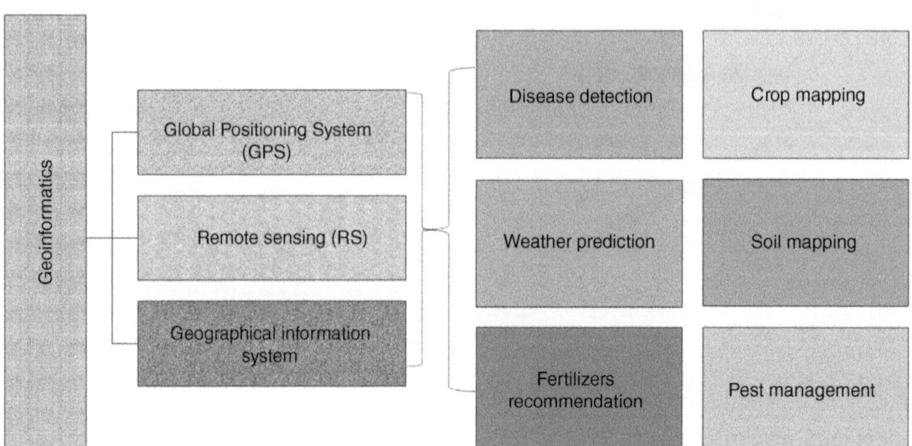

Figure 17.3 Diagrammatic representation of geoinformatics type are their application for management of abiotic and biotic stress.

17.4 Role of Geoinformatics in Abiotic and Biotic Stress

Leaves of the plants contain various pigment systems, and these systems change during abiotic and biotic stress. These pigments absorb visible (red) light wavelengths and significantly reflect infrared light, which is not visible to human eye. As a plant canopy matures and senesces from early spring growth to late-season maturity and senescence, its reflectance properties change. VIs (Vegetation Indices) are excellent approaches for quantitative and qualitative evaluations of vigor, vegetation cover, and growth dynamics generated from RS-based canopies, among other applications (Xue and Su 2017). Mapping vegetation is an important approach in the creation, planning, and maintenance of natural and polluted protected areas (Munier et al. 2018; Fabre et al. 2020). Vegetation maps are required for land use planning, management, and development of protected areas or monitoring of natural and polluted sites (Fabre et al. 2020). Half of the total area of the canopy's green parts per unit horizontal ground area is termed as "Land Area Index" (Chen and Black 1992). The total green LAI of all canopy levels including the understory is referred as satellite-derived number and it makes a large difference, mainly in forests (Figure 17.4) (GCOS 2011;

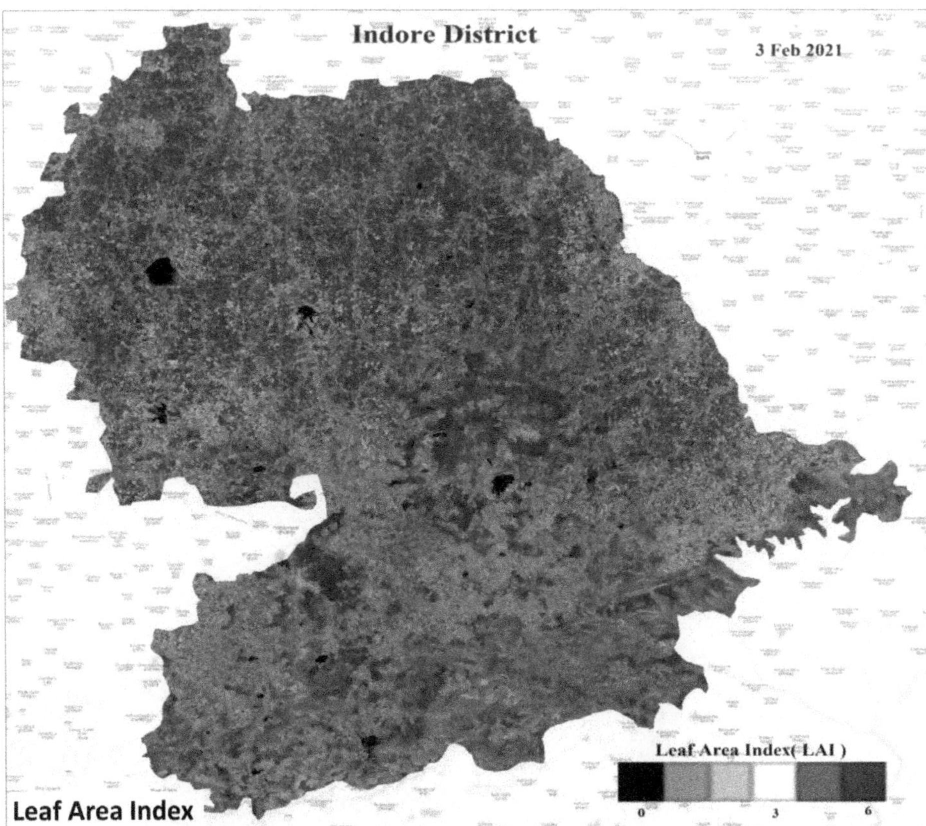

Figure 17.4 The LAI, total chlorophyll content of various types of plants in Indore district Madhya Pradesh, India. (*See insert for color representation of the figure.*)

Turner et al. 1999). In practice, the LAI measures the vegetation cover thickness (Zhang et al. 2005). LAI has been designated as essential climate variable (ECV) by the Global Climate Observing System (GCOS). A variety of LAI models are used in agriculture, ecology, carbon cycling, climate change, and other research to accurately depict radiation, heat, water, and other gas exchanges with the overlying atmosphere or underlying soil (Chase et al. 1996; Buermann et al. 2001). LAI is most commonly used in Soil-Vegetation-Atmosphere-Transfer schemes, biogeochemical cycle models, and agro-meteorology from crop assessment, all of which require long time series at various temporal and geographic scales (Fang et al. 2013; Nunes and Auge 1999).

Comparison of the present NDVI to a range of previous years' values for the same period is done by Vegetation Condition Index (VCI) (Figure 17.5). The VCI is reported in percent and indicates where the measured value falls between the previous year's extreme values (minimum and maximum) (Fabre et al. 2020). Poor and good vegetative state conditions are indicated by lower and higher values, respectively. VCI is primarily used in the detection and spatially outlining abnormalities in vegetation condition and growth, in terms of

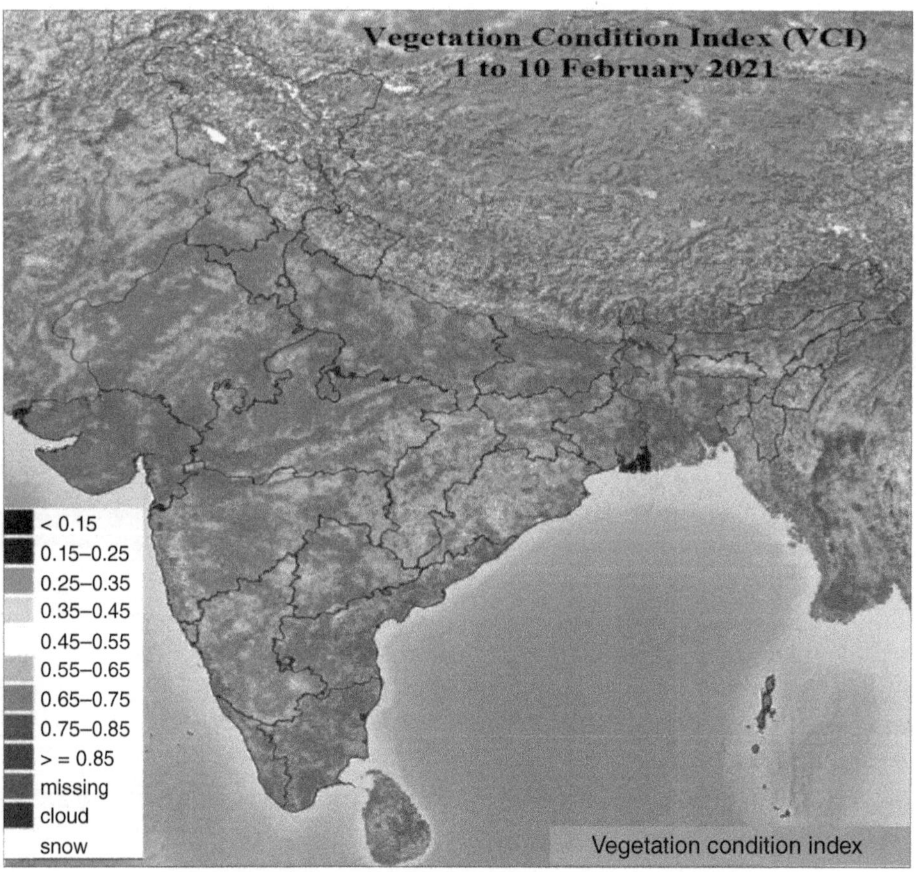

Figure 17.5 Diagrammatic representation of Vegetation Condition Index (VCI) of India from 1 February to 10 February 2021.

both augmentation and intensity. This is very beneficial for keeping track of the current growth season (e.g. early warning purposes). Another finest performing tool to estimate the canopy chlorophyll content is the IRECI (Figure 17.6).

17.4.1 Geoinformatics Tools for Abiotic and Biotic Stress Management

17.4.1.1 Global Positioning System (GPS)

The GPS is a navigation system based on satellites that lets people locate objects anywhere on the planet (Panda et al. 2010). It is provided continuously 24 hours per day worldwide data like real-time, 3-D positioning, navigation, and earth timing, etc. Signals are broadcasted by GPS satellites that allow GPS receivers to compute their position in real-time. This real-time positioning information can aid with agricultural, soil, and water mapping studies. Thus, GPSs are used in smart farming to detect the location of farm equipment within a 3-m range of an actual site in the field, and it is mainly used in agriculture for yield mapping and variable application rates of fertilizers and pesticides (Panda et al. 2010).

17.4.1.2 Remote Sensing (RS)

RS is an essential process of information science that supports the management of crop health, soil, weather, temperature, humidity, at the micro-level, and large scale (Singh

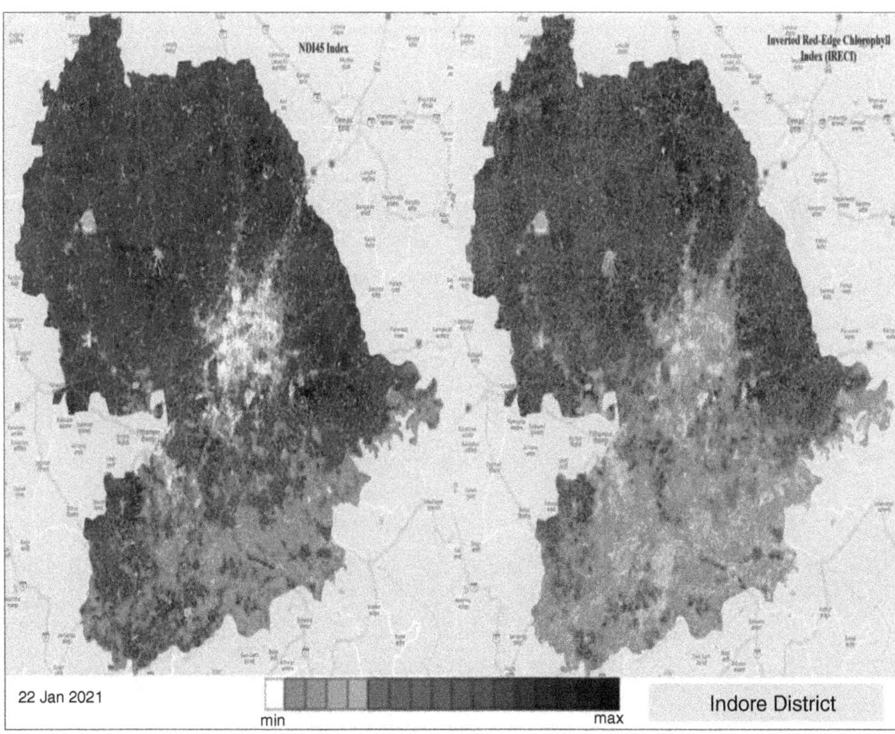

Figure 17.6 Diagrammatic representation of IRECI of Indore district, Madhya Pradesh, India. (*See insert for color representation of the figure.*)

et al. 2020). It is a scientific discipline, which examines the characteristics of earth's surface by collecting various forms of data from various sensors like balloon, rocket, aircraft, satellite, or a ground-based sensor (Singh et al. 2020). The RS technique uses wavelengths between 0.4 and 2.4 nm for land surveys (Ojo and Ilunga 2018).

17.4.1.3 Geographical Information System (GIS)

GIS is another technique based on computer utilization that helps in mapping, analyzing, and monitoring the things and event happened on earth. GIS computerized system can store data related to crop production and agronomic aspects that are used for sustainable agriculture. Mapping for a good understanding of crop yield, weeds, pests, fertility, or some other factors can be done with the help of stored information of GIS (Panda et al. 2010). This decision-making information helps for a better understanding of farmers for sustainable agriculture.

17.5 Nanoparticles

Nanotechnology is an emerging field of research and innovation concerned with making small materials on the scale of atoms and molecules called nanoparticles (Chinnamuthu and Murugesa Boopathi 2009). Size of nanoparticles range from 1 to 100 nm. These are the functional units of a bigger structure known as a nanotube. The nanotube is a novel carbon form that resembles a rolled-up two-dimensional graphene sheet. They offer unique

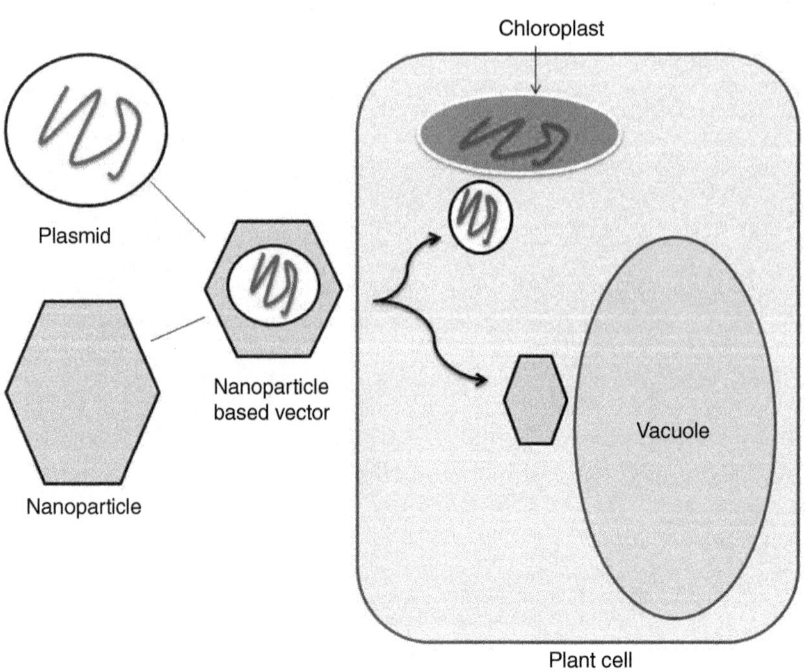

Figure 17.7 Nanoparticle-mediated transformation into the plant cell.

mechanical, electrical, thermal, optical, elastomeric, and other features that make them ideal for agri-field applications and the establishment of long-term agricultural conditions (Camilli et al. 2014; Chahine et al. 2014).

17.6 Role of NPs in Abiotic and Biotic Stress

Pest stress, microbial infections, weeds, natural catastrophes, soil fertility deficiencies, and other factors are responsible for around one-third of crop production being lost each year. In order to combat this issue, every day new methods and techniques are being developed, but most of these technical and related strategies have their results (Baker et al. 2017). Nanoparticles are created by using a broad class of materials such as magnetic materials, ceramics, metal oxides, quantum dots (Androvitsaneas et al. 2016; Bakalova et al. 2004; Das et al. 2015), semiconductors, dendrimers, lipids, emulsions (Anton and Vandamme 2009), and polymers (Bao et al. 2017; Bulovic et al. 2004). To counter these abiotic and biotic stresses, these nanoparticles are applied variously in agriculture (Das and Das 2019).

17.6.1 NPs in Abiotic Stresses

Abiotic stress means a particular environment that limits plant growth, development, and productivity (Boyer 1982). As plants cannot move from one place to another because of which they face abiotic stresses over their life cycle. Abiotic stresses in plants include salinity, drought, heat, flood, metals, chilling, etc. To overcome these environmental stresses, plants develop certain adaptations that modify their biochemical, morphological, gene expressions, and physiological pathways. Unfortunately, all these adaptation processes are not enough to neutralize the advert effect of environmental stresses. For example, osmotic potential of the soil decreases due to salinity stress resulting in reduced uptake of nutrients from the soil that causes a nutritional imbalance. This imbalance also creates ionic toxicity into plant cells that directly develop negative effects in various vital pathways such as photosynthesis, protein synthesis, lipid metabolism or other physiological and biochemical processes. As discussed earlier, plants make some adaptations in the presence of abiotic stresses. But these adaptations are not as effective for the long-term survival of the crop (Liang and Ritter 2014). This may lead to problems related to food security in the future due to low yield. One of the best solutions to this problem is nanotechnology. Conversion of heavy material into tiny pieces gives NPs some unique physicochemical properties like low volume-to-surface ratio because of which they attain more reactivity. These nanoparticles upregulate the antioxidant enzyme (CAT, SOD, and POD) activity, thereby indirectly catalyzing the biochemical pathways (Laware and Raskar 2014). Some of these biochemical pathways involved in the synthesis of antioxidant molecules are upregulated by nanoparticles. Various studies have shown that nanoparticle application can influence growth and development during various abiotic stresses on the plant body as discussed below.

17.6.1.1 Drought Stress

It occurs due to the unavailability of water in the plant. Higher temperature, low rainfall, dry wind, high light intensity, and fast evaporation rate of water from soil are some of the major causes of drought. The plant, therefore, is unable to obtain enough water for its growth and development. However, plant tolerance toward drought stress has been observed to be improved by the application of nanoparticles at different concentrations. Due to drought, leaf size is reduced, stem extension and root proliferation are extensively enhanced, and plant water relations are critically disturbed resulting in reduced water uptake efficiency. Positive effects on relative water content, photosynthesis, electrolyte leakage channel (ELI) of the membrane, and other components involved in drought stress like chlorophyll, carotenoid, carbohydrate, proline have been observed by pretreating the plants with SNPs (Asua 2002; Blankenship et al. 2011). On applying silicon nanoparticle, the adaptation mechanism that Hawthron plants have in maintaining critical physiological and biochemical conditions for avoiding stress due to drought is not well understood (Ashkavand et al. 2015). In addition to SiO_2 nanoparticle into a plant medium, various defense mechanisms were performed to protect the plant for example,

1) Loss of lipid peroxidation in the leaf cells by reduction of plasma wall penetrability.
2) Drought and heat stress resistance of plant cell walls (Zhu et al. 2004).

On comparing common silica fertilizer and silica nanoparticle, the amount of proline increased after silica nanoparticle application (Kalteh et al. 2014). Si nanoparticle acts as a catalyst for two enzymes, POD (peroxidase), CAT (catalase) in the plant leaves in case of stress. Silica nanoparticle also helps in increasing water efficiency during stress. TiO_2 application in plants causes increase in chlorophyll formation, rate of germination, RuBisco, photosynthesis (Hong et al. 2005; Zheng et al. 2005).

17.6.1.2 Salinity Stress

Salinity stress is a result of excess salt accumulation in the soil. It affects plant growth resulting in decreased production. Salt stress affects more than 20% of farmed land in the world today, and the number is growing by the day. Increased salinity in the soil has a negative impact on plants, causing increased osmotic pressure, accumulation and production of ROS (reactive oxygen species), and lowering the leaf water potential, all of which cause physiological and morphological changes, as well as ion toxicity, which alters biochemical processes (Rajput et al. 2015). Applying silicon fertilizer and silicon nanoparticles under salt stress enhanced growth and development and also increased chlorophyll and proline amount in *Ocimum basilicum* (basil) (Kalteh et al. 2014). Chlorophyll content, leaf dry weight, leaf fresh weight, proline quantity, and the regulation of the gene that produces antioxidant enzyme under salt stress all increased when SiO_2 nanoparticles were applied (Haghighi et al. 2012). The antioxidant defense system is activated, and osmolyte synthesis is induced when Si ions and silica nanoparticles are applied to rice (Abdel-Haliem et al. 2017; Hatami et al. 2016). Application of SiO_2 NPs to tomatoes during salinity stress can enhance seed, plant dry and fresh weight, chlorophyll, and proline content (Gowayed et al. 2017; Mushtaq et al. 2019). Sunflower cultivars show tolerance against salinity stress by applying a foliar spray of iron sulfate ($FeSO_4$) NPs. $FeSO_4$ nanoparticles provide plants some physiological adaptation against salinity like increased leaf area result in an enhanced net rate of assimilation of CO_2 (carbon dioxide). This raises the dry weight of the shoot, the CO_2 concentration in the sub-stomata, and photosystem II's maximum photochemical efficiency (Fv/Fm), but decreases the significant level of sodium (Na) in the leaves (Torabian et al. 2018). Application of SiO_2 to plants during salinity stress may improve chlorophyll content, antioxidant enzyme activity, leaf fresh weight, and proline accumulation (Haghighi et al. 2012; Kalteh et al. 2014; Siddiqui et al. 2014). The use of Cerium oxide nanoparticles (CeO-NPs) on *Glycine max*, which under the salinity can stimulate their growth by upregulating the rate of photosynthesis by controlling the activity of enzyme, Rubisco (Cao et al. 2017). On treating *Brassica napus L.* with CeO-NPs during salinity stress, it shows higher plant biomass and higher photosynthesis rate, but salt uptake from the soil is inhibited without changing the nutritional value (Rossi et al. 2016). A certain amount of zinc oxide nanoparticle generally enhances the growth and photosynthetic pigments, antioxidant responses in *Luminus termis* to protect it from salinity stress (Latef et al. 2017).

17.6.1.3 Metal Stress

The presence of excess metal nutrients in plants causes metal stress resulting in oxidative stress and toxicity (Rajput et al. 2019a). Some necessary micronutrients and macronutrients are zinc, copper, and manganese (Rajput et al. 2019b). These are required for plants' normal growth and development. All living organisms including plants require zinc as a micronutrient. It has a number of critical functions to perform in the cell. They prevent plants from absorbing toxic heavy metals like cadmium (Cd) (Baybordi 2005; Rajput et al. 2020; Venkatachalam et al. 2017). The effects of magnetic (Fe_3O_4) nanoparticles (nano-Fe_3O_4) to reduce heavy metal toxicity (Pb, Zn, Cu, Cd) in white were investigated. In a 1 mM solution of each heavy metal, the use of magnetic (Fe_3O_4) nanoparticles (2000 mg/l) drastically reduced growth inhibition and boosted protective mechanisms in white seedlings to reduce oxidative stress caused due to heavy metals. It is possible that the positive effects of nano-Fe_3O_4 under heavy metal stress are related to greater antioxidant enzyme activity. The reduction in MDA content verified their mitigating effect. The relieving effects of nano-Fe_3O_4 were thought to be linked to their heavy metal adsorption potential, and this could be credited to an electrostatic interaction between negatively changed adsorption surface and ions of heavy metals (Mingshu et al. 2021). In an experiment in *Brassica juncea*, iron oxide nanoparticles (Fe_3O_4 NP) were found to be effective in mitigating the detrimental consequences of Arsenic. Plant stress-related indicators decreased, which may be due to the plant's lower ability to absorb arsenic in the presence of Fe_3O_4 NP (Parveen et al. 2017). Zinc oxide nanoparticles were reported to increase wheat chlorophyll content, gas exchange properties, antioxidant enzymes, zinc absorption, and yield lowering Cd concentrations under cadmium stress (Hussain et al. 2018).

17.6.1.4 Ultraviolet (UV) Radiation Stress

UV radiation belongs to the nonionizing portion of the EM spectrum. There are three types of UV radiations based on their wavelength: UV-A, UV-B, UV-C. The most dangerous of all is UV-C having wavelength of 200–280 nm. UV-A is less hazardous and it represents 6.3% of incoming solar radiations having wavelength 320–400 nm. UV-B constitutes 1.5% of incoming solar radiation and is the most dangerous radiation that causes damage in plants, especially the photosynthesis apparatus. In plants, UV stress is induced by a variety of mechanisms including excessive formation of free radicals and other reactive species, as well as other related reasons. They have a negative impact on the plant's growth and development due to their extreme responsiveness. TiO_2 NPs have been found to play an important role in reducing light stress. When exposed to light, TiO_2 NPs act as catalyst in the oxidation–reduction reaction, resulting in the generation of superoxide anion radicals and hydroxide (Baiazidi-Aghdam et al. 2016). UV-B exposure, on the other hand, is another element that causes reduced photosynthesis and aberrant leaf structure. By blocking the actions of SOD and APX, it causes production of ROS in plant cells. According to Tripathi et al. (2017), UV-B exposure causes H_2O_2 and superoxide radical production leading to peroxidation of lipids and electrolyte leakage. Si NPs have been discovered to improve antioxidant activity in wheat, and thereby oxidative damage caused due to UV exposure is reduced. The method of action is unknown at this time, although it is thought that Si NPs protect the plant by an antioxidant defense system, which counteract photosynthetic damage produced by ROS. Thus, NPs are important in eliciting plant abiotic stress management responses. They can be employed as instruments in the abiotic stress management process in crops because they affect abiotic stress-induced responses at several levels. To assure low or no toxicity in agricultural applications, however, suitable tuning of NPs' physiochemical, optical, electrical, and biological properties is essential in NP engineering.

17.6.2 NPs in Biotic Stress

Biotic stress can be defined as the damage caused to plant cells due to other living things such as fungi, parasites, insects, weeds, bacteria, or virus (Atkinson and Urwin 2012). In contemporary times, sprays were the most widely used source of agricultural fertilizers, herbicides, insecticides, antibiotics, pesticides, and nutrients (Begum et al. 2010). The chemicals or mixture of chemicals that are used to prevent, eradicate, or control pests are called pesticides. Pests include disease vectors, undesired plant or animal species, and pests that disrupt or obstruct the production, processing, storage, transportation, or marketing of food, agricultural commodities, wood and wood products, or animal feedstuffs. According to WHO, to control insects, arachnids or other pests to grow inside or on the animals' bodies, insecticides, arachnoiditis, and other pesticides can be given to animals. NPs-based insecticides, pesticides, and herbicides are a new product that has provided potential power to cope up the problem of biotic stress (Rajput Vishnu et al. 2017)

17.6.2.1 Nanoinsecticides

When it comes to pesticide application, water solubility is crucial. The majority of insecticides on the market today have very low water solubility. This necessitates the use of organic solvents to thoroughly dissolve the herbicide in water, raising the pesticide's cost and toxicity, while the use of nanoparticles can increase pesticide solubility, thus reducing toxicity. Less water-soluble pesticides have been made using nanoparticles like modified chitosan and porous silica and have been successfully loaded into pesticides (Worrall Elizabeth et al. 2018). Azadirachtin with modified chitosan nanoparticles (hydrophobic insecticide) showed a good reduction of cell proliferation and prolonged drug release in *Spodoptera litura* ovarian cell lines (Lu et al. 2013). Intercalation of anacardic acid into LDH (layered double hydroxides) nanoparticles in insects has shown increased mortality (Nguyen et al. 2014). A direct spray onto the mustard leaves or skin of *S. litura* increases the mortality rate. These findings emphasize the potential advantages of employing nanoparticles to increase solubility. Another key issue, post-insecticide application evaporation or volatilization, has been handled utilizing solid lipid nanoparticles (SLNs) (Lai et al. 2006). Nanoparticles containing insecticides can help to reduce insecticide toxicity by delaying the release of active chemicals.

17.6.2.2 Nanofungicides

Fungicides are biological or chemical substances that are used to kill fungi. Fungicides are divided into two categories: fungicides that kill fungi and fungicides that kill fungi.

1) **Preventive fungicides** – These include compounds like sulfur, dichloro-carbamates, organometallics, phthalimides, and benzamides that prevent fungal infections in plants.
2) **Curative fungicides** – These are the compounds that migrate to the area infected by plant pathogen and prevent it from developing further, such as acetimides, dicarboximide, and sterol inhibitor.

Polymer mixes, silica, and chitosan are the examples of the most commonly investigated and useful nanoparticle carriers. The efficiency of the nanofungicide was checked on a very broad range of fungal species. But recently, a few nanofungicides were tested on the plant, and not toxic evidence was observed. Due to low water solubility, volatilization, and poor stability, generic insecticides on the application do not show the desired effect. Nanoparticle-based fungicide, pyrachlostrobin, which is a fungicide having lower water solubility, has proven to be a solution to these problems. Chitosan-lactide copolymer is loaded onto nanoparticles in various concentrations to inhibit *Colletotrichum gossypii* (Xu et al. 2014). Nanofungicides are an eco-friendly alternative to conventional fungicides. Copper nanoparticles including Cu-NPs and CuO-NPs, silver nanoparticles, i.e Ag-NPs, and zinc nanoparticles all have inhibitory effects. These characteristics are beneficial against economically significant foliar and soilborne plant diseases caused by a variety of pathogens. The characteristics of these metal nanoparticles were tested in vitro. Cu-NPs inhibited mycelial development most effectively with EC50 values ranging from 162 to 310 g/ml, followed by ZnO-NPs having mean inhibition rates ranging from 235 to 848 g/ml. CuO-NPs were ineffective against fungus. Ag-NPs had a substantial inhibitory effect solely on *Botrytis cinerea* where CuO-NPs were practically insensitive to all fungal species. ZnO-NPs were more toxic to all fungal species tested than $ZnSO_4$, whereas CuNPs were more fungi toxic to CuSO4 in all cases except *B. cinerea, Alternaria alternata*, and *Monilinia fructicola*, according to comparison of mycelial growth fungi toxicity experiments between the nanoparticles and their bulk-sized counterparts. The toxicity of Cu-NPs and CuO-NPs is positively correlated, whereas the toxicity of NPs and their bulk metal counterparts is unrelated. This suggests that bulk and nanosized metals may have different modes of action. Despite considerable heterogeneity among fungal species, all nanoparticles on application to spores instead of fungal hyphae showed a significant increase in fungi toxicity. As a result, such coated nanoparticles have a lot of potentials to be exploited as antifungal protective agents.

17.6.2.3 Nanoherbicide

Herbicides are the chemicals used to kill the unwanted plants and grasses, called weeds. The conventional herbicides that are sprayed also affect the crop (Deva and Kadiri 2016). They have a high toxicity level and a lengthy half-life, and they are unaffected by the conventional wastewater treatment plants (Camilli et al. 2014). This increases the level of soil pollutants, as well as water pollutants. As the weeds grow along with the crops in the fields, these herbicides harm the crops as well. The herbicides are used to kill the weeds showing harmful effects on the crops also. Here, nanoherbicides come into play as they can be used to solve this problem. Because of their small size (approximately 1–100 nm), these particles can easily dissolve in soil particles and kill weeds from their roots without harming the primary crops. Herbicides based on nanoparticles may be able to destroy weeds while increasing yields (Bickel and Killorn 2001). When herbicides like triazine, ametrine, and atrazine are coupled with nanoparticles, the efficiency of these nanoparticle-based herbicides can rise by as much as 84% (Grillo et al. 2012). Atrazine is the frequently used herbicide. As a result, the crops can more effectively utilize the soil nutrients that were previously distributed by weeds. This also increases the disease resistance of the crops. The nanoparticles not only act as compound carriers but also as a catalyst in degradation reactions of various compounds. Nanoparticles such as modified silver and carboxymethyl cellulose combination can further enhance the efficiency of degradation of herbicide.

17.6.3 NPs and Crop Improvement: Nanoparticle-Mediated Transformation

Nanoparticle-mediated gene transformation has several advantages as compared to the traditional transformation method. Traditional transformation method has certain limitations like some methods are useful for monocotyledons plants only and some for dicotyledons. But nanoparticle-based transformation is applicable for both monocotyledons, as well as dicotyledons. Nanoparticle mediated transformation can also overcome transgenic gene silencing. The efficiency of nanoparticle-mediated gene transformation is higher than the traditional method of transformation. It has the capacity to transfer multigene without any genomic barrier. Ceramics, silicates, metal oxides, magnetic materials, dendrimers, and liposomes are among the nanoparticles used for transformation (Holister et al. 2003). Mesoporous silica can also be used (Torney et al. 2007). For a successful genetic transformation, nanomaterials should be biodegradable, less genotoxic and cytotoxic (Buzea and Pacheco 2017; Czapar and Steinmetz 2017). *Agrobacterium tumefaciens*-mediated transformation has a capacity of 54.4%, while the efficiency of nanoparticle-mediated transformation is around 80.7%. Only 8% of naked DNA is used in nanoparticle-mediated transformations. These encouraging findings suggest that CaP nanoparticles are a more effective transforming vector in plants than *Agrobacterium tumefaciens* (Naqvi et al. 2012). Silicon carbide-mediated transformation has been successfully used to transform tobacco, soyabean, cotton, and rice (Armstrong and Green 1985; Wang et al. 1995; Serik et al. 1996; Asad and Arsh 2012; Latef et al. 2017).

17.7 Conclusion

The combination of two innovative technologies, geoinformatics and nanotechnology, have the potential power to coping up the problem related to abiotic and biotic stress caused by climate change that directly affected agriculture production. These two technologies have developed a concept known as the "3-T" concept. The first T stands for Trace, tracing, or real-time monitoring of yield, soil, water, and weather using geoinformatics; second T for Test, after getting geoinformatics data. It is essential to validate and confirm with ground data for this using various sensors for obtaining ground data, for example UGV, UAV, and nanosensors. After getting the test result, we examine the crop and soil at the micro-level. Third T stands for Treat that is directly related to pre-diagnosis of crop and soil before they are affected by various types of disease. For pre-diagnosis, plants and soil can use nano fertilizer pesticides, herbicides, and insecticides, which are more effective than conventional fertilizer pesticides, herbicides, and insecticides.

Acknowledgments

The authors are thankful to Prof. Minkina, and Dr. Rajput are grateful to the Russian Foundation for Basic Research, project no. 19-05-50097.

References

Abdel-Haliem, M.E.F., Hegazy, H.S., Hassan, N.S., and Naguib, D.M. (2017). Effect of silica ions and nano silica on rice plants under salinity stress. *Ecol. Eng.* 99: 282–289.

Androvitsaneas, P., Young, A.B., Schneider, C. et al. (2016). Charged quantum dot micropillar system for deterministic light matter interactions. *Phys. Rev. B* 93: 241409. https://doi.org/10.1103/physrevb.93.241409.

Anton, N. and Vandamme, T.F. (2009). The universality of low-energy nano-emulsification. *Int. J. Pharm.* 377: 142–147. https://doi.org/10.1016/j.ijpharm.2009.05.014.

Armstrong, C.L. and Green, C.E. (1985). Establishment and maintenance of friable, embryogenic maize callus and the involvement of L-proline. *Planta* 164: 207–214. https://doi.org/10.1007/BF00396083.

Asad, S. and Arsh, M. (2012). Silicon carbide whisker-mediated plant transformation. In: *Properties and Applications of Silicon Carbide* (ed. R. Gerhardt), 1–16. Rijeka: BoD-Books on Deman https://doi.org/10.5772/15721.

Ashkavand, P., Tabari, M., Zarafshar, M. et al. (2015). Effect of SiO_2 nanoparticles on drought resistance in hawthorn seedlings. *LeśnePraceBadawcze/Forest Research Papers Grudzień* 76 (4): 350–359.

Asua, J.M. (2002). Mini-emulsion polymerization. *Prog. Polym. Sci.* 27: 1283–1346. https://doi.org/10.1016/S0079-6700(02)00010-2.

Atkinson, N.J. and Urwin, P.E. (2012). The interaction of plant biotic and abiotic stresses: from genes to the field. *J. Exp. Bot.* 63 (10): 3523–3543. https://doi.org/10.1093/jxb/ers100.

Baiazidi-Aghdam, M.T., Mohammadi, H., and Ghorbanpour, M. (2016). Effects of nanoparticulateanatase, titanium dioxide on physiological and biochemical performance of *Linumusitatissimum* (Linaceae) under well watered and drought stress conditions. *Braz. J. Bot.* 39: 139–146.

Bakalova, R., Zhelev, Z., Ohba, H. et al. (2004). Quantum dots as photosensitizers? *Nat. Biotechnol.* 22: 1360–1361. https://doi.org/10.1038/nbt1104-1360.

Baker, S., Volova, T., Prudnikova, S.V. et al. (2017). Nanoagroparticles emerging trends and future prospect in modern agriculture system. *Environ. Toxicol. Pharmacol.* 53: 10–17.

Bao, W., Wan, Y., and Baluška, F. (2017). Nanosheets for delivery of biomolecules into plant cells. *Trends Plant Sci.* 22 (6): 445–447. https://doi.org/10.1016/j.tplants.2017.03.014.

Bargaz, A., Lyamlouli, K., Chtouki, M. et al. (2018). Soil microbial resources for improving fertilizers efficiency in an integrated plant nutrient management system. *Front. Microbiol.* 9: 1606.

Batukaeva, A., Endovitsky, A., Kalinichenko, V. et al. (2017). Cadmium status in chernozem of the Krasnodar Krai (Russia) after the application of phosphogypsum. *Proc. Estonian Acad. Sci.* 66 (4): 1–15. https://doi.org/10.3176/proc.2017.4.17.

Baybordi, A. (2005). Effect of zinc, iron, manganese and copper on wheat quality under salt stress conditions. *J. Water Soil* 140: 150e170.

Begum, N., Sharma, B., and Pandey, R.S. (2010). Evaluation of insecticidal efficacy of *Calotropis Procera* and *Annona Squamosa* ethanol extracts against *Musca Domestica*. *J. Biofertil. Biopestic.* 1: 101.

Beneduzi, A., Ambrosini, A., and Passaglia, L.M.P. (2012). Plant growth-promoting rhizobacteria (PGPR): their potential as antagonists and biocontrol agents. *Genet. Mol. Biol.* 4: 44–51.

Bickel, A. and Killorn, R. (2001). Spatial response of corn to banded zinc sulfate fertilizer in Iowa. *North central Extension Industry Soil Fertility Conference*, Potash and Phosphate Institute, 17, 14–15.

Blankenship, R.E., Tiede, D.M., Barber, J. et al. (2011). Comparing photosynthetic and photovoltaic efficiencies and recognizing the potential for improvement. *Science* 332: 805–809.

Boyer, J.S. (1982). Plant productivity and environment. *Science* 218 (4571): 443–448.

Buermann, W., Dong, J., Zeng, X. et al. (2001). Evaluation of the utility of satellite-based vegetation leaf area index data for climate simulations. *J. Clim.* 14: 3536–3550.

Bulovic, V., Mandell, A., and Perlman, A. (2004). Molecular memory device. US 20050116256, A1.

Buzea, C. and Pacheco, I. (2017). Nanomaterial and nanoparticle: origin and activity. In: *Nanoscience and Plant–Soil Systems. Soil Biology*, vol. 48 (eds. M. Ghorbanpour, K. Manika and A. Varma), 45. Cham: Springer https://doi.org/10.1007/978-3-319-46835-8_3.

Camilli, L., Pisani, C., Gautron, E. et al. (2014). A three-dimensional carbon nanotube network for water treatment. *Nanotechnology* 25: 065701. https://doi.org/10.1088/0957-4484/25/6/065701.

Cao, Z., Stowers, C., Rossi, L. et al. (2017). Physiological effects of cerium oxide nanoparticles on the photosynthesis and water use efficiency of soybean (*Glycine max* (L.) Merr.). *Environ. Sci. Nano* 4 (5): 1086e1094.

Chahine, N.O., Collette, N.M., Thomas, B.C. et al. (2014). Nanocomposite scaffold for chondrocyte growth and cartilage tissue engineering: effects of carbon nanotube surface functionalization. *Tissue Eng. A* 20: 2305–2315. https://doi.org/10.1089/ten.TEA.2013.0328.

Chase, T.N., Pielke, R.A., Kittel, T.G. et al. (1996). Sensitivity of a general circulation model to global changes in leaf area index. *J. Geophys. Res, D* 101: 7393–7408.

Chen, J.M. and Black, T.A. (1992). Defining leaf area index for non-flat leaves. *Plant Cell Environ.* 15: 421–429.

Chen, H. and Yada, R. (2011). *Trends Food Sci. Technol.* 22: 585.

Chinnamuthu, C.R. and Murugesa Boopathi, P. (2009). Nanotechnology and agroecosystem. *Madras Agric. J.* 96: 1–6.

Czapar, A.E. and Steinmetz, N.F. (2017). Plant viruses and bacteriophages for delivery in medicine and biotechnology. *Curr. Opin. Chem. Biol.* 38: 108–116. https://doi.org/10.1016/j.cbpa.2017.03.01.

Das, A. and Das, B. (2019). Nanotechnology a potential tool to mitigate abiotic stress in crop plants. In: *Abiotic and Biotic Stress in Plants* (ed. A.B. de Oliveira). IntechOpen https://doi.org/10.5772/intechopen.83562.

Das, S., Wolfson, B.P., Tetard, L. et al. (2015). Effect of N-acetyl cysteine coated CdS:Mn/ZnS quantum dots on seed germination and seedling growth of snow pea (*Pisum sativum* L.): imaging and spectroscopic studies. *Environ. Sci.* 2: 203–212. https://doi.org/10.1039/c4en00198b.

Davis, K.F., Odorico, P.D., and Rulli, M.C. (2014). Moderating diets to feed the future. *Earth's Future* 2: 559–565. https://doi.org/10.1002/2014EF000254.

Deva, S. and Kadiri, L. (2016). A review on the impact of nanoparticles on plant growth and development. *Adv. Life Sci.* 5 (1): 16–21.

Di Giacomo, R., Daraio, C., and Maresca, B. (2015). Plant nanobionic materials with a giant temperature response mediated by pectin-Ca2+. *Proc. Natl. Acad. Sci. U. S. A.* 112 (15): 4541–4545. https://doi.org/10.1073/pnas.

Ebrahimi, N. and Mansoori, G.A. (2014). Reliability for drug targeting in cancer treatment through nanotechnology. *Int. J. Med. Nano Res.* 1 (1) ISSN 2378-3664.

Fabre, S., Gimenez, R., Elger, A., and Rivière, T. (2020). Unsupervised monitoring vegetation after the closure of an ore processing site with multi-temporal optical remote sensing. *Sensors* 20 (17): 4800. https://doi.org/10.3390/s20174800.

Fang, H., Jiang, C., Li, W. et al. (2013). Characterization and intercomparison of global moderate resolution leaf area index (LAI) products: analysis of climatologies and theoretical uncertainties. *J. Geophys. Res. Biogeosci.* 118: 529–548.

FAO (2019). *Agriculture and Climate Change – Challenges and Opportunities at the Global and Local Level – Collaboration on Climate-Smart Agriculture*, 52. Rome: Food and Agriculture Organization (FAO) of the United Nations.

Frame, B.R., Shou, H., Chikwamba, R.K. et al. (2002). *Agrobacterium tumefaciens* mediated transformation of maize embryos using a standard binary vector system. *Plant Physiol.* 129: 13–22.

GCOS (2011). *Systematic Observation Requirements for Satellite-Based Products for Climate: 2011 Update; GCOS-154.* World Meteorological Organization: Geneva, Switzerland.

Ghasabkolaei, N., Choobbasti, A.J., Roshan, N., and Ghasemi, S.E. (2017). Geotechnical properties of the soils modified with nanomaterials: a comprehensive review. *Arch. Civil Mech. Eng.* 17 (3): 639–650.

Ghorbanpour, M. (2015). Major essential oil constituents, total phenolics and flavonoids content and antioxidant activity of Salvia officinalis plant in response to nano-titanium dioxide. *Indian J. Plant Physiol.* 20 (3): 249–256.

Ghorbanpour, M. and Hadian, J. (2015). Multi-walled carbon nanotubes stimulate callus induction, secondary metabolites biosynthesis and antioxidant capacity in medicinal plant *Saturejakhuzestanica* grown in vitro. *Carbon* 94: 749–759.

Ghorbanpour, M. and Hatami, M. (2014). Spray treatment with silver nanoparticles plus thidiazuron increases anti-oxidant enzyme activities and reduces petal and leaf abscission in four cultivars of geranium (*Pelargonium zonale*) during storage in the dark. *J. Hortic. Sci. Biotechnol.* 89 (6): 712–718.

Ghorbanpour, M., Fahimirad, S., Xue, Y., and Mansoori, G.A. (2010). Self-assembly of diamonded molecules and derivatives (MD simulations and DFT calculations). *Int. J. Mol. Sci.* 11 (1): 288–303. https://doi.org/10.3390/ijms11010288.

Giraldo, J.P., Landry, M.P., Faltermeier, S.M. et al. (2014). Plant nanobionics approach to augment photosynthesis and biochemical sensing. *Nat. Mater.* https://doi.org/10.1038/nmat3890.

Glick, B.R. (2012). Plant growth-promoting bacteria: mechanisms and applications. *Scientifica (Cairo)* 2012: 963401. https://doi.org/10.6064/2012/963401. Epub 2012 Sep 19. PMID: 24278762; PMCID: PMC3820493.

Gowayed, M.H., Al-Zahrani, H.S., and Metwali, E.M. (2017). Improving the salinity tolerance in potato (*Solanum tuberosum*) by exogenous application of silicon dioxide nanoparticles. *Int. J. Agric. Biol.* 19: 1.

Grillo, R., dos Santos, N.Z.P., Maruyama, C.R. et al. (2012a). The effect of N-Si on tomato seed germination under salinity levels. *J. Biol. Environ. Sci.* 6 (16): 87–90.

Grillo, R., dos Santos, N.Z.P., Maruyama, C.R. et al. (2012b). Poly (ε-caprolactone) nanocapsules as carrier systems for herbicides: physico-chemical characterization and genotoxicity evaluation. *J. Hazard. Mater.* 231: 1–9.

Grillo, R., Abhilash, P.C., and Fraceto, L.F. (2016). Nanotechnology applied to bio-encapsulation of pesticides. *J. Nanosci. Nanotechnol.* 16: 1231–1244.

Gupta, A., Eral, H.B., Hatton, T.A., and Doyle, P.S. (2016). Nanoemulsions: formation, properties and applications. *Soft Matter* 12: 2826–2841. https://doi.org/10.1039/ e5sm02958a.

Gutiérrez, J.M., González, C., Maestro, A. et al. (2008). Nano-emulsions: new applications and optimization of their preparation. *Curr. Opin. Colloid Interface Sci.* 13: 245–251. https://doi.org/10.1039/C5SM02958A.

Haghighi, M., Afifipour, Z., and Mozafarian, M. (2012). The effect of N-Si on tomato seed germination under salinity levels. *J. Biol. Environ. Sci.* 6 (16): 87–90.

Hajirostamlo, B., Mirsaeedghazi, N., Arefnia, M. et al. (2015). The role of research and development in agriculture and its dependent concepts in agriculture (short review). *Asian J. Appl. Sci. Eng.* 4 https://doi.org/10.1016/j. cocis.2008.01.005.

Han, J.H. et al. (2010). Excitation antennas and concentrators from core-shell and corrugated carbon nanotube filaments of homogeneous composition. *Nat. Mater.* 9: 833–839.

Hatami, M., Ghorbanpour, M., and Salehiarjomand, H. (2014). Nano-anatase TiO_2 modulates the germination behavior and seedling vigority of the five commercially important medicinal and aromatic plants. *J. Biol. Environ. Sci.* 8 (22): 53–59.

Hatami, M., Kariman, K., and Ghorbanpour, M. (2016). Engineered nanomaterial-mediated changes in the metabolism of terrestrial plants. *Sci. Total Environ.* 571: 275–291.

Hatami, M., Hadian, J., and Ghorbanpour, M. (2017). Mechanisms underlying toxicity and stimulatory role of single-walled carbon nanotubes in *Hyoscyamus niger* during drought stress simulated by polyethylene glycol. *J. Hazard. Mater.* 324: 306–320.

Holister, P., Weener, J.W., Román, C.V., and Harper, T. (2003). Nanoparticles. *Technol. White Papers* 3: 1–11.

Hong, F., Zhou, J., Liu, C. et al. (2005). Effect of nano-TiO on the photochemical reaction of chloroplasts of spinach. *Biol. Trace Elem. Res.* 105: 269–279.

Hu, Y., Li, J., Ma, L. et al. (2010). High efficiency transport of quantum dots into plant roots with the aid of silwet L-77. *Plant Physiol. Biochem.* 48: 703–709. https://doi.org/10.1016/j.plaphy.2010.04.001.

Hussain, A., Ali, S., Rizwan, M. et al. (2018). Zinc oxide nanoparticles alter the wheat physiological response and reduce the cadmium uptake by plant. *Environ. Pollut.* 242: 1518–1526.

Iranpour, B. and Haddad, A. (2016). The influence of nanomaterials on collapsible soil treatment. *Eng. Geol.* 205: 40–53.

Kah, M. (2015). Nanopesticides and nanofertilizers: emerging contaminants or opportunities for risk mitigation? *Front. Chem.* 3: 64.

Kalteh, M., Alipour, Z.T., Ashraf, S. et al. (2014). Effect of silica nanoparticles on basil (*Ocimum basilicum*) under salinity stress. *J. Chem. Health Risks* 4: 49–55.

Kalteh, M., Zarrin, T.A., Shahram, A. et al. (2014). Effect of silica nanoparticles on basil (*Ocimum basilicum*) under salinity stress. *J. Chem. Health Risks* 4 (3): 49e55.

Khiari, R. (2017). Valorization of agricultural residues for cellulose nanofibrils production and their use in nanocomposite manufacturing. *Int. J. Polym. Sci.* 9: 1–10.

Konstantatos, G. and Argent, E.H. (2009). Solution-processed quantum dot photodetectors. *Proc. IEEE* 97: 1666–1683. https://doi.org/10.1109/JPROC.2009.2025612.

Kumar, K.M., Hanumanthappa, M., Marimuthu, S., and Meenambigai, C. (2018). A review on enhancing the fertilizers use efficiency to minimize environmental impacts. *Int. J. Chem. Stud.* 6 (3): 2167–2174.

Kwon, K.C. and Daniell, H. (2015). Low-cost oral delivery of protein drugs bioencapsulated in plant cells. *Plant Biotechnol. J.* 13 (8): 1017–1022. https://doi.org/10.1111/pbi.12462.

Lai, F., Wissing, S.A., Müller, R.H., and Fadda, A.M. (2006). *Artemisia arborescens* L essential oil-loaded solid lipid nanoparticles for potential agricultural application: preparation and characterization. *AAPS PharmSciTech* 7: 10.

Lal (2014). Climate strategic soil management. *Challenges* 5 (1): 43–74.

Latef, A.A.H.A., Alhmad, M.F.A., and Abdelfattah, K.E. (2017). The possible roles of priming with ZnO nanoparticles in mitigation of salinity stress in lupine (*Lupinustermis*) plants, *J. Plant Growth Regul.* 36 (1): 60e70.

Laware, S.L. and Raskar, S. (2014). Effect of titanium dioxide nanoparticles on hydrolytic and antioxidant enzymes during seed germination in onion. *Int. J. Curr. Microbiol. App. Sci.* 3 (7): 749–760.

Liang, T. and Ritter, T. (2014). *Synthesis of Fluorides, Comprehensive Organic Synthesis II*, 2e, vol. 210–238. Elsevier https://doi.org/10.1016/B978-0-08-097742-3.00607-8.

Lu, W., Lu, M.L., Zhang, Q.P. et al. (2013). Octahydrogenated retinoic acid-conjugated glycol chitosan nanoparticles as a novel carrier of azadirachtin: synthesis, characterization, and in vitro evaluation. *J. Polym. Sci. A Polym. Chem.* 51: 39323940.

Luca, M. (2018). Nanotechnology in agriculture: new opportunities and perspectives. *New Visions Plant Sci.* 122–140 https://doi.org/10.5772/intechopen.74425.

Malik, M., Ahmad, A.A., Siddiqui, S., and Kamaluddin, A.M.Z. (2017). Methods in transgenic technology. *Plant Biotechnol. Princ. Appl.*: 93–115. https://doi.org/10.1007/978-981-10-2961-5.

Mansoori, G.A. (2017). An introduction to nanoscience and nanotechnology. In: *Nanoscience and Plant–Soil Systems. Soil Biology*, vol. 48 (eds. Ghorbanpour et al.). Springer International Publishing AG https://doi.org/10.1007/978-3-319-46835-8_13.

Mansoori, G.A., Mohazzabi, P., McCormack, P., and Jabbari, S. (2007). Nanotechnology in cancer prevention, detection and treatment: bright future lies ahead. *World Rev. Sci. Technol. Sustain. Dev.* 4 (2/3): 226–257.

Mansoori, G.A., Brandenburg, K.S., and Shakeri-Zadeh, A. (2010). A comparative study of two folate-conjugated gold nanoparticles for cancer nanotechnology applications. *Cancer* 2 (4): 1911–1928.

Mariano, M., El Kissi, N., and Dufresne, A. (2014). Cellulose nanocrystals and related nanocomposites: review of some properties and challenges. *J. Polym. Sci. B Polym. Phys.* 52 (12): 791–806.

Mason, T.G., Wilking, J.N., Meleson, K. et al. (2006). Nanoemulsions: formation, structure, and physical properties. *J. Phys. Condens. Matter* 18: R635–R666. https://doi.org/10.1088/0953-8984/18/41/r01.

Mingshu, L., Zhang, P., Adeel, M. et al. (2021). Physiological impacts of zero valent iron, Fe3O4 and Fe2O3 nanoparticles in rice plants and their potential as Fe fertilizers. *Environ. Pollut.* 269: 116134. https://doi.org/10.1016/j.envpol.2020.116134.

Mitter, N., Worrall, E.A., Robinson, K.E. et al. (2017). Clay nanosheets for topical delivery of RNAi for sustained protection against plant viruses. *Nat. Plants* 3: 16–207.

Munier, S., Carrer, D., Planque, C. et al. (2018). Satellite leaf area index: global scale analysis of the tendencies per vegetation type over the last 17 years. *Remote Sens.* 10 (3): 424. https://doi.org/10.3390/rs10030424.

Mushtaq, A.Y., Rizwan, S.A., Jamil, N.E. et al. (2019). Influence of silicon sources and controlled release fertilizer on the growth of wheat cultivars of Balochistan under salt stress. *Pak. J. Bot.* 51 (5): 1561–1567.

Naqvi, S., Maitra, A., Abdin, M. et al. (2012). Calcium phosphate nanoparticle mediated genetic transformation in plants. *J. Mater. Chem.* 22 https://doi.org/10.1039/C2JM11739H.

Nayana, A.R., Jerin, J.B., Ashitha, J., and Radhakrishnan, E.K. (2020). Nanotechnological advances with PGPR applications. *Sustain. Agric. Rev.* 41: 163–180.

Nazem, A. and Mansoori, G.A. (2008). Nanotechnology solutions for Alzheimer's disease: advances in research tools, diagnostic methods and therapeutic agents. *J. Alzheimers Dis.* 13 (2): 199–223.

Nazem, A. and Mansoori, G.A. (2014). Nanotechnology building blocks for intervention with Alzheimer's disease pathology: implications in disease modifying strategies. *J. Bioanal. Biomed.* 6 (2): 9–14.

Nguyen, T.N.Q., Le, V.A., Hua, Q.C., and NguÅn, T.N. (2014). Enhancing insecticide activity of anacardic acid by intercalating it into MgAl layered double hydroxides nanoparticles. *J. Vietnamese Environ.* 6: 208–211.

Nima, A.Z., Lahiani, M.H., Watanabe, F. et al. (2014). Plasmonically active nanorods for delivery of bio-active agents and high-sensitivity SERS detection in planta. *RSC Adv.* 4: 64985–64993. https://doi.org/10.1039/C4RA10358K.

Noji, T., Kamidaki, C., Kawakami, K. et al. (2011). Photosynthetic oxygen evolution in mesoporous silica material: adsorption of photosystem II reaction center complex into 23 nm nanopores in SBA. *Langmuir* 27 (2): 705–713.

Nunes, C. and Auge, J.I. (1999). Land-Use and Land-Cover Change (LUCC): Implementation Strategy; IGBP Report-48. University of North Texas Libraries, Digital Librar, Stockholm, Sweden.

Ojo, O.I. and Ilunga, F. (2018). Geospatial analysis for irrigated land assessment, modeling and mapping. *Multi-Purposeful Application of Geospatial Data* 9: 65–84.

Panda, S.S., Hoogenboom, G., and Paz, J.O. (2010). Remote sensing and geospatial technological applications for site-specific management of fruit and nut crops: a review. *Remote Sens.* 2 (8): 1973–1997. https://doi.org/10.3390/rs2081973.

Parisi, C., Vigani, M., and Rodríguez-Cerezo, E. (2015). Agricultural nanotechnologies: what are the current possibilities? *Nano Today* 10: 124–127.

Parveen, R., Shamsi, T.N., and Fatima, S. (2017). Nanoparticles-protein interaction: role in protein aggregation and clinical implications. *Int. J. Biol. Macromol.* 94: 386–395. https://doi.org/10.1016/j.ijbiomac.2016.10.024.

Porter, A.E., Gass, M., Muller, K. et al. (2007). Direct imaging of single-walled carbon nanotubes in cells. *Nat. Nanotechnol.* 2: 713–717. https://doi.org/10.1038/nnano.2007.347.

Prasad, R., Bhattacharyya, A., and Nguyen, Q.D. (2017). Nanotechnology in sustainable agriculture: recent developments, challenges, and perspectives. *Front. Microbiol.* 8: 1014.

Praveen, A., Khan, E., Ngiimei, D.S. et al. (2018). Iron oxide nanoparticles as nano-adsorbents: a possible way to reduce arsenic phytotoxicity in Indian mustard plant (*Brassica juncea* L.). *J. Plant Growth Regul.* 37: 612–624. https://doi.org/10.1007/s00344-017-9760-0.

Pyrzynska, K. (2011). Carbon nanotubes as sorbents in the analysis of pesticides. *Chemosphere* 83: 1407–1413. https://doi.org/10.1016/j.chemosphere.2011.01.057.

Rai, M. and Ingle, A. (2012). Role of nanotechnology in agriculture with special reference to management of insect pests. *Appl. Microbiol. Biotechnol.* 94 (2): 287–293.

Rajput Vishnu, D., Minkina, T., Sushkova, S. et al. (2017). Effect of nanoparticles on crops and soil microbial communities. *J. Soils Sediments* 18: 179–187.

Rajput, V.D., Chen, Y., and Ayup, M. (2015). Effects of high salinity on physiological and anatomical indices in the early stages of Populuseuphratica growth. *Russian J. Plant Physiol.* 62 (2): 229–236. https://doi.org/10.1134/S1021443715020168.

Rajput, V.D., Minkina, T., Sushkova, S. et al. (2019a). Toxicity assessment of metal oxide nanoparticles on terrestrial plants. *Compr. Anal. Chem.* 87: 189–207.

Rajput, V., Minkina, T., Ahmed, B. et al. (2019b). Interaction of copper based nanoparticles to soil, terrestrial and aquatic systems: critical review of the state of the science and future perspectives. *Rev. Environ. Contam. Toxicol.* 252: 51–96. https://doi.org/10.1007/398_2019_34.

Rajput, V., Minkina, T., Sushkova, S. et al. (2020). ZnO and CuO nanoparticles: a threat to soil organisms, plants, and human health. *Environ. Geochem. Health* 42: 147–158. https://doi.org/10.1007/s10653-019-00317-3.

Raju, P.L.N. (2006). Fundamentals of geographical information system. *Satellite Remote Sensing and GIS Applications in Agricultural Meteorology*, 103–120.

Raliya, R., Tarafdar, J.C., Gulecha, K. et al. (2013). Review article; scope of nanoscience and nanotechnology in agriculture. *J. Appl. Biol. Biotechnol.* 1: 041–044.

Rossi, L., Zhang, W., Lombardini, L., and Ma, X. (2016). The impact of cerium oxide nanoparticles on the salt stress responses of *Brassica napus L. Environ. Pollut.* 219: 28e36.

Scholes, D.G. and Sargent, H.E. (2014). Boosting plant biology. *Nat. Mater.* 13: 329–331. PMID 24651425.doi:https://doi.org/10.1038/nmat3926.

Serik, O., Ainur, I., Murat, K. et al. (1996). Silicon carbide fiber-mediated DNA delivery into cells of wheat (*Triticum aestivum* L.) mature embryos. *Plant Cell Rep.* 16: 133–136. https://doi.org/10.1007/BF01890853.

Service RF (2003). American Chemical Society meeting: nanomaterials show signs of toxicity. *Science* 300: 243.

Shah, F. and Wu, W. (2019). Improving crop yield and nutrient use efficiency via biofertilization-a global meta-analysis. *Sustain. For.* 11: 1485.

Siddiqui, M.H., Al-Whaibi, M.H., Faisal, M., and Al Sahli, A.A. (2014). Nano-silicon dioxide mitigates the adverse effects of salt stress on *Cucurbita pepo L. Environ. Toxicol. Chem.* 33 (11): 2429.

Siddiqui, M., Al-Whaibi, M., Firoz, M., and Al-Khaishany, M. (2015a). Role of nanoparticles in plants. In: *Nanotechnology and Plant Sciences Nanoparticles and their Impact on Plants.* Springer.

Siddiqui, M.H., Al-Whaibi, M.H., and Mohammad, F. (2015b). Nanotechnology and plant sciences: nanoparticles and their impact on plants. In: *Nanotechnology and Plant Sciences Nanoparticles and their Impact on Plants*, 1–303.

Singh, A., Sengar, R.S., Singh, A.K. et al. (2019). Impact of genetic engineering on agriculture food crops. *Biotech Today* 9 (1): 17–24.

Singh, A., Vishnu, R., Sapna, R. et al. (2020). Monitoring soil salinity and recent advances in mechanism of salinity tolerance in plants. *Biogeosyst. Tech.* 7 (2): 66–87. https://doi.org/10.13187/bgt.2020.2.66.

Singh, A., Rajput, V., Singh, A.K. et al. (2021). Transformation techniques and their role in crop improvements: a global scenario of GM crops. In: *Policy Issues in Genetically Modified Crops* (eds. P. Singh, A. Borthakur, A.A. Singh, et al.), 515–542. Academic Press https://doi.org/10.1016/b978-0-12-820780-2.00023-6.

Tadros, T.F., Izquierdo, P., Esquena, J., and Solans, C. (2004). Formation and stability of nanoemulsions. *Adv. Colloid Interf. Sci.* 108–109: 303–318. https://doi.org/10.1016/j.cis.2003.10.023.

Thome, A., Eddy, K.R., Reginatto, C., and Cecchin, I. (2015). Review of nanotechnology for soil and groundwater remediation: Brazilian perspectives. *Water Air Soil Pollut.* 226 (4): 121.

Torabian, S., Farhangi-Abriz, S., and Zahedi, M. (2018). Efficacy of $FeSO_4$ nano formulations on osmolytes and antioxidative Journal of Nanomaterials 11 enzymes of sunflower under salt stress. *Indian J. Plant Physiol.* 23 (2): 305–315.

Torney, F., Trewyn, B.G., Lin, V.S.Y., and Wang, K. (2007). Mesoporous silica nanoparticles deliver DNA and chemicals into plants. *Nat. Nanotechnol.* 2: 295–300. https://doi.org/10.1038/nnano.2007.108.

Tripathi, D.K., Singh, S., Singh, V.P. et al. (2017). Silicon nanoparticles more effectively alleviated UV-B stress than silicon in wheat (*Triticum aestivum*) seedlings. *Plant Physiol. Biochem.* 110: 70e81.

Turner, D.P., Cohen, W.B., Kennedy, R.E. et al. (1999). Relationships between leaf area index, FAPAR and net primary production of terrestrial ecosystems. *Remote Sens. Environ.* 70: 52–68.

Venkatachalam, P., Jayaraj, M., Manikandan, R. et al. (2017). Zinc oxide nanoparticles (ZnONPs) alleviate heavy metal-induced toxicity in *Leucaena leucocephala* seedlings: a physiochemical analysis. *Plant Physiol. Biochem.* 110: 59e69.

Vishwakarma, K., Upadhyay, N., Kumar, N. et al. (2018). Potential applications and avenues of nanotechnology in sustainable agriculture. In: *Nanomaterials in Plants, Algae, and Microorganisms*. https://doi.org/10.1016/B978-0-12-811487-2.00021-9.

Wan, Y., Li, J., Ren, H. et al. (2014). Physiological investigation of gold nanorods toward watermelon. *J. Nanosci. Nanotechnol.* 14: 6089–6094. https://doi.org/10.1166/jnn.2014.8853.

Wang, K., Drayton, P., Frame, B. et al. (1995). Whisker mediated plant transformation: an alternative technology. *in vitro Cell. Dev. Biol.* 31: 101–104. https://doi.org/10.1007/BF02632245.

Wong, M.H., Giraldo, J.P., Kwak, S.Y. et al. (2016). Nitroaromatic detection and infrared communication from wild-type plants using plant Nanobionics. *Nat. Mater.* 16 (2): 264–272. https://doi.org/10.1038/nmat4771.

Worrall Elizabeth, A., Hamid, A., ModyKarishma, T. et al. (2018). Nanotechnology for plant disease management. *Agronomy* 8: 2–24.

Xu, L., Cao, L.D., Li, F.M. et al. (2014). Utilization of chitosan-lactide copolymer nanoparticles as controlled release pesticide carrier for pyraclostrobin against *Colletotrichum gossypi*. *Southw. J. Dispers. Sci. Technol.* 35: 544–550.

Xue, J. and Su, B. (2017). Significant remote sensing vegetation indices: a review of developments and applications. *J. Sens.* 2017: 1353691.

Zhang, P., Anderson, B., Tan, B. et al. (2005). Potential monitoring of crop production using a satellite-based climate-variability impact index. *Agric. For. Meteorol.* 132: 344–358.

Zhang, J., Boghossian, A.A., Barone, P.W. et al. (2010). Single molecule detection of nitric oxide enabled by d(AT)(15) DNA adsorbed to near infrared fluorescent single-walled carbon nanotubes. *J. Am. Chem. Soc.* 20: 567–581.

Zhao, C., Liu, B., Piao, S. et al. (2017). Temperature increase reduces global yields of major crops in four independent estimates. *Proc. Natl. Acad. Sci. U. S. A.* 114: 9326–9331.

Zheng, L., Hong, F., Lu, S., and Liu, C. (2005). Effect of Nano-TiO_2 on strength of naturally aged seeds and growth of spinach. *Biol. Trace Elem. Res.* 104 (1): 83–89.

Zhu, J., Wei, G., Li, J. et al. (2004). Silicon alleviates salt stress and increases antioxidant enzymes activity in leaves of salt-stressed cucumber (*Cucumis sativus* L.). *Plant Sci.* 167: 527–533.

Zhu, X.G., Long, S.P., and Ort, D.R. (2010). Improving photosynthetic efficiency for greater yield. *Annu. Rev. Plant Biol.* 61: 235–261.

Index

Sustainable Agriculture Systems and Technologies, First Edition. Edited by Pavan Kumar, A.K. Pandey, Susheel Kumar Singh, S.S. Singh, and V.K. Singh.

g

k

l

t

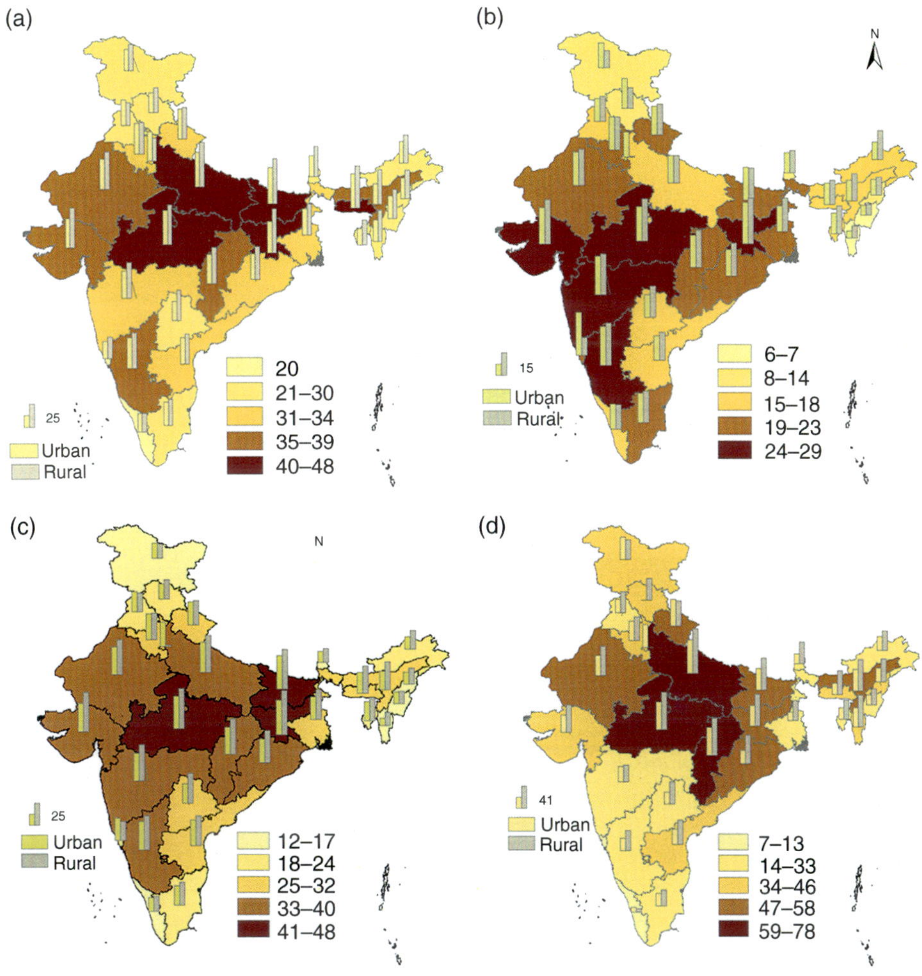

Figure 1.2 Status of undernutrition in India. (a) Prevalence of stunted children. (b) Prevalence of child wasting. (c) Prevalence of underweight children. (d) Under-five mortality rate. *Source:* Based on NFHS 2015–16.

Sustainable Agriculture Systems and Technologies, First Edition. Edited by Pavan Kumar, A.K. Pandey, Susheel Kumar Singh, S.S. Singh, and V.K. Singh.

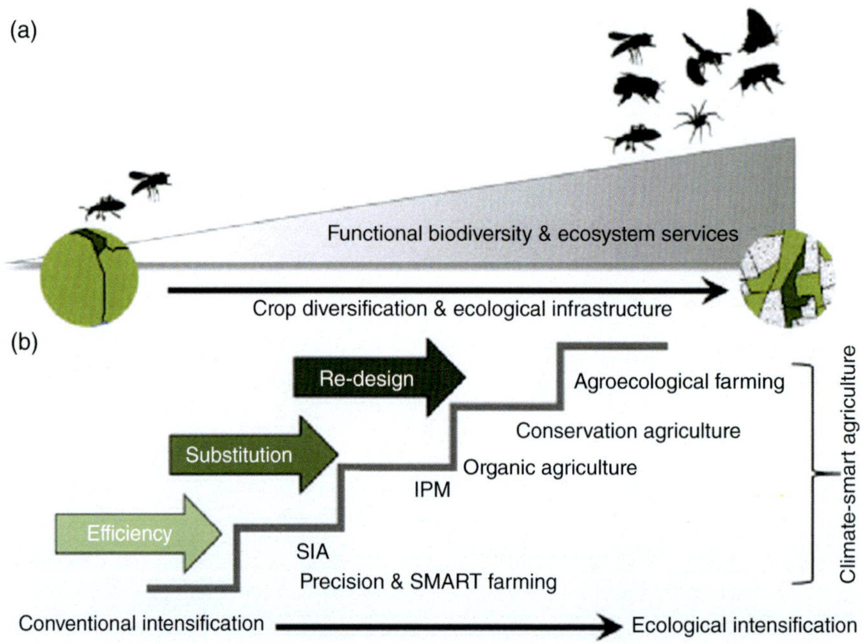

Figure 2.2 (a) Functional biodiversity delivering ecosystem services (e.g. pollination, natural biocontrol) in support of crop yields along a gradient of crop diversification and ecological infrastructure and (b) alternative options (efficiency, substitution, redesign) for farm management systems situated on a continuum from conventional agricultural intensification to ecological intensification.

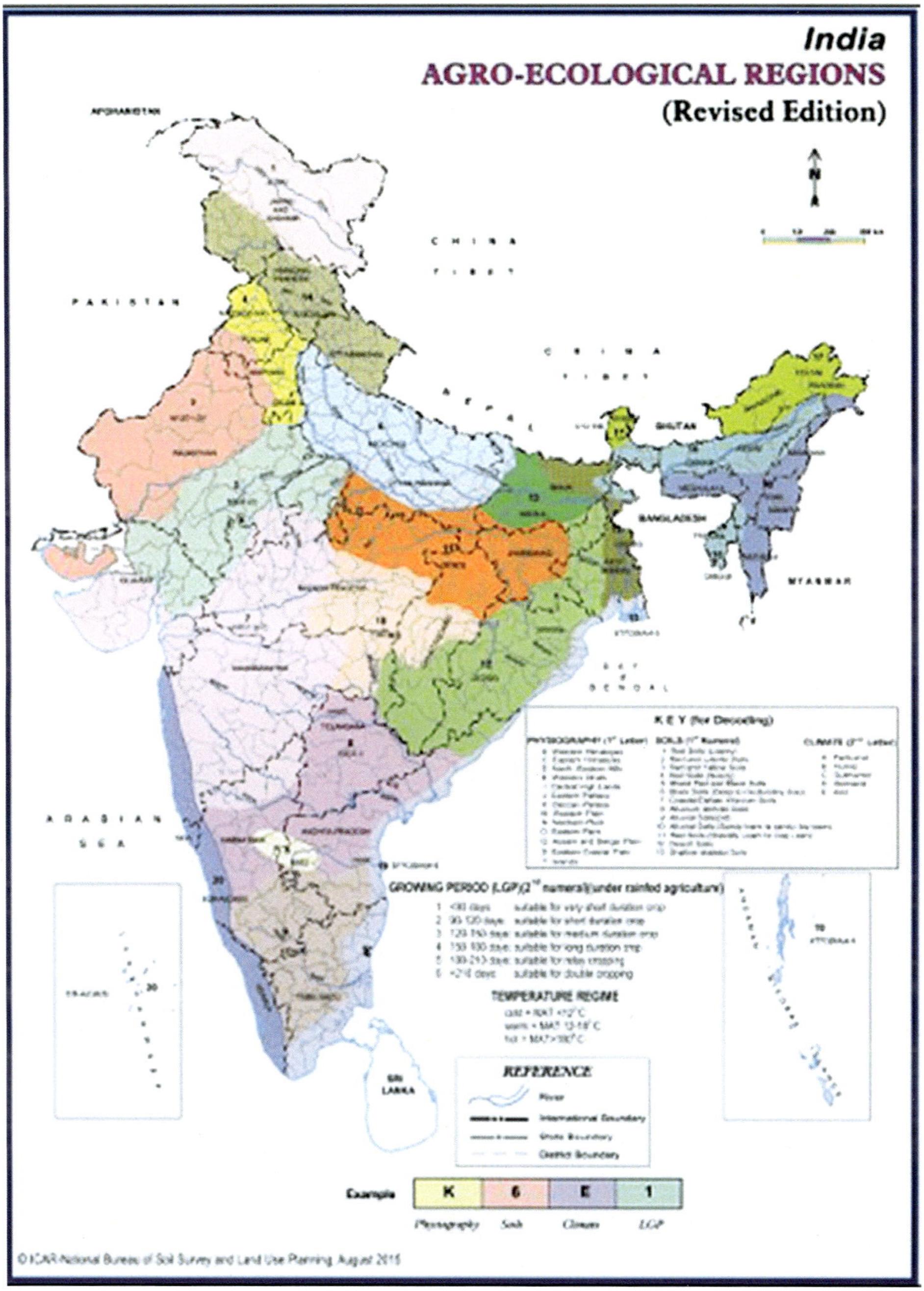

Figure 2.3 Agro-ecological regions map of the country (2015) and revised Bio-climates of India (2015) (e-Publication: ICAR-NBSS&LUP Technologies).

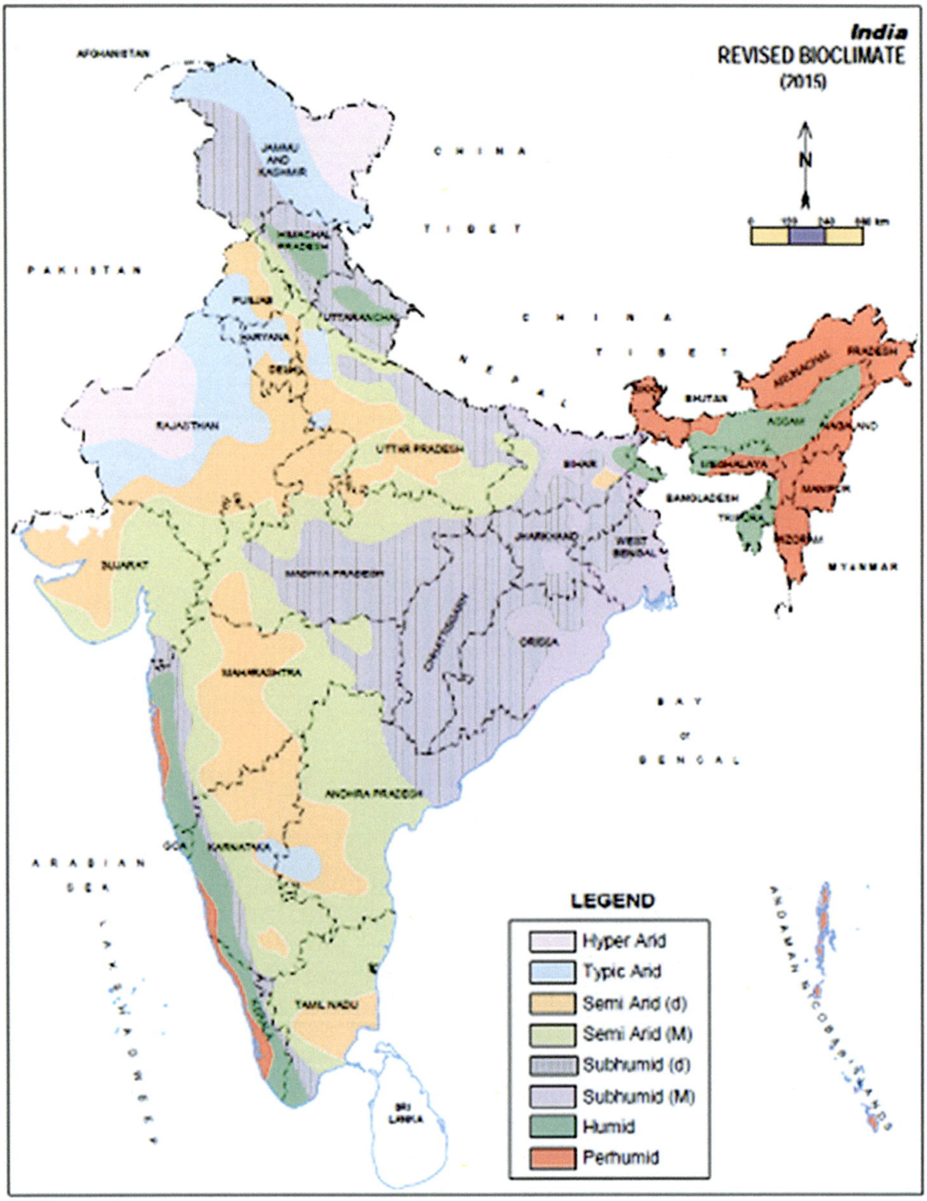

Figure 2.3 (Continued)

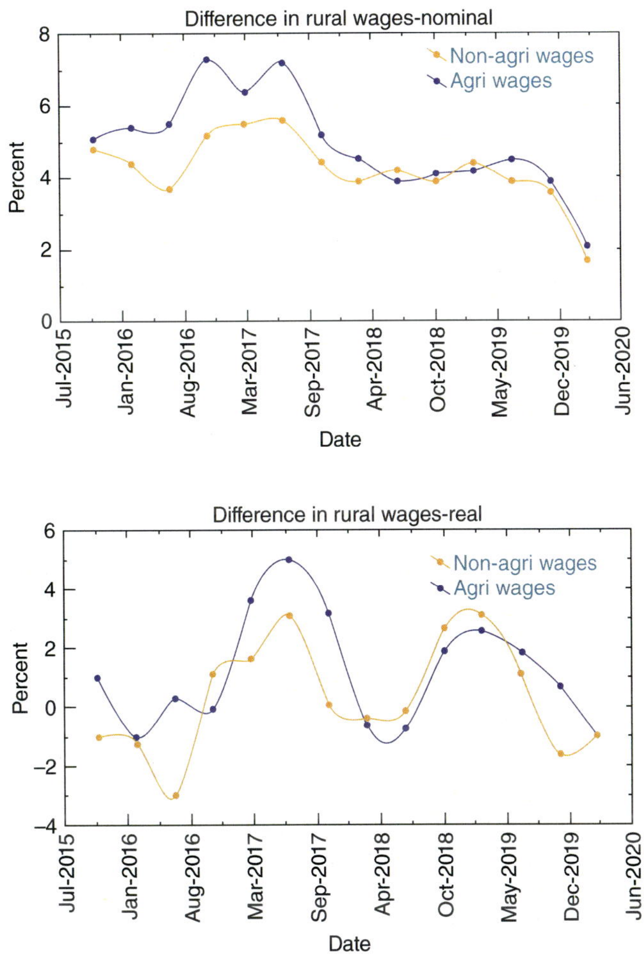

Figure 4.4 (Top panel) Variations in the average nominal wage (agriculturally related vs. nonagriculturally related in rural areas). (Bottom panel) Variations in the average real wage in rural areas based on data from RBI, accessed 16 May 2020.

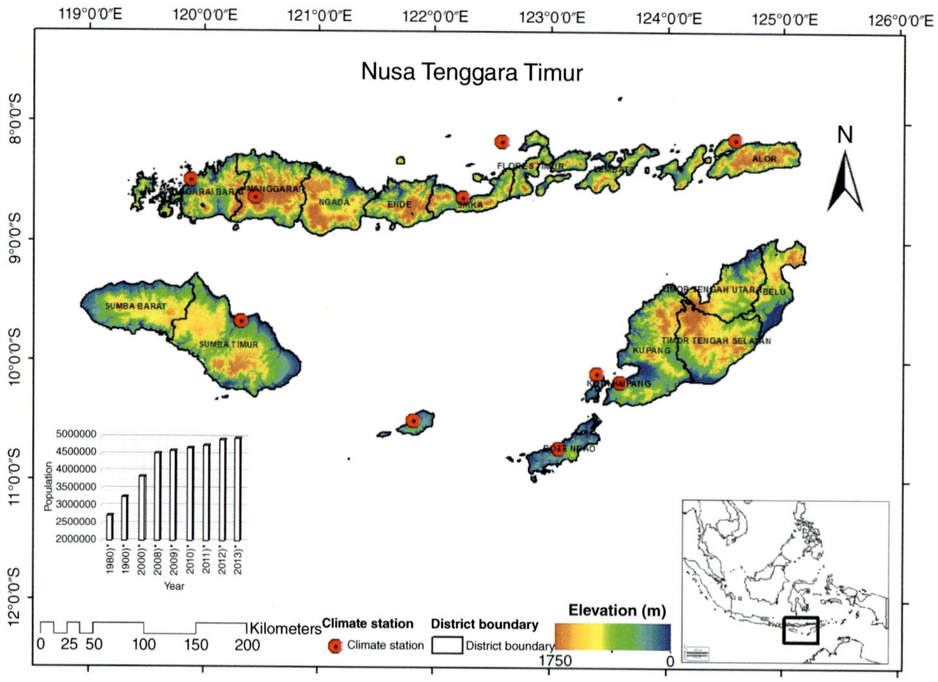

Figure 6.1 Study area. *Source:* Boubacar Siddighi Balde, Martiwi Diah Setiawati, I Wayan Nampa, Mohamed Esham.

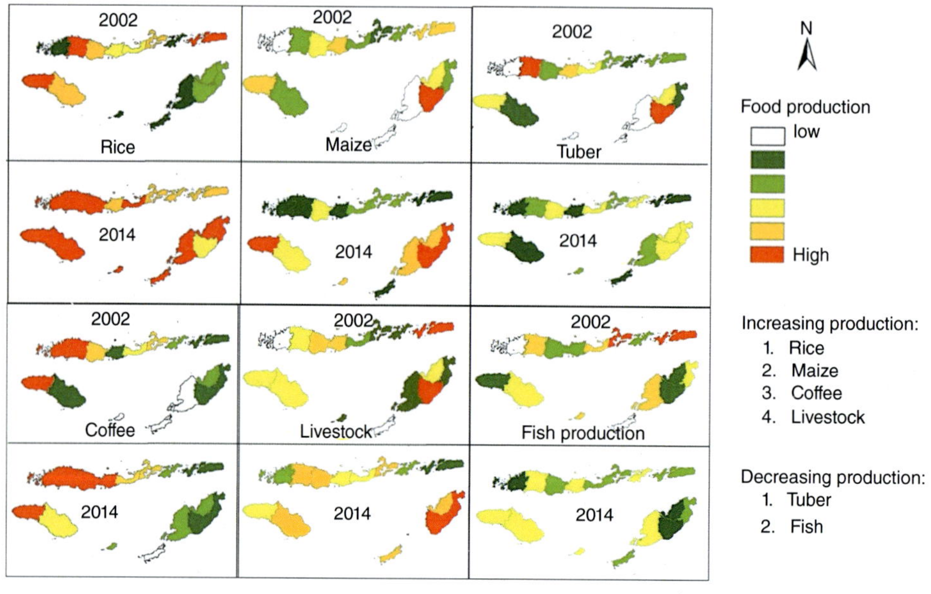

Figure 6.6 Food production between 2002 and 2014 in NTT provinces. *Source:* Boubacar Siddighi Balde, Martiwi Diah Setiawati, I Wayan Nampa, Mohamed Esham.

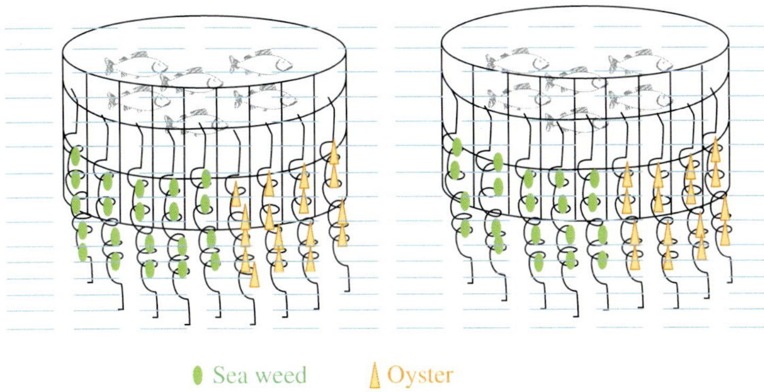

Figure 8.6 IMTA: Coculture of fish, sea weed, and oyster in cage system.

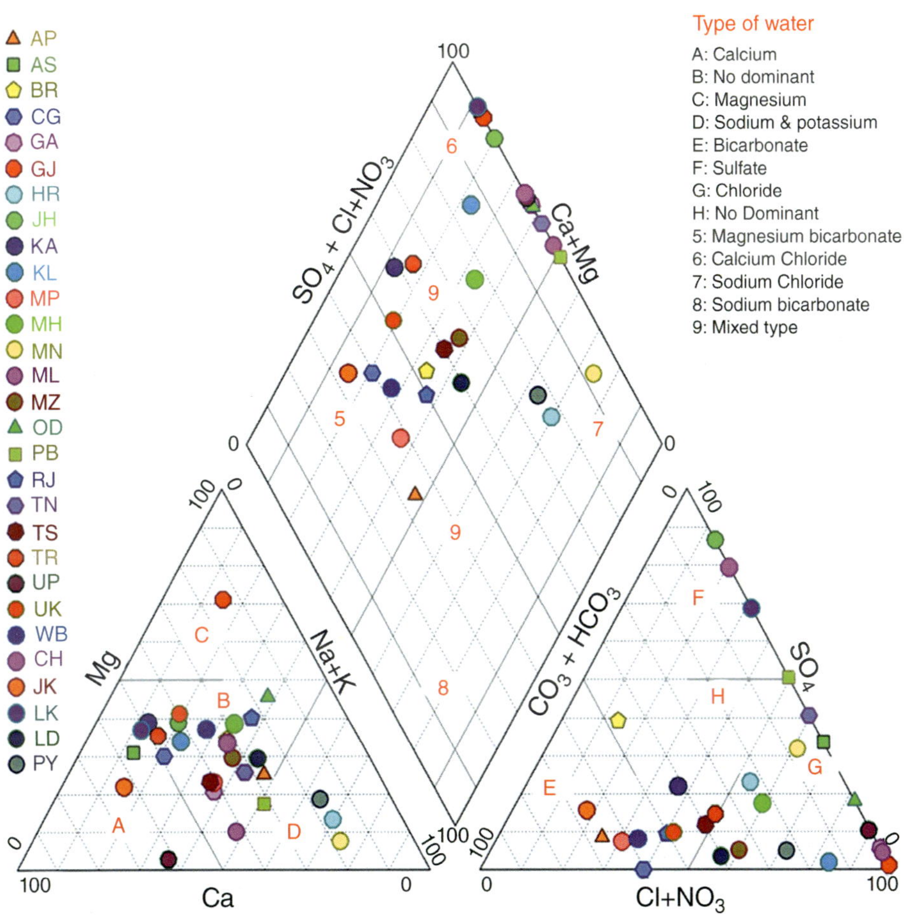

Figure 10.4 Ground water hydro chemistry of different states of India.

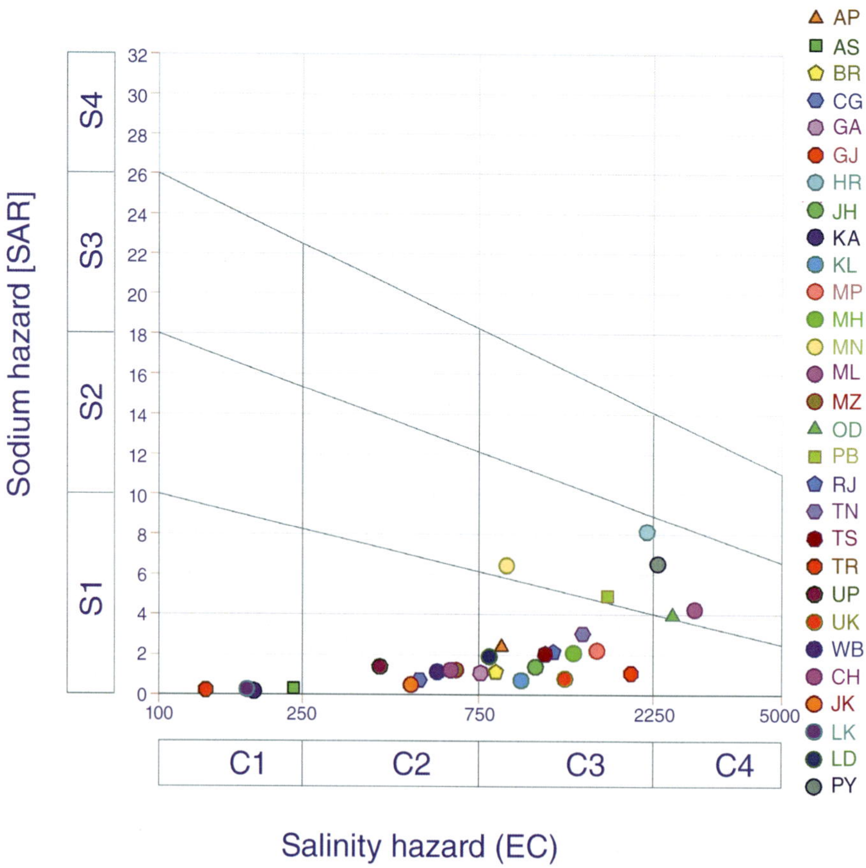

Figure 10.5 Salinity and sodium hazards of the Indian groundwater.

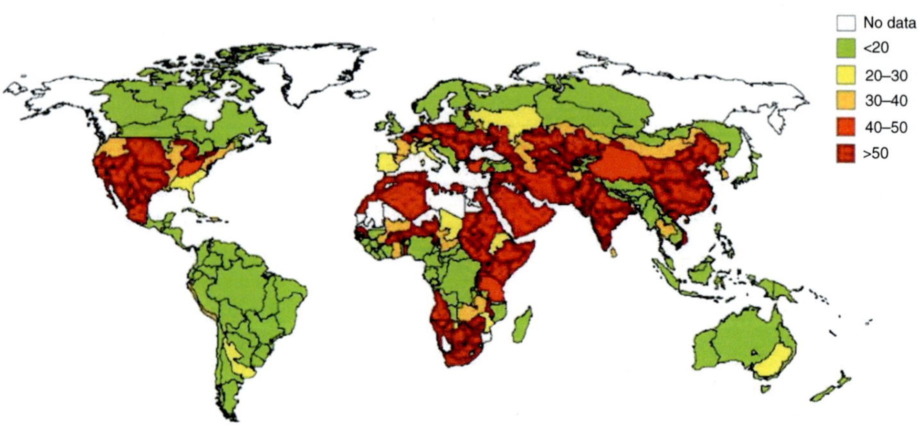

Figure 11.3 Water scarcity faced by the world by 2050. *Source:* Modified from Hoekstra and Chapagain (2007). © John Wiley & Sons.

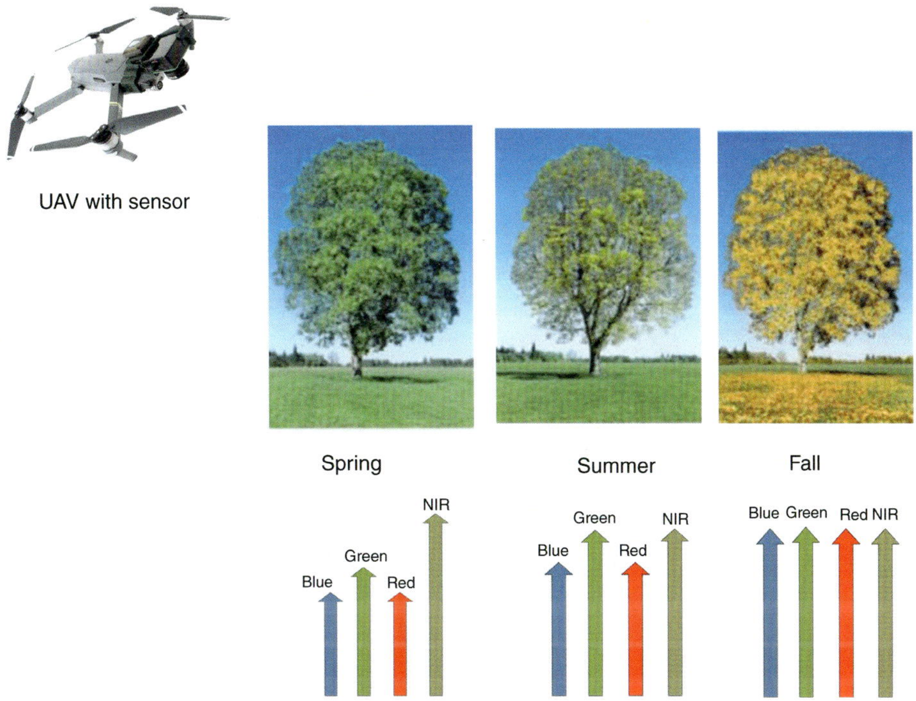

Figure 14.3 Green and healthy vegetation reflects a large portion of the near-infrared light in comparison to sparse vegetation that reflects more visible light.

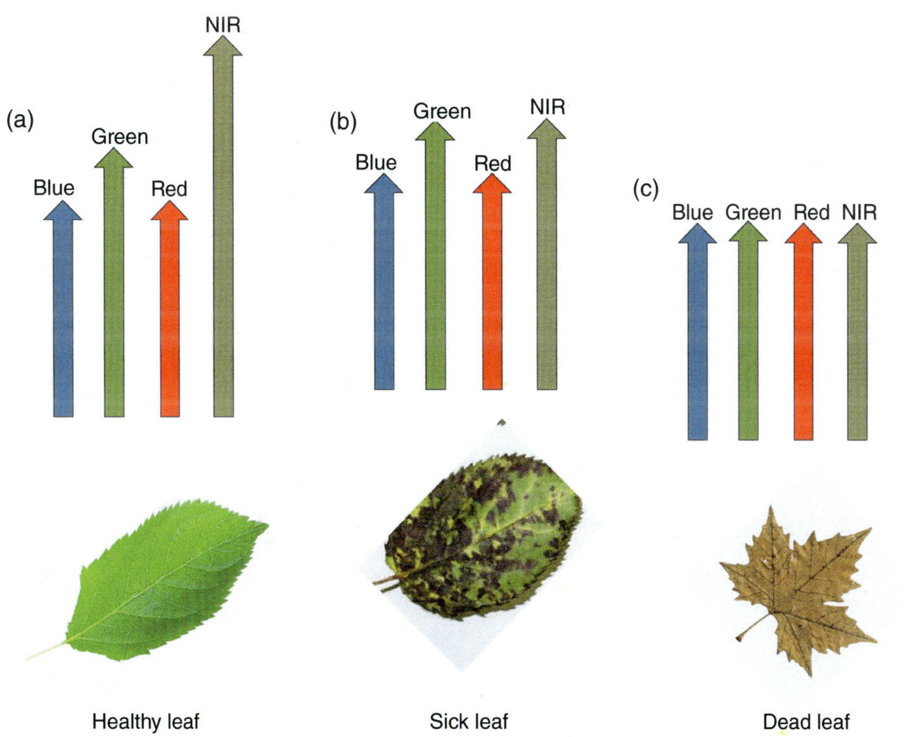

Figure 14.4 Reflection of light by healthy, sick, and dead leaves. Healthy leaf reflects more near infrared light. (a) Healthy leaf. (b) Sick leaf. (c) Dead leaf.

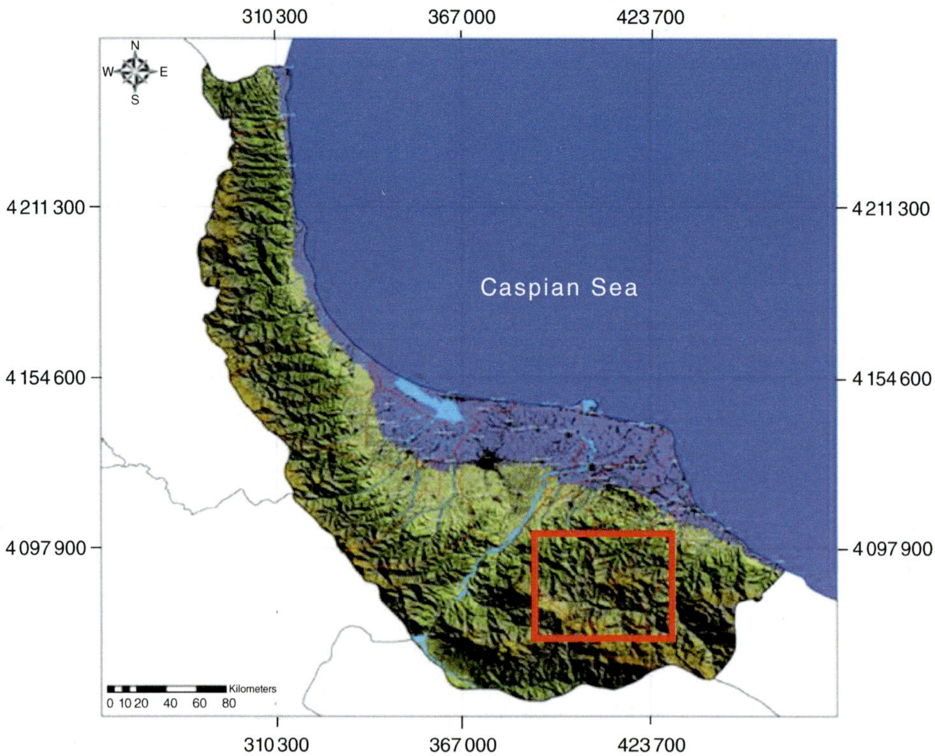

Figure 15.1 Location of study area.

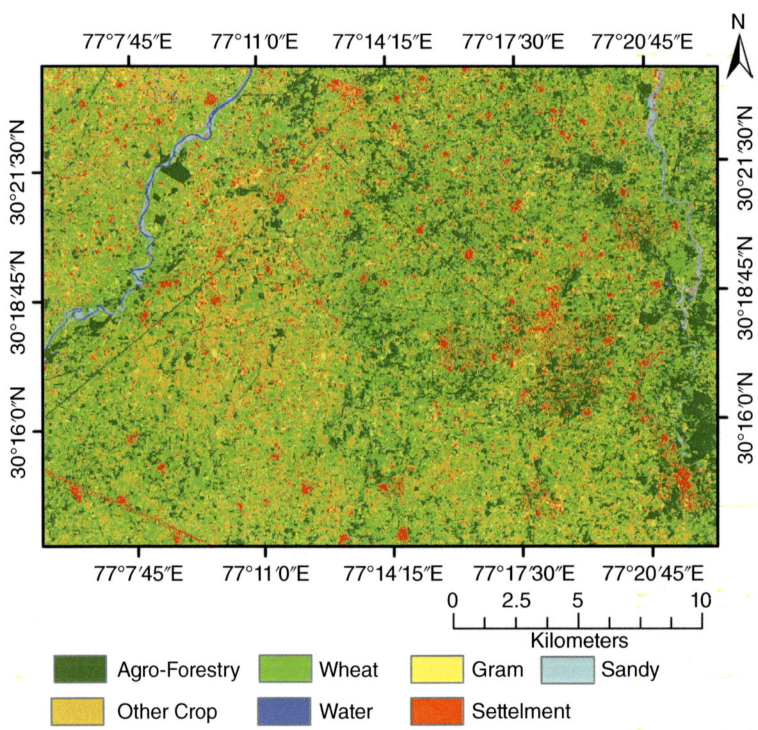

Figure 16.3 Random forest machine learning-based mapped cropland cover of the year 2019.

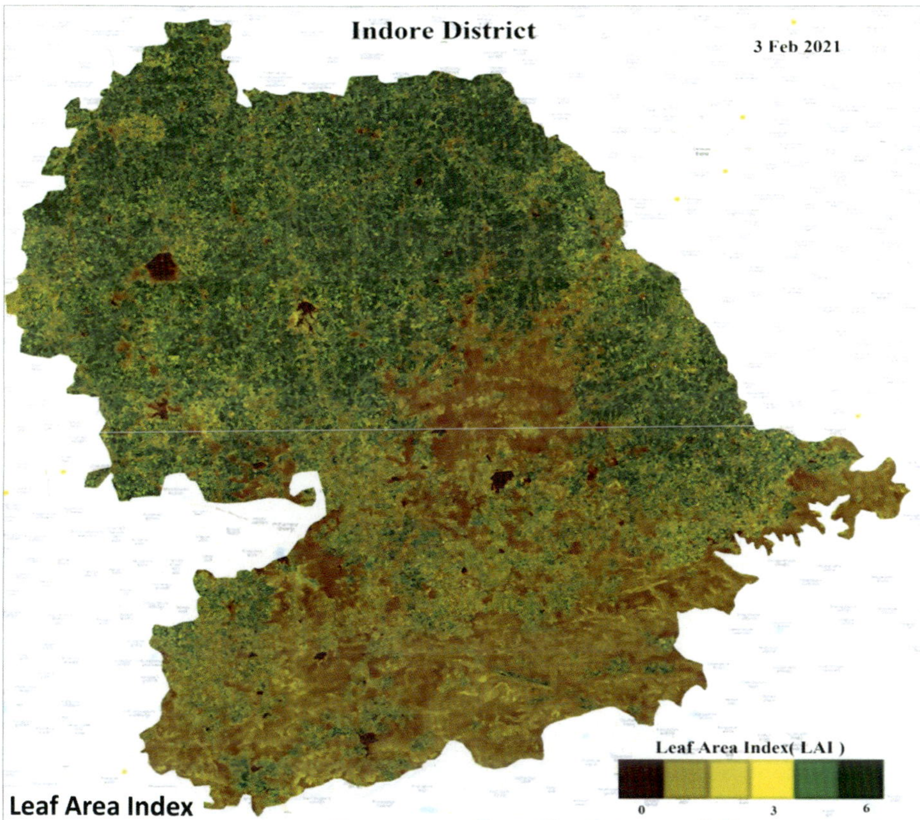

Figure 17.4 The LAI, total chlorophyll content of various types of plants in Indore district Madhya Pradesh, India.

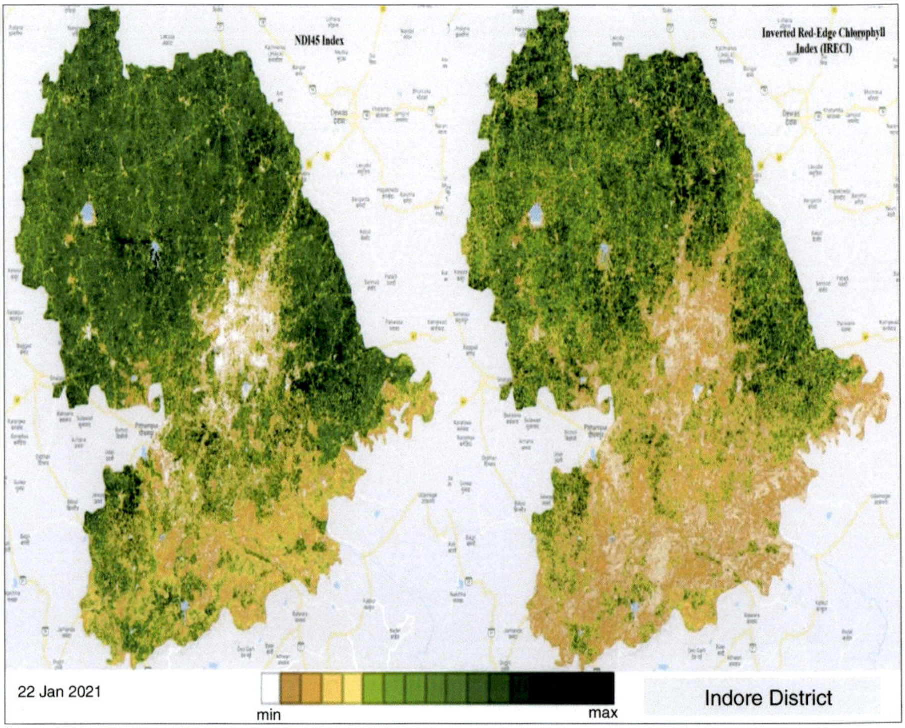

Figure 17.6 Diagrammatic representation of IRECI of Indore district, Madhya Pradesh, India.